Lecture Notes in Mathematics 1874

Editors:
J.-M. Morel, Cachan
F. Takens, Groningen
B. Teissier, Paris

Lecture Notes in Mathematics

M. Émery · M. Yor (Eds.)

In Memoriam
Paul-André Meyer

Séminaire
de Probabilités XXXIX

 Springer

Editors

Michel Émery
Institut de Recherche Mathématique Avancée
Université Louis Pasteur
7, rue René Descartes
67084 Strasbourg Cedex, France
E-mail: emery@math.u-strasbg.fr

Marc Yor
Laboratoire de Probabilités et Modèles Aléatoires
Université Pierre et Marie Curie
Boîte Courrier 188
4, place Jussieu
75252 Paris Cedex 05, France

Library of Congress Control Number: 2006920071

Mathematics Subject Classification (2000): 60Gxx, 60Hxx, 60Jxx, 91B28

ISSN print edition: 0075-8434
ISSN electronic edition: 1617-9692
ISSN Séminaire de Probabiltés, print edition: 0720-8766
ISBN-10 3-540-30994-2 Springer Berlin Heidelberg New York
ISBN-13 978-3-540-30994-9 Springer Berlin Heidelberg New York
DOI 10.1007/b128398

Springer is a part of Springer Science+Business Media
springer.com
© Springer-Verlag Berlin Heidelberg 2006
Printed in The Netherlands

Typesetting: by the authors and SPI Publisher Services using a Springer LATEX package

Cover design: *design & production* GmbH, Heidelberg

Printed on acid-free paper   SPIN: 11601395   VA 41/3100/SPI   5 4 3 2 1 0

Father was a mathematician, and when he
was not able to solve a problem he would turn
to Sanskrit Grammar.

RAJA RAO. *The Serpent and the Rope.*

This volume XXXIX of Séminaire de Probabilités
is dedicated to the memory of Paul-André Meyer,
who unexpectedly passed away on January 30th 2003, in Strasbourg.

Readers of the *Séminaire* need not be reminded how deeply Paul-André
Meyer's research has modified the landscape of Probability Theory during
three decades, from 1962 and 1963, when his articles in the *Illinois Journal
of Mathematics* extended to continuous time Doob's decomposition of supermartingales, to 1993, when his *Quantum Probability for Probabilists* appeared
in the Springer Lecture Notes.

Most important is probably the way his work continues to live on, sometimes in ways different from those he would have thought. It is used "as one
breathes" (to quote his "Titres et Travaux : Postface", which can be found
at the very beginning of this volume), that is, in a way both trite and fundamental, similar to the way Fubini's theorem is used . . .

The May-June 2005 issue of the *Annales de l'Institut Henri Poincaré*
(Vol. 41, n° 3), also dedicated to the memory of Meyer, was prepared at
the same time as this volume, under the auspices of a joint committee,
consisting of J. Azéma, S. Attal, D. Bakry, M.T. Barlow, G. Ben Arous,
J. Bertoin, P. Biane, C. Dellacherie, H. Föllmer, R.K. Getoor, J. Jacod,
T. Jeulin, M. Ledoux, J.-F. Le Gall, Y. Le Jan, D. Lépingle, J. Neveu,
K.R. Parthasarathy, E.A. Perkins, R. Rebolledo, A.N. Shiryaev, C. Stricker,
J.B. Walsh, S. Watanabe, K.-A. Yan and ourselves. Both volumes exemplify,
but do not exhaust, the broad spectrum of mathematical themes which inter-
ested Meyer.

We take this opportunity to recall to the reader's attention the tribute
paid to P.A. Meyer and J. Neveu in 1994, on occasion of their 60th birthdays;
it was published in 1996, in n° 236 of *Astérisque* and in n° 68 of *La Gazette
des Mathématiciens*.

The present volume opens with a slightly abridged version of Meyer's auto-
biographical "Titres et Travaux : Postface", followed by an English version

of Marc Yor's "Un modèle pour nous tous" (the French version has appeared in the *Notices de l'Académie des Sciences*) and by Stéphane Attal's "Disparition de Paul-André Meyer", already published in n° 96 of *La Gazette des Mathématiciens*.

Ten other personal *témoignages* then follow. Among them is the speech Jacques Neveu delivered at Meyer's funeral in Paris; the other nine were written for this memorial volume.

Very sadly, one of these *témoins*, Catherine Doléans-Dade, also untimely left us in 2004. She had been Meyer's first student, and her name remains associated to a most important object in stochastic calculus; our warmest thoughts go to Everett and their children. Her death followed shortly that of Joseph L. Doob, who was from the very beginning one of Meyer's permanent sources of inspiration ...

We thank the Académie des Sciences and the Société Mathématique de France for their kind permission to reprint Marc Yor and Stéphane Attal's témoignages, Anthony Phan for the drawing reproduced on the front cover, and Philip Protter for his help with the English language.

M. Émery and M. Yor

Contents

Titres et Travaux : Postface

Paul André Meyer

2 janvier 1998

Il y a maintenant juste trois ans que je suis à la retraite. Le moment est favorable pour écrire quelque chose sur mon expérience de mathématicien, avec un peu de recul et pas trop d'oubli. Prendre du recul n'a pas diminué mon estime pour les mathématiques, au contraire : avoir d'autres activités me permet de mieux apprécier l'extraordinaire concentration qu'exige le travail mathématique.

Les mathématiciens forment un monde un peu à part, aussi vaut-il la peine de raconter avec quelques détails comment j'y suis entré et j'y ai vécu. Cela ne ressemble pas tout à fait à la trajectoire de l'enfant doué.

Personne dans ma famille n'a fait de mathématiques avant moi. Tout le monde était dans le commerce, petit ou plus grand, achetant et revendant quelque chose. J'ai d'ailleurs gardé, à voir l'intelligence et le charme que mon père pouvait mettre dans le fonctionnement de sa boutique, une grande estime pour cette activité que l'on juge couramment stupide. Une légende familiale veut qu'un arrière grand-oncle paternel (ou quelque chose de ce genre) ait été un calculateur prodige, mais rien ne la garantit, et quant à moi j'ai continué jusqu'au lycée à compter sur mes doigts.

Mes premiers souvenirs mathématiques sont justement des souvenirs de grandes difficultés avec les opérations, particulièrement la division. C'est ma mère qui m'a tiré de là, m'apprenant à faire les divisions comme on le lui avait appris à l'école anglaise, c'est à dire en posant toutes les opérations intermédiaires, multiplications du côté droit, soustractions du côté gauche. Ce manque de sécurité, ce besoin de tout « poser », je peux dire que je l'ai ressenti toute ma vie.

J'ai le souvenir d'une composition de 8e complètement ratée, où j'avais eu 4 sur 20, et d'un moment de désespoir. C'est encore ma mère qui m'a tiré d'affaire. Elle m'a dit « Je ne sais pas faire tes problèmes comme ils le veulent, mais on m'a appris à l'école anglaise à les faire par l'algèbre. Je vais te montrer. Tu appelles ça x et ça y, alors cette phrase veut dire que $x + y = 3$, celle-là que $3x + 2y = 7$, tu remplaces y par $3 - x$ dans la seconde égalité, et tu trouves que $x = 1$, donc $y = 2$ ». J'étais émerveillé. J'emportai le cahier dans ma chambre, je refis le problème, puis *tous* les problèmes, et je finis même par comprendre la ruse qui permet, dans certains cas, de présenter la solution sans algèbre. J'ai rattrapé les autres sans peine. Ensuite, je me suis arrangé pour être toujours en avance sur ma classe, appliquant sans le savoir le principe que Spitzer devait m'expliquer en 1965 : « always tell the NSF that you plan to do what you just have done ».

J'ai eu jusqu'en 6ᵉ un délicieux et invraisemblable professeur, qui nous a appris, certes, à extraire des racines carrées et cubiques, mais qui était d'autant plus heureux devant une figure de géométrie qu'elle comportait plus de traits de toutes les couleurs. J'avais de bonnes notes. Entre la 6ᵉ et la 5ᵉ, arrivant à Paris au printemps, j'ai pris quelques leçons de mathématiques et de latin avec un professeur du Cours Hattemer qui se nommait M. Bonhoure ; on disait qu'il avait été jésuite. Il barra le premier dessin multicolore et me dit « Une figure ne doit pas comporter une lettre ou un trait inutile pour la démonstration ». J'avalai la leçon. Il me donna aussi quelques cours de latin, et lorsque mes parents me demandèrent (cela m'émerveille, mais c'est ainsi) ce que je voulais faire, je dis que le latin me plaisait, mais que je voulais faire le plus de mathématiques possible. J'abandonnai donc le latin. Hélas, comment aurions-nous pu savoir qu'en ce temps-là, pour faire « le plus de mathématiques possible » il fallait faire aussi du grec ?

J'étais en pension au premier trimestre de 5ᵉ. Je ne sais pas comment un formulaire de mathématiques me tomba entre les mains. Il commençait par des identités remarquables : $(a + b)^2$, $(a + b)^3$... que je vérifiai facilement. Puis il continuait : dérivées, différentielles... équation de Laplace, équations de Maxwell... Les identités remarquables exceptées, je ne compris rien, sauf que tout cela existait. Mon idéal mathématique devint pour des années de comprendre l'équation de Laplace et les équations de Maxwell. Je suis arrivé jusqu'à Laplace, mais jamais vraiment jusqu'à Maxwell.

Au second trimestre de 5ᵉ j'étais de nouveau en famille. Le cabinet de toilette à côté de ma chambre était devenu mon laboratoire, où j'avais une petite machine électrostatique, une bobine de Ruhmkorff. Je m'y suis construit un poste à galène (je me revois casque aux oreilles écoutant le *Martyre de Saint Sébastien,* la galène étranglant d'Annunzio tout en gardant le plus possible de Debussy). Quant aux mathématiques, nous avions (en 1947, déjà, et à Janson !) un professeur remplaçant qui peinait beaucoup. Un jour il se trompa au tableau et resta court. Je levai la main et lui dis où il se trompait. Il discuta un peu, je répondis « C'est logique ! » Alors un garçon dans les premiers rangs dit à voix haute « Ça n'a rien à voir avec la logique, ici on est en maths ».

A partir de la 4ᵉ, j'eus un vrai professeur, M. Heilbronn, qui était même docteur ès-sciences — sa thèse portait sur « les équations aux dérivées partielles selon Jules Drach », si ma mémoire est bonne. Par les hasards des changements de classe, je restai avec lui quatre ans. A vrai dire, ce n'était un bon professeur que pour les bons élèves, car les autres le mettaient dans des rages folles. Mais pendant quatre ans je n'eus avec lui aucune note inférieure à 18. J'aimais beaucoup la géométrie, la droite de Simpson et le cercle des neuf points, et je me posais tout seul le genre de problèmes puérils que je pouvais imaginer, imitant les exercices de fins de chapitre, tels que construire un triangle connaissant les rayons du cercle inscrit, du cercle circonscrit, et d'un cercle exinscrit. Je calculais mieux aussi (sur les lettres à défaut des chiffres), et j'aimais même la trigonométrie.

La géométrie plane, branche morte des mathématiques, avait cet avantage d'être concrète, et de mettre à la portée des enfants des énoncés non-évidents. Sa suppression en tant que théorie « inutile », fut l'œuvre de gens intelligents et honnêtes, et je pense que ce fut une sottise — l'une de ces sottises logiques et irréparables, sur lesquelles on ne peut que pleurer.

Je crois que pendant cette période de la 5e à la 2e — de 12 à 15 ans, mettons — j'ai été plus intelligent que pendant tout le reste de ma vie. J'ai l'impression que maintenant l'intelligence des adolescents est gaspillée.

En classe de seconde se produisit un événement important pour moi : mon cousin venait de passer son bac, dans la classe de Sciences Expérimentales (il a fait des études de médecine), et il me passa son livre de maths. Or dans ce mince livre, il y avait les dérivées et différentielles, primitives, logarithmes et exponentielles. Cela me transporta. L'idée que la fonction $1/x$ avait une primitive *pour laquelle il fallait inventer un nouveau nom* me stupéfia ; je ne voulus pas le croire, et je partis à la recherche de cette primitive. J'y passai beaucoup de temps, transformant la relation $y' = 1/x$ en diverses équations différentielles que j'essayais de résoudre. Enfin, j'en parlai à M. Heilbronn, qui m'affirma que vraiment il n'y avait rien à faire — et que j'aurais dû plutôt m'étonner que la *dérivée* d'une fonction algébrique soit algébrique.

A cette époque, je passais beaucoup de jeudis après-midi au Palais de la Découverte. A vrai dire, j'en passais autant au Musée Guimet, dont la bibliothèque me fascinait. Je tournais autour, mais jamais je n'ai osé demandé à y entrer.

Le résultat de tout cela fut que j'allai acheter le *Cours de Mathématiques Spéciales* de A. Decerf, qui avait l'avantage d'être mince. Ce n'était pas un livre extraordinaire, mais il était clair et je l'ai beaucoup aimé. Je ne m'en suis séparé que lorsqu'il est vraiment devenu impossible de le recoller. Le premier chapitre était le plus difficile, l'algèbre linéaire, avec la définition classique des déterminants, si bizarre. Je me suis passionné pour l'Analyse Combinatoire, et j'ai passé beaucoup de temps à compter les différentes manières de ranger ceci ou cela.

Je me rappelle un autre écueil sur lequel j'ai buté : je m'étonnais que l'on pût couper une intégrale en deux, dans la première moitié poser $s = t$, dans la seconde $s = 1 - t$, et les remettre ensemble. Il me semblait que s ne pouvait être égal *à la fois* à t et à $1 - t$. Je n'ai pu avancer *qu'en comprenant qu'il n'y avait rien à comprendre.* Cela me rappelle un passage de Stendhal (dans la *Vie de Henry Brulard* au chapitre 34) où il dit avoir été arrêté dans son

développement mathématique par l'impossibilité de « comprendre » comment
− multiplié par − peut donner +. Voici ce passage : « *Mon enthousiasme
pour les mathématiques avait peut-être eu pour base principale mon horreur
pour l'hypocrisie... Que devins-je quand je m'aperçus que personne ne pouvait m'expliquer comment il se faisait que : moins par moins donne plus ?...
On faisait bien pis que ne pas m'expliquer cette difficulté (qui sans doute
est explicable car elle conduit à la vérité), on me l'expliquait par des raisons
évidemment peu claires pour ceux qui me les présentaient... J'en fus réduit à
ce que je me dis encore aujourd'hui : il faut bien que − par − donne + soit vrai,
puisque évidemment, en employant à chaque instant cette règle dans le calcul,
on arrive à des résultats vrais et indubitables.* » Ainsi le progrès consiste à comprendre que l'imagination concrète doit lâcher prise et laisser le langage faire
son travail. Le langage mathématique va *beaucoup plus loin* que l'imagination
— pourquoi, c'est un mystère. Cela se retrouve à tous les niveaux, par exemple
on ne peut (à mon avis) « comprendre » la transformation de Lorentz ou les
fondements de la mécanique quantique, on peut seulement s'y habituer — et
une intuition se reconstitue à partir de cette habitude. Dans ce cas précis, bien
sûr, il aurait été facile de satisfaire aussi l'imagination, et l'obstacle a été créé
par de mauvais professeurs, qui ne comprenaient pas ce qu'ils enseignaient.

L'« absence d'hypocrisie » dont parle Stendhal me semble être un élément
important de l'attrait des mathématiques.

Cette opposition entre ce que dit le langage mathématique et ce qui est
compris par images se rencontre partout : rien dans notre imagination ordinaire ne nous prépare à concevoir la vitesse de la lumière comme vitesse limite,
et la transformation de Lorentz peut être (me semble-t-il) manipulée, mais
non « comprise ». C'est pire encore pour la mécanique quantique. Bien entendu, l'esprit d'un physicien qui manipule ces choses tous les jours finit par
les connaître parfaitement, mais je ne crois pas qu'il réalise plus qu'un court-
circuit du langage mathématique, une constitution sommaire d'images à partir
de celui-ci. Feynman n'a-t-il pas écrit que la mécanique quantique restait toujours aussi stupéfiante au bout de quarante ans d'expérience ? Ou que la question « Qu'est-ce qui fait que les masses s'attirent » est sans réponse jusqu'à
maintenant, et sans doute pour toujours ? En ce sens, la science n'explique
rien. Elle nous dit « c'est comme ça », de la même façon que mon père, à
toutes les questions que je lui posais à l'âge des questions, répondait « à cause
des mouches ».

Je reviens à ma formation. J'ai d'abord été intéressé par les nombres
et leurs propriétés, comme tous les débutants me semble-t-il, en raison du
caractère élémentaire de ce sujet. J'ai dû lire deux démonstrations fausses du
théorème de Fermat, une démonstration (obscure) de la transcendance des
nombres e et π. J'ai perdu maintenant l'intérêt pour les « nombres » en eux-
mêmes, et je ne peux m'empêcher de trouver puéril l'attachement que leur
portent beaucoup de mathématiciens. J'éprouve presque aussi peu d'intérêt
pour le théorème de Fermat que pour la coupe du monde de football.

M. Heilbronn avait été élevé dans la tradition de Borel, Lebesgue, Hada-
mard (au séminaire duquel il avait assisté autrefois), et il m'a conseillé de
lire les *Cours d'Analyse* de Goursat et Valiron, et des livres de la Collection
Borel : les *Leçons sur la Théorie des Fonctions,* les *Leçons sur l'Intégration* de
Lebesgue, puis de Sierpinski les *Leçons sur les Nombres Transfinis.* Mais il m'a
prévenu aussi que tout cela était vieilli, et qu'un jour je devrais lire Bourbaki
— en passant, et puisqu'il est à la mode à présent de décrier Bourbaki, je note
mon émerveillement lorsque je l'ai lu (en classe préparatoire sans doute), avec
le fascicule de résultats de Théorie des Ensembles, et les admirables premiers
volumes de l'Algèbre et de la Topologie Générale, supérieurs par leur sobriété
à ceux qui sont venus par la suite. Je pense que toute une génération a été
enthousiasmée par cette perfection d'écriture — au détriment peut-être de
l'imagination créatrice, mais je n'en suis pas sûr.

La fascination de Bourbaki dans ces années-là ne s'exerçait pas seulement
sur les mathématiciens. Ce texte à la fois limpide et difficile a suscité la jalousie
des philosophes. Une partie des textes que contient le sottisier de Sokal peut
se comprendre comme une imitation de Bourbaki, à la façon dont un bébé
imite le bruit du français. Mais Bourbaki n'y est pour rien.

J'ai beaucoup profité de mes années de lycée, où j'ai appris à rédiger
clairement et sobrement. *J'ai surtout immensément aimé l'école.* J'avais le
sentiment que l'on pouvait tout savoir, et que le savoir menait à tout. Je n'ai
pas retrouvé chez mes enfants cet amour de l'école, et il me semble que quelque
chose s'est perdu d'une génération à l'autre.

Lorsque j'étais en Terminale, j'ai eu mes premiers contacts avec des
mathématiciens en activité, à qui je fus envoyé par le physicien P. Grivet,
un ami d'amis et un homme très bienveillant. Ce furent Raphaël Salem, qui
me prêta la première édition du traité d'intégration de Saks, Henri Cartan,
qui me conseilla de lire les *Fonctions de Variables Réelles* de Bourbaki, et
Laurent Schwartz, qui me recommanda son livre sur les distributions. J'ai
acheté ces livres, ils m'ont servi plus tard, mais bien sûr je n'y compris rien en
ce temps-là. Plus important, P. Grivet persuada mes parents que mon projet
de préparer le seul concours de la rue d'Ulm n'était pas une folie.

Je me suis trouvé parfaitement heureux en analyse : la théorie descriptive
des ensembles, l'intégration (et la théorie des fonctions d'une variable complexe). Cependant, je n'avais pas d'aversion pour la géométrie. Mes premiers
contacts avec la géométrie différentielle (dans le vieux manuel de Lainé) ont
été heureux. Je me rappelle avoir réfléchi en ce temps-là sur la définition des
différentielles secondes « complètes » dans le livre de Goursat. Je me souviens
aussi d'avoir été émerveillé en Terminale et en Hypotaupe par le petit traité de
Duporcq *Premiers Principes de Géométrie Moderne,* c'est à dire la géométrie
projective complexe à la Poncelet.

En Hypotaupe et Taupe, je travaillais beaucoup et je réussissais bien. Je
n'étais pas le meilleur. Celui-ci, qui avait fait du grec, est entré à Polytechnique ; j'ai aperçu son nom dans les journaux (il y a déjà longtemps) comme

fondé de pouvoir d'une grande banque dans quelque opération financière international, et il me semble qu'il y a eu là un gaspillage d'intelligence. Il me semble trop distrait et trop gentil pour avoir fait un excellent banquier. En tout cas, je n'ai pas vu mentionner son nom à propos du Crédit Lyonnais.

Je détestais en idée Polytechnique depuis l'adolescence, où j'avais lu (dans un livre datant de 1880) que les polytechniciens marchaient au pas et qu'un planton les empêchait d'entrer et de sortir. Si je ne pouvais entrer à l'Ecole Normale en une fois, j'irais à la Sorbonne ou aux USA.

L'année du concours de l'Ecole Normale, je quittai le lycée au milieu de l'année, car mon professeur de physique–chimie me persécutait. Je fus reçu au concours (parmi les derniers d'une toute petite fournée) grâce à un 20/20 en physique et à la lucide bonté de J. Deny, le second examinateur de maths.

Le jour où j'ai vu mon nom affiché à la porte de l'Ecole Normale a été l'un des plus lumineux de ma vie (il faisait d'ailleurs très beau). En y entrant, je ne savais pas si je choisirais les maths ou bien la physique, et les splendides possibilités de manipulations offertes aux élèves me firent hésiter un moment. Mais l'inadaptation pédagogique des cours de physique, comparée à la lumineuse perfection de Cartan, me firent choisir les maths. L'effet sur d'autres fut exactement contraire. Je me rends compte à présent que l'aptitude à tirer de l'information d'un exposé obscur ou incomplet fait partie de l'entraînement d'un physicien, et je regrette d'avoir méprisé ces bons maîtres.

Entré en 1954 rue d'Ulm, j'ai passé l'Agrégation en 1957 (à nouveau dernier sur une courte liste). Ce fut une année mémorable, où les candidats furent autorisés à passer l'oral sans veste — mais non sans cravate — car il faisait si chaud que le Préfet de Police avait autorisé les agents à enlever les leurs. J'ai eu une quatrième année d'Ecole, et je suis entré au CNRS avec un dossier contenant seulement deux lignes de Cartan « M. Meyer me semble être l'un des bons éléments de sa promotion ». J'ai jeté un coup d'œil à ce dossier par curiosité, alors que j'étais membre de la Commission du CNRS — à une époque où pour entrer au CNRS il fallait avoir une thèse de 3e cycle achevée, et des publications.

D'abord j'ai suivi le chemin tout tracé : rédiger le cours de Cartan sur l'homologie, lire de l'algèbre ; puis je me suis senti incapable de supporter une compétition aussi vive, et lorsque Cartan m'a proposé d'étudier les fibrés holomorphes je me suis jeté dans les probabilités. C'était alors une branche basse des mathématiques, bien qu'exerçant une certaine fascination d'ordre esthétique (et je reste fasciné par le hasard, quarante ans après). A cet égard, on peut comparer le hasard aux « nombres ». Les cours de la Sorbonne en probabilités étaient nuls à un point que l'on ne peut imaginer. Le seul grand probabiliste français, Paul Lévy, y était interdit d'enseignement ; son disciple Loève — un moins grand mathématicien, mais un admirable professeur — avait été écarté par xénophobie. La seule personnalité qui, sans être très utile, n'était du moins pas nuisible, était R. Fortet. Sans volonté de puissance personnelle, il accueillait généreusement les débutants, et nous sommes nombreux à penser à lui avec gratitude. J'ai d'abord travaillé seul, essayant de lire les livres de Lévy — mais celui-ci n'était pas mon genre. J'ai vraiment été formé par le grand traité de Doob, alors tout récent. J'avais une solide formation en théorie de la mesure, et Doob a fait de moi un probabiliste.

Comme Choquet avait été mon professeur, comme Deny m'avait sauvé par les cheveux au concours, je suivais le Séminaire de Théorie du Potentiel, et je me rappelle bien quand Brelot m'a dit « Il paraît qu'il y a un grand travail qui vient de sortir, dû à Hunt ; il démontre des résultats importants de théorie du potentiel par des méthodes purement probabilistes ». C'était une occasion de devenir probabiliste tout en restant un mathématicien « pur », chose qui jusqu'alors n'avait été possible qu'en Russie. Justement cette année-là (58/59), Loève est venu à Paris. L'année suivante, je l'ai suivi aux USA — en partie pour mettre l'Atlantique entre la guerre d'Algérie et moi, car on parlait de résilier les sursis. J'étais aussi entré en rapport avec Doob, à l'occasion d'une première note (assez stupide) sur la « séparabilité des processus stochastiques ». J'ai donc passé une année universitaire, moitié à Berkeley avec Loève, moitié à Urbana chez Doob. En fait, à Noël de cette année-là j'ai transmis par l'intermédiaire de Loève une note aux Comptes Rendus, qui la veille du Jour de l'An s'est révélée complètement fausse ; Loève a eu la gentillesse de téléphoner à Fréchet pour qu'il la retire. Mais en revenant en juin j'avais déjà une demi-thèse sur les processus de Markov. Elle n'avait rien à voir avec Loève et assez peu avec Doob : j'avais trouvé un problème intéressant en piochant Hunt, et lu les travaux des élèves de Dynkin sur ce sujet. À vrai dire, j'ai assez peu lu en ce temps-là Dynkin lui-même, qui me semblait trop abstrait.

La note retirée me donne l'occasion de dire que j'ai fait pas mal d'erreurs dans mon travail mathématique, et que j'ai vu à l'œuvre sur moi-même les mécanismes de l'erreur, et dans certains cas l'utilité de l'erreur. Parfois, si l'on avait au départ une idée juste de la difficulté d'une démonstration, on perdrait courage. Paul Lévy a dit quelque part qu'il avait progressé « d'erreur en erreur vers la vérité ».

Surtout, j'ai une chose importante à dire à propos de ces premiers travaux. On décrit la formation des étoiles par l'effondrement d'un nuage de gaz, après quoi naît spontanément une réaction nucléaire (ainsi, Jupiter aurait dû être le compagnon du soleil, mais s'est trouvé trop petit pour s'allumer). L'image convient aux mathématiciens : il y a d'abord accumulation de connaissances, après quoi ou bien on « s'allume » pour devenir un mathématicien et produire des résultats nouveaux, ou bien on reste « froid » : un érudit, nullement méprisable, mais privé de la grâce. C'est un phénomène propre aux mathématiques, me semble-t-il, dû au fait que l'activité y est entièrement mentale, sans les occupations annexes qu'apporte la pratique du laboratoire. Les gens du dehors ne comprennent pas bien cela.

Au retour des Etats-Unis, j'ai fait à Paris un séminaire sur les processus de Markov, dans le cadre du Séminaire Brelot-Choquet-Deny, qui a été très bien accueilli — je me rappelle Dixmier sortant d'un exposé sur la théorie de la mesure abstraite en disant « ce n'est pas si dégueulasse après tout ! ». J'ai soutenu ma thèse en 1961, le premier des normaliens de mon année. Après mon service militaire, j'ai pris mon premier poste dans l'Académie de Strasbourg (précisément à Mulhouse). Nous sommes arrivés à Strasbourg par un froid terrible au début de l'année 1964. Nous avons tant aimé Strasbourg, l'Alsace, les Vosges et la Forêt Noire, que je n'en ai plus bougé, bien que des occasions de « remonter » à Paris se soient manifestées presque aussitôt. Nous n'avons quitté Strasbourg que pour un séjour de deux ans de l'autre côté du Rhin, à Freiburg. J'aurais eu sans doute une vie mathématique plus longue, si j'étais retourné à Paris ; nous avons connu ici des années creuses, sans le moindre étudiant. J'ai toujours donné à ma famille une certaine priorité sur mon travail, et j'ai eu ma récompense lorsqu'une de mes filles m'a dit « Quand nous étions petits, nous nous disions que tu ne faisais rien, et nous avions un peu honte ».

Les mathématiciens sont friands d'anecdotes, mais je n'en raconterai pas. Les souvenirs imprimés dans ma mémoire sont pour la plupart de mauvais souvenirs, où je joue un rôle ridicule, et personne ne peut m'obliger à les raconter. En voici une tout de même : j'ai été invité une fois à dîner chez L.C. Young à Madison avec deux personnages illustres : Littlewood, âgé de 90 ans (sa moustache lui donnait l'air d'un phoque, et il mâchait en silence), et Mark Kac, qui m'a pris à partie pendant tout le dîner, en se moquant de ces Français toujours si abstraits, alors que ce qui compte en probabilités, c'est le concret, etc., etc. Des gens comme Arnold répètent maintenant ce genre de bêtises. J'ai gardé le souvenir d'un bon dîner, car au fond j'étais d'accord avec lui, et certain d'avoir autant que lui travaillé dans le cambouis (après tout, je suis né à Boulogne-Billancourt).

• • •

Le moment est venu de parler de mon travail mathématique, en essayant de n'être pas trop ennuyeux.

J'y vois trois périodes. La théorie des processus de Markov d'abord, puis la théorie des martingales et ce que l'on appelle la « théorie générale des processus », sans doute la période la plus heureuse et la plus fructueuse de ma vie mathématique. Après cela, j'ai changé de sujet tous les deux ou trois ans, sans approfondir vraiment les choses, mais avec des enthousiasmes de brève durée : applications des martingales à l'analyse, géométrie différentielle stochastique, « mécanique stochastique » à la Nelson, « calcul de Malliavin ». Mes toutes dernières années d'activité ont été consacrées aux probabilités quantiques de Hudson et Parthasarathy, sujet pour lequel j'ai mené une propagande efficace, me semble-t-il, mais sans démontrer de résultats personnels.

En théorie des processus de Markov, j'ai démontré dans la ligne de Doob, Hunt et Dynkin des résultats qui à l'époque ont été considérés comme « importants », mais maintenant dans cette théorie l'herbe pousse entre les rails. Un théorème de représentation des fonctions excessives, que j'ai démontré en m'appuyant en grande partie sur des travaux préliminaires russes (Volkonskii, Shur), a pu être traduit presque littéralement deux ou trois ans plus tard, comme le théorème de décomposition des surmartingales de la classe (D), qui est sans doute le seul résultat de moi que le « grand public » des probabilistes purs et appliqués puisse citer. En fait, il ne s'agissait que d'une traduction !

La théorie générale des processus est l'étude des filtrations, des temps d'arrêt, des martingales et semimartingales. Elle a été fondée par Doob, développée par Doob, Chung et l'école russe. Le groupe strasbourgeois est venu ensuite, et nous avons beaucoup fait pour la simplifier et en faire un langage simple et commode — si commode qu'il a été adopté par les mathématiciens appliqués, démentant ainsi les accusations de Kac. Nous, c'est à dire pas mal de monde à Strasbourg : Dellacherie (qui écrivait aussi bien que moi, mais moins vite), Catherine Doléans, Michel Weil, Maisonneuve, J.A. Yan, Giorgio Letta — et bien sûr beaucoup d'autres ailleurs ou plus tard. Ces travaux sont destinés à rester, j'en suis persuadé, et de la meilleure manière qui soit : en devenant des « trivialités », que l'on utilise comme on respire.

Cette période a amené des publications : celle de mon livre « Probabilités et Potentiel », repris plus tard en collaboration avec Dellacherie (puis ralenti par son départ et ma paresse au point que les derniers volumes en sont passés inaperçus). Surtout, celle des Séminaires de Probabilités de Strasbourg, encore vivants après trente ans.

Après la théorie générale est venue la période de calcul stochastique, c'est-à-dire des intégrales stochastiques et de leurs applications — aux équations différentielles stochastiques d'une part, et à l'analyse d'autre part. J'y ai travaillé mais avec Catherine Doléans (qui a fait la découverte principale, mais n'en a pas tiré de bénéfice, à cause de l'antiféminisme du Middle West), puis en suivant la voie ouverte par la thèse d'Emery.

Le troisième sujet de recherches que je considère comme important est une étude des « chaos » de Wiener (rien à voir avec la théorie du chaos dont tout le monde parle), commencée en essayant de comprendre une application qu'en avait donnée Malliavin. En utilisant des méthodes apprises chez Stein, j'ai pu démontrer une inégalité intéressante (qui a été aussi le point de départ des travaux beaucoup plus vastes de Bakry, puis Bakry-Emery). J'ai travaillé sur les chaos de Wiener, selon d'autre points de vue, avec Y.Z. Hu et J.A. Yan. Enfin je les ai retrouvés dans mon enthousiasme pour les « probabilités quantiques ».

A partir des années 70, je me suis tenu à l'écart des courants nouveaux des probabilités. Je n'avais plus assez de temps à consacrer à un travail de fond, ou même à la lecture. J'avais toujours devant moi un article à lire comme referee, des lettres d'évaluation à écrire pour ces terribles organismes de recherche américains, des rapports pour la commission du CNRS[1]. Mon intérêt personnel s'est tourné vers des sujets marginaux, où la compétition n'était pas trop vive. J'ai le regret d'avoir été un peu touche-à-tout, laissant tomber un sujet

---

[1] Encore pires, les années d'après 68 passées à rédiger des statuts et constitutions à la place d'un pouvoir politique défaillant, incapable (il l'est toujours) soit de savoir ce qu'il veut, soit de l'imposer.

au moment où les choses devenaient intéressantes et un peu difficiles. À trente ans je me serais sans doute comporté autrement.

A côté de la recherche et des publications, d'autres choses occupent la vie d'un mathématicien. Je n'ai fait de l'enseignement élémentaire que pendant les premières années de ma carrière : j'aimais surtout enseigner en premier cycle, mais les examens m'horrifiaient. J'avais une grande pitié pour cette foule de jeunes gens, filles et garçons, qui se faisaient massacrer, et pour lesquels je ne pouvais presque rien. Il n'aurait été possible de les tirer de là qu'au prix d'un effort que ni l'État, ni nous-mêmes n'étions prêts à fournir.

Dans les années 80, je me suis pas mal occupé des relations avec les probabilistes chinois et la Chine. Nous avons eu des collaborations très fructueuses avec J.A. Yan, W.A. Zheng, Y.Z. Hu et plusieurs autres. Cela m'a valu d'aller plusieurs fois en Chine, et aussi à Taiwan.

Je n'ai pas manqué d'honneurs : j'ai eu dans ma jeunesse le prix Peccot et le prix Maurice Audin, dans mon âge plus mûr le prix Ampère, j'ai été élu correspondant de l'Académie des Sciences. J'ai été nommé deux fois au Comité National du CNRS, et une fois au Jury d'Admission. J'ai été invité dans de grands congrès à l'étranger. Si j'en avais désiré davantage, j'aurais sans doute pu l'obtenir — et je ne ferai pas la fine bouche, tout cela m'a réjoui et encouragé, particulièrement les occasions de voyager en Inde, en Chine, au Japon. Mais j'éprouvais aussi un sentiment de malaise, dont je reparlerai.

Je pense avoir été un directeur de recherches à la fois bienveillant et tyrannique. Il est très difficile de se juger soi-même à cet égard, presque autant que de se juger en tant que père. Le départ de mes anciens élèves a été relativement difficile à supporter, particulièrement celui de Dellacherie — pourtant je suis trop lent pour travailler volontiers avec d'autres. Je n'ai pas eu beaucoup d'élèves ; sans doute parce que j'étais à Strasbourg, mais aussi parce que je manquais de sujets à leur donner. Quand j'avais une « bonne question », je la traitais moi-même. Dans les meilleures années, il y avait à la fois suffisamment de monde à Strasbourg pour justifier cours et séminaires, et suffisamment de postes d'assistants pour nourrir les étudiants, qui pouvaient alors trouver leur sujet eux-mêmes, sans pression excessive. C'est alors que je pouvais les aider.

● ● ●

Que j'aie été ainsi reconnu comme un mathématicien par le milieu mathématique, sous la forme de distinctions objectives, est d'autant plus important que *je ne suis jamais parvenu à me reconnaître moi-même comme tel.*

Je devrais être un adhérent des *Social Sciences* à la mode, tant j'ai été conscient pendant toute ma vie du caractère collectif de la science, du mouvement souterrain d'informations qui se trouvent à point nommé là où il faut (mes propres emprunts à Blackwell, à Volkonskii et Shur à des moments cruciaux), du caractère injuste de beaucoup d'attributions, du caractère fragile de la connaissance (des choses bien connues dans ma jeunesse sont maintenant oubliées), etc. En ce qui concerne les « dons », je ne me reconnais que le goût du travail (le premier de tous les dons, il est vrai), une excellente mémoire, et une certaine habileté manuelle qui m'a permis de taper des kilomètres de papier, pour moi et pour d'autres. Tout cela, en somme, des qualités de bon élève. Je n'en vois qu'une qui dépasse ce niveau : une certaine aptitude à distinguer l'essentiel de l'accessoire.

J'ai mis une certaine insistance à souligner plus haut que j'ai été reçu à quelques concours, mais dans les derniers rangs. Il me semble que je n'aurais pu réussir les épreuves d'un quelconque « Rallye Mathématique » destiné à sélectionner, à la manière des Russes, de futurs mathématiciens.

En gros, tant que j'ai eu la force de lire des articles difficiles, de les digérer et de les exposer, je suis arrivé à découvrir des choses nouvelles. Je sais par expérience que si l'on prend son temps pour lire, en cherchant vraiment à comprendre, on en est souvent (mais pas toujours : il y a un risque à prendre) récompensé bien au delà de ce qu'on espérait. Il est tout à fait normal qu'un mathématicien qui découvre un théorème utilise pour cela des chemins compliqués ou artificiels, et ne comprenne pas ce qui est vraiment au fond des choses. Chung m'a dit une fois « Monsieur Meyé, people say that you don't write your own papers, you rewrite the papers of other mathematicians ». Je peux accepter ce jugement en partie.

Le travail de lecture est peu glorieux, mais il est à la base de l'activité scientifique. Les historiens des sciences ne font attention qu'au processus de découverte, alors que le contrôle exercé sur celle-ci par des collègues (surtout s'ils sont mus par la jalousie) est essentiel. La vérification des travaux de de Branges sur la conjecture de Bieberbach, de Wiles sur le théorème de Fermat, a été un travail énorme — et personne ne tresse de couronnes à ceux qui passent un an à vérifier une démonstration et la trouvent fausse. C'est un peu comme si l'on réduisait une fourmilière à la reine. Le système d'avancement des scientifiques ignore trop cette activité. J'ai rencontré en Californie, lorsque j'étais étudiant, un vrai lecteur et écouteur qui se nommait John Woll. Comme il ne publiait pas suffisamment il a dû partir et j'ai perdu sa trace. Dans les séminaires, il avait l'air de tout comprendre tout de suite, et faisait des remarques intéressantes. Il a lu mon premier article et a découvert une partie des erreurs qui s'y cachaient.

J'ai beaucoup lu, mais j'ai aussi beaucoup été lu pendant quelques années, et je n'en suis pas peu fier.

• • •

La phrase relative aux *Social Sciences* est restée en suspens. Il me faut expliquer pourquoi je n'y adhère pas. Je sais parfaitement ce qu'on peut dire de la Science. Les gens lui demandent beaucoup trop, mais moi non. Non seulement j'ai depuis longtemps renoncé à tout savoir, mais je me suis persuadé que si je savais tout, je ne saurais pas grand chose. Cependant, mon enthousiasme pour la science n'a pas diminué, et je suis heureux d'avoir participé à la vie scientifique, même de façon relativement marginale.

L'histoire des sciences telle qu'elle est racontée dans les *Social Sciences* (voir par exemple *A qui appartient la découverte de l'électron*, dans *La*

Recherche de novembre 1997) est dépourvue de toute grandeur humaine. Elle ne s'exprime qu'en termes de pressions sociales, réseaux d'influences, copinages et hostilités. Je me rappelle avoir voyagé entre Paris et Strasbourg dans le même compartiment que trois profs. Pendant tout le trajet ils n'ont parlé qu'avancement : proviseurs, syndicats, commissions, concours. C'est l'univers des *Social Sciences*. Moi qui ai été un étudiant–2CV des années 50, j'y reconnais les étudiants-Austin des années 70, nés dans un monde pauvre en postes. Les *Social Sciences* ne sont que l'expression des classes moyennes frustrées.

Je veux bien qu'on soit corrosif. Un livre corrosif comme *La double hélice* de Watson laisse sa place à la grandeur, c'est-à-dire à l'admirable conjonction de l'intelligence et du travail. Pour ma part, au cours de ma vie de mathématicien, j'ai rencontré deux formes de grandeur. D'abord la grandeur intellectuelle, qui n'est pas forcément arrogante ou écrasante (je pourrais citer ici des noms en bien et en mal). J'ai beaucoup admiré des mathématiciens, pas tous célèbres, que possédait véritablement le désir de comprendre, et qui n'auraient pu vivre sans un beau problème difficile.

L'extrême concentration que j'ai rencontrée chez les « vrais mathématiciens », je ne l'ai éprouvée que de rares fois dans ma vie. Mon expérience la plus proche de celles que rapportent les grands créateurs concerne un travail mineur, où je voulais démontrer que la théorie du potentiel décrite axiomatiquement par Brelot entre dans la théorie des semi-groupes. Je n'y arrivais pas malgré beaucoup d'efforts, mais j'avais réduit le problème à la construction d'une certaine fonction sousharmonique, et je me suis couché en plein travail ; le matin, je me suis réveillé en pensant « considère une somme de carrés de fonctions harmoniques », et cela marchait.

The Life and Scientific Work of Paul André Meyer (August 21st, 1934 - January 30th, 2003) "Un modèle pour nous tous"

Marc Yor

Laboratoire de Probabilités et Modèles Aléatoires, Université Pierre et Marie Curie, Boîte courrier 188, F-75252 Paris Cedex 05, France

Summary. The life and scientific works of Paul André Meyer are presented, with some special emphasis on the contents of the Séminaires de Probabilités and the treatise Probabilités et Potentiel, both of which were key achievements in the career of Paul André Meyer.

Key words: General Theory of Processes, Séminaire de Probabilités, Stochastic Integration, Stochastic Differential Geometry, Quantum Probability

1 Introduction and a short vita

Paul André Meyer was born on August 21st, 1934 in Boulogne-Billancourt, a suburb of Paris. He entered the École Normale Supérieure de la rue d'Ulm in 1954, and was a student there between 1954 and 1958. After preparing and defending his doctoral thesis, under the supervision of Jacques Deny, he became Professor at the Université de Strasbourg, a position he held from 1964 until 1970, before becoming Directeur de Recherches in CNRS from 1970 onwards. He was elected a corresponding member of the Académie des Sciences de Paris in 1978. He retired completely from scientific activities in 1994.

Between roughly 1962 and 1992, Meyer's publications were constantly at the forefront of developments of research in probability theory, and more particularly in the study of stochastic processes.

In the following, I present the main results of Paul André Meyer, I hope in not too technical a manner, and I sketch a portrait of the admirable person he was, most certainly worthy of the accolade given to Laurent Schwartz by Alain Connes: "un modèle pour nous tous".

2 The genesis of Paul André Meyer's work

In order to understand the genesis of the work of Paul André Meyer, it is useful to consider the situation of probability theory in France in 1957, as Paul André Meyer was about to start his 4th year in the École Normale. The following lines are borrowed from a text of G. Choquet: *"Paul Lévy, who was considered abroad as one of the founding fathers of modern probability, was little known in France. France, which had been the cradle of probability with Pascal, was running the risk of seeing the line of her famous theoreticians (Borel, Fréchet, Lévy ...) cut short. The Parisian mathematicians, being aware of this danger, invited M. Loève, a former disciple of P. Lévy, to come from Berkeley and spend a year in Paris, in order to sow the good seed by giving a course and a seminar ...*

By the end of the year, Meyer had assimilated the probabilistic techniques and was ready to begin his own research; this was the time when G. Hunt was publishing, in the United States, two fundamental memoirs which were renewing at the same time potential theory and the theory of Markov processes by establishing a precise link, in a very general framework, between an important class of Markov processes and the class of kernels in potential theory which French probabilists had just been studying. Meyer, who was in close relation with these potentialists, and who was quite knowledgeable about convex functional analysis, was particularly well positioned to explore Hunt's theory.

In 1961, under the guidance of Jacques Deny, Meyer defended his thesis on multiplicative and additive functionals of Markov processes, which established him at once as a top researcher among probabilists and potentialists."

3 The post-doctoral period (1961-1966)

(3.1) Following his thesis, Meyer (1962-1963) published some fundamental papers in probability theory which demonstrated his technical virtuosity, and the depth of his understanding of supermartingales, i.e.: processes which model "unfair games". More precisely, in these papers Meyer shows that, as a deep extension of Doob's results in the discrete time case, any supermartingale of class (D) may be decomposed as the difference of a martingale and a predictable increasing process.

(3.2) A first—very structured and pedagogical—evaluation of the results obtained so far by P.A. Meyer appeared in 1966, with the publication of his book: Probabilités et Potentiel (1966) which is divided into 3 sections:

A. Elements of Probability Calculus
B. Martingale theory
C. Analytical tools in Potential theory.

This book, which was immediately translated into English (Blaisdell, 1966) had a huge impact. It became the reference book for a large number of probabilists, roughly for the period 1967–1975, at the beginning of their research in probability. In particular, for myself and probabilists of my generation, it opened the door to graduate study (leading to the Diplôme d'études approfondies) and the preparation for the Agrégation (an exam which, in the French system, enables one to take up a teaching position in Classes préparatoires aux grandes écoles ...).

Throughout his career, Meyer always came back to the contents of his book, adding successive layers of knowledge to it; it was, indeed, the cornerstone of his work.

4 The first fifteen years of the Séminaire de Probabilités

Not only was Paul André Meyer a virtuoso of probability theory, potential theory, and so on, but he also had his own deep vision of establishing an easily accessible body of knowledge, essentially by concentrating his own publications, as well as those of his students, collaborators, etc..., in Séminaire Notes. Here he was following the tradition of the Brelot-Choquet-Deny mimeographed volumes.

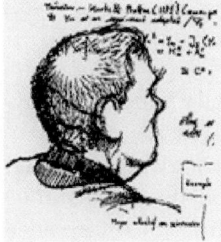

This led to the birth of the Séminaire de Probabilités, which appeared annually from 1967 onwards, each volume totalling between 200 and 500 pages, within the Springer Lecture Notes in Mathematics series. In each Séminaire, Meyer systematically discussed the main developments of the moment, at an international level, of research in probability.

This really was a titanic work, into which Meyer threw himself, with all his intellectual strength and enthusiasm, together with the help of C. Dellacherie and M. Weil. He often presented much clearer versions than the original works, he extended the results of many authors, from the continuous to the discontinuous case, from the integrable to the non-integrable case, and so on ..., so that the Séminaire became an indispensable working tool for any researcher concerned with stochastic processes. In fact, Meyer always insisted on this aspect of the Séminaires, and often compared them to the "working book" of a general medical doctor, the pages of which need to be replaced as medicine evolves. With each volume, Meyer provided some comments, improvements, and corrections to previous articles, always closely related to the most recent developments in the subject.

The list of the fundamental results to which Meyer gave, in the Séminaires, an enlightened view and an essential personal contribution, is indeed quite lengthy; after making some difficult choices, I list the following:

- in Séminaire I (1967), a sequence of four papers about stochastic integrals, where the spaces of additive functionals - martingales for a general Markov process - are thoroughly studied, following the fundamental paper of Kunita-Watanabe, which also appeared in 1967;
- in Séminaire II (1968), the detailed guide to the general theory of stochastic processes, which culminated in 1972 with the publication of the book of C. Dellacherie: Capacités et Processus Stochastiques;
- in Séminaire III (1969), the Burkholder inequalities in martingale theory;
- in Séminaire IV (1970), the discussion of Stroock-Varadhan's martingale problem formulation, which constitutes a very powerful and elegant method of constructing Markov processes with a given (pre-)generator, thus completing in an original manner Itô's pathwise construction of diffusion processes with the help of the resolution of stochastic differential equations;
- in Séminaire V (1971), a wonderful presentation of Itô's excursion theory, which had just appeared in the Berkeley Proceedings (1970), as well as a simplified proof of Knight's theorem about continuous orthogonal martingales (Meyer had just heard the lecture by F. Knight in the Math. Institute of Oberwolfach). I should not forget Meyer's presentation of the results about time reversal for Markov processes, obtained by Chung and Walsh. In fact, Meyer discussed this topic (namely, time reversal) a number of times, starting with Nagasawa's original results (1964) and progressing towards those of Azéma (1972) which he discussed in Séminaire VIII, to finish with Kuznetsov's contributions (see later).
- in Séminaire VI (1972), Meyer's presentation of Föllmer's measure associated to a supermartingale (the original article appeared the same year) thanks to which the tendency of these processes (supermartingales) to decrease may be measured in a global way; the same Séminaire also contains a renewed discussion (see Séminaire III) of the inequalities of Burkholder-

Davis-Gundy for martingales; these inequalities played, a few years later, an essential role in L^p estimates of semimartingales;

- Séminaire VII (1973), contains an impressive collection of 11 articles by Meyer, on very different subjects, among which may be found: surgery of Markov processes, Azéma's work about the general theory of processes, as well as a paper which became quite famous: "On a problem of filtration", in which Meyer was deeply astonished by the fact that the probability theory he had developed might be used in applied questions! What happened next (see below!) can only have astonished him further!

- in Séminaire VIII (1974), besides a lecture about time-reversal, one finds a big paper written with Maisonneuve concerning homogeneous Markovian random sets;

- again, in Séminaire IX (1975), the contributions of Paul André Meyer are very important, roughly about fifteen; they were written on his own, or in collaboration, or holding the hand of a researcher ... There, one finds some complements to the general theory of processes, some studies of the theory of flows (7 papers!), some studies of time reversal, work on predictability, as well as a discussion of the marvellous book by A. Garsia about the (inequalities for) discrete time martingales. With hindsight, one realizes that Meyer is gathering his ammunition for the fireworks which are about to occur in the next Séminaire ...

- Séminaire X, which appeared in 1976, and contains the "Cours sur les Intégrales stochastiques" (150 pages) by Meyer, constitutes an "explosion", both of the contents of the Séminaire and of Meyer's creativity: this period was probably one of the magical moments in the research career of Paul André Meyer, a period when the efforts and tenacity shown during many years finally bore fruit. More precisely, here, with the help of the general theory of processes, and that of the many papers written by Meyer, C. Dellacherie and C. Doléans in the previous Séminaires, Meyer was now able to develop a theory of stochastic integration with respect to a general semimartingale. One of the strong points of this Course is the derivation of a Tanaka formula for the local times, and their existence, related to a general semimartingale.

It does not seem too much of an exaggeration to say that this "Cours" was the result of about thirty years of research by the international probabilistic community, culminating in a presentation as general and efficient as possible of the theory of stochastic integrals. This theory was initially created by Itô as early as 1944 for Brownian motion, and became known by probabilists outside of Japan as late as 1969, with the wonderful, although very concise, book of H.P. McKean: *Stochastic Integrals*, but which only discusses diffusions (i.e.: Markov processes with continuous trajectories).

The richness of the "Cours" is largely acknowledged by the impressive crop of results and publications which it generated in the years following its publication, that is, roughly between 1976 and 1980:

- the Séminaires XI to XIV contain many complements to the "Cours"; a whole new generation of French probabilists (Emery, Jacod, Jeulin, Lenglart, Lépingle, Stricker ...) and also those abroad (Pratelli, Yan, Yoeurp) became accustomed to changing—probabilities, filtrations, and so on—, in a general framework, by having the essential techniques developed in the Cours. Local times became a crucial element in the study of stochastic processes: a little later, J.F. Le Gall used them to simplify considerably the arguments of the Japanese school on fine results for stochastic differential equations; a little later again, a systematic exploitation of local times of intersections of planar Brownian motion were made, in order to study its multiple points (see Le Gall's Cours de St-Flour (1990); LNM 1527);
- the richness of the Cours is also shown by the publication, in the following years, of a number of treatises on stochastic calculus (with, or without, applications ...). In one of the first such books, R. Durrett, then a professor at UCLA, wrote that his own book was mainly a translation from the Alsatian to the American language ...

All of these developments led P.A. Meyer, at the end of 1979, to ask J. Azéma and myself to be in charge of the yearly editions of the Séminaires, from Séminaire XIV onwards, a task which we accepted, although we felt strongly that Meyer did us an honor of which we were unworthy; however, this may have facilitated Meyer's deep involvement in the developments—by then in full blossom—of Stochastic Differential Geometry, with the essential and independent contributions at that time of J.M. Bismut, D. Elworthy, P. Malliavin and L. Schwartz.

In order to accommodate the abundance of new results in this field, which were extremely clearly exposed by Meyer in Séminaire XV, it was necessary to add an additional volume (the whole Séminaire comprised more than 500 pages, to the horror of Springer!). Among these papers, I would like to cite the celebrated: "Géométrie stochastique sans larmes", as well as a very beautiful synthesis given at the Durham Congress (July 1980), which constituted a key moment for the international probabilistic community.

5 The five volume Treatise: Probabilités et Potentiel (1975–1992) by C. Dellacherie and Paul André Meyer (and B. Maisonneuve, for vol. V)

(5.1) At this point in the evocation of the works of Paul André Meyer, it seems judicious to leave aside for a moment the litany of the Séminaires (to return to it later) in order to present the main characteristics of the huge scientific undertaking which the writing of the Treatise Probabilités et Potentiel, published in 5 volumes by Hermann between 1975 and 1992, represented.

(5.2) To write of the Treatise that it is a completely new edition of the book of Paul André Meyer published in 1966 (see Section 3 above) gives only a very vague idea of the corpus of knowledge which is contained within it. It represents, in fact, a deep synthesis of the main results obtained by Dellacherie and Meyer, and more generally by the group of probabilists formed by, educated by, or linked with, one way or another, the "Strasbourg school of probability", which was simultaneously, quite virtual (before the advent of the Internet), and also truly "real"!

(5.3) I now discuss, volume by volume, chapter by chapter, the contents of the Treatise:

- Volume I: (Chapters I to IV). Published in 1975. In this volume, the authors present the foundations of the treatise: the two first chapters contain the complete theory of integration and its diverse variants for probabilists. Chapter III presents the theory of analytic sets, and Choquet capacities, which are indispensable tools for the section theorems; the latter play a central role in the general theory of processes. Chapter IV is devoted to an introduction to stochastic processes: their filtrations, stopping times, optional, predictable, progressive processes ...

- Volume II: (Chapters V to VIII). Martingale theory; published in 1980. This second volume gives an exhaustive presentation of the theory of martingales, beginning with the discrete case (Chapter V) and then continuing in the case of continuous time (Chapter VI), where the full power of the general theory of processes, including "direct" and dual projections, is utilized. The decomposition theorems of super- and semi- martingales are presented in Chapter VII where one can also find a study of H^p-spaces of semimartingales, and the H^1-BMO duality. Finally, the theory of stochastic integrals is developed in Chapter VIII. Apart from the general Itô-Tanaka formula, this chapter contains the characterization of semimartingales as "good integrators" taking values in L^0 due to Dellacherie and Bichteler (following the pioneering works of Métivier and Pellaumail). The full strength of functional analysis (vector measures taking values in L^0; works of Maurey and Nikishin) is put to use ... All of Meyer's virtuosity was necessary to remain in such close contact with the fascinating new developments of that period. Although it says, in the Preface to this volume, that the presentation does not pretend to be definitive because the theory progresses too fast, it transpires that, more than twenty years later, it still gives a very coherent and up-to-date discussion of martingale theory. This discussion was further completed in volume V.

- Volume III: (Chapters IX, X, XI). Discrete potential theory; published in 1983. This third volume, which consists of 3 chapters, provides a first introduction to potential theory: more precisely, in Chapter IX, potential theory with respect to a discrete time semigroup (which amounts to a discussion relative to a single kernel) is presented; Chapter X presents the notion of réduite in the set-up of the "gambling houses" of Dubins and Savage, whose beautiful book: *How to gamble if you must* is too little known; this chapter also contains some beautiful applications, which go far beyond those to be found in the 1966 book (Bernstein and Bochner's theorems, and so on ...). Finally, Chapter XI presents new methods in capacity theory, and their applications to gambling houses.

- Volume IV: (Chapters XII to XVI). Potential theory associated with a resolvent; theory of Markov processes; published in 1987. This sub-title indicates quite well the nature of the two parts of the volume: the first part (Chap. XII and XIII) is devoted to the study of the notion of a submarkovian resolvent, and goes as far as the representation theorem of excessive functions in terms of extremal ones; the second part discusses the theory of Markov processes, presented here in the form of Feller processes, which suffices to present Ray processes. Besides general discussions—which are unavoidable on this subject—one also notices that Chapter XV contains a deep study of local times associated with closed random sets and, in particular, it contains Itô's excursion theory, and some of its applications to the study of Brownian motion. Finally, Chapter XVI shows how the most general class of Markov processes, namely the "right processes", may be reduced to Ray processes. I should also indicate that this volume contains a deep discussion of the opérateur "carré du champ", a notion due to H. Kunita (1969) and rediscovered by J.P. Roth in his thesis (1976) in potential theory; this notion then played an important role in the derivation of the log-Sobolev inequalities of Bakry-Emery. This volume also presents the Lévy systems of Markov processes, which had already been thoroughly studied by P.A. Meyer in Séminaire I (see above).

- Volume V: (Chapters XVII to XXIV). Markov processes (the end); Complements of Stochastic Calculus. This fifth and final volume begins with Chapter XVII, in which the authors gathered some important complements to the so-called "right processes", which constitute the ultimate "good" class of Markov processes and whose definition emerged little by little, as Meyer's works on Markov processes developed. The second part of the Chapter is devoted to the study of excessive measures. The whole Chapter is very much correlated to the monographs of Sharpe (1988) on Markov processes, and of Getoor (1990) on excessive measures.

Chapter XVIII discusses in depth the important subject of time-reversal for Markov processes: this operation is made difficult by the fact that if one reverses a Markov process at a fixed time, the reversed process is no longer time-homogeneous; on the other hand, the time-homogeneity is preserved when reversing at a "return time". To discuss this result, the authors presented all the necessary notions and techniques, which were often borrowed from Azéma's work in 1972–73.

Chapter XIX discusses the theory of Markov processes with random birthing, a theory which culminated in the works and, in particular, the measures, of Kuznetsov (1974), which are presented in the first section.

This is followed by a second section on the theory of flows and Palm measures. The Chapter concludes with some applications to last exit decompositions for Markov processes.

Chapter XX offers a definitive treatment of filtrations and martingales related to random sets, and is partly inspired by Azéma's works on random closed sets. The same chapter also contains an excellent presentation of the theory of enlargements of filtrations, a theory which although quite natural, the "French school of probability" has had a lot of difficulty in exporting, despite repeated efforts ...

Chapter XXI is devoted to instances of chaotic decompositions, which were then very much studied, following the discovery by M. Emery in Séminaire XXIII of the remarkable fact that Azéma's martingale enjoys the chaotic decomposition property, which brings this martingale to the forefront of the important Markov processes, almost with the same fundamental status as Brownian motion and the Poisson process. The same Chapter also contains a beautiful introduction to analysis on Wiener space, in other words, to Malliavin Calculus, which is close to the presentation of D. Williams in his "To begin at the beginning", in the Proceedings of the 1980 Durham Conference (Springer LNM 851), which was mentioned earlier on.

In Chapter XXII, the works of Meyer on the Littlewood-Paley inequalities and the Riesz transforms are taken up in detail. Meyer contributed immensely to the popularization of the works of Burkholder-Gundy-Silverstein, so that they should be better known in both the probabilistic and analystic communities, by going back and forth between the purely analytical studies of harmonic functions on one hand, and, on the other hand, Brownian studies, and/or studies involving certain Lévy processes obtained from the subordination of Brownian motion.

Chapter XXIII provides yet another pedagogical presentation of the martingale inequalities, as they were developed in, and around, the Strasbourg Séminaires (an excellent presentation is given by Lenglart-Lépingle-Pratelli in Séminaire XIV). One also finds, in the same Chapter, a deep discussion of multiplicative representations of submartingales, a theme which was dear to Meyer, and was inherited from the early works of Itô, Watanabe, Kunita, etc. Stochastic differential equations are also studied in a general framework.

Chapter XXIV, which closes this fifth volume, and the whole treatise, is devoted to deep results about the descriptive theory of sets, in a form which may be used by probabilists.

(5.4) To summarize, the Treatise represents a wonderful synthesis of the works of Paul André Meyer, starting from the publication of his book in 1966, and decanting the twenty five volumes of the Séminaire which were published up to 1992, and are described as a treasure by the three authors of volume V. It is really sad that this volume is certainly among the less disseminated ones; I hesitate to formulate an explanation, (but perhaps ...).

(5.5) This sketch of the contents of the Treatise should not prevent us from noticing that Paul André Meyer wrote several Lecture Notes in Mathematics (Springer LNM), of high quality:

- LNM 26: *Markov processes* (1967)
- LNM 77: *Markov processes: Martin's boundary* (1968)
- LNM 284: *Martingales and Stochastic Integrals I* (1972)
- LNM 307: *Presentation of Markov processes* (1973) In: École d'Été de Saint-Flour, 1971

without forgetting the remarkable:

- LNM 1538: *Quantum Probability for Probabilists* (1993)

which, despite the insistence of Springer-Verlag and of many readers, he always refused to publish as a book.

6 Back to the Séminaires de Probabilités

(6.1) Mentioning Meyer's Quantum Probability Lecture Notes brings us back quite naturally to Meyer's contributions to the Séminaires de Probabilités (the discussion of which we left aside after Séminaire XVI (1980)). Two "periods" appear quite clearly: the first one between 1980 and 1991, and the second one, between 1992 and 1994, during which Meyer published only a little.

(6.2) Concerning the first period (1982–1991), one should cite, first of all, some important papers of Meyer about Nelson's stochastic mechanics (Séminaires XVIII (1984) and XIX (1985)), which is really a renaissance from the famous Princeton Lecture Notes of Nelson (1967): Dynamical Theories of Brownian Motion. These Séminaire papers were written by P.A. Meyer and W. Zheng, and are very often quoted. But the main contributions of Meyer from this period are the publications, in the form of a series, of his Éléments de Probabilités Quantiques (vol. XX (1986): Chap. I to V; vol. XXI (1987): Chap. VI to VIII; vol. XXII (1988): Chap. IX and X) which, of course, constitute a first draft of what ultimately became the volume: *Quantum Probability for Probabilists* (1993), already cited above.

Next, in Séminaire XXIII (1989), Meyer constructs the solutions of structure equations for martingales, and highlights their relationships with quantum probability. This was an occasion for Meyer to popularize the astonishing properties of Azéma's martingale, which may be obtained as the projection of a real-valued Brownian motion on the filtration generated by its signs, and which has already been mentioned in the above discussion of Chap. XXI of the Treatise.

All of this work, and these publications, bore fruit: they kindled the interest of French probabilists in Quantum Probability. I remember that, during those years, Meyer used to give informal lectures on Saturday mornings in the Laboratoire de Probabilités of Paris VI, on his way to visit his parents who lived in the Paris region; it was during these lectures that Ph. Biane assimilated very quickly the topic of Quantum Probability and became, a few years later, a leading expert on Voiculescu's free probability theory.

The Séminaires during these years are also strongly enriched with the answers given by "Meyer's listeners" to his incessant interrogations about quantum probabilities. Finally, about the end of this "first period", I should like to mention the 3 papers by Meyer about Quantum diffusions in Séminaire XXIV (1990).

(6.3) Concerning the second period, one should mention four articles of Meyer in Séminaire XXVII (1993), which still concern Quantum Probability (written while his LNM 1538 was appearing).

To this day—that is in 2005—the Séminaire de Probabilités still appears annually, under the guidance of M. Emery (Strasbourg) and M. Ledoux (Toulouse), until 2004, and, after 2005, with a new editorial team. Let us hope that the original spirit of the Séminaire, originating of course from Meyer, is still easily recognizable: that the Séminaires contain many papers from young researchers, they also feature specialized Courses which are quite close to current developments, and try to imitate the role of Meyer's Very Quick Writing Machine (sorry, in English, VQWM does not look at all like TGV; I wonder why?) ...

The general theory of stochastic processes (GTP, for short) has become an indispensable tool in (the foundation of) Mathematical Finance, as W. Schachermayer (Vienna) (see the beginning of his St-Flour Summer School, 2000) and F. Delbaen (ETH Zürich) have demonstrated very clearly. Immediately after P.A. Meyer's death, Delbaen wrote, (in substance): *"We constantly use the GTP, which we call the 'Meyereries'; it looks as if Paul André did all he could to open the gate for us ..."*. I wonder what Meyer would think about this? I think I understand that, a little similarly to D. Williams in Great Britain, J. Pitman in Berkeley, and many others ... he was somewhat skeptical, and a little ironic, even saddened, as far as this very first role of the GTP in Math. Finance is concerned, but the history of Mathematics made us familiar with a number of these role reversals which may happen between "the most theoretical" research, and the "most practical" applications. But, is this really the case here?

Whatever the answer to this question may be, this domain of applications of the GTP and, hopefully, others to come, illustrate the pertinence of the Ariadne thread of the GTP, of which Meyer was, simultaneously, the founder and the working craftsman ...

This Ariadne thread of the GTP consisted, for Meyer, in highlighting "the martingale backbone" inside studies done in the Markovian framework (following thus the founding paper of Kunita-Watanabe (1967)) and then eradicating, as much as possible, the Markov property ... a property which is open to discussion when one examines closely many random phenomena (although, conversely, any process may be considered as a Markov process, under the condition that it is "immersed" in a sufficiently big space of measures (e.g., Knight's prediction theory (1975), expounded in book form by its author (1991)).

7 Paul André Meyer, the man

In this final section, I would like, at the same time, to synthesize the activity of the researcher, and to evoke other facets of Paul André Meyer, at least the ones I was fortunate enough to get a glimpse of. I find it most difficult to choose the right words, as he disliked very much every pretentious label. Should I say that he was an invaluable "Directeur de Recherches", in the full sense of the term, that is knowing in which direction he was going, and warmly inviting his students and, more generally, everyone around him to follow? This is indeed true, but the feeling which all who worked with him and lived around him got was much deeper than that, i.e. that of being with an elder member of "the family"—he was in fact my so-called "godfather" when I was in CNRS—in whom we could have complete confidence and for whom mathematical rigor went hand in hand with a noble spirit of mind and a moral rectitude.

I have also been asked—by several colleagues, among whom Nick Bingham—to make a list of the (direct) students of Paul André Meyer: to answer this query, I may of course write down the names of C. Dellacherie, M. Weil, B. Maisonneuve, D. Bakry, M. Emery, C. Stricker, but I find this task almost impossible, as every researcher—young and not so young—who ever came in contact with Paul André Meyer and submitted a paper to the Séminaire knows very well how much he/she owes to Paul André, who invariably rewrote the paper, added his own comment, and so on ... Should not these researchers be considered as students of Paul André?

I would also like to present three remarks[1] which were made to me by Yves Meyer, a great friend of Paul André, an expert on wavelets, as well as the thesis supervisor of Martin, Paul André's son:

- the first remark is the reason P.A. Meyer studied probability theory: "to understand the world around us" (P.A. Meyer dixit). However, by Paul André's own confession, this was not enough: to understand this world, one needs either many more probabilistic tools, or other tools! This brings us to the work of Paul André in Stochastic Differential Geometry, as well as in Quantum Probability;

[1] See also the Témoignage of Yves Meyer: "Un chemin vers la lumière", in this volume.

- the second remark is close to the first one: P.A. Meyer was always extremely interested, if not passionate, about other cultures, and in particular, those of the Far East. P.A. Meyer was inhabited by a fever and a spiritual questioning which brought him to an open and erudite knowledge of cultures and religions;

- the third remark concerns the very high regard which P.A. Meyer had, contrary to many mathematicians, for statistics as an intellectual discipline. As he wanted to "form" my background, he kept explaining to me that statistics is not a trivial implementation of intellectual tools elaborated by probabilists. On the contrary, statistics builds itself from a never-ending dialogue between mathematical models and problems posed by practitioners. Thanks to his infinite humility and respect for others, P.A. Meyer conveyed to me how wrong the arrogant pretension of "pure mathematicians" that any probabilist could be a good teacher of statistics is ... These were discussions in the 60's; nowadays this opinion is widely held ...

Returning to some more personal memories, I keep as a treasure in my mind some stays I had the privilege to make in his family apartment, where, surrounded by his wife Geneviève, and sometimes their grown-up children, we exchanged some simple thoughts, and mostly silence, sometimes looking at the impressive bookshelves where mathematics and physics competed with the best literature from the entire world.

I still kick myself, and so do several colleagues, for not daring to disturb him as soon as he retired, in 1994; Paul André had given us so much, we could not keep bothering him with our little problems ...

This permanent position of withdrawal and this discretion were essential components of the personality of Paul André Meyer; if some ultimate proof were necessary, it is provided by ten pages which Paul André wrote in January 1998, entitled: "Titles and Works: Postface"[2], all of them filled with a good dose of humor and humility, and which were made public in the Journées de Probabilités de Toulouse, in September 2003.

As a conclusion, may the French educational system and, more generally, university teaching throughout the world, continue to favor the development of such figures.

---

[2] This is the first item in this volume.

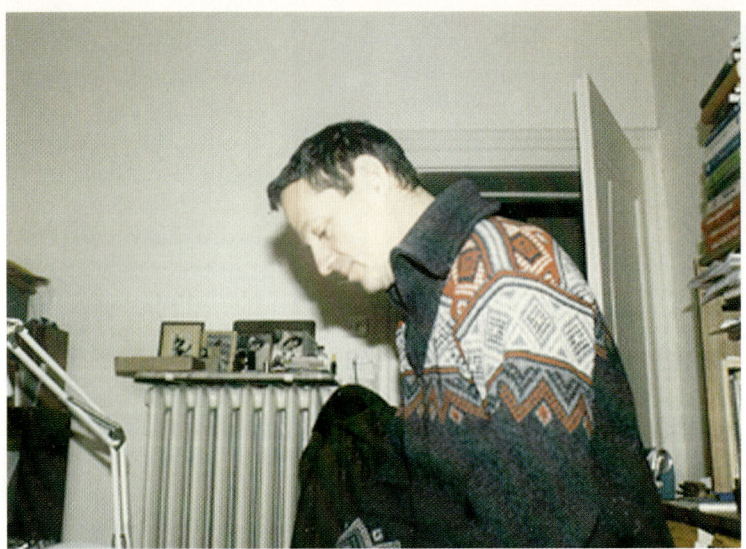

Acknowledgements: I would like to thank Madame Geneviève Meyer who sent me the four photographs of Paul André, and to assure her of the affection in which the probabilistic community held Paul André and his family. Anthony Phan, for the two drawings of P.A. Meyer attending some Séminaire sessions, Yves Meyer for his remarks, and finally Christina Goldschmidt and Nick Bingham for their efforts to eradicate my gallicisms.

Disparition de Paul-André Meyer

Stéphane Attal

Institut Camille Jordan, Université Claude Bernard, 21 avenue Claude Bernard,
F-69 622 Villeurbanne Cedex, France
e-mail: attal@math.univ-lyon1.fr

Jeudi 30 janvier 2003, Paul-André Meyer est décédé d'un infarctus foudroyant.
Cette disparition est celle d'un grand mathématicien, qui a transformé le
paysage de la théorie des processus stochastiques en France et dans le monde.

Souvenirs

Lorsque j'ai appris la nouvelle le lendemain matin, par un courrier électronique
de sa fille Thérèse, j'ai bien entendu été très triste, assommé par le caractère
complètement inattendu de cet événement. Il y a quelques semaines à peine
je conversais encore avec lui et il était comme toujours vif et gai.

Mais c'est surtout lorsque je me suis rendu à Strasbourg pour la cérémonie
que j'ai été profondément bouleversé. En arrivant à Strasbourg même, au
département de mathématiques ou en refaisant le chemin à pied jusque chez
lui, tous les souvenirs sont revenus avec énormément de poids.

Je repensais à ce séminaire du mardi matin. Toutes les semaines nous nous
retrouvions devant le tableau du quatrième étage de la tour I.R.M.A. Nous
étions parfois 10 personnes, parfois 3 personnes, et nous écoutions Meyer. Il
venait chaque fois avec un nouveau texte dactylographié : il avait démontré
un résultat nouveau, ou alors il avait lu et entièrement réécrit un article d'un
autre, ou encore il nous dispensait le 10e chapitre de son cours sur les algèbres
de von Neumann, les algèbres de Lie ou la théorie quantique des champs.

Je repensais à l'intérêt et à l'enthousiasme constant qu'il avait pour tout
ce que je faisais. Je devais surement être la millième personne à lui montrer
fièrement ses résultats, mais je crois bien que son intérêt était sincère, intact.
Une joie permanente à voir les mathématiques avancer, à voir un jeune pro-
gresser.

Je me souvenais de ce chemin que nous faisions souvent ensemble à midi,
je l'accompagnais à pied vers chez lui et nous parlions. Nous parlions de
mathématiques bien sûr, mais aussi de voyages (en particulier de l'Inde), des
langues, de musique classique.

Je repensais à ces séminaires où il écoutait l'invité pendant les cinq premières minutes, puis fermait les yeux ou lisait son courrier pendant le reste de l'exposé, et, à la fin, posait de nombreuses questions très pertinentes, proposait des directions de développement qui émerveillaient l'orateur.

Je me souvenais du premier congrès où je fus invité pour présenter mes résultats, il a tenu à m'accompagner. On a fait une marche en forêt et il m'a raconté ses débuts. Je crois qu'il était ému comme moi.

Je me souvenais de la fois où je n'ai pas été retenu comme professeur sur un poste et il m'a dit « C'est une très bonne nouvelle ! Tu ne dois pas quitter le C.N.R.S. avant d'avoir effectué encore de nombreux voyages dans des départements de mathématiques ou de physique à l'étranger ! »

Je me disais que tout cela avait été un bagage extraordinaire pour un jeune chercheur et que je ne le revivrai sans doute plus jamais.

Je me sentais orphelin en ce mardi matin pluvieux devant l'église St Maurice de Strasbourg.

Parcours

Je vais maintenant retracer les grandes lignes du parcours personnel et scientifique de Paul-André Meyer. Je ne l'ai moi-même côtoyé que durant sa période « Probabilités Quantiques ». Pour les autres périodes je me suis appuyé sur des renseignements de Michel Émery, Claude Dellacherie et Michel Weil, mais aussi de Martin Meyer pour les étapes personnelles. Je les en remercie. J'ai aussi eu à ma disposition un texte privé de Paul-André Meyer lui-même, retraçant son parcours.

Paul-André Meyer est né en 1934 dans la région parisienne, dans une famille de négociants qui s'appelaient à l'époque Meyerowitz. Entre 1940 et 1946 sa famille se réfugie en Argentine. Il gardera toute sa vie une connaissance parfaite et beaucoup d'amour pour la langue espagnole. Sa femme Geneviève a aussi vécu son enfance à Buenos Aires et a fréquenté le même lycée français. Mais ils ne se sont rencontrés que beaucoup plus tard sur les bancs de la Sorbonne.

De retour à Paris, il entre au lycée Janson de Sailly et découvre un vrai goût pour les mathématiques au contact d'un professeur qui l'a marqué, M. Heilbronn. Il passe le bac en 1952 et à la même époque son père fait changer le nom de leur famille en Meyer. Il entre à l'E.N.S. en 1954, en s'y préparant seul.

Meyer, dans ses notes personnelles, ne se présente pas comme un élève particulièrement brillant, reçu parmi les derniers à l'E.N.S. et à l'Agreg ; mais il est au moins certain que sa progression, à partir du moment où il a entamé sa thèse, a été fulgurante.

À ses débuts de normalien, il suit le séminaire de Théorie du Potentiel avec ses professeurs Choquet et Deny. Sur le conseil de Brelot, il s'intéresse à des travaux récents de Hunt qui montre des résultats importants de théorie du potentiel par des méthodes probabilistes. Meyer rencontre Loève qui passe une année à Paris, puis le suit aux États-Unis (Berkeley). Il entre en contact avec Doob, dont il avait lu le grand traité, alors tout récent. Il s'agit là d'une période qui va vraiment déclencher sa carrière.

Ses premiers résultats importants dans les années 60 furent de caractériser les fonctions excessives qui se représentent comme potentiel d'une fonctionnelle additive. Puis en appliquant ces méthodes à la théorie des martingales, il caractérise les surmartingales qui sont différences d'une martingale et d'un processus croissant. C'est la fameuse « Décomposition de Doob-Meyer ». En appliquant cela au carré d'une martingale la voie est ouverte à toute la théorie de l'intégration stochastique pour les martingales de carré intégrable (en particulier grâce à la définition des crochets droits).

Dans ces années sa vie personnelle s'accélère : conversion au catholicisme puis mariage en 1957, naissance de 3 enfants entre 1959 et 1962 (la quatrième naîtra en 1969), prix et cours Peccot en 1963 et surtout installation de la nouvelle famille dans la région de Strasbourg en 1964.

Meyer dit qu'ils ont tellement aimé Strasbourg, l'Alsace, les Vosges et la Forêt Noire, qu'ils n'ont plus bougé. Et cela malgré de très nombreuses propositions qui lui seront faites de revenir à Paris (ce qui n'empêchera pas Meyer de garder des contacts très étroits avec les collègues probabilistes parisiens, en particulier avec Neveu). C'est le début de « l'École strasbourgeoise des probabilités ».

Très vite, Meyer va développer une équipe avec ses étudiants, en particulier Doléans, Dellacherie, Maisonneuve, Weil. Il va recevoir de nombreux visiteurs Doob, Ito, Chung, Spitzer, Walsh, Knight, Getoor, etc. En quelques années, Paul-André Meyer et son groupe vont étendre tous les résultats décrits ci-dessus et vraiment créer la « Théorie générale des processus ».

Ils posent la théorie définitive de l'intégration stochastique avec intégrand prévisible ct intégrateur semi-martingale (Schwartz a un jour suggéré qu'elles étaient bien mal nommées et qu'elles devraient s'appeler « meyergales »).

La théorie des temps d'arrêt va y jouer un rôle crucial et offre un langage naturel qui n'a pas d'équivalent en analyse. Associée à ces temps d'arrêt est la théorie des différentes mesurabilités pour les processus (progressifs, optionnels, prévisibles). C'est une théorie extrêmement subtile (cf. les fameux théorèmes de « Début » ou de « Section ») mais maintenant tellement clarifiée que l'on a du mal à imaginer que des notions aussi simples que la tribu prévisible engendrée par les processus continus à gauche, ou les temps d'arrêt prévisibles annoncés par une suite de temps d'arrêt, ont demandé énormément de temps et d'efforts pour arriver à une telle efficacité. Meyer écrit : « Ces travaux sont destinés à rester, j'en suis persuadé, et de la meilleure manière qui soit : en devenant des trivialités, que l'on utilise comme on respire. »

C'est aussi l'époque du premier livre de Meyer (1966), « *Probabilités et Potentiel* », qui sera ensuite repris et enrichi en 5 volumes avec C. Dellacherie (le 5ᵉ volume en collaboration avec B. Maisonneuve aussi). Véritable bible de la théorie générale des processus stochastiques, de l'intégration stochastique, des processus de Markov..., ces volumes ont connu un succès décroissant avec le numéro du volume. Le premier volume est dans le bureau de presque tout probabiliste, le cinquième n'a pas dépassé 100 ventes. C'est certainement une bien grande injustice pour une série inégalée encore aujourd'hui, et il semble que Paul-André Meyer en avait gardé une certaine amertume.

C'est aussi l'époque où démarre la série des « Séminaires de Probabilités ». Commencée en 1967 cette série de volumes chez Springer continue encore aujourd'hui avec le volume XXXVI qui vient de paraître.

Cette édition annuelle a constitué à une époque une véritable référence. Certains articles ou cours qui y figurent ont influencé toute une génération de mathématiciens. C'est une publication, originale à plus d'un titre, à laquelle Meyer était très attaché. D'abord parce que c'était son enfant, mais aussi pour le vent de liberté qui y soufflait. Ces volumes contenaient bien entendu des articles fondamentaux qui restent des références 30 ans plus tard, mais aussi des remarques, des mises au point, des cours, et tout un joyeux mélange qui a été le ferment de la progression des idées et de la communication entre chercheurs de toute une époque.

Ce mélange a aussi porté du tort à l'image de cette publication auprès des mathématiciens non probabilistes, qui l'ont considérée comme une publication pas très sérieuse. On connaît des exemples de commissions de spécialistes qui ne comptent pas les articles publiés dans cette série comme de véritables publications.

Il est certainement dommage que l'esprit scientifique au sens large de cette revue, cette liberté d'écrire ce que l'on jugeait important, loin de la pression des commissions, du comptage des publications, aient été contraints de reculer.

Cette période pour l'équipe strasbourgeoise culmine dans les années 70 avec le fameux « Cours sur les Intégrales Stochastiques » auquel participaient Chou, Dellacherie, Émery, Pratelli, Stricker, Yan, Yœurp mais aussi, à distance, Azéma, El Karoui, Jacod, Lenglart, Lépingle, Mémin, Métivier, Pellaumail, Yor.

Parallèlement à ces travaux en théorie générale des processus, Meyer poursuit les applications de la théorie des martingales en analyse. En théorie des espaces homogènes et des intégrales singulières, les martingales lui donnent des résultats à la fois profonds et valables sur des espaces très généraux. Par exemple, 10 ans après ses apports à la théorie de Littlewood-Paley, les analystes travaillaient encore à des extensions à $\mathbf{R}^n$, ignorant, parce que sa modestie lui interdisait de le leur signaler, que Meyer avait obtenu les inégalités avec des constantes universelles, indépendantes de la dimension, ou même de la nature euclidienne ou non de l'espace.

Vers la fin des années 70 Meyer s'intéresse aux martingales continues sur les variétés différentiables et à l'intégration stochastique intrinsèque dans ce contexte. Il a beaucoup discuté avec Schwartz sur ce sujet, mais ils ont rédigé séparément. On doit à Meyer sur ce sujet l'introduction et l'étude des martingales dans une variété, riemannienne ou non, mais pourvue d'une connexion, la classification des transports stochastiques.

En 1980 il fait connaissance avec le calcul de Malliavin, au congrès de Durham. Ce thème rejoint son goût pour les « chaos de Wiener ». Meyer sera au départ de nombreux développements. En particulier ses équivalences de normes Sobolev en dimension infinie donneront lieu à toute une série de travaux fameux de Bakry, et aussi dans un autre registre de Hu et Yan.

Ce goût pour les chaos de Wiener trouvera un écho très fort en 1984 quand Meyer écoute un exposé de Parthasarathy à Pise sur le tout nouveau calcul stochastique quantique. Meyer est tellement enthousiasmé qu'il décide fermement de consacrer le reste de sa carrière à ce sujet. Pendant plus de 14 ans Meyer va explorer le monde des probabilités quantiques de fond en comble. Il a contribué à de nombreux développements importants (bien qu'il s'en défendît). J'en cite quelques uns qui me viennent à l'esprit, sans être exhaustif : introduction et développement de la théorie des noyaux de Maassen à 3 arguments, contre-exemple célèbre à la représentation en intégrales stochastiques quantiques des opérateurs bornés (avec Journé), une nouvelle définition des intégrales stochastiques quantiques (avec moi-même) permettant de s'abstraire du domaine des vecteurs cohérents de Hudson et Parthasarathy. Il a aussi beaucoup fait de propagande et de travail de réécriture sur ces sujets, donnant lieu au Lecture Notes « *Quantum probability for probabilists* » qui reste un succès et une référence.

Ce travail de réécriture permanent est une autre caractéristique du travail de Paul-André Meyer. Il disait lui-même qu'il ne pouvait rien lire sans le rédiger à sa manière et l'exposer. Ainsi, en plus de ses travaux originaux considérables, Meyer a réécrit de très nombreux articles. Mais il faut savoir que les articles repris par Meyer pouvaient être très différents de l'article original (nouvelles démonstrations, simplifications, extensions, rédaction enfin lisible) et devenaient parfois la référence plutôt que l'article original.

Ce travail de réécriture et de compilation est une composante aussi importante du travail de Meyer que ses articles originaux. Mais parfois ce double rôle a été mal compris et Meyer a été parfois plus perçu comme un encyclopédiste que comme un mathématicien original. Il est bien évident que cela est totalement faux et que son œuvre dépasse même sûrement ce que l'on en connaît. En effet sa très grande modestie naturelle l'a souvent conduit à s'effacer derrière d'autres personnes. Son œuvre originale est considérable et elle a changé le paysage des mathématiques de ces 40 dernières années.

D'ailleurs, je me souviens d'un jour où je me plaignais devant lui du manque de reconnaissance dont souffraient les probabilités quantiques en France. Il me répondit : « J'ai passé tout le début de ma carrière à faire des probabilités en m'abritant sous le parapluie de la théorie du potentiel, ça m'a ouvert des portes. Tu n'as qu'à faire de même et faire des probabilités quantiques en t'abritant sous le parapluie de la théorie des probabilités ! »

Comme les temps ont changé ! La théorie des probabilités et des processus stochastiques est passée en 40 ans du statut de théorie mineure, appliquée et inavouable, à un statut de théorie à part entière, reconnue et développée. Et ça, Paul-André Meyer y est pour beaucoup.

Durant sa carrière de Paul-André Meyer a été souvent récompensé : prix Peccot, prix Maurice Audin, puis le fameux prix Ampère. Il fut deux fois nommé au Comité National du C.N.R.S et élu correspondant de l'Académie des Sciences. Mais il n'a jamais fait grand cas de toutes ces distinctions, allant même jusqu'à partager financièrement un de ses prix avec un co-auteur, ou encore ne jamais mentionner à sa famille l'obtention de certaines distinctions.

Caractère

Je voudrais terminer ce portrait en parlant de l'homme, de son caractère, de l'impression qu'il laissait à ceux qui le côtoyaient.

Paul-André Meyer était un mathématicien très important, reconnu dans le monde entier, une véritable référence et pourtant il était d'une incroyable modestie. Je parle là d'une modestie sincère et non pas des habituels discours « Mais non, mais non, je n'y suis pas pour grand chose ! » que l'on entend souvent. Un souvenir personnel illustre ce trait de caractère.

Il s'agit de la seule fois où il s'est (un peu) fâché contre moi. Je finissais de rédiger ma thèse et récupérais un chapitre à l'imprimante. Il passe et voit que j'y parle de *noyaux de Maassen-Meyer*. Il me dit : « Je ne veux pas que tu les appelles ainsi, appelle-les *noyaux de Maassen* ! » Je lui explique que les opérateurs à noyaux sur l'espace de Fock quand ils sont à deux arguments sont effectivement de Maassen mais que c'est bien à lui, Meyer, que l'on doit avoir montré la nécessité d'introduire le troisième argument et que depuis tout le monde dit *noyaux de Maassen-Meyer*. Il me répond qu'il ne veut pas que des objets mathématiques soient nommés d'après son nom. Je lui demande alors comment il fait pour parler de la *décomposition de Doob-Meyer*. Alors il se retourne, part et me lance en criant « Je n'ai JAMAIS appelé cela *décomposition de Doob-Meyer* ! » et il claque violemment la porte derrière lui.

Tous ses anciens étudiants témoignent que Meyer s'est beaucoup occupé d'eux. Il a toujours été très présent, tapant souvent lui-même des articles, avec sa machine à écrire que tout le monde savait reconnaître (avant l'arrivée de TeX). Il portait aux travaux des autres, en particuliers de ses étudiants, un enthousiasme communicatif. Il débordait d'idées et de commentaires. Et puis, il mettait lui-même une telle énergie dans tout ce qu'il faisait, il avait une telle capacité de travail et de concentration, que ceux qui travaillaient avec lui ne pouvaient qu'être portés, poussés par cet élan.

Dans toute sa carrière Paul-André Meyer a attaché relativement peu d'importance aux applications que pouvaient avoir ses travaux (comme en physique ou en mathématiques financières). Je le cite :

« Rien dans notre imagination ordinaire ne nous prépare à concevoir la vitesse de la lumière comme une vitesse limite, et la transformation de Lorentz peut être manipulée mais non *comprise*. C'est pire encore pour la mécanique quantique. Bien entendu, l'esprit d'un physicien qui manipule ces choses tous les jours finit par les connaître parfaitement, mais je ne crois pas qu'il réalise plus qu'un court-circuit du langage mathématique, une constitution sommaire d'images à partir de celui-ci. Feynman n'écrit-il pas que la mécanique quantique restait toujours aussi stupéfiante au bout de 40 ans d'expériences ? Ou que la question : Qu'est-ce qui fait que les masses s'attirent ? est sans réponse jusqu'à maintenant et sans doute pour toujours. En ce sens la science ne nous explique rien, elle nous dit : C'est comme ça ! »

Pourtant, dans les toutes dernières années de sa carrière, il semble avoir regardé la physique d'un autre œil. En effet, nous sommes nombreux à l'avoir vu faire des efforts en direction de la physique quantique. Il m'a répété sans cesse que faire des probabilités quantiques n'avait pas de sens sans s'intéresser aussi à la physique qu'il y avait derrière. Il me disait que les probabilistes (quantiques ou non) avaient sûrement beaucoup de choses à dire aux physiciens mais que pour cela il fallait faire l'effort d'aller vers eux, de comprendre leurs problèmes, leurs besoins.

Je terminerai en évoquant les passions de Meyer en dehors des sciences. Les quatre principales étaient je crois sa famille, la musique, les langues et la littérature.

Paul-André Meyer était en effet très proche de sa famille, qu'il faisait souvent passer avant tout. Il écrit : « J'ai toujours donné à ma famille une certaine priorité sur le travail, et je me suis senti un jour récompensé quand une de mes filles m'a dit : Quand nous étions petits, nous avions un peu honte, car nous pensions que tu ne faisais rien ! »

Paul-André Meyer était un grand mélomane et un très bon musicien. Il jouait lui-même du violon, de l'alto et surtout de la flûte traversière. Toute sa famille a grandi dans cette passion de la musique (deux filles musiciennes professionnelles). Le soir, la « petite famille Meyer » jouait à la maison des sonates, des trios, des quatuors avec piano.

Paul-André Meyer était aussi un impressionnant connaisseur de langues étrangères. Il parlait très bien espagnol, anglais, allemand, mais aussi chinois, bengali. Il avait de bonnes connaissances de japonais, hindi, sanscrit, russe. Cet amour des langues s'accompagnait naturellement d'un amour des voyages. Sa carrière de mathématicien lui a d'ailleurs permis d'en effectuer beaucoup. Il a par exemple beaucoup travaillé aux relations scientifiques entre la France et la Chine, faisant de nombreux voyages et cours, formant de nombreux étudiants chinois en France.

Paul-André était beaucoup plus qu'un grand mathématicien ou même qu'un grand scientifique. Toutes les personnes qui le côtoyaient ne pouvaient qu'être frappées par son immense culture. Parmi les auteurs qu'il a le plus aimés, il faut citer : Pouchkine, Saint Simon (le mémorialiste), Borges, Tchekhov, les poètes de la Chine des Tang et des Song. Ses livres de chevet étaient les Essais de Montaigne et le Zhuangzi, texte taoïste de l'antiquité chinoise.

Il avait aussi une vie spirituelle intense, une réflexion profonde sur son engagement religieux, tout en gardant une grande ouverture d'esprit. Il a en particulier voué une véritable passion à l'hindouisme. Allant même jusqu'à prendre sa retraite pour achever une œuvre qui le « travaillait depuis l'enfance » : apprendre le bengali, le hindi et le sanscrit, pour traduire en français les œuvres du disciple de Ramakrishna, Mahendranath Gupta, en particulier le volumineux Kathamrita. Cette œuvre immense (environ 950 pages de traduction), qu'il a achevée juste avant de nous quitter, représentait un jardin secret, une passion de toute sa vie, qui n'étaient connus que de ses très proches, mais qui montre encore une fois combien Paul-André Meyer pouvait être déterminé et fidèle dans tout ce qu'il faisait.

Paul-André Meyer part en laissant derrière lui près de 200 publications, une dizaine de livres, de nombreux élèves, « petits-élèves » et même « arrière-petits-élèves ». Il en est de certains hommes comme des grandes œuvres, ils ne disparaissent jamais vraiment.

Témoignages

Jacques Azéma, Claude Dellacherie, Catherine Doléans-Dade, Michel Émery, Yves Le Jan, Bernard Maisonneuve, Yves Meyer, Jacques Neveu, Nicolas Privault and Daniel Revuz

Que dire ? Il avait une telle horreur des hommages... Au moins nous sera-t-il permis de partager quelques souvenirs. Celui de cette machine à écrire, d'abord, dont les caractères « non indispensables » avaient été remplacés par des symboles mathématiques ; on reconnaissait sa frappe à l'autre bout du monde. La fierté secrète éprouvée quand une de nos idées se retrouvait dans ses lignes. À l'heure des conference-call et des show minutés des grands congrès internationaux, il est devenu difficile d'imaginer que cette machine, dans le silence d'un village d'Alsace, ait pu à elle seule révolutionner les probabilités françaises.

Si l'on savait lire entre les formules, on pouvait discerner, cachés par la profondeur des mathématiques et la beauté du texte, le plaisir de se retrouver entre amis, la joie peut-être aussi d'avoir dévoilé, à l'aide d'une idée simple, quelques-uns des petits secrets du monde.

L'hôtel Gutenberg, avait-il réellement choisi ce nom par hasard, lui qui pouvait, des heures durant, parler des fontes de l'A.M.S. ou que l'on croisait dans Paris à la recherche d'un pinceau destiné à calligraphier des idéogrammes ? L'hôtel Gutenberg, d'où s'échappaient une ou deux fois l'an une vingtaine de mathématiciens mal fagotés venus des quatre coins du globe, désorientés dans les rues de Strasbourg.

Une fois passés la cathédrale et les quais, au pied de la tour de l'I.R.M.A., un sourire clair d'enfant posé sur un vieux duffle-coat, Paul André Meyer.

Jacques Azéma

★ ★ ★ ★ ★

Nostalgie du fossoyeur

Je suis arrivé à l'âge où régulièrement on perd un être proche, un être cher, une part de sa chair. Meyer, puis Doob, puis Doléans ; Joe, André, Catherine dans l'ordre des générations.

Doob avait atteint un grand âge, et nous nous savons mortels. Par un heureux hasard nous étions en même temps, il y a plus de trente ans, à l'Institut de Princeton. Hebdomadairement il écoutait ma conférence, nous mangions ensemble et devisions ; c'est ainsi que j'appris que « y » ne se dit pas « aie grec » en anglais, et beaucoup d'autres choses. Un grand homme humain. C'est bien cela, pense chacun, et beaucoup plus, mais comment s'exprimer convenablement sur un homme ?

Meyer, André, il devait parler de moi aux Journées de Rouen pour mes 60 ans, sa mort a brutalement arrêté une correspondance renouvelée, et c'est moi qui ai dû parler de lui, de notre collaboration, en ouverture des Journées probabilistes de Toulouse. Ce que je sais dire, je ne sais pas forcément l'écrire, pas encore en tout cas. Lors de ma première année à Strasbourg, alors qu'il m'avait donné à lire un article de Getoor, je suis arrivé dans son bureau avec la nouvelle « j'ai trouvé une erreur dans l'article de M. Getoor (prononcé à la française) » ; il a ri et j'ai fini par comprendre. André aimait beaucoup les jeux de mots, et nous en rivalisions plus tard dans notre correspondance.

Et Catherine, son sourire, ses grands yeux, sa rudesse et sa gentillesse. Vue pour la dernière fois, il y a plus de vingt ans, chez elle à Urbana puis chez moi à Rouen. Brillante étudiante à Paris, premier élève de Meyer à Strasbourg, elle fut mathématicienne puis mère, essaya de concilier les deux dans l'indifférence d'un monde trop masculin. Nous étions les chouchous de la bibliothécaire, Mlle Frenkel.

Ich sinke, ihr Lieben, ich komme, ich komm'

Claude Dellacherie

★ ★ ★ ★ ★

Il est difficile d'exprimer en quelques mots l'influence que Paul-André Meyer eut dans ma vie mathématique. Lorsque j'arrivai à Strasbourg en automne 1965, je sortais de quelques années de quasi-isolement à l'université de Paris et je fus heureuse de trouver à Strasbourg, sous la direction de Paul-André Meyer, un groupe probabiliste beaucoup plus restreint et plus humain. André venait de finir d'écrire son premier livre Probabilités et Potentiel ; chaque semaine il nous en exposait une partie, et c'est ainsi que je commençai à m'intéresser aux processus stochastiques.

Maintenant que j'ai depuis longtemps quitté la France et le monde des mathématiciens, quels souvenirs me reste-t-il de ces années strasbourgeoises ? La bienveillance d'André ; sa disponibilité pour ses élèves ; la clarté de ses cours et l'atmosphère amicale qui y régnait ; les petites remarques humoristiques qui les parsemaient ; le souvenir d'un cours entier passé à démontrer la mesurabilité d'une fonction ; une ironie qu'il savait faire partager en trois mots prononcés lentement, un regard et un demi-sourire ; les tous premiers séminaires de probabilités auxquels venaient assister trois ou quatre probabilistes de Paris.

Lorsque je quittai le monde des Mathématiques, André ne fut plus pour moi un ex-directeur de thèse, mais tout simplement un ami. Durant toutes ces années, il fut le seul avec qui je restai en correspondance ; correspondance annuelle et banale, dans laquelle nous échangions nos vœux et nous nous donnions des nouvelles de nos familles. Avec lui a disparu mon dernier lien avec l'ancien groupe de Strasbourg.

Catherine Doléans-Dade

★ ★ ★ ★ ★

Tribut du passé

> Je trouve très heureux le projet de demander
> à tous ceux qui l'ont fréquenté d'écrire à son
> sujet ; mon témoignage sera peut-être le plus
> bref et sans doute le plus pauvre, mais non le
> moins impartial du volume que vous éditerez.
> J. L. BORGES, *Funes ou la mémoire.*

Machine à écrire. Dans les années soixante-dix, imitant le courrier administratif et ses lettres référencées JP/AN027, où JP étaient les initiales de l'auteur de la lettre et AN celles de la dactylo, Meyer envoyait des lettres d'allure très officielle, du style : « Vos intéressantes remarques ont été tapées par le secrétariat du Séminaire comme note pour le prochain volume », avec : Notre référence AM/ME012, où AM signifiait évidemment André Meyer, et ME ... machine à écrire.

Pour reconnaître, dans les Séminaires, ce qu'il avait tapé, les caractères de ses machines étaient un indice bien plus fiable que la signature de l'auteur. Lors de l'arrivée des micro-ordinateurs,

> ces écrans qui, dit-on, « traitent » les textes (comme on traiterait les
> eaux usées),[1]

inquiet de l'impersonnalité des articles composés par TeX, il imite à l'ordinateur la machine à écrire. Mais il se rassure bientôt : TeX est tellement malléable que, avec les mêmes caractères que tout le monde, chaque page porte la marque de l'auteur, repérable d'un coup d'œil. Il nous a quittés juste après le volume XXXVI, le dernier encore formaté à façon par chaque auteur...

---

[1] E. Orsenna, *Grand Amour.*

Sainteté. Au séminaire, l'orateur se bat avec le symbole d'un opérateur pseudo-différentiel. Meyer me glisse un billet : « Le graal aussi, c'est un saint bol ».

Une autre fois, vers 1990, entrant dans mon bureau, il extrait solennellement de son vieux cartable râpé un paquet de lessive Saint-Marc et une bouteille d'eau Saint-Yorre : « Voici les patrons que tous les probabilistes devraient révérer ». Sur la bouteille, une petite étiquette, hélas perdue depuis, tapée en TEX : « Tested for probability by the Berkeley Probability Lab. – Guaranteed content: 100 % probabilistic water ».

Une autre fois encore : « Voici pourquoi le Gange est sacré : avec les cadavres qui s'y décomposent, c'est un vrai miracle que les pélerins qui y font leurs ablutions en ressortent, sinon saints, du moins sains ! »

Sanscrit. Au sortir de son cours de sanscrit : « Aujourd'hui, à propos du linguiste bien connu Hermann Grassmann, auteur d'un excellent dictionnaire de sanscrit, notre professeur a eu ce commentaire : "Il paraît qu'il aurait fait aussi un peu de mathématiques". »

Chine. Un peu avant ses soixante ans, il surprend par hasard des échanges entre Dellacherie et moi sur la façon de fêter ça. « Vous ne ferez rien pour moi ! Ces célébrations ! Tu imagines ces malheureux, rivés à leur siège, sous les éloges ? De toute façon, je m'arrangerai pour être ailleurs, en Chine, par exemple. »

À la fin des années soixante-dix, jouant, le regard pétillant, à imiter la langue de bois de la révolution culturelle, il se plaisait à commenter, dans le style du Quotidien du Peuple, la dernière réunion de Commission des Enseignants. Le document ci-dessous date de cette époque.

Séminaire de Probabilités
Université de Syldavie

DIFFUSION RESTREINTE

1979/1980

GÉOMÉTRIE DIFFERENTIELLE STOCHASTIQUE

ou

REJETONS LES MENSONGES EHONTES DE L'ENNEMI

par P.A. Meyer

Depuis plusieurs années nous voyons se développer une insidieuse entre-
prise pour saper notre moral, en faisant croire à nos combattants probabilistes qu'ils ont besoin, pour remporter la victoire, d'avoir dans leur
sac autre chose que Halmos et Loève, Doob et Chung, Lévy et Ito - autre
chose même que le Lecture Note de Jacod ! Ce lâche complot dissimule son
visage sous le nom de << Géométrie Stochastique >> : dénonçons sans relâche
les agresseurs idéologiques ! Restons indéfectiblement attachés aux vrais
principes probabilistes-diffusionnels et nous vaincrons !

Michel Émery

★ ★ ★ ★ ★

⋆ ⋆ ⋆ ⋆ ⋆

J'avais commencé à écrire quelque chose sur Meyer, mais ce que j'avais écrit ne me plaisait pas beaucoup, alors je l'ai perdu. Finalement je me suis dit que si j'en ressentais trop vivement la vanité, ce genre n'était pas fait pour moi. Et puis, comme il fallait « boucler » le numéro de l'IHP qui lui est dédié, j'ai pensé à lui et je me suis aperçu que je commençais à oublier certaines choses. Alors ça m'a donné envie, finalement, d'essayer... pour moi en fait, je crois.

La dernière fois que je l'ai vu, c'était à Strasbourg, il y a trois ans environ. Je donnais un séminaire et il était passé nous dire bonjour. Il m'a dit qu'il devrait partir avant la fin. Malgré tout, ça m'a stimulé, je me suis senti un bref moment face au maître, comme cela m'était arrivé certainement en quelques occasions, il y a vingt ans. Mais assez vite il a fermé les yeux.

Donc il y a cela bien sûr, cette position magistrale qu'il occupait quand j'ai commencé à travailler. Mais qui reposait, je crois, moins sur l'autorité, que sur la séduction et l'abondance de la parole. Il aimait répandre des idées, des problématiques, magnifier un peu leur importance, entraîner derrière lui une petite foule, voyager dans les champs de l'analyse et du calcul des probabilités. Potentiels, Markov, martingales, semigroupes, espaces de Fock, tout s'enchaîne bien sûr, mais ça fait de grands paysages et il y bâtissait de grandes architectures. On voyageait avec lui, ou on lisait les relations dans le séminaire. Ça avait quelque chose des récits des grandes explorations dont je me régalais enfant.

Mais ce n'est pas si simple. Un homme qui après avoir écrit des milliers de pages, entraîné toute une école à sa suite, se retire presque brutalement et presque totalement du monde académique, sans amertume aucune, je pense, a dû garder ou mûrir en lui toutes ces années ce projet. Les farces de potache qu'il glissait ici et là en étaient peut être un signe mineur. Son goût des langues orientales (terrain où je m'étais un peu aventuré), un indice plus significatif. Travailleur acharné des mathématiques, il m'a toujours semblé qu'il échappait pourtant parfaitement au mortel esprit de sérieux, et posait souvent un sourire sur tout ça. Il pouvait le partager et on sentait alors de l'amitié. Là sans doute résidait son attrait qui demeure, et nourrit le regret. Plusieurs visages ne font pas une statue.

Yves Le Jan

⋆ ⋆ ⋆ ⋆ ⋆

C'est en 67, alors que je cherchais à échapper au Service des Études et Recherche sur la Circulation routière (ma fascination pour les probabilités enseignées par Neveu à l'IHP était devenue trop forte), que je suis allé avec J.L. Vermeulen faire une petite virée à Strasbourg, pour prendre contact avec un jeune professeur à la renommée déjà grande. Notre entrevue avec Meyer fut fort sympathique et très encourageante : comme nous nous informions sur les centres d'intérêt de la recherche strasbourgeoise, il nous dit que lui et son équipe travaillaient *un petit peu sur les martingales* et nous en fûmes enchantés, car nous avions présenté à Bonitzer, un ingénieur des Ponts et Chaussées plein de verve et de curiosité, une partie du livre de Doob et croyions être bien au courant ! Toutefois l'atmosphère ne s'était vraiment détendue que lorsque nous avions évoqué notre passé polytechnicien (quelques années plus tard une notation qui ne lui plaisait pas me vaudra pourtant un « Si tu écris ça, je vais te traiter de polytechnicien »). Ensuite j'ai pris une ou deux fois le Paris-Strasbourg, pour assister au Séminaire de Strasbourg et me laisser séduire par son ambiance familiale, par la simplicité avec laquelle Meyer interrompait l'orateur, suggérait, s'émerveillait ou soulevait des questions avec une générosité qui n'appartenait qu'à lui. J'avais attrapé le virus.

En 67-68 j'obtins un poste d'assistant à Strasbourg, où je suis resté six ans. La première année (67-68), Meyer faisait un cours sur la frontière de Martin ; Claude Dellacherie disait qu'il n'y comprenait rien, mais cela ne me rassurait pas : lui faisait semblant. Suivirent les années Dacunha(-Castelle), Bretagnolle (dans un exposé du second sur les PAI, André faisait circuler un petit rébus constitué de : Bretagne, Olive, A, Fort). Et puis nous avons eu Kai Lai Chung et ses sticky points, John Walsh et ses *ex is pour is dans É* (traduction : Xs, s, A) et discussions « à bâtons rompus » (ça faisait rigoler Badrikian). Meyer, lui, vivait l'après-mai 68 près de Freiburg en Forêt Noire, mettant doublement en pratique le dicton alsacien : « l'intérieur est à l'extérieur », et c'est là, tout près du petit téléski qui frôlait son jardin, que je suis venu lui présenter mes processus en dents de ski descendantes (résultats de renouvellement pour processus de Markov généraux). C'est surtout dans le cadre sympathique de l'Ecole de St-Flour (juste avant ma thèse, puis quelques années plus tard) que j'ai vraiment rencontré personnellement André et sa famille.

Pour le dernier volume de la 2e mouture de Probabilités et Potentiel, André a fait appel à moi pour la partie concernant les processus de Markov, car je venais avec Pat Fitzsimmons de renouveler un peu le point de vue, grâce aux mesures de Kuznetsov. J'ai alors eu une abondante correspondance mathématique avec André. C'est lui qui faisait tout le boulot de frappe (il disait que ça l'empêchait de réfléchir) et j'ai eu souvent des scrupules à demander vraiment ce que je désirais. De plus, n'étant à l'époque pas initié aux ordinateurs et les évitant à cause de mes yeux, nous dépendions de la Poste pour nos correspondances et certaines de mes rédactions sont arrivées trop tard ! Il me disait que cela n'était pas grave, puisque je devais être amené à les exposer dans un cours d'été de Saint-Flour.

En décembre 94 André écrit : « ça fait un temps fou (des mois) que je veux t'écrire, mais j'ai remis de jour en jour — chaque jour apportant sa charge de minuscules obligations qui suffisent à occuper tout le temps ». Puis après quelques nouvelles tristes, il me fait part de son prochain départ à la retraite en ces termes : « Avec tout cela, qui ne représente d'ailleurs que les soucis ordinaires dans chaque famille, je me suis bien détaché des mathématiques actives — je passe mes journées à essayer de ne pas travailler (ce n'est pas difficile). Pourtant, je suis encore attiré par les mathématiques *passives* —

par exemple, j'ai commencé un cours sur les Algèbres de Lie, qui m'amuse beaucoup. J'ai demandé ma mise à la retraite au 1er janvier prochain, pour toutes sortes de raisons. Ce ne sera pas un grand changement. »

André avait pris l'habitude de me parler de l'état de santé de ses parents, de Geneviève ou encore du violoncelle de Thérèse ; il me parlait aussi volontiers de ses activités extra-mathématiques et surtout de musique. En cela il cherchait, je suppose, à me mettre à l'aise, car je m'étais éloigné des mathématiques « actives » depuis mes difficultés oculaires, qui ont été attribuées pendant 15 ans à une myopathie de Steinert (ce qui a été finalement démenti !). J'avais été amené, à un moment où mon moral flanchait et pour expliquer une certaine mollesse dans ma collaboration au livre, à lui parler de cette myopathie et des soucis qu'elle causait à Anne et moi-même au sujet de nos filles. Il m'a alors apporté un soutien inestimable à travers quelques lettres bouleversantes, auxquelles j'ai répondu trop parcimonieusement, cherchant à tout prix à oublier mes problèmes, plutôt qu'à les ressasser. Qu'il me pardonne et que me pardonnent aussi tous les mathématiciens amis qui représentent son héritage (souvent partagé avec Jacques Neveu) et à qui je n'ai plus donné de nouvelles depuis si longtemps.

À force de concentration je viens de parvenir à faire remonter à la surface quelques bribes des belles années strasbourgeoises et des plus difficiles années récentes. Merci, André, d'avoir tant cru en les mathématiques et de nous avoir entraînés dans tes parages.

Bernard Maisonneuve

★ ★ ★ ★ ★

Un chemin vers la lumière

> Nous sommes tous menés, menés mais non contraints et forcés, menés comme par la lumière.
>
> JEAN CAVAILLÈS

En 1963 les locaux du département de mathématique de l'Université de Strasbourg se réduisaient à cinq bureaux prêtés par la Faculté de théologie protestante. Les professeurs étaient entassés dans quatre petits bureaux tandis que nous, les assistants, étions regroupés dans le cinquième qui était spacieux, bruyant, enfumé, mais très convivial. Paul-André Meyer venait d'être nommé professeur et moi assistant. Nous partagions une même boîte aux lettres, ce qui nous amenait à causer en ramassant notre courrier. La délicieuse amabilité de Paul-André fit le reste. Nous devînmes amis. Nous organisâmes un séminaire commun dont l'un des thèmes était de relier la théorie de Littlewood-Paley à celle des martingales. Nous anticipions ainsi les travaux de Burkholder et Gundy qui eurent, par la suite, une grande influence sur la pensée de Paul-André.

Dès lors, à sa demande, je l'appelai André. André fut pour moi un maître vénéré et, bien qu'il n'ait pas été mon directeur de recherche, il m'a profondément influencé. Les autres mathématiciens strasbourgeois disaient « le grand Meyer » en parlant d'André ; « le petit Meyer » c'était moi. Mais cela ne me gênait pas, car l'œuvre d'André en probabilité était déjà éclatante.

Hier, le 11 mars 2005, à Evry, je participais à un jury de thèse et le seul « Meyer » à être mentionné dans la bibliographie de l'impétrant (Sadek Gala) fut Martin Meyer, fils d'André et ensuite mon élève en thèse. Cela eût comblé de joie André et je désespère de ne plus pouvoir le lui dire.

Revenons aux années soixante. J'étais marié, ma femme était provisoirement sans emploi et nous avions un bébé. André et Geneviève nous invitaient souvent chez eux. J'y fis la connaissance de leurs enfants, c'est-à-dire de Catherine, de Martin (qui plus tard deviendrait mon élève) et d'Alice. Thérèse n'était pas encore née. Après le repas, Geneviève enfilait ses gants de caoutchouc et allait faire la vaisselle en poussant de légers soupirs. Je ne suis jamais arrivé à l'en empêcher et à faire la vaisselle à sa place. Elle respectait et admirait l'activité intellectuelle d'André et lui évitait tout souci ménager. La vaisselle terminée, Geneviève revenait discuter avec nous. J'aimais André, j'aimais son ironie et sa douceur, et je l'écoutais avec vénération. En me citant des textes que je ne connaissais pas encore, André souriait de mon admiration pour Simone Weil.

Pour aller chez André et Geneviève, il nous suffisait de traverser quelques rues. Leur appartement était rempli de livres ; c'était une vraie bibliothèque. Geneviève et André lisaient sans arrêt, ils lisaient ensemble et ces lectures étaient une sorte de prière. André travaillait à la maison, chez lui. Il n'établissait pas de frontière entre son travail de chercheur, son activité de lecteur, sa vie spirituelle et sa vie affective. Tout se mêlait ; une même ferveur et une même ascèse le conduisaient vers la lumière.

André conservera toute sa vie son amour de la lecture. Lire, relire et finalement comprendre un grand texte lui permettait de prendre son élan pour aller plus loin. Le mathématicien argentin Alberto Calderón me conseillait de ne pas lire et d'attaquer un problème d'une façon personnelle et originale, sans être influencé par ce que d'autres avaient essayé de faire. Antoni Zygmund se plaignait d'avoir trop lu, d'avoir été encombré par ses lectures. André a lu toute sa vie. Cette façon de travailler et l'attention qu'André porta à la pensée des autres mathématiciens est, à mon avis, l'une des manifestations de son exquise attention aux autres. Il voyait venir un temps où les scientifiques dédaigneraient de lire les travaux de leurs collègues et se contenteraient de publier des textes dont ils seraient les seuls lecteurs.

D'autres fois, André et Geneviève nous invitaient à un pique-nique dans les Vosges. Il était prudent de prendre un second petit-déjeuner avant de partir pour ces aventures, car la nourriture ne se méritait qu'après une longue marche et le pique-nique ne débutait que vers deux heures de l'après-midi. Ces pique-niques avaient pour origine un usage introduit par Joseph Doob à Urbana. Ceci mérite quelques explications et un retour en arrière.

André entra à l'Ecole Normale Supérieure, en 1954, parmi les derniers d'une toute petite fournée. Ensuite André entra directement au CNRS, après une quatrième année à l'ENS. Ses maîtres furent alors Gustave Choquet, Jacques Deny et Marcel Brelot. Ce dernier l'orienta vers les travaux de Hunt et Doob où des résultats importants de théorie du potentiel étaient obtenus par des méthodes purement probabilistes. André m'a souvent parlé de ce moment historique où les probabilités acquièrent enfin leurs lettres de noblesse auprès des analystes et autres « mathématiciens purs ». Mais André m'a confié que ce lien avec la théorie du potentiel n'était pas la seule explication de son intérêt pour les probabilités. *Il étudiait les probabilités afin de pouvoir comprendre le monde qui nous entoure.* Cet aveu m'a bouleversé et a influencé mes propres recherches en traitement du signal et de l'image.

André passa un an aux Etats-Unis. Ce voyage à Urbana lui permit de rencontrer Doob qui l'initia à l'art du pique-nique. Mais ce voyage ne fut pas facile, car André était déjà marié et sa fille Catherine était encore un bébé. Geneviève a alors tout fait pour que le travail d'André ne soit pas trop perturbé par les problèmes matériels et les cris de Catherine. De retour en France, André soutenait une thèse d'Etat (l'équivalent de notre habilitation), effectuait son service militaire et obtenait son premier poste à Mulhouse, puis à Strasbourg. A Strasbourg, André créa un centre de recherche en probabilités qui allait très rapidement acquérir une réputation internationale. André n'a plus quitté Strasbourg.

André nous dit qu'il a eu peu d'élèves et qu'il en aurait eu davantage s'il n'était pas resté à Strasbourg. Mais peu de maîtres ont eu une influence intellectuelle et morale aussi forte sur leurs disciples. Après avoir donné, en avril 2002, une conférence à Toulouse, j'accompagnai Michel Ledoux dans son bureau et Dominique Bakry nous y rejoignit. Nous étions heureux d'évoquer le passé et de parler de ce que nous devions à André. De retour à Paris, je téléphonai à Geneviève et André pour leur raconter mon séjour à Toulouse. Geneviève fut heureuse d'entendre mon témoignage, tandis qu'André était un peu gêné.

La vie et l'œuvre d'André ne se limitèrent pas aux mathématiques et à sa famille. André a étudié et a admiré les cultures chinoises et indiennes. Son extrême pudeur, son exquise politesse et son attention aux autres l'apparentent aux traditions de l'Orient. Il construisit avec soin et opiniâtreté des relations scientifiques avec la Chine. Après avoir pris sa retraite, il apprit le bengali et traduisit et publia un texte qui, tout comme le fut sa vie, est une invitation, un appel à l'amour, à la sagesse et au dépouillement.

Yves Meyer

★ ★ ★ ★ ★

★ ★ ★ ★ ★

J'ai connu Paul-André Meyer en 1959 à l'Institut Henri Poincaré alors qu'il avait commencé sa thèse avec Jacques Deny. Je n'ai donc pas rencontré Paul-André lorsqu'il était à l'Ecole Normale, où il est entré en 1954, car j'étais à cette époque à New-York et à Berkeley. Nous avons d'emblée sympathisé, mûs par notre intérêt commun pour les Probabilités.

De cette année 1959, je me souviens notamment de Michel Loève, alors professeur à Berkeley et en année sabbatique à Paris, qui nous disait son espoir que nous poursuivions l'œuvre magistrale de Paul Lévy et que nous renouvelions tous les deux l'étude des Probabilités en France après les développements importants que notre discipline avait connus depuis les années 1950 d'une part aux Etats-Unis sous la conduite de W. Feller et de J.L. Doob et d'autre part en URSS à l'inspiration de A.N. Kolmogorov. A juste titre, Paul-André admirait d'ailleurs beaucoup Doob auprès duquel il avait passé une année.

Paul-André probabiliste bien que normalien, n'aurait pu en ces années 60 avoir de meilleur directeur de thèse que Jacques Deny et par voie d'osmose que Brelot et Choquet. En ces temps les Probabilités avaient besoin de plus en plus de mathématiques pour se développer. Dans cet esprit G.A. Hunt, professeur à Princeton, venait de publier des longs mémoires approfondissant les liens entre la théorie du potentiel et celle des processus de Markov ; invité à Orsay pour un an, il participait au séminaire B.C.D. (Brelot-Choquet-Deny) et eut une grande influence sur nous tous.

Après ces années d'apprentissage, Paul-André entré au C.N.R.S. décida de s'installer à Strasbourg et d'y créer un séminaire. Je n'ai certainement pas été le seul à regretter qu'il refuse ensuite les propositions qui lui furent faites de revenir à Paris mais les jeunes probabilistes lorsqu'ils ne choisirent pas de partir à Strasbourg, établirent rapidement un axe Paris-Strasbourg.

Le séminaire de Probabilité de Strasbourg connut un très grand succès. Comme le rappelait Michel Emery dans un hommage à Paul-André, celui-ci ne pouvait lire ou exposer un article sans en réécrire instantanément une version personnelle ; ces textes avec les travaux personnels très importants de Paul-André le conduisirent à l'écriture avec C. Dellacherie d'un grand traité intitulé « Probabilités et Potentiel ». Ainsi naquirent les « Probabilités strasbourgeoises »...

D'autres que moi évoqueront les travaux de recherche exceptionnels qui ont établi la renommée de Paul-André. Mais je ne puis terminer cet hommage sans évoquer ses très grandes qualités humaines : urbanité constante dans ses rapports avec autrui doublée d'une grande modestie, un très grand dévouement reconnu par ses nombreux élèves et collaborateurs, un enthousiasme communicatif pour sa discipline.

Paul-André a disparu sans que nous nous y attendions. Nous garderons longtemps le souvenir de ce très grand Probabiliste et Mathématicien ainsi que de l'homme merveilleux qu'il fut.

Jacques Neveu

★ ★ ★ ★ ★

Cela se passe au milieu d'un séjour de deux ans en Chine alors que j'essayais d'écrire une thèse dans un certain isolement. Un jour de septembre 1991, retournant au bâtiment des étrangers et m'interrogeant une fois de plus sur l'utilité de poursuivre ces travaux, je trouvai un courrier venant de France, dont l'adresse était parfaitement rédigée en caractères chinois. L'enveloppe contenait cinq pages dactylographiées intitulées « Notes sur un travail de N. Privault », consistant en une série de remarques et conseils détaillés de P.A. Meyer sur un manuscrit que j'avais pu lui transmettre via S. Paycha. C'est probablement grâce à cet encouragement que j'ai continué ma thèse.

Nicolas Privault

★ ★ ★ ★ ★

Les tributs payés à Meyer mettent en lumière la variété des domaines qu'il a ouverts ou popularisés. Mais il me semble que l'on oublie parfois celui des processus de Markov et de la théorie du potentiel. Jusqu'à la parution du livre de Blumenthal et Getoor ceux qui travaillaient dans ce domaine disposaient de très peu de textes facilement accessibles sur les travaux de Hunt. L'un de ceux-ci était le volume du séminaire Brelot-Choquet-Deny consacré à la théorie probabiliste du potentiel, tome à la rédaction duquel Meyer avait contribué de façon essentielle. Il y avait aussi sa thèse « Fonctionnelles multiplicatives et additives de Markov » dont le titre était tout un programme et a fourni du travail aux chercheurs pendant plusieurs années.

Dans sa thèse il donnait dans le cadre markovien un exemple général de décomposition des surmartingales. Cela allait le conduire au fameux théorème de décomposition dont la partie « unicité » était quelque chose de radicalement neuf et sans doute son résultat le plus remarquable. Je me risque à penser que c'était aussi son opinion. Dans une conversation que j'eus avec lui il y a bien longtemps, malgré sa modestie habituelle (« Nos travaux seront vite oubliés », « Ne prenons pas trop au sérieux ce que nous faisons ») il admit que là, il avait eu une idée (la seule bien sûr qu'il ait jamais eue !).

Daniel Revuz

★ ★ ★ ★ ★

Kernel and Integral Representations of Operators on Infinite Dimensional Toy Fock Spaces

Yan Pautrat

Université Paris-Sud, Bâtiment 425, F-94 405 Orsay Cedex, France
e-mail: yan.pautrat@math.u-psud.fr

Summary. We study conditions for the existence of a Maassen-Meyer kernel representation for operators on infinite dimensional toy Fock spaces. When applied to the toy Fock space of discrete quantum stochastic calculus, this condition gives a criterion for the existence of a representation as a quantum stochastic integral, as well as explicit formulas for deriving the coefficients of this representation.

Key words: Fock space, toy Fock space, Maassen–Meyer representation, quantum stochastic integration

Introduction

The representation of operators on Fock space over $L^2(\mathbb{R}_+)$ or $L^2(\mathbb{R}_+^d)$ as Maassen–Meyer kernel operators—that is, as series of iterated integrals of scalars with respect to quantum noises—is of particular interest. Indeed, this type of representation has been heavily studied because kernel operators naturally arise as solutions of quantum stochastic differential equations (see [5], [6]); what is more, composing kernel operators gives a kernel operator which can be explicitly computed. Nevertheless, there is no satisfactory criterion for representability of an operator, as existing results are scarce (see [1], [3]).

The purpose of this article is to give very general representability criteria in the simpler case of infinite dimensional toy Fock space, that is, antisymmetric Fock space over $\ell^2(\mathbb{N})$ or $\ell^2(\mathbb{N}^d)$. The search for such results was motivated by Attal's rigorous method to approximate boson Fock space over $L^2(\mathbb{R}_+)$ (which we denote by Φ) by its "discrete-time" counterpart: infinite dimensional toy Fock space $T\Phi$ (see [2]). Since the approximation method is explicit, finding criteria for the representability of operators on $T\Phi$ could lead to representations as kernel operators (and, as we shall see later, as quantum stochastic integrals) of approximations of operators on Φ.

Furthermore, the toy Fock space TΦ conspicuously has interesting features regarding representability as kernel operators: indeed it is straightforward to see that the von Neumann subalgebra of $\mathcal{B}(\mathrm{T}\Phi)$ (the set of bounded operators on TΦ) generated by all fundamental operators a_i^+, a_i^- a_i°, $i \geqslant 0$, is $\mathcal{B}(\mathrm{T}\Phi)$ itself. This leads to some kind of kernel representability result for bounded operators of TΦ. Another simple approach, presented in section 2, gives a more precise result on the representation of a bounded operator; yet this result is still unsatisfactory both in its range of application (the operators considered in physical practice are rarely bounded) and in the underlying meaning given to a kernel representation. In this paper we discuss a more satisfactory definition of such a representation. We then search after equivalent formulations of that definition; we will see that, with minor domain assumptions, these formulations lead to explicit formulas for the coefficients involved in the kernel and yield a very general sufficient condition for the existence of a kernel representation. The proofs use discrete-time analogues of continuous-time methods developed by Lindsay in [4]; we point out along the way a mistake contained in one of the propositions of that paper, give a counterexample and make a tentative correction. Later on we apply our results on kernel representations to obtain criteria and formulas for representations of operators as quantum stochastic integrals.

This paper is organized as follows: in section 1 we give all the needed notations. In section 2 we fulfill the above program on kernel representations. In section 3 we apply our kernel representation theorems to the study of quantum stochastic integral representations.

1 Notations

In the purpose of generalizing our representability theorems to criteria that apply in any antisymmetric Fock space over an infinite-dimensional, separable Hilbert space, we introduce all notations in a general framework. Denote by $\mathcal{A}$ an arbitrary infinite countable set, and by $\mathcal{P}$ the set of all finite subsets of $\mathcal{A}$. The space we will work on, throughout this paper, is the antisymmetric Fock space over $\ell^2(\mathcal{A})$, which, by Guichardet's interpretation, can be seen as $\ell^2(\mathcal{P})$, that is, the space of all maps $f : \mathcal{P} \mapsto \mathbb{C}$, such that

$$\sum_{M \in \mathcal{P}} |f(M)|^2 < +\infty.$$

The most familiar case is $\mathcal{A} = \mathbb{N}$, where $\ell^2(\mathcal{P})$ is the infinite dimensional toy Fock space.

Let us denote by X_A the indicator function of $A \in \mathcal{P}$; the set of all X_A's constitutes a (Hilbertian) basis of $\ell^2(\mathcal{P})$. This particular basis being fixed, one defines for all $B \in \mathcal{P}$ three operators by

$$a_B^+ X_A = \begin{cases} X_{A \cup B} & \text{if } B \cap A = \varnothing \\ 0 & \text{otherwise,} \end{cases}$$

$$a_B^- X_A = \begin{cases} X_{A \setminus B} & \text{if } B \subset A \\ 0 & \text{otherwise,} \end{cases}$$

$$a_B^\circ X_A = \begin{cases} X_A & \text{if } B \subset A \\ 0 & \text{otherwise.} \end{cases}$$

Those operators are closable and we keep the same notations for their closures. They are called respectively the *creation*, *annihilation* and *conservation* operators. From now on, for two subsets A, B of $\mathcal{A}$, we write

- $A + B$ for $A \cup B$ if $A \cap B = \varnothing$,
- $A - B$ for $A \setminus B$ if $B \subset A$,

using the convention that any quantity in which $A + B$ (respectively $A - B$) appears as a variable or as an index is null if $A \cap B \neq \varnothing$ (respectively $B \not\subset A$). We will write for example

$$a_B^+ X_A = X_{A+B},$$
$$a_B^- X_A = X_{A-B}.$$

Practically this will turn out to be a short notation to restrict the range of summation of sums in which the variables are elements of $\mathcal{P}$.

Let us write the general expression of the action of an operator $a_A^+ a_B^\circ a_C^-$ on a vector f of $\ell^2(\mathcal{A})$. First, if A, B, C are not mutually disjoint then $a_A^+ a_B^\circ a_C^-$ is null; otherwise for all M in $\mathcal{P}$,

$$a_A^+ a_B^\circ a_C^- \, f(M) = f(M + C - A) \tag{1}$$

if $B \subset M$, and zero otherwise.

Remark that we have not included here a precise definition of what we will call a kernel representation. We postpone this and the preparative discussion to the next section.

2 Kernel representation theorems on $\ell^2(\mathcal{P})$

Let us describe a tentative approach to representations of bounded operators; to that end let us restrict our framework to the (ordered) case where $\mathcal{A} = \mathbb{N}$, and denote by p_i the orthogonal projection on the subset $\ell^2(\mathcal{P}_i)$, where $\mathcal{P}_i$ is the set of subsets of $\{0, \dots, i-1\}$. In that case, for any bounded operator K on $\ell^2(\mathcal{P})$, $p_i K p_i$ is an operator on $\ell^2(\mathcal{P}_i)$, and the sequence $(p_i K p_i)_{i \geq 0}$ converges strongly to K. It is known from the case of finite dimensional toy Fock spaces that every one of these $p_i K p_i$ coincides with a kernel operator $p_i \sum_{A,B,C < i} k_i(A, B, C) a_A^+ a_B^\circ a_C^- p_i$. It is easy to see that the k_i's are compatible in the sense that there exists a kernel $k : \mathcal{P}^3 \mapsto \mathbb{C}$ which extends all kernels k_i.

This k is such that for any vector $f \in \ell^2(\mathcal{P})$, any $M \in \mathcal{P}$,

$$(Kf)(M) = \lim_{i \to \infty} \sum_{U+V+W=M} \sum_{N \in \mathcal{P}_i} k(U, V, N) f(V + W + N), \qquad (2)$$

so that, in some sense, the function k satisfies an analogue of the formula which defines kernel operators in continuous time (see [6], [5]). Nevertheless, let us discuss what we want a kernel representation to be. Heuristically, such a representation should be a series $\sum_{(A,B,C) \in \mathcal{P}^3} k(A, B, C) a_A^+ a_B^\circ a_C^-$, the meaning of the sum being taken in a weak sense: we expect that, for every f in some domain, the formal computation $\sum_{A,B,C} k(A, B, C) a_A^+ a_B^\circ a_C^- \sum_N f(N) X_N$, gives the right result at every M in $\mathcal{P}$, that is, that the equality

$$(Kf)(M) = \sum_{U+V+W=M} \sum_{N \in \mathcal{P}} k(U, V, N) f(V + W + N), \qquad (3)$$

holds for every N in $\mathcal{P}$. It is a sensible demand, both from a pragmatic point of view and from an intuitive one (the sums $\sum_{A,B,C}$ and $\sum_N$ above should be independent of the order of summation, so the final one also should) that this series be *absolutely convergent*. In equation (2) the series a priori lacks that property. It is therefore natural to look for conditions upon which that property holds.

Returning to the general case (where $\mathcal{A}$ is any infinite countable set), we will solve this problem and obtain further representation theorems. The strategy of our proof will be completely different from the above; our basic tools will be the following:

- the transform $k \mapsto k'$ similar to the one defined by Lindsay (see [4]) in the regular Fock space; we give its definition below.
- The additional feature, specific to the case of discrete-time, of equivalence of two variables and three variables representation. This feature is the simple fact that, since any a_B° is $a_B^+ a_B^-$, one should (and does, as Proposition 1 will prove) obtain equivalent actions for the formal series $\sum_{(A,B,C) \in \mathcal{P}^3} k(A, B, C) a_A^+ a_B^\circ a_C^-$ and $\sum_{(A,B) \in \mathcal{P}^2} k(A, B) a_A^+ a_B^-$ if we use the correspondence

$$k(A, B) = k(A \backslash B, A \cap B, B \backslash A) \quad \text{and} \quad k(A, B, C) = k(A \cup B, B \cup C). \quad (4)$$

Definition 1. *For a function $k : \mathcal{P}^3 \mapsto \mathbb{C}$, let us define $k' : \mathcal{P}^3 \mapsto \mathbb{C}$ as*

$$k'(A, B, C) = \sum_{V \subset B} k(A, V, C),$$

and for a function $k : \mathcal{P}^2 \mapsto \mathbb{C}$, let us define $k' : \mathcal{P}^2 \mapsto \mathbb{C}$ as

$$k'(A, B) = \sum_{V \subset A \cap B} k\big((A \backslash B) \cup V, (B \backslash A) \cup V\big).$$

Remark. For subsets A, B, C which are not mutually disjoint, the operator $a_A^+ a_B^\circ a_C^-$ is null; therefore, when considering kernels with *three* arguments, the function k needs only to be defined on mutually disjoint triples for our purpose, and similarly for k'.

Properties of the transform

- The correspondence $k \mapsto k'$ is bijective for both three-arguments and two-arguments kernels thanks to the Moebius inversion formula which yields

$$k(A, B, C) = \sum_{V \subset B} (-1)^{|B-V|} k'(A, V, C) \qquad (5)$$

and

$$k(A, B) = \sum_{V \subset A \cap B} (-1)^{|A \cap B - V|} k'\big((A \setminus B) \cup V, (B \setminus A) \cup V\big). \qquad (6)$$

- The correspondence defined by (4) is bijective between the set of functions defined on the subset of $\mathcal{P}^3$ of mutually disjoint triples of $\mathcal{P}$ and the set of functions on $\mathcal{P}^2$.
- If, in the next few lines, we add indices and denote kernels by k_2, k_2', k_3, k_3', depending on the number of variables, the following diagram is commutative:

$$\begin{array}{ccc} k_2 & \longleftrightarrow & k_3 \\ \updownarrow & & \updownarrow \\ k_2' & \longleftrightarrow & k_3' \end{array}$$

where arrows are either the correspondence in (4) or the transformation in Definition 1. That means in particular that notations like k_2', k_3' would be unambiguous.
- Our choice for the transform $k \mapsto k'$ extends the equalities in (4) to the functions k':

for mutually disjoint $(A, B, C) \in \mathcal{P}^3,$ $k'(A, B, C) = k'(A \cup B, B \cup C).$

Because of these properties we will not distinguish anymore notations between k_2 and k_3, nor between k_2' and k_3'.

The following proposition contains the first properties that link the four different forms of a kernel.

Proposition 1. *Let f be a fixed vector in $\ell^2(\mathcal{A})$. Define the four assumptions:*

- *for all mutually disjoint U, V, W in $\mathcal{P}$*

$$\sum_{N \in \mathcal{P}} |k(U, V, N) f(V + W + N)| < +\infty \qquad (7)$$

- *for all disjoint U, V in $\mathcal{P}$*

$$\sum_{N \in \mathcal{P}} |k(U, N)f(V + N)| < +\infty \tag{8}$$

- *for all disjoint U, V in $\mathcal{P}$*

$$\sum_{N \in \mathcal{P}} |k'(U, V, N)f(V + N)| < +\infty \tag{9}$$

- *for all U in $\mathcal{P}$*

$$\sum_{N \in \mathcal{P}} |k'(U, N)f(N)| < +\infty. \tag{10}$$

Then the conditions on two-arguments kernels are equivalent to their three-arguments counterparts, that is, (7) and (8) are equivalent, (9) and (10) are equivalent. What is more, the conditions on the kernels imply the conditions on their transforms, that is, (7), (8) imply (9), (10).

Besides, if all conditions are satisfied, then the following are defined and equal for all $M \in \mathcal{P}$:

$$\sum_{U+V+W=M} \sum_N k(U, V, N)f(V + W + N) \tag{11}$$

$$\sum_{U+V=M} \sum_N k'(U, V, N)f(V + N) \tag{12}$$

$$\sum_{U+V=M} \sum_N k(U, N)f(V + N) \tag{13}$$

$$\sum_N k'(M, N)f(N) \tag{14}$$

Proof. Let us start with the proof that (7) implies (8): first fix U_0, V_0;

$$\sum_N |k(U_0, N)f(V_0 + N)|$$

$$= \sum_N |k(U_0 \setminus N, U_0 \cap N, N \setminus U_0)f(V_0 + U_0 \cap N + N \setminus U_0)|$$

$$\leqslant \sum_{U+V=U_0} \sum_N |k(U, V, N \setminus U_0)f(V + V_0 + N \setminus U_0)|$$

$$\leqslant 2^{|U_0|} \sum_{U+V=U_0} \sum_{N \text{ disjoint from } U,V} |k(U, V, N)f(V + V_0 + N)|$$

$$\leqslant 2^{|U_0|} \sum_{U+V=U_0} \sum_N |k(U, V, N)f(V + V_0 + N)|$$

$$< +\infty$$

where the $2^{|U_0|}$ arises because any N can be written as at most $2^{|U_0|}$ different "$N \setminus U_0$".

Now prove that (8) implies (7):

$$\sum_N |k(U, V, N)f(V + W + N)| = \sum_N |k(U + V, N + V)f(W + (N + V))|$$

$$= \sum_{N \supset V} |k(U + V, N)f(W + N)|$$

$$< +\infty.$$

The equivalence of (9) and (10) is shown exactly in the same way.

To show that the above conditions on k imply those on the transform k', we prove that (7) implies (9):

$$\sum_N |k'(U, V, N)f(V + N)| \leqslant \sum_N \sum_{\alpha \subset V} |k(U, \alpha, N)f(\alpha + (V \setminus \alpha) + N)|$$

and the right-hand side is just a finite sum of series of the type (7) with $(U, V, W) = (U, \alpha, V \setminus \alpha)$.

The equalities are obvious once the summability assumptions allow all manipulations on the sums. $\qquad\square$

Remark. It is not true in general that the conditions on k' imply their counterpart on k. Here is a counterexample: let k be of the form

$$k(U, V, W) = (-1)^{|V|} j(U, W)$$

for some function j of two disjoint finite subsets of $\mathbb{N}$, which is not to be confused with the kernel k expressed as a function of two variables.

Then $k'(U, V, W) = \mathbb{1}_{V=\varnothing} j(U, W)$ and (9) simply becomes

$$\text{for all } U \in \mathcal{P}, \qquad \sum_N |j(U, N)f(N)| < +\infty, \tag{15}$$

whereas (7) is

$$\text{for all } (U, V) \in \mathcal{P}^2, \qquad \sum_N |j(U, N)f(V + N)| < +\infty. \tag{16}$$

Now if one considers

- a function j such that $j(U, W) = 0$ if the cardinality of W is different from 1,
- a vector f null on sets of cardinality one,

then (15) is trivial while (16) becomes

$$\text{for all } (U, V) \in \mathcal{P}^2, \qquad \sum_{n \geqslant 0} |j(U, \{n\}) f(V + \{n\})| < +\infty. \qquad (17)$$

and still implies a condition on the values of j and f, so that many counterexamples exist.

The same type of counterexamples holds in continuous time for the equivalence described by Lindsay in [4]. The next theorem describes a class of vectors for which the equivalence of (7), (8), (9), (10) holds; once again this class translates in continuous time to a class of vectors for which the equivalence described by Lindsay holds.

Theorem 1. *Let f be a vector in $\ell^2(\mathcal{P})$ for which there exists a function $\phi : \mathcal{A} \mapsto \mathbb{R}_+$ such that*

$$\text{for all } (A, B) \in \mathcal{P}^2, \qquad |f(A + B)| \leqslant |f(A)| \prod_{i \in B} \phi(i).$$

Then assumptions (7), (8), (9), (10) are equivalent for that f.

Proof. What is left to prove is that (9) implies (7). Using the particular hypothesis on f, one has for all U, V, W,

$$\sum_N |k(U, V, N) f(V + W + N)| \leqslant \prod_{i \in W} \phi(i) \sum_{N \text{ disjoint from } V, W} |k(U, V, N) f(V + N)|$$

$$\leqslant \prod_{i \in W} \phi(i) \sum_N |k(U, V, N) f(V + N)|$$

so we can reduce the proof to the case where $W = \varnothing$. But in that case, using the inverse Moebius transform,

$$\sum_N |k(U, V, N) f(V + N)| \leqslant \sum_{\alpha \subset V} \sum_N |k'(U, \alpha, N) f(V + N)|$$

$$\leqslant \sum_{\alpha \subset V} \sum_N |k'(U, \alpha, N) f(\alpha + N)| \prod_{i \in V \setminus \alpha} \phi(i)$$

$$< +\infty. \qquad \qquad \square$$

Definition 2. *The vectors which satisfy the property mentioned in the previous theorem are called subexponential. The set of all subexponential vectors, which we denote by $s\mathcal{E}$, contains all linear combinations of exponential vectors or vectors X_A, $A \in \mathcal{P}$.*

The set $s\mathcal{E}$ is not a vector space; nevertheless, it has the properties that, if f, g are in $s\mathcal{E}$ and λ, μ are in $\mathbb{C}$, then $|\lambda| |f| + |\mu| |g|$ is in $s\mathcal{E}$.

Here is a precise definition, following our earlier discussion, of what we call a kernel operator:

Definition 3. *A (possibly unbounded) operator K on $\ell^2(\mathcal{P})$ is said to have a kernel representation if there exists a function k such that:*

- *Dom K is exactly the set of $f \in \ell^2(\mathcal{P})$ that satisfy one of the conditions (7) or (8),*
- *the equalities in (11), (12), (13) or (14) define a square integrable function of M,*
- *$Kf(M)$ is equal to the corresponding expression for all $M \in \mathcal{P}$.*

Now, up to an additional simple assumption, the kernel decomposition takes a clear meaning: suppose that an operator K has such a representation and that the basis $\{X_A\}$ is in Dom K, then writing $\sum_N k'(M,N)\mathbb{1}_A(N) = (KX_A)(M)$ yields the fundamental formula

$$\langle X_M, KX_A \rangle = k'(M,A). \tag{18}$$

Thanks to this formula, (10) becomes

$$\forall M, \qquad \sum_N |\langle X_M, KX_N \rangle f(N)| < +\infty,$$

and the other two assumptions (square-integrability and equality of expressions) simply mean that one can write rigorously

$$\forall M, \qquad \left\langle X_M, K \sum_N f(N)X_N \right\rangle = \sum_N \langle X_M, KX_N \rangle f(N).$$

What this means indeed is that kernel representations are just another way to write the above expansion. Of course there are conditions for this expansion to be meaningful and conditions for the obtained representation to actually represent the original operator. We discuss these conditions after the following proposition:

Proposition 2. *Let K be an operator with domain such that*

$$\{X_A,\ A \in \mathcal{P}\} \subset \text{Dom}\, K \subset s\mathcal{E}.$$

Then K can be extended to a kernel operator if and only if for every $f \in \text{Dom}\, K$ and every M in $\mathcal{P}$,

$$\begin{cases} \sum_N |\langle X_M, KX_N \rangle f(N)| < +\infty \ \text{and} \\ \sum_N \langle X_M, KX_N \rangle f(N) = Kf(M). \end{cases}$$

In that case, the associated kernel is given by

$$k'(A,B) = \langle X_A, KX_B \rangle.$$

Proof. Thanks to Proposition 1, Theorem 1 and formula (18), this is a simple rephrasing of our definition. □

It is clear that, if the domain of the operator K is not contained in the subexponential subset, then the obtained kernel operator can be such that $\{X_A\} \subsetneq \operatorname{Dom} K_k$ and $\{X_A\} = \operatorname{Dom} K \cap \operatorname{Dom} K_k$.

It is also clear that, even if the conditions

$$\sum_N \left| \langle X_M, K X_N \rangle f(N) \right| < +\infty$$

$$\sum_M \left| \sum_N \langle X_M, K X_N \rangle f(N) \right|^2 < +\infty$$

hold for any f in $\operatorname{Dom} K$, so that the associated kernel operator K_k is well defined on $\operatorname{Dom} K$ and coincides with K on $\{X_A, A \in \mathcal{P}\}$, one needs a kind of closability assumption to make sure that the kernel operator is indeed an extension of K: this assumption is exactly the condition

$$\sum_N \langle X_M, K X_N \rangle f(N) = K f(M). \tag{19}$$

It does not seem that a more concise formulation can be found: the usual closability property would be that if a sequence $(u_n)_{n \geqslant 0}$ converges to zero and is such that the sequence of images $(K u_n)_{n \geqslant 0}$ is convergent, then its limit is zero.

Assumption (19) is weaker than closability in the sense that it only considers approximating sequences $(u_n)_{n \geqslant 0}$ made of partial sums of $\sum_N f(N) X_N$, but also stronger than closability in the sense that convergence to zero of $(u_n)_{n \geqslant 0}$ with the weak convergence assumption on the images that $\langle X_M, K u_n \rangle$ converges for all M and defines a square-integrable function of M must imply that that limit is zero.

On the other hand, it is clear that these properties are satisfied if one assumes that K has an adjoint defined on all vectors X_M, M in $\mathcal{P}$. Indeed in that case, we have the following result. Note that no assumption of the type $\operatorname{Dom} K \subset s\mathcal{E}$ is needed; actually the proof of this theorem proves the stronger summability assumptions on the kernels k themselves and not on the transforms k'.

Theorem 2. *Let K be an operator on $\ell^2(\mathcal{P})$ such that the set of all X_A is in $\operatorname{Dom} K \cap \operatorname{Dom} K^*$. Then the kernel operator defined by (18) is a closed extension of K.*

Proof. Let us define the kernel k by (18). We will show that assumption (7) holds for *any* vector f of $\ell^2(\mathcal{P})$. Indeed, let us fix U, V, W; then from the Moebius inversion formula, one has for all N disjoint from U, V, W:

$$|k(U, V, N)| \leqslant \sum_{\alpha \subset V} |k'(U + \alpha, N + \alpha)|$$

$$\leqslant \sum_{\alpha \subset V} |\langle X_{U+\alpha}, K X_{N+\alpha} \rangle|$$

$$\leqslant \sum_{\alpha \subset V} |\langle K^* X_{U+\alpha}, X_{N+\alpha} \rangle|,$$

which is a finite sum of square-summable terms. Besides, $N \mapsto f(V+W+N)$ is also square-summable, and therefore $\sum_N |k(U, V, N) f(V + W + N)|$ is finite.

The condition on f appears in the sequel: the domain of the kernel operator $K_k f$ is the set of vectors f in TΦ such that

$$K_k f(M) = \sum_N \langle X_M, K X_N \rangle f(N)$$

defines a function belonging to $\ell^2(\mathcal{P})$. That quantity is equal to

$$\sum_N \langle K^* X_M, X_N \rangle f(N) = \langle K^* X_M, f \rangle,$$

so that the domain of K_k is the set of vectors f such that $M \mapsto \langle K^* X_M, f \rangle$ is square-integrable. For all f in Dom $\overline{K}$, $\langle K^* X_M, f \rangle = \langle X_M, \overline{K} f \rangle$, so $\overline{K} \subset K_k$.

We now prove that K_k is closed: let $(f_n)_{n \geqslant 0}$ be a sequence in Dom K_k that converges to some f in TΦ and such that $(K_k f_n)_{n \geqslant 0}$ converges to some ϕ in TΦ. Then

$$\sum_M |\langle K^* X_M, f \rangle|^2 = \sum_M \lim |\langle K^* X_M, f_n \rangle|^2$$

$$\sum_M \lim |\langle K^* X_M, f_n \rangle|^2 \leqslant \liminf \sum_M |\langle K^* X_M, f_n \rangle|^2$$

$$\liminf \sum_M |\langle K^* X_M, f_n \rangle|^2 = \liminf \|K_k f_n\|^2$$

$$\liminf \|K_k f_n\|^2 = \|\phi\|^2$$

$$\|\phi\|^2 < +\infty,$$

so that f lies in Dom K_k; besides

$$K_k f(M) = \langle K^* X_M, f \rangle$$
$$\langle K^* X_M, f \rangle = \lim \langle K^* X_M, f_n \rangle$$
$$\lim \langle K^* X_M, f_n \rangle = \lim \langle X_M, K_k f_n \rangle$$
$$\lim \langle X_M, K_k f_n \rangle = \phi(M)$$

holds for every M in $\mathcal{P}$ so that $\phi = K_k f$ and the proof is complete.   □

Remark. The inclusion $\overline{K} \subset K_k$ is a priori not an equality.

This theorem gives numerous examples of operators that admit a kernel representation.

3 Integral representation of operators on toy Fock space

In this section we restrict ourselves to operators on the toy Fock space TΦ of quantum stochastic calculus, for which $\mathcal{A} = \mathbb{N}$; indeed we wish to consider integrals of operators, and this demands a natural ordering on our set $\mathcal{A}$. Let us suppose that an operator K on $\mathrm{T}\Phi$ has a kernel representation in the sense of Definition 3. This kernel representation is formally an expression of K as $\sum_{A,B,C} k(A,B,C) a_A^+ a_B^\circ a_C^-$. Let us go on with formal expressions and apply careless manipulations to this sum: we write for all $(A,B,C) \neq (\varnothing,\varnothing,\varnothing)$ $a_A^+ a_B^\circ a_C^- = a_{A\setminus i}^+ a_B^\circ a_C^- a_i^+$ when e.g. the largest element i in $A \cup B \cup C$ is in A, and regroup terms. We obtain

$$k(\varnothing,\varnothing,\varnothing) + \sum_i \sum_{A,B,C<i} k(A+i,B,C) a_A^+ a_B^\circ a_C^- a_i^+$$

$$+ \sum_i \sum_{A,B,C<i} k(A,B+i,C) a_A^+ a_B^\circ a_C^- a_i^\circ + \sum_i \sum_{A,B,C<i} k(A,B,C+i) a_A^+ a_B^\circ a_C^- a_i^-,$$

that is, we obtain an integral representation of K with the following integrands:

$$\begin{cases} k_i^+ = \sum_{A,B,C<i} k(A+i,B,C) a_A^+ a_B^\circ a_C^- \\ k_i^\circ = \sum_{A,B,C<i} k(A,B+i,C) a_A^+ a_B^\circ a_C^- \\ k_i^- = \sum_{A,B,C<i} k(A,B,C+i) a_A^+ a_B^\circ a_C^- \end{cases} \tag{20}$$

Giving a more rigorous meaning to that demands, of course, a definition of the integrals. The definitions are actually quite simple (see [7]): the domain of a sum $\sum_i h_i$ is the set of vectors f such that for all $A \in \mathcal{P}$, the series $\sum_i h_i f(A)$ is summable and defines a square-integrable function of A. An integral $\sum_i h_i^\epsilon a_i^\epsilon$ is such a sum with $h_i = h_i^\epsilon a_i^\epsilon$; on the other hand, one can see from Definition 3 and from the expression (1) that a kernel representation is simply a series $\sum_{A,B,C} \left(k(A,B,C) a_A^+ a_B^\circ a_C^- \right)$ in the above sense. This justifies the above manipulations except for domain properties; what it exactly yields is

$$K \subset k(\varnothing,\varnothing,\varnothing) + \sum_i (k_i^+ a_i^+ + k_i^\circ a_i^\circ + k_i^- a_i^-),$$

which does not prove in our sense that our operator has an integral representation, since the right-hand side is itself an extension of the integral

$$k(\varnothing,\varnothing,\varnothing) + \sum_i k_i^+ a_i^+ + \sum_i k_i^\circ a_i^\circ + \sum_i k_i^- a_i^-. \tag{21}$$

What we have shown is that on some domain (the intersection of the domain of this integral and the domain of the operator), the operator coincides with the integral. Moreover, if a vector X_A is in the domain of K, then it is also in the domain of the integral (21), since only one of the three series $\sum_i k_i^\epsilon a_i^\epsilon X_A$ has an infinite number of nonzero terms.

The interesting aspect of this formal computation is that the integrands defined by (20) are everywhere defined as finite sums of everywhere defined operators, and are described as simple kernel operators with explicit kernels.

Using the formula (18) then gives the action of the operators k_i^ϵ expressed as functions of K:

$$\begin{cases} \langle X_A, k_i^+ X_B \rangle = \mathbb{1}_{A \triangle B < i} \langle X_{A_i[+i}, K X_{B_i[} \rangle \\ \langle X_A, k_i^- X_B \rangle = \mathbb{1}_{A \triangle B < i} \langle X_{A_i[}, K X_{B_i[+i} \rangle \\ \langle X_A, k_i^\circ X_B \rangle = \mathbb{1}_{A \triangle B < i} \left(\langle X_{A_i[+i}, K X_{B_i[+i} \rangle - \langle X_{A_i[}, K X_{B_i[} \rangle \right) \end{cases}$$

Using the fundamental operators of abstract Ito calculus on toy Fock space p_i, d_i, these operators have much more interesting expressions. We recall briefly from [7] the definitions of these operators: for $f \in T\Phi$ and $i \in \mathbb{N}$, $p_i f$ and $d_i f$ are defined as the vectors of $T\Phi$ such that, for $M \in \mathcal{P}$,

$$p_i f(M) = \mathbb{1}_{M < i} f(M),$$
$$d_i f(M) = \mathbb{1}_{M < i} f(M + i).$$

In particular,

$$p_i X_A = \mathbb{1}_{A < i} X_A,$$
$$d_i X_A = \mathbb{1}_{A < i} X_{A - i},$$

from which one sees immediately that one has

$$\begin{cases} k_i^+ \ p_i = d_i K p_i \\ k_i^- \ p_i = p_i K a_i^+ p_i \\ k_i^\circ \ p_i = d_i K a_i^+ - p_i K p_i \end{cases} \tag{22}$$

The above discussion proves the following result:

Theorem 3. *Let K be an operator on $T\Phi$ such that the set of all X_A is in $\mathrm{Dom}\, K \cap \mathrm{Dom}\, K^*$. Then the operator*

$$\lambda + \sum_i k_i^+ a_i^+ + \sum_i k_i^\circ a_i^\circ + \sum_i k_i^- a_i^- - K,$$

where the operators k_i^ϵ are defined by (22) and $\lambda = \langle \mathbb{1}, K\mathbb{1} \rangle$, is a restriction of the zero process and its domain contains the set $\{X_A, A \in \mathcal{P}\}$.

The p_i's on the right of expressions in (22) are here only for the sake of symmetry; if one notices that $d_i = p_i a_i^-$ and that a_i^+, a_i^- are mutually adjoint, these formulas show that $(k^*)_i^+ = k_i^-$, $(k^*)_i^- = k_i^+$ and $(k^*)_i^\circ = k_i^\circ$, where the $(k^*)_i^\epsilon$ are the integrands in the integral representation of K^*.

References

1. S. Attal: *Non-commutative chaotic expansion of Hilbert–Schmidt operators on Fock space*, Comm. in Math. Phys. Vol. **175**, Springer, Berlin, 1996, pp. 43–62.

2. S. Attal: *Approximating the Fock space with the toy Fock space*, Sém. de Probabilités, Vol. XXXVI, Lecture Notes in Mathematics 1801, Springer, Berlin, 2002, pp. 477–491.
3. V.P. Belavkin, and J.M. Lindsay: *The kernel of a Fock space operator II*, Quantum probability and related topics VIII, World scientific, Singapore, 1993, pp. 87–94.
4. J.M. Lindsay: The kernel of a Fock space operator I, Quantum probability and related topics VIII, World scientific, Singapore, 1993, pp. 271–280.
5. H. Maassen: Quantum Markov processes on Fock space described by integral kernels, Quantum probability and related topics II, Lecture Notes in Mathematics Vol. 1136, Springer, Berlin, 1985, pp. 361–374.
6. P.-A. Meyer: Quantum probability for probabilists, Lecture Notes in Mathematics 1538, Springer, Berlin, 1993.
7. Y. Pautrat: From Pauli matrices to quantum Ito formula, Mathematical Physics, Analysis and Geometry, 8, pp. 121–155, 2005.

Le Théorème de Pitman, le Groupe Quantique $SU_q(2)$, et une Question de P. A. Meyer

Philippe Biane

CNRS, Département de Mathématiques et Applications, École Normale
Supérieure, 45, rue d'Ulm, F-75005 Paris, France
e-mail: Philippe.Biane@ens.fr

Summary. On montre comment le théorème $2S - B$ de Pitman peut s'obtenir en
« cristallisant » une marche de Bernoulli quantique.

1 Introduction

Le but de cette note est de relier le théorème de Pitman, dans sa version
en temps discret, avec la théorie des représentations du groupe $SU(2)$, et
plus généralement des groupes quantiques $SU_q(2)$. On verra que ce théorème
apparaît naturellement lors de la « cristallisation », au sens de Kashiwara, i.e.
lorsque $q \to 0$.

L'idée de considérer les représentations de $SU(2)$ en liaison avec les processus de Bessel de dimension 3 provient d'une question de P. A. Meyer, qui avait
demandé, lors d'un exposé en 1989, comment calculer la loi de la partie radiale
d'une marche de Bernoulli quantique (voir la partie 3 pour plus de précisions).

Cette étude ne nous en apprendra pas plus sur le théorème de Pitman
lui-même, mais elle a le mérite de suggérer comment généraliser ce résultat à
des situations plus complexes, en dimensions supérieures. En fait la transformation de Pitman est reliée à des opérations introduites par Littelmann [12]
en théorie des représentations. Les liens entre ces objets sont étudiés plus en
détails dans [5].

L'article est organisé de la façon suivante : dans la partie 2 je rappelle
l'énoncé du théorème de Pitman ainsi que sa version discrète. La partie 3
contient des généralités sur les processus stochastiques non commutatifs,
ainsi que la définition des marches de Bernoulli quantiques, puis je donne
une interprétation de ces marches de Bernoulli quantiques en termes de

représentations du groupe $SU(2)$. Dans la partie 4 je montre comment, en passant de $SU(2)$ au groupe quantique $SU_q(2)$, puis en faisant tendre q vers 0, on obtient le théorème de Pitman de façon naturelle. Finalement on termine par des considérations sur les généralisations possibles de ces résultats.

2 Le théorème de Pitman

Soit $(X_t; t \geq 0)$ un mouvement brownien réel, issu de 0, le théorème de Pitman [16] affirme que le processus $(T_t := 2S_t - X_t; t \geq 0)$, où $S_t = \sup_{s \leq t} X_s$, est un processus de Bessel de dimension 3, i.e. a la même loi que la norme d'un mouvement brownien de dimension 3. Une fois énoncé, le théorème est facile à vérifier, il suffit de calculer la loi du processus $((X_t, S_t); t \geq 0)$, qui est un processus de Markov, pour en déduire le résultat. On obtient de cette façon un bonus, le fait que pour tout $t > 0$ la loi conditionnelle de la variable X_t, connaissant la trajectoire $(T_s, 0 \leq s \leq t)$, est uniforme sur l'intervalle $[-T_t, T_t]$. Faisons dès à présent une observation. Soit $(X_t, Y_t, Z_t; t \geq 0)$ un mouvement brownien de dimension trois, alors le processus $R_t = \sqrt{X_t^2 + Y_t^2 + Z_t^2}$ est un processus de Bessel de dimension trois, X_t est un mouvement brownien réel, et on vérifie facilement, en utilisant l'invariance par rotation de la loi du mouvement brownien, que la loi conditionnelle de X_t sachant la trajectoire $(R_s; 0 \leq s \leq t)$ est la loi uniforme sur l'intervalle $[-R_t, R_t]$. On voit ainsi que les processus $((X_t, T_t); t \geq 0)$ et $((X_t, R_t); t \geq 0)$ ont mêmes lois marginales de rang 1, les processus obtenus en prenant la première ou la deuxième coordonnée ont la même loi, mais les lois des couples de processus ne coïncident pas, car T est mesurable par rapport à la filtration de X, ce qui n'est pas le cas de R. La clé pour comprendre le théorème de Pitman va consister à trouver une interpolation naturelle entre les lois de ces deux couples de processus. Il est possible de faire cette interpolation de façon purement « classique », comme dans [6] mais ici nous allons la faire en introduisant des approximations en temps discret de ces processus, ce qui amène à considérer les groupes quantiques $SU_q(2)$, comme expliqué ci-dessous.

La démonstration originale de Pitman repose en effet sur un analogue en temps discret du théorème. Considérons une marche de Bernoulli, i.e. $X_n = x_1 + x_2 + \ldots + x_n$ où les x_j sont des variables de Bernoulli indépendantes, centrées ($P(x_j = \pm 1) = 1/2$). Le résultat est que le processus $(T_n := 2S_n - X_n; n \geq 1)$, où $S_n = \sup\{X_k; 0 \leq k \leq n\}$, est une chaîne de Markov sur $\mathbb{N}$ de probabilités de transition

$$p(k, k+1) = \frac{k+2}{2(k+1)}; \qquad p(k, k-1) = \frac{k}{2(k+1)}. \qquad (1)$$

De plus la loi conditionnelle de X_n sachant la trajectoire $(T_k; k \leq n)$ est uniforme sur l'ensemble $\{-T_n, -T_n+2, \ldots, T_n-2, T_n\}$. Le théorème de Pitman pour le mouvement brownien s'en déduit par un changement d'échelle et un passage à la limite. La démonstration de Pitman repose sur le fait que la transformation de Pitman peut s'inverser, c'est-à dire que l'on peut retrouver la trajectoire de X en connaissant celle de T, à condition de connaître en plus la valeur de X_n, en fait on a la formule explicite

$$X_k = \inf\{2T_l; k \leq l \leq n\} \wedge (X_n + T_n) - T_k.$$

Évidemment une formule analogue est valable dans le cas du temps continu. Proposons ici une approche équivalente mais plus adaptée à notre propos. On calcule les probabilités de transition de la chaîne de Markov $((S_n, X_n); n \geq 1)$, à valeurs dans l'ensemble $\{(s, k) \in \mathbb{N} \times \mathbb{Z} \mid s \geq k\}$, on trouve facilement

$$p((s,k),(s,k+1)) = \tfrac{1}{2}, \quad p((s,k),(s,k-1)) = \tfrac{1}{2} \qquad \text{si } s > k$$
$$p((s,s),(s+1,s+1)) = \tfrac{1}{2}, \quad p((s,s),(s,s-1)) = \tfrac{1}{2} \,,$$

ce qui donne les probabilités de transition de la chaîne de Markov $((T_n, X_n); n \geq 1)$, à valeurs dans $\{(t, k) \in \mathbb{N}^* \times \mathbb{Z} \mid k \in (-t, -t+2, \ldots, t-2, t)\}$,

$$p((t,k),(t-1,k+1)) = \tfrac{1}{2}, \quad p((t,k),(t+1,k-1)) = \tfrac{1}{2} \qquad \text{si } s > k \quad (2)$$
$$p((t,t),(t+1,t+1)) = \tfrac{1}{2}, \quad p((t,t),(t+1,t-1)) = \tfrac{1}{2} \,.$$

On vérifie alors, par récurrence sur n, que la loi conditionnelle de X_n, sachant $T_1, \ldots, T_n$, est la loi uniforme sur l'ensemble $\{-T_n, -T_n + 2, \ldots, T_n - 2, T_n\}$. Le fait que $(T_n; n \geq 0)$ soit une chaîne de Markov avec les bonnes probabilités de transition s'en déduit aisément. Notre stratégie pour retrouver de façon naturelle le théorème de Pitman va consister à exhiber un analogue discret du processus $((X_t, R_t); t \geq 0)$, ce que nous ferons en considérant des marches aléatoires quantiques. Ensuite nous déformerons la loi de ce processus au moyen de la théorie des groupes quantiques, par un paramètre réel $q \in]0, 1]$, de sorte que la loi de chaque composante reste constante, que la loi conditionnelle de X_n sachant $R_1, \ldots, R_n$ soit uniforme sur l'intervalle $\{-R_n, -R_n + 2, \ldots, R_n - 2, R_n\}$, mais que la loi du couple $((X_n, R_n); n \geq 1)$ converge vers celle du couple $((X_n, T_n); n \geq 1)$ lorsque $q \to 0$.

3 Marches de Bernoulli quantiques

Dans cette partie je vais faire des rappels sur les processus stochastiques non-commutatifs et les marches de Bernoulli quantiques. Ce sujet est traité dans [1] et dans le premier chapitre de [4], auxquels je renvoie pour plus de détails. Pour une introduction aux probabilités quantiques, on peut consulter le livre de référence de Meyer [11].

Dans la suite on considère des espaces de probabilités non-commutatifs. Un tel espace est une algèbre A sur $\mathbb{C}$, avec une unité, habituellement munie d'une involution antilinéaire $*$, et d'une forme linéaire η, qui joue le rôle d'espérance, qui vérifie $\eta(1) = 1$ et $\eta(a^*a) \geq 0$ pour tout $a \in A$. La donnée d'un élément autoadjoint $x = x^*$ détermine alors une suite de nombres réels $(\eta(x^n); n \geq 0)$, qui sont les moments d'une mesure de probabilités sur $\mathbb{R}$, non nécessairement unique. Si cette loi est unique on l'appelle la loi de l'élément x. Par exemple, lorsque l'algèbre est $M_n(\mathbb{C})$, et $\eta(M) = \frac{1}{n}Tr(M)$ alors la loi d'un élément autoadjoint $M \in M_n(\mathbb{C})$ est la mesure empirique sur les valeurs propres de M (comptées avec leur multiplicité). Évidemment dans ce cas la loi est à support compact et donc déterminée par ses moments.

Plus généralement si on se donne une famille d'éléments autoadjoints qui commutent $x_1, \ldots, x_k$, alors les nombres $\eta(x_1^{l_1} \ldots x_k^{l_k}); l_1, l_2, \ldots, l_k \geq 1$ sont les moments d'une mesure de probabilités sur $\mathbb{R}^k$, que l'on appelle la loi jointe des x_k si elle est unique. De nouveau, si l'algèbre est de dimension finie, la loi est à support compact (et même fini), et donc est déterminée par ses moments. C'est dans ce cadre que nous nous placerons dans toute la suite.

L'exemple le plus simple est celui de $M_2(\mathbb{C})$, les matrices de taille 2×2 sur $\mathbb{C}$, avec $\eta(M) = \frac{1}{2}Tr(M)$. On considère les matrices de Pauli

$$x = \begin{pmatrix} 0 & 1 \\ 1 & 0 \end{pmatrix}, \quad y = \begin{pmatrix} 0 & -i \\ i & 0 \end{pmatrix}, \quad x = \begin{pmatrix} 1 & 0 \\ 0 & -1 \end{pmatrix};$$

chacune d'elles suit une loi de Bernoulli symétrique, et elles vérifient les relations de commutation $[x, y] = 2iz$, $[y, z] = 2ix$ et $[z, x] = 2iy$. En particulier, elles ne commutent pas, on ne peut donc pas parler de leur loi jointe. L'algèbre $M_2(\mathbb{C})^{\otimes N}$ est isomorphe à l'algèbre des matrices $M_{2^N}(\mathbb{C})$, et le produit tensoriel des traces sur chacun des facteurs est la trace sur $M_{2^N}(\mathbb{C})$. Dans cette algèbre on considère les éléments (I désigne la matrice identité)

$$x_i = I^{\otimes(i-1)} \otimes x \otimes I^{\otimes(N-i)} \qquad y_i = I^{\otimes(i-1)} \otimes y \otimes I^{\otimes(N-i)}$$

$$z_i = I^{\otimes(i-1)} \otimes z \otimes I^{\otimes(N-i)}$$

pour $i = 1, \ldots, N$, et leurs sommes partielles

$$X_n = x_1 + \ldots + x_n, \quad Y_n = y_1 + \ldots + y_n, \quad Z_n = z_1 + \ldots + z_n.$$

Chacun des processus $(X_n; 1 \leq n \leq N), (Y_n; 1 \leq n \leq N)$ et $(Z_n; 1 \leq n \leq N)$ est constitué d'opérateurs qui commutent, on peut donc calculer leur loi, qui est, dans les trois cas, celle d'une marche de Bernoulli centrée. Ces opérateurs vérifient les relations de commutation $[X_n, Y_m] = 2iZ_{n \wedge m}$, ainsi que celles obtenues par permutation circulaire de X, Y, Z. On constate que ces relations de commutation entraînent que les opérateurs $\Sigma_n = \sqrt{I + X_n^2 + Y_n^2 + Z_n^2}$ commutent entre eux. On peut calculer la loi du processus $(\Sigma_n; n \geq 1)$ et on trouve par des calculs explicites (cf [4]) que c'est la loi de la chaîne de Markov sur $\mathbb{N}^*$, de lois de transition

$$p(k, k+1) = \frac{k+1}{2k}, \qquad p(k, k-1) = \frac{k-1}{2k}. \tag{3}$$

Autrement dit, le processus $(R_n := \Sigma_n - I, n \geq 1)$ a les mêmes probabilités de transition que (1). De plus, pour tout n, les opérateurs $R_1, \ldots, R_n$ et X_n

commutent et la loi conditionnelle de X_n sachant $R_1, \ldots, R_n$ est uniforme sur
$\{-R_n, -R_n + 2, \ldots, R_n - 2, R_n\}$.

Plutôt que de reproduire la preuve de [4], je vais expliquer comment on peut retrouver ce résultat par la théorie des représentations de $SU(2)$. Les relations de commutation entraînent que, pour chaque $n \geq 1$, les opérateurs X_n, Y_n et Z_n engendrent une représentation de l'algèbre de Lie $\mathfrak{sl}_2(\mathbb{C})$. Rappelons que cette algèbre de Lie est de dimension 3, engendrée par trois éléments x, y, z, qui vérifient les relations

$$[x, y] = 2iz, \quad [y, z] = 2ix, \quad [z, x] = 2iy.$$

D'autre part, l'opérateur $\Sigma_n^2 = I + X_n^2 + Y_n^2 + Z_n^2$, est l'image dans cette représentation de l'opérateur de Casimir $1 + x^2 + y^2 + z^2$, qui engendre le centre de l'algèbre enveloppante de $\mathfrak{sl}_2(\mathbb{C})$. On sait que les représentations irréductibles de $\mathfrak{sl}_2(\mathbb{C})$ sont indexées par les entiers $l \geq 1$; pour chacun de ces entiers il existe une unique (à isomorphisme près) représentation irréductible de dimension l. De plus dans cette représentation l'opérateur de Casimir est un multiple de l'identité de valeur propre l^2. On peut en déduire la loi du processus $(R_k; k \leq N)$. Pour cela supposons que nous avons décomposé la représentation $(\mathbb{C}^2)^{\otimes n}$ en composantes irréductibles. Alors pour trouver les valeurs propres et les sous-espaces propres de R_{n+1}, pour chaque copie $V_k \subset (\mathbb{C}^2)^{\otimes n}$ de la représentation de dimension k, il nous faut décomposer la représentation $V_k \otimes \mathbb{C}^2$. Les formules de Clebsch-Gordan donnent la réponse, cette représentation se scinde en somme directe de deux représentations $V_k \otimes V_2 \sim V_{k-1} \oplus V_{k+1}$. Les probabilités de transition sont proportionnelles aux dimensions respectives de ces représentations, ce qui redonne bien la formule cherchée. La loi conditionnelle de X_n sachant la trajectoire $R_1, \ldots, R_n$ s'obtient en remarquant que dans une représentation irréductible V_k, où R_n prend la valeur $k-1$, l'image de l'élement x de $\mathfrak{sl}_2(\mathbb{C})$ a un spectre formé des k valeurs propres $\{-k+1, -k+3, \ldots, k-3, k-1\}$, chacune ayant multiplicité 1.

Une fois ces observations faites nous allons donner une construction plus abstraite de la marche aléatoire ci-dessus. Nous aurons besoin de la notion de variable aléatoire à valeurs dans un espace non-commutatif. Un espace non-commutatif est donné par une algèbre qui joue le rôle d'algèbre de fonctions sur l'espace. La notion de variable aléatoire à valeurs dans un espace est remplacée par celle de morphisme d'algèbres, à valeurs dans un espace de probabilités non-commutatif. Le lien avec la situation classique est facile à comprendre, une variable aléatoire ordinaire $X : \Omega \to E$ donne lieu à un morphisme d'algèbres $\mathcal{X} : L^\infty(E) \to L^\infty(\Omega)$ défini par la formule $\mathcal{X}(f) = f(X)$, et réciproquement la donnée d'un tel morphisme détermine la valeur de la variable aléatoire X (presque sûrement). Dans la situation non-commutative, les points des espaces non-commutatifs n'existent plus, mais la notion de morphisme d'algèbres garde un sens. Nous allons donc construire les processus X, Y, Z, R ci-dessus et les considérer comme des coordonnées d'un processus à valeurs dans un espace non-commutatif. Cet espace non-commutatif sera le dual de $SU(2)$ ce qui signifie que l'algèbre de « fonctions non-commutatives » qui le définira sera l'algèbre du groupe $SU(2)$. Il y a plusieurs choix possibles d'algèbres de convolution sur un groupe, on peut convoler par exemple les fonctions L^1, ou les mesures finies, ou encore les fonctions continues ou les fonctions régulières si on est sur un groupe de Lie. Les subtilités liées au choix de la topologie, et in fine de la complétion ne sont pas très utiles pour comprendre les calculs algébriques, je renvoie à la discussion de [2] pour plus d'information, et je me contenterai de donner une description informelle de cette algèbre. Comme nous avons vu, il existe pour chaque entier $l \geq 1$ une unique représentation irréductible de dimension l du groupe $SU(2)$, ce qui suggère de prendre pour l'algèbre $\mathcal{A}(SU(2))$ le produit direct des algèbres de matrices $M_l(\mathbb{C})$ pour $l \geq 1$. Tout autre choix pourra se plonger dans cette grosse algèbre. Dans cette algèbre se trouvent des éléments x, y, z correspondant à la base de $\mathfrak{sl}_2(\mathbb{C})$, et dont les composantes sur chaque sous-algèbre $M_l(\mathbb{C})$ forment la représentation de dimension l de $\mathfrak{sl}_2(\mathbb{C})$. L'opérateur de Casimir $1 + x^2 + y^2 + z^2$ a pour composante dans chaque algèbre $M_l(\mathbb{C})$ un opérateur scalaire de valeur propre l^2, tandis que les composantes des opérateurs x, y et z sont diagonalisables, et leurs valeurs propres, de multiplicité 1, sont $\{-l+1, -l+3, \ldots, l-3, l-1\}$. On peut considérer ces opérateurs comme trois coordonnées sur l'espace non-commutatif sous-jacent. Chaque sous-algèbre $M_l(\mathbb{C})$ est une sorte de sphère non-commutative de dimension 3, de rayon $l-1$.

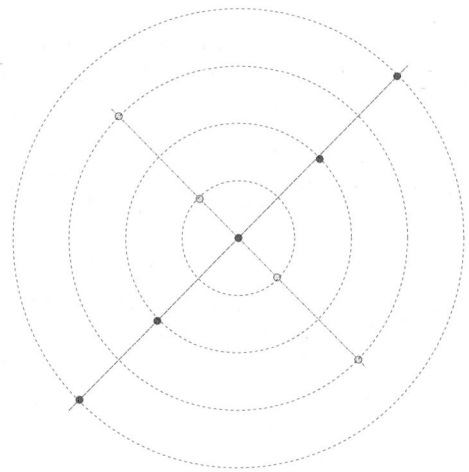

L'espace non-commutatif correspondant à $\mathcal{A}(SU(2))$

À chaque direction de l'espace à trois dimensions (α, β, γ), on peut faire correspondre l'opérateur $\alpha x + \beta y + \gamma z$. Dans une telle direction, la coordonnée peut prendre des valeurs entières. On peut dans cet espace déterminer simultanément le rayon de la sphère sur laquelle on se trouve et la coordonnée dans une des directions de l'espace, auquel cas on trouve que les valeurs possibles des coordonnées sont $-r, r+2, \ldots, r-2, r$ où r est le rayon de la sphère. Par contre on ne peut pas déterminer simultanément les coordonnées dans deux directions de l'espace qui ne sont pas parallèles, car les opérateurs correspondants ne commutent pas.

J'ai représenté de manière symbolique cet espace dans le dessin ci-dessus. Les algèbres $M_l(\mathbb{C})$ sont représentées par des sphères « floues » concentriques (j'ai représenté celles de rayons 0 à 4). J'ai aussi montré les coordonnées que l'on peut obtenir en faisant une mesure dans deux directions sur les sphères de rayons 3 et 4.

Nous allons maintenant construire un processus stochastique non-commutatif sur l'algèbre $\mathcal{A}(SU(2))$. Pour cela il nous faut construire des morphismes de $\mathcal{A}(SU(2))$ à valeurs dans un espace de probabilités non-commutatif. On va se servir de la structure d'algèbre de Hopf de $\mathcal{A}(SU(2))$.

L'algèbre $\mathcal{A}(SU(2))$ possède en effet une structure d'algèbre de Hopf (comme toute algèbre de groupe), ce qui correspond à la donnée d'un coproduit, qui est un morphisme $\Delta : \mathcal{A}(SU(2)) \rightarrow \mathcal{A}(SU(2)) \otimes \mathcal{A}(SU(2))$, ainsi que d'autres données, satisfaisant des axiomes que l'on peut trouver par exemple dans [11]. Pour comprendre la signification de cette structure, considérons un groupe G, et l'espace des fonctions (continues, ou essentiellement bornées, ou polynomiales, etc...) sur G, muni de la multiplication usuelle qui en fait une algèbre $A(G)$ (que l'on notera $C(G), L^\infty(G), \mathcal{P}(G), \dots$ suivant l'espace de fonctions choisi). La donnée de cet algèbre caractérise l'ensemble G en tant qu'espace, mais ne permet pas de trouver sa structure de groupe. Pour cela il faut considérer le produit $G \times G \rightarrow G$. Comme tout à l'heure on peut considérer le morphisme d'algèbre $\Delta : A(G) \rightarrow A(G \times G)$ induit par cette application, donné par la formule $\Delta f(g, h) = f(gh)$. Si on identifie $A(G) \otimes A(G)$ avec une algèbre de fonctions sur $G \times G$, alors Δ est un coproduit. Les propriétés qui font de G un groupe (associativité, élément neutre, existence d'un inverse) peuvent se traduire en terme de propriétés du coproduit. Par exemple l'associativité du produit dans le groupe est équivalente au fait que les deux morphismes $(\Delta \otimes I) \circ \Delta, (I \otimes \Delta) \circ \Delta : A(G) \rightarrow A(G) \otimes A(G) \otimes A(G)$ sont égaux. La notion d'algèbre de Hopf est « autoduale », ce qui signifie que si H est une algèbre de Hopf alors l'espace dual H^* est encore une algèbre de Hopf, il suffit de renverser les flèches dans tous les morphismes considérés (modulo les problèmes usuels de passage au dual si la dimension est infinie). Ainsi si G est un groupe abélien, alors il a un groupe dual $\hat{G}$, et les algèbres de Hopf correspondantes sont en dualité. Dans la situation que nous considérons, le groupe $SU(2)$ n'est pas commutatif, par conséquent l'algèbre du groupe, dont la structure d'algèbre de Hopf est duale de celle de l'algèbre des fonctions sur G, n'est pas l'algèbre des fonctions sur un espace ordinaire, mais nous pouvons la considérer comme l'algèbre des fonctions sur un espace non-commutatif. Nous pouvons facilement écrire des formules pour le coproduit d'éléments de cette algèbre suffisamment simples. En fait on a

$$\Delta(\alpha x + \beta y + \gamma z) = \alpha(x \otimes I + I \otimes x) + \beta(y \otimes I + I \otimes y) + \gamma(z \otimes I + I \otimes z)$$

(I désigne l'identité de $\mathcal{A}(SU(2))$) ce qui suffit à caractériser ce coproduit.

Passons maintenant à la construction de la marche de Bernoulli quantique. Soit un groupe G, muni d'une mesure de probabilités μ, nous pouvons construire une marche aléatoire dont la loi des accroissements est μ en considérant l'espace produit $(G^N, \mu^{\otimes N})$ et en définissant la suite de variables aléatoires $(X_n(g_1, \ldots, g_N) = g_1 g_2 g_3 \ldots g_n; n = 1, 2, \ldots, N)$. Si nous prenons le point de vue des algèbres, les variables X_n induisent des morphismes $\mathcal{X}_n : A(G) \to A(G)^{\otimes N}$ donnés au moyen du coproduit Δ par les formules

$$\mathcal{X}_1(x) = x \otimes I^{\otimes(N-1)}$$
$$\mathcal{X}_2(x) = \Delta(x) \otimes I^{\otimes(N-2)}$$
$$\mathcal{X}_3(x) = ((I \otimes \Delta) \circ \Delta(x)) \otimes I^{\otimes(N-3)}$$

et ainsi de suite jusqu'à

$$\mathcal{X}_n(x) = ((I^{n-2} \otimes \Delta) \circ \mathcal{X}_{n-1}(x)) \otimes I^{\otimes(N-n)}.$$

On peut faire la même construction avec n'importe quelle algèbre de Hopf H munie d'une involution $*$, (en fait il suffit que H soit munie d'un coproduit coassociatif, les autres propriétés ne jouant pas de rôle dans la construction), à condition de disposer d'un état η sur l'algèbre de Hopf, c'est-à-dire une forme linéaire positive, qui vaut 1 sur l'identité. On considère alors l'algèbre $H^{\otimes N}$, munie de l'état $\eta^{\otimes N}$ et on définit les $\mathcal{X}_k : H \to H^{\otimes N}$ par la formule ci-dessus. La marche de Bernoulli quantique construite plus haut correspond à cette construction pour l'algèbre $\mathcal{A}(SU(2))$, lorsque l'état est la trace normalisée sur la composante $M_2(\mathbb{C})$.

Dans le cas général, une telle marche aléatoire quantique satisfait la propriété de Markov suivante : soit $f \in H$ et ξ un élément de l'algèbre engendrée par les $(\mathcal{X}_k(H); k \leq n - 1)$, alors on a

$$\eta^{\otimes N}[\mathcal{X}_n(f)\xi] = \eta^{\otimes N}[\mathcal{X}_{n-1}(P(f))\xi]$$

où $P : H \to H$ est défini par

$$Pf = (I \otimes \eta) \circ \Delta(f).$$

Dans le cas d'une marche aléatoire ordinaire sur un groupe abélien, en identifiant H avec l'algèbre des fonctions sur $\hat{G}$, et η avec la mesure μ, on voit que P est l'opérateur de convolution par la mesure μ

$$Pf(x) = \int_{\hat{G}} f(xy) d\mu(y).$$

Dans le cas général cet opérateur possède une propriété de positivité : si f est positif dans l'algèbre H alors $P(f)$ l'est également. Il est même complètement positif, une propriété plus forte, mais nous ne nous servirons pas de cette notion (cf [11], [15]). Nous l'appellerons l'opérateur de transition de la marche aléatoire quantique.

Si on a construit une telle marche aléatoire quantique on peut en déduire des processus stochastiques « classiques » à valeurs dans des espaces ordinaires, pour cela il suffit de trouver une sous-algèbre commutative $B \subset H$ de l'algèbre de Hopf, telle que les images $(X_n(B); n \geq 0)$ commutent dans l'espace de probabilités non-commutatif. L'algèbre B, étant commutative, s'interprète comme une algèbre de fonctions sur un espace $\hat{B}$ (son spectre), de même, il existe un espace de probabilités Ω dont l'algèbre des variables aléatoires bornées soit l'algèbre engendrée par les $(X_n(B); n \geq 0)$, et les morphismes $\mathcal{X}_n$ proviennent de variables aléatoires définies sur Ω et à valeurs dans $\hat{B}$.

Dans le cas qui nous intéresse, il y a deux sortes de sous-algèbres intéressantes, d'une part pour chaque sous-groupe commutatif de $SU(2)$, on peut considérer la sous-algèbre de $\mathcal{A}(SU(2))$ qu'il engendre. Cette algèbre est isomorphe à l'algèbre des fonctions sur le dual de ce groupe commutatif, et on obtient ainsi une marche aléatoire sur ce groupe dual. La loi des accroissements de cette marche aléatoire est obtenue en prenant la restriction de l'état à la sous-algèbre engendrée par le groupe. Par exemple si on considère le groupe à un paramètre engendré par un élément $v \in SU(2)$, alors on obtient une marche aléatoire sur le groupe dual, qui est isomorphe à $\mathbb{Z}$. C'est ainsi que l'on peut interpréter les marche aléatoires de Bernoulli X_n, Y_n, Z_n obtenues au début, elles correspondent aux sous-groupes à un paramètre de $SU(2)$ de générateurs ix, iy et iz respectivement.

Une autre sous-algèbre commutative intéressante est le centre de l'algèbre $\mathcal{A}(SU(2))$. Celui-ci est engendré par l'opérateur de Casimir, et il est isomorphe à l'algèbre des fonctions sur $\mathbb{N}$. On peut montrer que ses images par les $\mathcal{X}_n$ engendrent une algèbre commutative. Comment, dans ce contexte, calculer la loi de ce processus ? On utilise la propriété de Markov énoncée plus haut. En effet si une sous-algèbre $B \subset H$ est stable par l'opérateur P, alors elle vérifie la propriété de Markov avec pour opérateur de transition la restriction de P à B. Il est facile de voir que si l'algèbre engendrée par les $\mathcal{X}_k(B)$ est commutative et si B est stable par P, alors les restrictions des $\mathcal{X}_k$ à B définissent un processus de Markov classique sur $\hat{B}$ de noyau de transition donné par P. Dans le cas du centre de l'algèbre $\mathcal{A}(SU(2))$, on peut calculer ce noyau de transition explicitement ce qui revient comme précédemment à calculer la décomposition en composantes irréductibles de la représentation $V_k \otimes V_2$. Le formalisme abstrait ci-dessus a un avantage, il nous permet de découvrir, caché dans la marche de Bernoulli quantique, un processus de Markov non-commutatif à valeurs dans un « vrai » espace commutatif. Considérons en effet l'algèbre engendrée par un sous-groupe à un paramètre de $SU(2)$ et par le centre de $\mathcal{A}(SU(2))$. Cette algèbre est évidemment commutative et on peut vérifier qu'elle est stable par l'opérateur P. La restriction de P à cette algèbre correspond donc à un noyau markovien sur un espace « commutatif ». Commençons par déterminer quel est cet espace. Pour le trouver, diagonalisons la composante du générateur du groupe à un paramètre dans un des espaces $M_l(\mathbb{C})$. Cet opérateur admet l valeurs propres distinctes. On voit alors que la sous-algèbre engendrée par ce sous-groupe à un paramètre et par le centre est isomorphe au produit des sous-algèbres des matrices diagonales dans $\prod_{l \geq 1} M_l(\mathbb{C})$, (en particulier c'est une algèbre abélienne maximale). Il est commode d'utiliser les valeurs propres du générateur du groupe à un paramètre et le rayon (qui vaut $l-1$ dans la représentation de dimension l) pour paramétrer les points de cet ensemble. On voit donc qu'il est donné par

$$\{(r,k) \in \mathbb{N} \times \mathbb{Z} \mid k \in \{-r, -r+2, \ldots, r-2, r\}\} \,.$$

Calculons le noyau de transition obtenu en restreignant P à cette sous-algèbre. Pour cela il est utile de considérer l'algèbre de Hopf duale de $\mathcal{A}(SU(2))$. L'opérateur dual de P envoie $\mathcal{A}^*(SU(2))$ dans elle-même, et est donné par $u \mapsto \chi u$ où χ est le caractère normalisé de la représentation de dimension 2 de $SU(2)$, et où on a identifié $\mathcal{A}^*(SU(2))$ avec l'algèbre des fonctions sur $SU(2)$, munie du produit ordinaire. Choisissons pour tout r une base orthonormale $(e^r_j; j \in -r, -r+2 \ldots, r-2, r)$ de la représentation V_{r+1} dans laquelle le générateur du sous-groupe à un paramètre est diagonal, alors il faut comprendre comment les fonctions coefficients $u^r_{jj}(g) = \langle ge^r_j, e^r_j \rangle$ sont transformées par P^*. La réponse est donnée par la formule classique de Clebsch-Gordan

$$u^r_{jj} u^1_{-1-1} = \frac{r-j+2}{2(r+1)} u^{r+1}_{j-1j-1} + \frac{r+j}{2(r+1)} u^{r-1}_{j-1j-1}$$

$$u^r_{jj} u^1_{11} = \frac{r+j+2}{2l} u^{r+1}_{j+1j+1} + \frac{r-j}{2(r+1)} u^{r-1}_{j+1j+1}.$$

Comme $u^r_{jj}\chi = \frac{1}{2}(u^2_{-1-1} + u^2_{11})u^r_{jj}$ on en déduit les probabilités de transition de la chaîne

$$p((r,k),(r+1,k+1)) = \frac{r+k+2}{2(r+1)}$$

$$p((r,k),(r+1,k-1)) = \frac{r-k+2}{2(r+1)}$$

$$p((r,k),(r-1,k+1)) = \frac{r-k}{2(r+1)}$$

$$p((r,k),(r-1,k-1)) = \frac{r+k}{2(r+1)}.$$

On a donc obtenu une sous-algèbre commutative stable par P, toutefois la marche de Bernoulli quantique ne définit pas un processus classique associé sur l'espace correspondant car les images de cette sous-algèbre par les morphismes $\mathcal{X}_k; k \geq 1$ ne commutent pas entre elles, comme on peut le voir en remarquant que X_m ne commute pas avec R_n si $n > m$. Le processus défini par les restrictions des $\mathcal{X}_k$ à cette sous-algèbre est donc un processus de Markov non-commutatif, à valeurs dans un espace commutatif, avec un noyau de transition commutatif.

4 Groupes quantiques

L'algèbre de Hopf de $SU(2)$ peut être déformée en introduisant un paramètre réel q. Je renvoie au premier chapitre de [10] pour une présentation claire et concise des principales propriétés de cette construction. On obtient une algèbre de Hopf $\mathcal{A}(SU_q(2))$, dont la structure d'algèbre ne dépend pas de q, mais dont le coproduit, lui, en dépend. Pour définir cette algèbre on part des générateurs suivants de $\mathfrak{sl}_2(\mathbb{C})$

$$e = \begin{pmatrix} 0 & 1 \\ 0 & 0 \end{pmatrix}, \quad f = \begin{pmatrix} 0 & 0 \\ 1 & 0 \end{pmatrix}, \quad h = \begin{pmatrix} 1 & 0 \\ 0 & -1 \end{pmatrix}$$

vérifiant les relations de commutation

$$[h, e] = 2e, \quad [h, f] = -2f, \quad [e, f] = h.$$

Ces dernières formules définissent l'algèbre enveloppante de $\mathfrak{sl}_2(\mathbb{C})$. On définit l'algèbre enveloppante quantique de $SU_q(2)$, q étant un paramètre (dans notre cas $q \in]0, 1]$) au moyen de générateurs t, e, f, qui vérifient les relations

$$tet^{-1} = q^2 e, \quad tft^{-1} = q^{-2} f, \quad ef - fe = \frac{t - t^{-1}}{q - q^{-1}}$$

avec un coproduit donné par les formules suivantes :

$$\Delta(t) = t \otimes t, \quad \Delta(e) = e \otimes t^{-1} + 1 \otimes e, \quad \Delta(f) = f \otimes 1 + t \otimes f .$$

On vérifie que si l'on pose formellement $t = q^h$ et que l'on fait tendre q vers 1 dans ces formules on retrouve les relations de commutation définissant l'algèbre enveloppante de $\mathfrak{sl}_2(\mathbb{C})$, ainsi que son coproduit, par exemple la relation $[h, e] = 2e$ s'obtient en dérivant en $\epsilon = 0$ la relation $q^{\epsilon h} e q^{-\epsilon h} = q^{2\epsilon} e$.

On montre que les représentations irréductibles de dimension finie de cette algèbre sont des déformations de celles de $\mathfrak{sl}_2(\mathbb{C})$. Pour chaque entier $r \geq 0$ il existe deux représentations dans des espaces V_{r+1}^+ et V_{r+1}^-, munis de bases $v_k^{r\pm}; k \in \{-r, -r+2, \ldots, r-2, r\}$, données par les formules

$$tv_j^{r\pm} = \pm q^j v_j^{r\pm}$$

$$ev_j^{r\pm} = \pm \sqrt{\left[\frac{r-j}{2}\right]_q \left[\frac{r+j+2}{2}\right]_q} \, v_{j+2}^{r\pm}$$

$$fv_j^{r\pm} = \sqrt{\left[\frac{r-j+2}{2}\right]_q \left[\frac{r+j}{2}\right]_q} \, v_{j-2}^{r\pm}.$$

où on a posé, conformément à une tradition bien établie,

$$[n]_q = \frac{q^n - q^{-n}}{q - q^{-1}}.$$

Le coproduit permet de définir le produit tensoriel (au sens de Kronecker) de deux représentations. On montre que ce produit tensoriel obéit aux mêmes règles que celui des représentations de $SU(2)$, i.e. on a une décomposition

$$V_{r_1+1}^{\epsilon_1} \otimes V_{r_2+1}^{\epsilon_2} = \bigoplus_{r=|r_2-r_1|,|r_2-r_1|+2,\ldots,r_1+r_2} V_{r+1}^{\epsilon_1\epsilon_2}.$$

Désormais nous ne nous intéressons qu'aux représentations du type V_l^+, et on considère l'algèbre de Hopf $\mathcal{A}^+(SU_q(2)) = \prod_{r\geq 0} End[V_{r+1}^+]$, qui en tant qu'algèbre est isomorphe à $\mathcal{A}(SU(2))$, mais dont le coproduit est déformé par les formules ci-dessus. Nous allons déformer la marche aléatoire quantique considérée au paragraphe précédent, et calculer le noyau markovien associé à ses restrictions. On considère donc l'état $\eta = \frac{1}{2}Tr$ sur la composante de dimension 2 de $\mathcal{A}^+(SU_q(2))$. On peut alors calculer les probabilités de transition obtenues en restreignant l'opérateur de transition à la sous-algèbre engendrée par t et par le centre de $\mathcal{A}^+(SU_q(2))$. Si l'on choisit une base dans chaque représentation, formée des vecteurs propres de l'opérateur t, cette algèbre est encore la sous-algèbre abélienne maximale de $\mathcal{A}^+(SU_q(2))$ formée des éléments dont toutes les composantes sont des matrices diagonales. Dans la représentation de dimension $r+1$ les valeurs propres de t sont $q^r,\ldots,q^{-r}$. On note alors $(e_{jk}^r; j,k \in \{-r,-r+2,\ldots,r-2,r\})$ la base des matrices élémentaires correspondantes dans chaque $M_{r+1}(\mathbb{C})$. On note aussi u_{jk}^r la base duale dans $(\mathcal{A}^+(SU_q(2)))^*$. En utilisant encore les formules de Clebsch-Gordan (cette fois-ci pour le cas quantique, cf. Klimyk et Vilenkin [11], formules (6) et (9), §14.4.3, mais la normalisation n'est pas la même que celle du présent article), qui s'écrivent

$$u_{jj}^r u_{-1-1}^1 = q^{-(r+j)/2} \frac{\left[\frac{r-j+2}{2}\right]_q}{[r+1]_q} u_{j-1j-1}^{r+1} + q^{(r-j+2)/2} \frac{\left[\frac{r+j}{2}\right]_q}{[r+1]_q} u_{j-1j-1}^{r-1}$$

$$u_{jj}^r u_{11}^1 = q^{(r-j)/2} \frac{\left[\frac{r+j+2}{2}\right]_q}{[r+1]_q} u_{j+1j+1}^{r+1} + q^{-(r+j+2)/2} \frac{\left[\frac{r-j}{2}\right]_q}{[r+1]_q} u_{j+1j+1}^{r-1}$$

on constate que l'algèbre engendrée par t et par le centre de $\mathcal{A}^+(SU_q(2)$ (c'est-à-dire par les e_{jj}^r) est stable par l'opérateur de transition. On en déduit les probabilités de transition de la chaîne de Markov

$$p((r,k),(r \mid 1,k+1)) = q^{(r-k)/2} \frac{\left[\frac{r+k+2}{2}\right]_q}{[r+1]_q} - \frac{q^{r+1}-q^{-k-1}}{2(q^{r+1}-q^{-r-1})} \qquad (4)$$

$$p((r,k),(r+1,k-1)) = q^{-(r+k)/2} \frac{\left[\frac{r-k+2}{2}\right]_q}{2[r+1]_q} = \frac{q^{-k+1}-q^{-r-1}}{2(q^{r+1}-q^{-r-1})}$$

$$p((r,k),(r-1,k+1)) = q^{-(r+k+2)/2} \frac{\left[\frac{r-k}{2}\right]_q}{2[r+1]_q} = \frac{q^{-k-1}-q^{-r-1}}{2(q^{r+1}-q^{-r-1})}$$

$$p((r,k),(r-1,k-1)) = q^{(r-k+2)/2} \frac{\left[\frac{r+k}{2}\right]_q}{2[r+1]_q} = \frac{q^{r+1}-q^{-k+1}}{2(q^{r+1}-q^{-r-1})} .$$

On peut vérifier sur ces probabilités de transition que la coordonnée k suit une marche de Bernoulli, en effet $p((r,k),(r+1,k+1))+p((r,k),(r-1,k+1)) = \frac{1}{2}$, qui ne dépend pas de k, mais que la coordonnée r n'est pas markovienne (car la probabilité $p((r,k),(r+1,k+1)) + p((r,k),(r+1,k-1))$, de passer de r à $r+1$ dépend de k). Par contre on voit facilement sur ces probabilités de transition, par récurrence sur n, que si la chaîne part de $(R_0 = 0, X_0 = 0)$, alors la loi conditionnelle de $(X_n$ sachant $R_1, \ldots, R_n)$ est uniforme sur $\{-R_n, -R_n + 2, \ldots, R_n\}$. On en déduit que le processus $(R_n; n \geq 1)$ est bien une chaîne de Markov de probabilités de transition (3).

En faisant tendre q vers 0, on constate que les probabilités de transition (4) convergent vers (2), ce qui redonne bien le théorème de Pitman.

Pour terminer avec $SU_q(2)$, faisons une dernière remarque sur la propriété de Markov de la chaîne de Bernoulli quantique déformée. On aurait pu, à la place de l'état η, utiliser l'état ρ, obtenu à partir du caractère de Woronowicz [17], qui est défini par ($A^{(2)}$ désignant la composante de A dans $M_2(\mathbb{C})$)

$$\rho(A) = \frac{2}{[2]_q} \eta(tA) = \frac{1}{[2]_q} Tr \left(A^{(2)} \begin{pmatrix} q & 0 \\ 0 & q^{-1} \end{pmatrix} \right) .$$

Cet état a été utilisé par Izumi [9], qui a montré que l'opérateur associé laisse invariant le centre de $\mathcal{A}^+(SU_q(2))$ (voir aussi [7],[14]). Pour le vérifier il suffit de calculer les probabilités de transition associées, on trouve

$$\tilde{p}((r,k),(r+1,k+1)) = \frac{2q}{[2]_q} p((r,k),(r+1,k+1)) = \frac{q^{r+2} - q^{-k}}{[2]_q(q^{r+1} - q^{-r-1})}$$

$$\tilde{p}((r,k),(r+1,k-1)) = \frac{2q^{-1}}{[2]_q} p((r,k),(r+1,k-1)) = \frac{q^{-k} - q^{-r-2}}{[2]_q(q^{r+1} - q^{-r-1})}$$

$$\tilde{p}((r,k),(r-1,k+1)) = \frac{2q}{[2]_q} p((r,k),(r-1,k+1)) = \frac{q^{-k} - q^{-r}}{[2]_q(q^{r+1} - q^{-r-1})}$$

$$\tilde{p}((r,k),(r-1,k-1)) = \frac{2q^{-1}}{[2]_q} p((r,k),(r-1,k-1)) = \frac{q^{r} - q^{-k}}{[2]_q(q^{r+1} - q^{-r-1})}.$$

Cela permet de voir que la restriction à l'algèbre engendrée par t donne lieu à la marche de Bernoulli décentrée de lois d'accroissements

$$p(k,k+1) = \frac{q}{[2]_q}, \qquad p(k,k-1) = \frac{q^{-1}}{[2]_q}$$

alors que la restriction au centre donne la chaîne de Markov de lois de transition

$$p(k,k+1) = \frac{q[k+1]_q}{[2]_q[k]_q}, \qquad p(k,k-1) = \frac{q^{-1}[k-1]_q}{[2]_q[k]_q}.$$

La chaîne de Markov sur le centre s'obtient donc par une transformation de Doob à partir de la marche aléatoire sur le dual du groupe engendré par t. Pour retrouver la loi de la marche de Bernoulli quantique de tout-à-l'heure, on doit utiliser la formule $\eta^{\otimes N}(.) = \frac{[2]_q^N}{2^N} \rho^{\otimes N}(.t^{\otimes N})$. La preuve de la markovianité de la restriction au centre peut alors se faire suivant les lignes de [3], Theorem 4.1.

Évidemment il n'y a pas besoin d'introduire autant d'objets sophistiqués pour comprendre le théorème de Pitman, toutefois ces constructions donnent l'idée de généralisations naturelles. Il suffit de remplacer le groupe quantique $SU_q(2)$ par un autre groupe quantique, et de regarder comment se comporte la restriction des marches de Bernoulli à la sous-algèbre engendrée par le centre et par le tore maximal canonique, lorsque $q \to 0$. On espère alors obtenir des formules explicites, à la Pitman, qui permettent de construire le mouvement brownien à valeurs dans la chambre de Weyl du groupe compact sous-jacent, à partir d'un mouvement brownien de même dimension. En pratique cette approche a des limites, car il est difficile de calculer les coefficients de Clebsch-Gordan dans le cas général (voir toutefois l'article [8], qui est à l'origine des travaux de Kashiwara sur les cristaux). Néanmoins, il existe d'autres approches plus directes qui permettent d'obtenir de telles formules, et qui sont développées dans [5].

Références

1. P. Biane, *Marches de Bernoulli quantiques*. Séminaire de Probabilités, XXIV, 1988/89, 329–344, Lecture Notes in Math., 1426, Springer, Berlin, 1990.
2. P. Biane, *Équation de Choquet-Deny sur le dual d'un groupe compact*. Probab. Theory Related Fields 94 (1992), no. 1, 39–51.
3. P. Biane, *Minuscule weights and random walks on lattices*. Quantum probability & related topics, 51–65, QP-PQ, VII, World Sci. Publishing, River Edge, NJ, 1992.
4. P. Biane, *Calcul stochastique non-commutatif*. Lectures on probability theory (Saint-Flour, 1993), 1–96, Lecture Notes in Math., 1608, Springer, Berlin, 1995.
5. P. Biane, P. Bougerol, N. O'Connell *Littelmann paths and Brownian paths*. Duke Math. J. 130(1), 127–167, (2005).
6. P. Bougerol, T. Jeulin, *Paths in Weyl chambers and random matrices*. Probab. Theory Related Fields 124 (2002), no. 4, 517 543.
7. B. Collins, *Intégrales matricielles et probabilités non-commutatives*. Thèse, Université Paris 6, Janvier 2003.
8. E. Date, M. Jimbo, T. Miwa, *Representations of $U_q(\mathfrak{gl}(n, C))$ at $q = 0$ and the Robinson-Schensted correspondence*. Physics and mathematics of strings, 185–211, World Sci. Publishing, Teaneck, NJ, 1990.
9. M. Izumi, *Non-commutative Poisson boundaries and compact quantum group actions*. Adv. Math. 169 (2002), no. 1, 1–57.
10. M. Kashiwara, *Bases cristallines des groupes quantiques*. Rédigé par C. Cochet. Cours Spécialisés, 9. Société Mathématique de France, Paris, 2002.
11. A.U. Klimyk, N.Ja. Vilenkin, *Representation of Lie groups and special functions. Vol. 2. Class I representations, special functions, and integral transforms*. Mathematics and its Applications (Soviet Series), 74. Kluwer Academic Publishers Group, Dordrecht, 1993.

12. P. Littelmann, *Paths and root operators in representation theory*. Ann. of Math. (2) 142 (1995), no. 3, 499–525.

13. P.A. Meyer, *Quantum probability for probabilists*. Lecture Notes in Mathematics, 1538. Springer-Verlag, Berlin, 1993.

14. S. Neshveyev, L. Tuset, *The Martin boundary of a discrete quantum group*. math.OA/0209270.

15. K.R. Parthasarathy, *An introduction to quantum stochastic calculus*. Monographs in Mathematics, 85. Birkhäuser Verlag, Basel, 1992.

16. J.W. Pitman. *One-dimensional Brownian motion and the three-dimensional Bessel process*. Adv. Appl. Probab. 7 (1975) 511-526.

17. S.L. Woronowicz, *Compact matrix pseudogroups*. Comm. Math. Phys. 111 (1987), no. 4, 613–665.

A Simple Proof of Two Generalized Borel-Cantelli Lemmas

Dedicated to the memory of Paul André Meyer

Jia-An Yan[*]

Academy of Mathematics and Systems Science, Chinese Academy of Sciences,
Beijing 100080, China
e-mail: jayan@mail.amt.ac.cn

Summary. Kochen and Stone [2] have proposed a generalization of the Borel-Cantelli lemma. Recently Petrov [3,4] further generalized their result. In this note we first use the Chung-Erdös inequality to give a simple proof of Kochen-Stone's result and then show that the latter implies Petrov's generalized Borel-Cantelli lemma.

Keywords: Chung-Erdös inequality, Generalized Borel-Cantelli Lemma

For the reader's convenience we recall from [1] the Chung-Erdös inequality (see (1) below) and give its proof.

Lemma 1. *Let* $\{A_k, 1 \leq k \leq n\}$ *be a sequence of events. Then*

$$P\left(\bigcup_{k=1}^{n} A_k\right) \geq \frac{\left(\sum_{k=1}^{n} P(A_k)\right)^2}{\sum_{i,k=1}^{n} P(A_i A_k)}. \tag{1}$$

Proof. Let $X_k = I_{A_k}$. By Schwarz's inequality, we have

$$\left(E\left(\sum_{k=1}^{n} X_k\right)\right)^2 \leq P\left(\sum_{k=1}^{n} X_k > 0\right) E\left[\left(\sum_{k=1}^{n} X_k\right)^2\right].$$

Since $P(\sum_{k=1}^{n} X_k > 0) = P(\bigcup_{i=k}^{n} A_k)$, we get (1).

The following generalized Borel-Cantelli lemma is due to Kochen-Stone [2]. We give here a new and simple proof which is based on the Chung-Erdös inequality.

---
[*] Work supported by the Ministry of Science and Technology, the 973 project on Mathematics, and the Knowledge Innovation Program of the Chinese Academy of Sciences.

Theorem 1. *Let $\{A_n, n \geq 1\}$ be such that $\sum_{n=1}^{\infty} P(A_n) = \infty$. Then*

$$P(A_n, \text{i.o.}) \geq \limsup_{n \to \infty} \frac{(\sum_{k=1}^{n} P(A_k))^2}{\sum_{i,k=1}^{n} P(A_i A_k)} = \limsup_{n \to \infty} \frac{\sum_{1 \leq i < k \leq n} P(A_i) P(A_k)}{\sum_{1 \leq i < k \leq n} P(A_i A_k)}.$$
(2)

In particular, if $\{A_n, n \geq 1\}$ are pairwise independent or negatively correlated (i.e., $P(A_i A_k) \leq P(A_i) P(A_k), \forall i \neq k$), then $P(A_n, \text{i.o.}) = 1$.

Proof. Let $a_n = (\sum_{k=1}^{n} P(A_k))^2$, $b_n = \sum_{i,k=1}^{n} P(A_i A_k)$. Then by the assumption $\lim_{n \to \infty} a_n = \infty$ and by (1), we have $\lim_{n \to \infty} b_n = \infty$. From (1) and the fact that $\sum_{i,k=m+1}^{n} P(A_i A_k) \leq b_n - b_m$ we get that

$$P\left(\bigcup_{k=m+1}^{\infty} A_k\right) = \lim_{n \to \infty} P\left(\bigcup_{k=m+1}^{n} A_k\right) \geq \limsup_{n \to \infty} \frac{(\sqrt{a_n} - \sqrt{a_m})^2}{b_n - b_m}$$

$$= \limsup_{n \to \infty} \frac{a_n}{b_n}.$$

Letting $m \to \infty$ gives the inequality in (2). Since $\sum_{k=1}^{\infty} P(A_k) = \infty$, and

$$\left(\sum_{k=1}^{n} P(A_k)\right)^2 \leq 2 \sum_{1 \leq i < k \leq n} P(A_i) P(A_k) + \sum_{k=1}^{n} P(A_k),$$

we have

$$\lim_{n \to \infty} \frac{\sum_{k=1}^{n} P(A_k)}{\sum_{1 \leq i < k \leq n} P(A_i) P(A_k)} = 0.$$

Thus the equality in (2) holds.

Recently, Petrov [3] found that if $\sum_{n=1}^{\infty} P(A_n) = \infty$ and $P(A_i A_k) \leq HP(A_i) P(A_k)$ for all $i, k > N$ such that $i \neq k$ and for some constants $H \geq 1$ and N, then $P(\limsup A_n) \geq 1/H$. Petrov [4] further generalized his result as follows:

Theorem 2. *Let $\{A_n, n \geq 1\}$ be such that $\sum_{n=1}^{\infty} P(A_n) = \infty$. Let H be a real constant. Put*

$$\alpha_H = \liminf_{n \to \infty} \frac{\sum_{1 \leq i < k \leq n} \left(P(A_i A_k) - HP(A_i) P(A_k)\right)}{\left(\sum_{k=1}^{n} P(A_k)\right)^2}.$$

Then $P(A_n, \text{i.o.}) \geq 1/(H + 2\alpha_H)$.

We will show that this result is a simple consequence of Kochen-Stone's result. In fact, as indicated by Petrov [4], from both equalities

$$2 \sum_{1 \leq i < k \leq n} P(A_i A_k) = \sum_{i,k=1}^{n} P(A_i A_k) - \sum_{k=1}^{n} P(A_k)$$

and

$$2 \sum_{1 \leq i < k \leq n} P(A_i) P(A_k) = \left(\sum_{k=1}^{n} P(A_k) \right)^2 - \sum_{k=1}^{n} P(A_k)^2 \,,$$

one gets

$$H + 2\alpha_H = \liminf_{n \to \infty} \left\{ \left(\sum_{i,k=1}^{n} P(A_i A_k) \right) \left(\sum_{k=1}^{n} P(A_k) \right)^{-2} \right.$$
$$\left. - \left(\sum_{k=1}^{n} P(A_k) \right)^{-1} + H \left(\sum_{k=1}^{n} P(A_k)^2 \right) \left(\sum_{k=1}^{n} P(A_k) \right)^{-2} \right\}$$

(see (12)-(14) of Petrov [4]). However, from the above equality we deduce that

$$H + 2\alpha_H = \liminf_{n \to \infty} \left(\sum_{i,k=1}^{n} P(A_i A_k) \right) \left(\sum_{k=1}^{n} P(A_k) \right)^{-2} .$$

Thus Theorem 1 implies Theorem 2.

References

1. K.L. Chung, P. Erdös: On the application of the Borel-Cantelli lemma. Trans. Amer. Math. Soc. **72** (1952), 179–186.
2. S. Kochen, C. Stone: A note on the Borel-Cantelli lemma. Ill. J. Math. **8** (1964), 248–251.
3. V.V. Petrov: A note on the Borel-Cantelli lemma. Stat. Prob. Lett. **58** (2002), 283–286.
4. V.V. Petrov: A generalization of the Borel-Cantelli lemma. Stat. Prob. Lett. **67** (2004), 223–239.

Natural Decomposition of Processes and Weak Dirichlet Processes

François Coquet[*], **Adam Jakubowski**[**,1]
Jean Mémin[***] **and Leszek Słomiński**[**,2]

[*]LMAH, Université du Havre, 25 rue Philippe Lebon, 76063 Le Havre Cedex,
France
e-mail: coquet@univ-lehavre.fr
[**]Faculty of Mathematics and Computer Science, N. Copernicus University, ul.
Chopina, 87-100 Toruń, Poland
e-mail: adjakubo@mat.uni.torun.pl; leszeks@mat.uni.torun.pl
[***]IRMAR, Université de Rennes 1, Campus de Beaulieu, 35042 Rennes Cedex,
France
e-mail: memin@univ-rennes1.fr

Abstract. A class of stochastic processes, called "weak Dirichlet processes", is introduced and its properties are investigated in detail. This class is much larger than the class of Dirichlet processes. It is closed under C^1-transformations and under absolutely continuous changes of measure. If a weak Dirichlet process has finite energy, as defined by Graversen and Rao, its Doob-Meyer type decomposition is unique. The methods developed here have been applied to study generalized martingale convolutions.

Mathematics Subject Classification: 60G48, 60H05

Keywords: weak Dirichlet processes, Dirichlet processes, semimartingales, finite energy processes, quadratic variation, mutual quadratic covariation, Ito's formula, generalized martingale convolution.

1 Introduction

The quadratic variation of a stochastic process as well as the mutual covariation of two stochastic processes have been well-known for a long time to be at the core of the theory of stochastic integration, if only because the quadratic variation appears explicitely in Ito's formula for semimartingales. And indeed, every attempt to generalize Ito's calculus to a wider class of integrators (for instance Dirichlet processes) or to functions less regular than C^2 functions has

---

[1] Supported in part by Komitet Badań Naukowych under Grant No 1 P03A 022
26 and completed while the author was visiting Université de Rennes I.
[2] Supported in part by Komitet Badań Naukowych under Grant No 1 P03A 022
26 and completed while the author was visiting Université de Rennes I.

to deal with quadratic variations or covariations of the processes that appear. In this aspect, the most enlightening work is perhaps Föllmer's paper ([9]).

On the other hand, it was proven by Graversen and Rao ([11]) that the existence of a Doob-Meyer type decompositon for a process X is narrowly linked to the fact that X has a finite energy, which is a somewhat weaker assumption than the existence of an integrable quadratic variation for X. The best known class of processes with finite energy (beyond the class of square integrable semimartingales) is the class of Dirichlet processes. A larger class has been recently introduced by Errami and Russo ([6]) under the name "weak Dirichlet processes". The present paper explores some desirable properties of such processes. Although our definition of a quadratic variation is different from Errami and Russo's one -it is in a way more classical- at any rate both coincide as far as semimartingales are concerned. The other noticeable difference is that we do not assume continuity.

In part 2, we give an as explicit as possible link between quadratic variation, energy, Dirichlet processes, weak Dirichlet processes, and "natural" (then, "Doob-Meyer type") decomposition.

Part 3 proves that any $\mathcal{C}^1$ function F of a weak Dirichlet process X is again a weak Dirichlet process. We use an Ito-type formula which gives explicitely the decomposition of $F(X)$ for $\mathcal{C}^2$ transformations, and in the general case we find an explicit formula for the martingale part only. This extends the known results: Errami, Russo and Vallois obtained related results with an Ito-type formula ([8]) under slightly more restrictive conditions on F.

Part 4 which is closest to Errami and Russo's work mentionned above, deals with processes X that may be written $X_t = \int_0^t G(t,s)\,dL_s$ where L is a quasi-left continuous—but not necessarily continuous—square-integrable martingale and G a deterministic function. We give two sets of hypotheses under which X is a weak Dirichlet process, and also give its natural decomposition. This section is illustrated with 3 examples, the third one giving furthermore a formula of Fubini type.

Last, the appendix gives some counter-examples related to the quadratic variation or to the regularity of paths of processes. Although such examples may be well-known, we could not find any in the literature, and we hope that they can enlight some of the technical problems we are confronted with here and there in the paper.

2 Basic notations and results about processes with finite energy and weak Dirichlet processes

In what follows, we are given a probability space $(\Omega, \mathcal{G}, \mathbf{P})$.

We also fix a positive real number T. Unless otherwise stated, every process or filtration will be indexed by $t \in [0, T]$. A filtration $(\mathcal{F}_t)_{t \leq T}$ is denoted by $\mathcal{F}$. All filtrations are assumed to be right-continuous and defined on $(\Omega, \mathcal{G}, \mathbf{P})$ with $\mathcal{F}_T \subset \mathcal{G}$.

We are also given a refining sequence D_n of subdivisions of $[0, T]$ whose mesh goes to 0 when $n \to \infty$. For every n, $D_n = \{0 = t_0^n, t_1^n, \ldots, t_{N(n)}^n = T\}$.

We work with processes having a.s. right-continuous trajectories with left limits (such a process is called càdlàg), null in 0 and, unless otherwise stated, admitting a finite energy in the sense defined below following Graversen and Rao ([11]):

Definition 2.1. *We say that X is a process of finite energy if*

$$\sup_n E \left[\sum_{t_i^n \in D_n} (X_{t_i^n} - X_{t_{i-1}^n})^2 \right] < +\infty. \tag{1}$$

This "sup" will be denoted $\mathcal{E}n(X)$.

Of course, if X has a finite energy, $|X_t|^2$ is integrable for every $t \leq T$ and also $\sum_{s \leq T} \Delta X_s^2$ is integrable.

We recall Graversen and Rao's main result in [11]:

Theorem 2.1. *If X is a process with finite energy, then it can be written as a sum $X = M + A$, where M is a square-integrable martingale and A is a predictable process such that there exists a subsequence (D_{n_j}) of (D_n) satisfying, for every $t \leq T$*

$$E \left[\sum_{t_i^{n_j} \in D_{n_j}, t_i^{n_j} \leq t} (A_{t_i^{n_j}} - A_{t_{i-1}^{n_j}})(N_{t_i^{n_j}} - N_{t_{i-1}^{n_j}}) \right] \longrightarrow 0 \tag{2}$$

as $j \to \infty$ for all square integrable martingales N.

If, moreover, $X = M' + A'$ is any other such decomposition, the process $A - A'$ is a continuous martingale.

Last, if we write

$$M_t^n = \sum_{t_i^n \in D_n, t_i^n \leq t} \left[X_{t_i^n} - E[X_{t_i^n} | \mathcal{F}_{t_{i-1}^n}] \right], \tag{3}$$

then for all $t \in \bigcup_n D_n$, M_t is the weak limit in $\sigma(\mathbf{L}^2, \mathbf{L}^2)$ of the sequence, $(M_t^{n_j})$.

Any process A satisfying (2) will be called a "natural" process, and any decomposition of X as $X = M + A$, satisfying (2) and (3) will be called a "natural decomposition".

This definition of a "natural" process coincides with P.A. Meyer's one if X is a submartingale of class D. Graversen and Rao assume that the filtration is "free of times of discontinuity", but this assumption is not necessary, The important thing is that, as in the classical Doléans' proof for integrable increasing processes, "natural" implies "predictable"; to prove that, one uses the characterization of predictable processes expressed for example in Dellacherie's book ([4] p. 85, **T31**). So, this decomposition of X is a Doob-Meyer-type decomposition. Theorem 2.2 will provide us with an important setting where the decomposition $X = M + A$ in Theorem 2.1 is unique.

We will use the notion of weak Dirichlet process introduced by Errami and Russo ([6]) in a slightly different context.

Definition 2.2. *We say that X is a weak Dirichlet process if it admits a decomposition $X = M + A$, where M is a local martingale and A is a predictable process such as $[A, N] = 0$ for all continuous local martingales N.*

In the above definition and in the sequel we use the notion of mutual covariation and quadratic variation in the following sense (taken from [2]).

Definition 2.3. (i) *Two processes X and Y admit a quadratic (mutual) covariation along (D_n) if there exists a càdlàg process denoted $[X, Y]$ such that for every $t \leq T$*

$$[X, Y]_t = [X, Y]_t^c + \sum_{s \leq t} \Delta X_s \Delta Y_s$$

and

$$S^n(X, Y)_t := \sum_{t_i^n \in D_n, t_i^n \leq t} (X_{t_{i+1}^n} - X_{t_i^n})(Y_{t_{i+1}^n} - Y_{t_i^n}) \xrightarrow{P} [X, Y]_t \ \ as \ n \to \infty.$$

$$(4)$$

(ii) *A process X admits a quadratic variation along (D_n) if there exists a càdlàg process denoted $[X, X]$ with for every $t \leq T$*

$$[X, X]_t = [X, X]_t^c + \sum_{s \leq t} \Delta X_s^2$$

and

$$S^n(X, X)_t := \sum_{t_i^n \in D_n, t_i^n \leq t} (X_{t_{i+1}^n} - X_{t_i^n})^2 \xrightarrow{P} [X, X]_t \ \ as \ n \to \infty. \quad (5)$$

Remark 2.1. (i) The decomposition $X = M + A$ of a weak Dirichlet process is unique.

To see this suppose that we have decompositions $X = M + A = M' + A'$ with A and A' predictable and verifying $[A, N] = [A, N'] = 0$ for every continuous local martingale N. Then $A - A'$ is a predictable local martingale, hence a continuous local martingale. Then

$$[A - A', A - A']_T = [A - A', A]_T - [A - A', A']_T = 0$$

and we deduce that $A = A'$.

(ii) A weak Dirichlet process X need not admit a quadratic variation. We know only that for every continuous martingale N the covariation $[X, N]$ exists.

(iii) Of course, in general, a decomposition $X = M + A$ with a martingale M and a predictable process A, does not imply that $[A, N] = 0$ for every continuous martingale N, when $[A, N]$ exists. For example, take A as a continuous martingale and $N = A$.

The class of weak Dirichlet processses is much larger than the class of Dirichlet processes. We recall:

Definition 2.4. *A Dirichlet process is the sum of a local martingale and a continuous process whose quadratic variation is identically zero.*

Remark 2.2. Note that a Dirichlet process admits a quadratic variation, which is equal to the quadratic variation of its martingale part. Our definition of a quadratic variation, which follows Föllmer's one in [9], is weaker than the definition in [10], and slightly different from Russo and Vallois' one in [16]. However the three of them coincide as far as semimartingales are concerned, and a Dirichlet process according to the definition in [10] is also Dirichlet according to both other ones.

The following notion of pre-quadratic variation is weaker than quadratic variation; however, both notions coincide under stronger assumptions, as will be seen below.

Definition 2.5. *A process X (not necessarily càdlàg) admits a pre-quadratic variation along (D_n) if there exists an increasing process denoted by $S(X, X)$ such that for every $t \leq T$*

$$S^n(X, X)_t := \sum_{t_i^n \in D_n, t_i^n \leq t} (X_{t_{i+1}^n} - X_{t_i^n})^2 \xrightarrow{\text{P}} S(X, X)_t \quad as \ n \to \infty. \quad (6)$$

Remark 2.3. We can find examples of continuous processes X such that $S(X, X)$ is defined but not continuous (see example in Appendix 5.2), hence X does not admit a quadratic variation.

For every $t \leq T$, denoting by π_t any subdivision of $[0, t]$, we consider the sum

$$S^{\pi_t}(X, X) := \sum_{t_i \in \pi_t, i > 0} (X_{t_i} - X_{t_{i-1}})^2.$$

Proposition 2.1. *Let X be a càdlàg process with a pre-quadratic variation S such that the following property (referred to as property (S)) holds:*

$S(X, X)$ is right continuous and for every $t \leq T$, $S^{\pi_t}(X, X) \xrightarrow{\text{P}} S(X, X)_t$ as the mesh of π_t tends to 0.

Then $S(X, X)$ is the quadratic variation of X along any sequence (D_n) of subdivisions of $[0, T]$ whose meshes tends to 0.

This result is proved in [6], Lemme (3.11).

Remark 2.4. (i) The class of Dirichlet processes is larger than the space of special semimartingales. Every continuous function admitting a quadratic variation equal to zero is a deterministic Dirichlet process.

(ii) Every continuous function is a deterministic weak Dirichlet process for any filtration:

Actually, let us consider a bounded continuous martingale N nul in 0, we have

$$E\left(\left|\sum_{t_i^n \in D_n} (f(t_{i+1}^n) - f(t_i^n))(N_{t_{i+1}^n} - N_{t_i^n})\right|^2\right) \leq \sup_{t_i^n} (f(t_{i+1}^n) - f(t_i^n))^2 E(N_T)^2.$$

and, from continuity of f this last term tends to 0 when $n \to \infty$.

We give in Section 4 nondeterministic examples of weak Dirichlet processes, which are not ordinary Dirichlet processes.

Remark 2.5. The family of processes with finite energy is clearly stable under addition, however we do not know if this stability holds for the family of processes admitting a quadratic variation. Of course this is true for the family of Dirichlet processes and for weak Dirichlet processes

Theorem 2.2. *Assume X is a process with finite energy. The following three conditions are equivalent:*

(i) *X is a weak Dirichlet process,*
(ii) *for every continuous local martingale N, the quadratic covariation $[X, N]$ is well-defined,*
(iii) *for every locally square integrable martingale N, the quadratic covariation $[X, N]$ is well-defined.*

If any of these conditions is fulfilled, the decomposition $X = M + A$ as a weak Dirichlet process is a natural decomposition of X as expressed in Theorem 2.1, and this last decomposition is unique.

Proof: (i) $\Rightarrow$ (iii) Let us write $X = M + A$ as in Definition 2.2 and consider the decomposition $N = N^c + N^d$, where N^c is the continuous and N^d the purely discontinuous part of N. By the definition of a weak Dirichlet process the covariation $[X, N^c]$ is well-defined. To prove the existence of $[X, N^d]$ we use the following lemma.

Lemma 2.1 *Assume that X has a finite energy, and that N is a locally square integrable martingale which is the compensated sum of its jumps. Then X and N admit a covariation such that*

$$[X, N]_t = \sum_{s \leq t} \Delta X_s \Delta N_s. \tag{7}$$

Proof of Lemma 2.1: By using a localizing sequence of stopping times, one can assume that N is a square integrable martingale. One can find a sequence $(N^p)_p$ of martingales having finite variation and only a finite number of jumps, such that $N^p \to N$ in $\mathbf{H^2}$ the space of square integrable martingales.

We have then, for fixed p,

$$\sum_{t_i^n \in D_n, t_i^n \le t} (X_{t_{i+1}^n} - X_{t_i^n})(N^p_{t_{i+1}^n} - N^p_{t_i^n}) \xrightarrow{\mathrm{P}} \sum_{s \le t} \Delta X_s \Delta N^p_s \qquad (8)$$

as $n \to \infty$.

On the other hand,

$$E \left| \sum_{t_i^n \in D_n, t_i^n \le t} (X_{t_{i+1}^n} - X_{t_i^n})(N^p_{t_{i+1}^n} - N^p_{t_i^n}) - \sum_{t_i^n \in D_n, t_i^n \le t} (X_{t_{i+1}^n} - X_{t_i^n}) \right.$$

$$\left. \times \left(N_{t_{i+1}^n} - N_{t_i^n} \right) \right|$$

$$\le E \left| \left(\sum_{t_i^n \in D_n, t_i^n \le t} (X_{t_{i+1}^n} - X_{t_i^n})^2 \right)^{1/2} \right.$$

$$\left. \times \left(\sum_{t_i^n \in D_n, t_i^n \le t} ((N_{t_{i+1}^n} - N^p_{t_{i+1}^n}) - (N_{t_i^n} - N^p_{t_i^n}))^2 \right)^{1/2} \right|$$

$$\le (\mathcal{E}n(X))^{1/2} E \left([N - N^p, N - N^p]_t \right)^{1/2}$$

which goes to 0 as $p \to \infty$ since $[N - N^p, N - N^p]_t \xrightarrow{\mathrm{P}} 0$.

Last,

$$E \left| \sum_{s \le t} \Delta X_s (\Delta N_s - \Delta N^p_s) \right| \le E \left(\left(\sum_{s \le t} \Delta X_s^2 \right)^{1/2} \left(\sum_{s \le t} (\Delta N_s - \Delta N^p_s)^2 \right)^{1/2} \right.$$

$$\left. \le (\mathcal{E}n(X))^{1/2} E \left[[N - N^p, N - N^p]_t \right]^{1/2} \right)$$

which goes to zero as p goes to infinity, hence $\sum_{s \le t} \Delta X_s \Delta N^p_s$ converges in L^1 to $\sum_{s \le t} \Delta X_s \Delta N_s$.

These three convergences give the lemma. $\square$

$(iii) \Rightarrow (ii)$ is obvious.

$(ii) \Rightarrow (i)$ Let $X = M + A$ be a decomposition from Theorem 2.1 and let N be a continuous local martingale. Define $T_p = \inf\{t : |N_t| \ge p\}$ then (T_p) is a localizing sequence of stopping times. We will prove that for every p, $[A, N^{T_p}] = 0$, which implies that also $[A, N] = 0$.

By hypothesis we have the convergence $S^n(X, N^{T_p})_t \xrightarrow{\mathrm{P}} [X, N^{T_p}]_t$.

Hence we deduce that also $S^n(A, N^{T_p})_t \xrightarrow{P} [A, N^{T_p}]_t$. We will show that in fact

$$S^n(A, N^{T_p})_T \to [A, N^{T_p}]_T \quad in \ L^1. \tag{9}$$

To see this, it is sufficient to check uniform integrability of the sequence $\{S^n(A, N^{T_p})_T\}$. Writing $|S^n(X, N^{T_p})_T| \le S^n(X, X)_T^{1/2} S^n(N^{T_p}, N^{T_p})_T^{1/2}$, and using the Hölder inequality, we get:

$$E|S^n(X, N^{T_p})_T|^{4/3} \le (E[S^n(X, X)_T])^{2/3}(E[S^n(N^{T_p}, N^{T_p})_T^2])^{1/3} < \infty.$$

A similar proof gives $E|S^n(M, N^{T_p})_T|^{4/3} < +\infty$.

As a consequence, we deduce uniform integrability of $\{(S^n(A, N^{T_p})_T\}$ and (9) holds true. Therefore, in particular

$$E[S^n(A, N^{T_p})_T] \to E[A, N^{T_p}]_T$$

and due to (2)

$$E[A, N^{T_p}]_T = 0. \tag{10}$$

Note that the process $[A, N^{T_p}]$ has a finite variation ; moreover, since N is continuous, $[A, N^{T_p}]$ is also a continuous process. Therefore, to get $[A, N^{T_p}] = 0$ it is sufficient to prove that $[A, N^{T_p}]$ is a local martingale

Consider a bounded stopping time $\tau \le T$; the same arguments as above show that $E[S^n(A, N^{T_p \wedge \tau})_T]$ converges to $E[[A, N^{T_p \wedge \tau}]_T]$ and (2) gives

$$E[[A, N^{T_p}]_\tau] = E[[A, N^{T_p \wedge \tau}]_T] = 0.$$

Hence, it follows easily that the stopped process $[A, N]^{T_p}$ is a martingale and $[A, N^{T_p}] = 0$. Since $P(T_p = T) \uparrow 1$, the proof of the last implication is completed.

Finally, take $X = M' + A'$ any natural decomposition of X. From the proof of $(ii) \Rightarrow (i)$, $X = M' + A'$ is a decomposition of X as a weak Dirichlet process, but from Remark 2.1, such a decomposition is unique. This proves the last assertion of theorem. □

We get easily the following

Corollary 2.1. *Let us consider a weak Dirichlet process X with finite energy.*

(i) *If Q is a probability measure absolutely continuous with respect to $\mathbf{P}$, with bounded Radon-Nikodým derivative, then X is a Q weak Dirichlet process.*

(ii) *For an $a > 0$ we define $\hat{X} = \sum_{s \le .} \Delta X_s 1_{\Delta|X_s|>a}$, then $X - \hat{X}$ is a weak Dirichlet process.*

Now, we will consider processes with finite energy X admitting additionally a quadratic variation $[X, X]$. Then of course $E[X, X]_T < \infty$.

Theorem 2.3. *Assume X is a weak Dirichlet process with finite energy, admitting a quadratic variation process.*

(i) *In the natural decomposition $X = M + A$, M is a square integrable martingale and A has an integrable quadratic variation.*

(ii) *The natural decomposition is minimal in the following sense. If $X = M' + A'$ is another decomposition with a local martingale M' and a predictable process A'*
then $[A', A']$ is well defined and:

$$[A', A'] = [M - M', M - M'] + [A, A].$$

Proof: (*i*) To begin with, we notice that

$$E\left[\sum_{s \leq T} \Delta A_s^2\right] < \infty;$$

actually, for every predictable stopping time S, $\Delta A_S = E[\Delta X_S | \mathcal{F}_{S-}]$, hence $E\left[\sum_{s \leq T} \Delta A_s^2\right] \leq E\left[\sum_{s \leq T} \Delta X_s^2\right] < \infty$. It follows that $E[\sum_{s \leq T} \Delta M_s^2] < \infty$ and M is a locally square integrable martingale.

Let us consider the decomposition $M = M^c + M^d$ where M^c is the continuous part of M and M^d its purely discontinuous part. Writing $[M^d, A] = [M^d, X] - [M^d, M^d]$, we get the existence of $[M^d, A]$. Using the property of decomposition $X = M + A$, we have $[M, A] = [M^d, A]$. From Lemma 2.1 we deduce:

$$[M, A] = \sum_{s \leq \cdot} \Delta M_s \Delta X_s - \sum_{s \leq \cdot} \Delta M_s^2 = \sum_{s \leq \cdot} \Delta M_s \Delta A_s.$$

Now, by the definition of quadratic variation of X and M one gets the existence of $[A, A]$:

$$[A, A] = [X, X] - 2[M, A] - [M, M].$$

Finally,

$$E([M, M]_T + [A, A]_T) \leq E[X, X]_T + 2E[\sum_{s \leq T} \Delta M_s^2]^{1/2} E[\sum_{s \leq T} \Delta A_s^2]^{1/2} < \infty.$$

(ii) Since $A' = M + A - M'$, by linearity $[A', A']$ is well defined and we can write:

$$[A', A'] = [M+A-M', M+A-M'] = [M-M', M-M'] + [A, A] - 2[A, M-M'].$$

But, $M - M'$ is a continuous local martingale and as A is taken from the natural decomposition of X, we get: $[A, M - M'] = 0$, hence the desired result. □

3 Stability of weak Dirichlet processes under C^1 transformations

Assume X is a process of finite energy. Let us denote by μ the jump measure of X. Then $\sum_{s\leq.}\Delta X_s^2$ can be written $\int_0^\cdot \int_{\mathbb{R}\setminus\{0\}} x^2\mu(ds,dx)$. Since $E\sum_{s\leq T}\Delta X_s^2 < \infty$, X admits a Lévy system ν which is the predictable compensator of μ; then the predictable increasing process $\int_0^\cdot \int_{\mathbb{R}\setminus\{0\}} x^2\nu(ds,dx)$ is well defined and

$$E\sum_{s\leq T}\Delta X_s^2 = E[\int_0^T \int_{\mathbb{R}\setminus\{0\}} x^2\mu(ds,dx)] = E[\int_0^T \int_{\mathbb{R}\setminus\{0\}} x^2\nu(ds,dx)] < \infty.$$

We begin with C^2 stability:

Theorem 3.1. *Let $X = M + A$ be a weak Dirichlet process of finite energy and F a C^2-real valued function with bounded derivatives f and f'. Then the process $(F(X_t)_{t\geq 0})$ is a weak Dirichlet process of finite energy and the decomposition $F(X) = Y + \Gamma$ holds with the martingale part*

$$Y_t = F(0) + \int_0^t f(X_{s-})dM_s$$

$$+ \int_0^t \int_{\mathbb{R}} \Big(F(X_{s-}+x) - F(X_{s-}) - xf(X_{s-})\Big)(\mu - \nu)(ds,dx)$$

and the predictable part

$$\Gamma_t = (S)\int_0^t f(X_{s-})dA_s - 1/2\sum_{s\leq t} f'(X_{s-})(\Delta A_s)^2 + 1/2\int_0^t f'(X_s)d[M,M]_s^c$$

$$+ \int_0^t \int_{\mathbb{R}\setminus\{0\}} \Big(F(X_{s-}+x) - F(X_{s-}) - xf(X_{s-})\Big)\nu(ds,dx),$$

where $(S)\int_0^\cdot f(X_{s-})dA_s$ is well defined as a limit in probability of Riemann sums. More precisely for every t

$$\sum_{t_i^n \in D_n, t_i^n \leq t} (f(X_{t_i^n})(A_{t_{i+1}^n} - A_{t_i^n}) + 1/2f'(X_{t_i^n})(A_{t_{i+1}^n} - A_{t_i^n})^2)$$

$$\xrightarrow{\ P\ } (S)\int_0^t f(X_{s-})dA_s.$$

Proof. Fix $t > 0$. We borrow arguments from Föllmer [9]. For $\epsilon > 0$ we define $J(1) = \{s \leq t; |\Delta X_s| > \epsilon\}$. In the following, the elements of D_n are, for short, written t_i instead of t_i^n. Then

$$\sum_{t_i \in D_n, t_i \leq t} (F(X_{t_{i+1}}) - F(X_{t_i})) = \sum_1 (F(X_{t_{i+1}}) - F(X_{t_i}))$$

$$+ \sum_2 (F(X_{t_{i+1}}) - F(X_{t_i})),$$

where $\sum_1$ denotes the sum (depending on $\omega \in \Omega$) over all $t_i \in D_n, t_i \leq t$ such that $(t_i, t_{i+1}] \cap J(1) \neq \emptyset$ and $\sum_2$ the sum over all other t_i's.

Then by Taylor's formula

$$\sum_2 (F(X_{t_{i+1}}) - F(X_{t_i})) = \sum_2 f(X_{t_i})(X_{t_{i+1}} - X_{t_i})$$

$$+ 1/2 \sum_2 f'(X_{t_i})(X_{t_{i+1}} - X_{t_i})^2 + \sum_2 r_2(X_{t_i}, X_{t_{i+1}}),$$

where $r_2(X_{t_i}, X_{t_{i+1}}) = C_i^\epsilon (X_{t_{i+1}} - X_{t_i})^2$ with $\max_2 |C_i^\epsilon| \leq ||f'||$ and

$$\lim_{\epsilon \downarrow 0} \limsup_{n \to \infty} P(\max_2 |C_i^\epsilon| > \delta) = 0, \quad \delta > 0. \tag{11}$$

Hence

$$\sum_{t_i \in D_n, t_i \leq t} (F(X_{t_{i+1}}) - F(X_{t_i})) = \sum_{t_i \in D_n, t_i \leq t} f(X_{t_i})(X_{t_{i+1}} - X_{t_i})$$

$$+ 1/2 \sum_{t_i \in D_n, t_i \leq t} f'(X_{t_i})(X_{t_{i+1}} - X_{t_i})^2 + \sum_2 r_2(X_{t_i}, X_{t_{i+1}})$$

$$+ \sum_1 \{F(X_{t_{i+1}}) - F(X_{t_i}) - f(X_{t_i})(X_{t_{i+1}} - X_{t_i}) - 1/2 f'(X_{t_i})(X_{t_{i+1}} - X_{t_i})^2\}$$

$$= \sum_{t_i \in D_n, t_i \leq t} f(X_{t_i})(M_{t_{i+1}} - M_{t_i}) + \sum_{t_i \in D_n, t_i \leq t} f(X_{t_i})(A_{t_{i+1}} - A_{t_i})$$

$$+ 1/2 \sum_{t_i \in D_n, t_i \leq t} f'(X_{t_i})(A_{t_{i+1}} - A_{t_i})^2 + 1/2 \sum_{t_i \in D_n, t_i \leq t} f'(X_{t_i})(M_{t_{i+1}} - M_{t_i})^2$$

$$+ \sum_{t_i \in D_n, t_i \leq t} f'(X_{t_i})(M_{t_{i+1}} - M_{t_i})(A_{t_{i+1}} - A_{t_i}) + \sum_2 r_2(X_{t_i}, X_{t_{i+1}})$$

$$+ \sum_1 \{F(X_{t_{i+1}}) - F(X_{t_i}) - f(X_{t_i})(X_{t_{i+1}} - X_{t_i}) - 1/2 f'(X_{t_i})(X_{t_{i+1}} - X_{t_i})^2\}$$

$$= I_1^n + I_2^n + I_3^n + I_4^n + I_5^n + I_6^{n,\epsilon} + I_7^{n,\epsilon}.$$

Now, note that by the definition of a stochastic integral we have $I_1^n \xrightarrow{P} \int_0^t f(X_{s-}) dM_s$.

The following simple lemma will be very useful in order to estimate the other terms.

Lemma 3.1 *Assume that two càdlàg processes X and Y admit a quadratic covariation $[X,Y]$ and that the sequence $\{Var(S^n(X,Y))_T\}$ is bounded in probability.*

To any two càdlàg processes Z and U we associate the sequences $\{Z^n\}$ and $\{U^n\}$ of processes, where Z^n and U^n are the respective discretizations of Z and U along D_n; precisely $Z_t^n = Z_{t_i^n}$, $U_t^n = U_{t_i^n}$ when $t \in [t_i^n, t_{i+1}^n[$. Then, for every continuous real function f, g and every t, the following convergence holds:

$$\int_0^t f(Z_{s-}^n)g(\Delta U_s^n)dS^n(X,Y)_s \xrightarrow{\ P\ } \int_0^t f(Z_{s-})g(\Delta U_s)d[X,Y]_s,$$

where these integrals are Stieltjes integrals with respect to the processes $S^n(X,Y)$ or $[X,Y]$.

Proof of Lemma 3.1 From the proof of [2, Lemma 1.3] one can deduce that $\int_0^\cdot f(Z_{s-}^n)g(\Delta U_s^n)dS^n(X,Y)_s \xrightarrow{\ P\ } \int_0^\cdot f(Z_{s-})g(\Delta U_s)d[X,Y]_s$ in the Skorokhod J_1 topology (see e.g. [13]). Since for every t

$$f(Z_{t-}^n)g(\Delta U_t^n)\Delta S^n(X,Y)_t \xrightarrow{\ P\ } f(Z_{t-})g(\Delta U_t)\Delta[X,Y]_t,$$

the desired result follows from properties of the Skorokhod J_1 topology.   □

It is clear by Lemma 3.1 that we have the convergences:

$$I_4^n \xrightarrow{\ P\ } 1/2 \int_0^t f'(X_{s-})d[M,M]_s,$$

$$I_5^n \xrightarrow{\ P\ } \int_0^t f'(X_{s-})d[M^d,A]_s,$$

where M^d denotes purely discontinuous part of M.

Since X is a process with finite energy, by (11)

$$\lim_{\epsilon \downarrow 0} \limsup_{n \to \infty} P(|I_6^{n,\epsilon}| > \delta) = 0, \quad \delta > 0.$$

We observe also that P-almost surely the following limit exists:

$$\lim_{\epsilon \downarrow 0} \lim_{n \to \infty} I_7^{n,\epsilon} = \sum_{s \leq t}\{F(X_s) - F(X_{s-}) - f(X_{s-})\Delta X_s - 1/2f'(X_{s-})\Delta X_s^2\}$$

$$= \sum_{s \leq t}\{F(X_s) - F(X_{s-}) - f(X_{s-})\Delta X_s\} - 1/2\sum_{s \leq t}\{f'(X_{s-})\Delta M_s^2$$

$$+ f'(X_{s-})\Delta A_s^2 + 2f'(X_{s-})\Delta M_s^d \Delta A_s\}.$$

On the other hand it is obvious that P-almost surely

$$\sum_{t_i \in D_n, t_i \leq t} (F(X_{t_{i+1}}) - F(X_{t_i})) \to F(X_t) - F(0)$$

and putting together all convergences, we deduce that $\{I_2^n + I_3^n\}$ is converging in probability; denote its limit as $(S) \int_0^t f(X_{s-})dA_s$. Observe also that

$$\int_0^t f'(X_{s-})d[M, M]_s^c = \int_0^t f'(X_{s-})d[M, M]_s - \sum_{s \leq t} f'(X_{s-})\Delta M_s^2$$

and

$$\int_0^t f'(X_{s-})d[M^d, A]_s = \sum_{s \leq t} f(X_{s-})\Delta M_s \Delta A_s.$$

As a consequence we obtain the formula

$$F(X_t) = F(0) + \int_0^t f(X_{s-})dM_s + (S) \int_0^t f(X_{s-})dA_s$$

$$+ 1/2 \int_0^t f'(X_{s-})d[M, M]_s^c - 1/2 \sum_{s \leq t} f'(X_{s-})\Delta A_s^2$$

$$+ \sum_{s \leq t} \{F(X_s) - F(X_{s-}) - \Delta X_s f(X_{s-})\}.$$

Now, writing

$$\sum_{s \leq t} \{F(X_s) - F(X_{s-}) - \Delta X_s f(X_{s-})\}$$

$$= \int_0^t \int_{\mathbb{R} \setminus \{0\}} (F(X_{s-} + x) - F(X_{s-}) - x f(X_{s-}))\mu(ds, dx)$$

and using the basic inequalities

$$|F(y + x) - F(y) - x f(y)| \leq \|f'\| x^2$$

and

$$|F(y + x) - F(y) - x f(y)| \leq 2\|f\| |x|,$$

we get the decomposition

$$\sum_{s \leq \cdot} \{F(X_s) - F(X_{s-}) - \Delta X_s f(X_{s-})\}$$

$$= \int_0^{\cdot} \int_{\mathbb{R} \setminus \{0\}} (F(X_{s-} + x) - F(X_{s-}) - x f(X_{s-}))(\mu - \nu)(ds, dx)$$

$$+ \int_0^{\cdot} \int_{\mathbb{R} \setminus \{0\}} (F(X_{s-} + x) - F(X_{s-}) - x f(X_{s-}))\nu(ds, dx)$$

where

$$\int_0^{\cdot} \int_{\mathbb{R}\backslash\{0\}} (F(X_{s-} + x) - F(X_{s-}) - xf(X_{s-})(\mu - \nu)(ds, dx)$$

is a square integrable purely discontinuous martingale, which we will denote by L and

$$\int_0^{\cdot} \int_{\mathbb{R}\backslash\{0\}} (F(X_{s-} + x) - F(X_{s-}) - xf(X_{s-}))\nu(ds, dx)$$

is an increasing predictable square integrable process.

Then we get the decomposition $F(X_{\cdot}) = F(0) + Y_{\cdot} + \Gamma_{\cdot}$, as written in the statement of Theorem 3.1.

It remains to prove that, for every continuous local martingale N, the equality $[\Gamma, N] = 0$ holds.

First note that

$$\sum_{t_i \in D_n, t_i \le t} (\Gamma_{t_{i+1}} - \Gamma_{t_i})(N_{t_{i+1}} - N_{t_i})$$

$$= \sum_{t_i \in D_n, t_i \le t} \int_{t_i}^{t_{i+1}} (f(X_{s-}) - f(X_{t_i}))dM_s(N_{t_{i+1}} - N_{t_i})$$

$$- \sum_{t_i \in D_n, t_i \le t} (L_{t_{i+1}} - L_{t_i})(N_{t_{i+1}} - N_{t_i})$$

$$+ \sum_{t_i \in D_n, t_i \le t} f(X_{t_i})(A_{t_{i+1}} - A_{t_i})(N_{t_{i+1}} - N_{t_i})$$

$$+ 1/2 \sum_{t_i \in D_n, t_i \le t} f'(X_{t_i})(A_{t_{i+1}} - A_{t_i})^2(N_{t_{i+1}} - N_{t_i})$$

$$+ 1/2 \sum_{t_i \in D_n, t_i \le t} f'(X_{t_i})(M_{t_{i+1}} - M_{t_i})^2(N_{t_{i+1}} - N_{t_i})$$

$$+ \sum_{t_i \in D_n, t_i \le t} f'(X_{t_i})(M_{t_{i+1}} - M_{t_i})(A_{t_{i+1}} - A_{t_i})(N_{t_{i+1}} - N_{t_i})$$

$$+ \sum_2 r_2(X_{t_i}, X_{t_{i+1}})(N_{t_{i+1}} - N_{t_i})$$

$$- \sum_1 \{F(X_{t_{i+1}}) - F(X_{t_i}) - f(X_{t_i})(X_{t_{i+1}} - X_{t_i})$$

$$- 1/2 f'(X_{t_i})(X_{t_{i+1}} - X_{t_i})^2\}(N_{t_{i+1}} - N_{t_i})$$

$$= I_1^n + I_2^n + I_3^n + I_4^n + I_5^n + I_6^n + I_7^{n,\epsilon} + I_8^{n,\epsilon}$$

Clearly,

$$|I_1^n| \le (\sum_{t_i \in D_n, t_i \le t} (\int_{t_i}^{t_{i+1}} (f(X_{s-}) - f(X_{t_i}))dM_s)^2)^{1/2} (\sum_{t_i \in D_n, t_i \le t} (N_{t_{i+1}} - N_{t_i})^2)^{1/2},$$

where by the definition of the stochastic integral

$$E \sum_{t_i \in D_n, t_i \le t} \left(\int_{t_i}^{t_{i+1}} (f(X_{s-}) - f(X_{t_i})) dM_s \right)^2$$

$$= E \sum_{t_i \in D_n, t_i \le t} \int_{t_i}^{t_{i+1}} (f(X_{s-}) - f(X_{t_i}))^2 d[M, M]_s \to 0.$$

On the other hand by Lemma 3.1

$$I_2^n \xrightarrow{P} [L, N]_t = 0,$$

$$I_3^n \xrightarrow{P} \int_0^t f(X_{s-}) d[A, N]_s = 0,$$

$$I_4^n \xrightarrow{P} 1/2 \int_0^t f'(X_{s-}) \Delta A_s d[A, N]_s = 0,$$

$$I_5^n \xrightarrow{P} 1/2 \int_0^t f'(X_{s-}) \Delta N_s d[M, M]_s = 0,$$

and

$$I_6^n \xrightarrow{P} \int_0^t f'(X_{s-}) \Delta M_s d[A, N]_s = 0.$$

Finally, for every $\epsilon > 0$

$$I_6^{n,\epsilon} \xrightarrow{P} 0 \qquad \text{and} \qquad I_7^{n,\epsilon} \to 0, \ P - a.s.$$

and the proof of Theorem 3.1 is completed. $\qquad\qquad\qquad\qquad\square$

Corollary 3.1. *Let $X = M + A$ be a weak Dirichlet process of finite energy admitting a quadratic variation and F be a C^2-real valued function with bounded derivatives f and f'. Then the process $(F(X_t)_{t \ge 0})$ is a weak Dirichlet process of finite energy admitting a quadratic variation and the decomposition $F(X) = Y + \Gamma$ holds with the martingale part*

$$Y_t = F(0) + \int_0^t f(X_{s-}) dM_s$$

$$+ \int_0^t \int_{\mathbb{R} \setminus \{0\}} \left(F(X_{s-} + x) - F(X_{s-}) - x f(X_{s-}) \right) (\mu - \nu)(ds, dx)$$

and the predictable part

$$\Gamma_t = \int_0^t f(X_{s-}) dA_s + 1/2 \int_0^t f'(X_s) d[X, X]_s^c$$

$$+ \int_0^t \int_{\mathbb{R} \setminus \{0\}} \left(F(X_{s-} + x) - F(X_{s-}) - x f(X_{s-}) \right) \nu(ds, dx).$$

Proof. By Theorem 1.7(i), A admits integrable quadratic variation $[A, A]$ and due to Lemma 3.1

$$1/2 \sum_{t_i \in D_n, t_i \leq t} f(X_{t_i})(A_{t_{i+1}} - A_{t_i})^2 \overset{\text{P}}{\longrightarrow} 1/2 \int_0^t f(X_{s-})d[A, A]_s.$$

Since $[X, X]^c = [M, M]^c + [A, A]^c$, it is clear that

$$1/2 \int_0^t f(X_{s-})d[A, A]_s - 1/2 \sum_{s \leq t} f(X_{s-})\Delta A_s^2 + 1/2 \int_0^t f(X_{s-})d[M, M]_s$$

$$= 1/2 \int_0^t f(X_{s-})d([A, A]_s^c + [M, M]^c) = 1/2 \int_0^t f(X_{s-})d[X, X]_s^c$$

and the decomposition $F(X_t) = F(0) + Y_t + \Gamma_t$ in the statement of Corollary 3.1 is a consequence of Theorem 3.1.

Finally, by the Theorem from [9, page 144] we obtain that also $F(X)$ admits a quadratic variation, which completes the proof. $\qquad\square$

Theorem 3.2. *Let $X = M + A$ be a weak Dirichlet process of finite energy and F a C^1-real valued function with bounded derivative f.*

Then the process $(F(X_t)_{t\geq 0})$ is a weak Dirichlet process of finite energy and the decomposition $F(X) = Y + \Gamma$ holds with the martingale part

$$Y_t = F(0) + \int_0^t f(X_{s-})dM_s$$

$$+ \int_0^t \int_{\mathbb{R}\backslash\{0\}} \Big(F(X_{s-} + x) - F(X_{s-}) - xf(X_{s-})\Big)(\mu - \nu)(ds, dx)$$

Remark 3.1. This theorem has formally almost the same statement as Theorem 3.1. However, we have here no explicit formula for Γ. The delicate point here is the behaviour of the sum

$$\sum_{s \leq t} \Big(F(X_s) - F(X_{s-}) - \Delta X_s f(X_{s-})\Big)$$

which is not necessarily absolutely convergent and which does not define a process with finite variation.

Proof of Theorem 3.2: We consider a sequence $(F^p)_{p \in \mathbb{N}}$ of C^2 real functions such that $\|F - F^p\| + \|f - f^p\| \to 0$, when $p \to \infty$. Using Theorem 3.1 we can write

$$F^p(X_t) = F^p(0) + Y_t^p + \Gamma_t^p$$

where

$$Y_t^p = \int_0^t f^p(X_{s-})dM_s + L_t^p$$

with

$$L_t^p = \int_0^t \int_{\mathbb{R}\setminus\{0\}} \Big(F^p(X_{s-} + x) - F^p(X_{s-}) - xf^p(X_{s-}) \Big)(\mu - \nu)(ds, dx).$$

The sequence $\{\int_0^{\cdot} f^p(X_{s-})dM_s + L_{\cdot}^p\}_{p\in\mathbb{N}}$ is a Cauchy sequence in the space H^2 of square integrable martingales, hence the limiting martingale exists and has the form $\int_0^{\cdot} f(X_{s-})dM_s + L_{\cdot}$:

Actually, for p, q integers

$$\|Y^p - Y^q\|_{H^2} \le E\Big(\int_0^T (f^p(X_{s-}) - f^q(X_{s-}))^2 d[M, M]_s\Big)$$

$$+ E[\int_0^T \int_{\mathbb{R}\setminus\{0\}} \Big(F^p(X_{s-} + x) - F^p(X_{s-}) - xf^p(X_{s-})$$

$$- F^q(X_{s-} + x) + F^q(X_{s-}) + xf^q(X_{s-}) \Big)^2 \nu(ds, dx)]$$

$$\le \|F^p - F^q\|^2 E[[M, M]] + 2\|f^p - f^q\|^2 E[[M, M]].$$

Now we write:

$$\Gamma_t^p = F^p(X_t) - F^p(0) - \int_0^t f^p(X_{s-})dM_s - L_t^p.$$

Clearly the sequence of predictable processes (Γ^p) converges uniformly in probability and its limit (i.e. the process Γ) has to be also predictable.

It remains to prove that $[\Gamma, N] = 0$ for every continuous local martingale N.

Fix t. In the sequel we use the notations from the proof of Theorem 3.1. By Taylor's formula

$$\sum_2 (F(X_{t_{i+1}}) - F(X_{t_i})) = \sum_2 f(X_{t_i})(X_{t_{i+1}} - X_{t_i}) + \sum_2 r_1(X_{t_i}, X_{t_{i+1}}),$$

where $r_1(X_{t_i}, X_{t_{i+1}}) = C_i^\epsilon(X_{t_{i+1}} - X_{t_i})$ satisfy $|C_i^\epsilon| \le \|f\|$ and

$$\lim_{\epsilon \downarrow 0} \limsup_{n\to\infty} P(\max_2 |C_i^\epsilon| > \delta) = 0, \quad \delta > 0.$$

Therefore,

$$\sum_{t_i \in D_n, t_i \le t} (\Gamma_{t_{i+1}} - \Gamma_{t_i})(N_{t_{i+1}} - N_{t_i})$$

$$= \sum_{t_i \in D_n, t_i \le t} \int_{t_i}^{t_{i+1}} f(X_{t_i}) - f(X_{s-})dM_s(N_{t_{i+1}} - N_{t_i})$$

$$- \sum_{t_i \in D_n, t_i \le t} (L_{t_{i+1}} - L_{t_i})(N_{t_{i+1}} - N_{t_i})$$

$$+ \sum_{t_i \in D_n, t_i \leq t} f(X_{t_i})(A_{t_{i+1}} - A_{t_i})(N_{t_{i+1}} - N_{t_i})$$

$$+ \sum_2 r_1(X_{t_i}, X_{t_{i+1}})(N_{t_{i+1}} - N_{t_i})$$

$$- \sum_1 \{F(X_{t_{i+1}}) - F(X_{t_i}) - f(X_{t_i})(X_{t_{i+1}} - X_{t_i})(N_{t_{i+1}} - N_{t_i})\}$$

$$= I_1^n + I_2^n + I_3^n + I_4^{n,\epsilon} + I_5^{n,\epsilon}.$$

Clearly, the first three sums tend to 0 analogously to the proof of Theorem 3.1. Next,

$$\lim_{\epsilon \downarrow 0} \limsup_{n \to \infty} P(|I_4^{n,\epsilon}| > \delta) = 0, \delta > 0$$

and for every $\epsilon > 0$

$$I_5^{n,\epsilon} \to 0, \ P - a.s.$$

which completes the proof of Theorem 3.2.   □

Corollary 3.2. *Let $X = M + A$ be a weak Dirichlet process of finite energy admitting a quadratic variation process and F a C^1-real valued function with bounded derivative f. Then the process $(F(X_t)_{t \geq 0})$ is a weak Dirichlet process of finite energy admitting a quadratic variation and the decomposition $F(X) = Y + \Gamma$ holds with the martingale part*

$$Y_t = F(0) + \int_0^t f(X_{s-})dM_s$$

$$+ \int_0^t \int_{\mathbb{R} \backslash \{0\}} \left(F(X_{s-} + x) - F(X_{s-}) - xf(X_{s-}) \right)(\mu - \nu)(ds, dx)$$

The quadratic variation process of $F(X_t)_t$ is given by

$$[F(X), F(X)]_t = \int_0^t (f(X_s))^2 d[M, M]_s^c + \int_0^t (f(X_s))^2 d[A, A]_s^c$$

$$+ \sum_{0 \leq s \leq t} (F(X_s) - F(X_{s-}))^2.$$

Proof. Follows easily from Theorem 3.2, [9, Theorem, page 144] and the equality $[X, X]^c = [M, M]^c + [A, A]^c$.   □

In the related work ([8]) one can see an Ito-type formula for $C^{1,\lambda}$ functions of processes admitting a quadratic variation.

We are able to prove a version of Theorem 3.2 for weak Dirichlet processes also with infinite energy. However, in this case we have restricted our attention to processes with a continuous predictable part.

Theorem 3.3. *Let $X = M + A$ be a weak Dirichlet process with continuous predictable part A and F a C^1 real-valued function with bounded derivative f.*

Then the process $(F(X_t)_{t\geq0})$ is a weak Dirichlet process and the decomposition $F(X) = Y + \Gamma$ holds with the martingale part

$$Y_t = F(0) + \int_0^t f(X_{s-})dM_s$$

$$+ \int_0^t \int_{\mathbb{R}\backslash\{0\}} \Big(F(X_{s-} + x) - F(X_{s-}) - xf(X_{s-}) \Big)(\mu - \nu)(ds, dx).$$

Proof. We consider a sequence $(F^p)_{p\in\mathbb{N}}$ of C^2 real functions such that locally on compact sets $\|F - F^p\| + \|f - f^p\| \to 0$, when $p \to \infty$. Let A^p be a sequence of continuous processes with finite variation such that

$$\sup_{t\leq T} |A_t^p - A_t| \xrightarrow{\ \mathrm{P}\ } 0.$$

Using the classical Itô's formula for the semimartingale $X^p = M + A^p$ we can write

$$F^p(X_t^p) = F^p(0) + Y_t^p + \Gamma_t^p$$

where

$$Y_t^p = \int_0^t f^p(X_{s-}^p)dM_s + L_t^p$$

with

$$L_t^p = \int_0^t \int_{\mathbb{R}\backslash\{0\}} \Big(F^p(X_{s-}^p + x) - F^p(X_{s-}^p) - xf^p(X_{s-}^p) \Big)(\mu - \nu)(ds, dx).$$

Similarly to the proof of Theorem 3.2, we check that

$$\sup_{t\leq T} |Y^p - Y_t| \xrightarrow{\ \mathrm{P}\ } 0.$$

On the other hand it is clear that

$$\sup_{t\leq T} |F^p(X_t^p) - F(X_t)| \xrightarrow{\ \mathrm{P}\ } 0,$$

which implies that Γ, a uniform limit of predictable processes, is also predictable. Finally, by the same arguments as in the proof of Theorem 3.2 we prove that $[\Gamma, N] = 0$ for every continuous local martingale N.

4 Weak Dirichlet processes and generalized martingale convolutions

In this section we deal with processes X such that

$$X_t = \int_0^t G(t, s)dL_s \tag{12}$$

where L is a quasi-left continuous square integrable martingale, and G a real valued deterministic function of (s,t).

Let us consider the following hypotheses on G.

(H_0): $(t,s) \to G(t,s)$ is continuous on $\{(s,t) : 0 < s \le t \le T\}$.

(H_1): For all s, $t \to G(t,s)$ has a bounded energy on $]s,T]$, that is

$$V_2^2(G)((s,T],s) = \sup_n \sum_{t_i \in D_n, t_i \ge s} (G(t_{i+1},s) - G(t_i,s))^2 < \infty.$$

(H_2) : $$E[\int_0^T V_2^2(G)(]s,T],s)d[L,L]_s] < \infty$$

(H_3) : $$E[\int_0^T \Gamma^2(s)d[L,L]_s] < \infty$$

where $\Gamma^2(s) = \sup_{t \le T} G^2(t,s)$.

Remark 4.1. Instead of (H_0), Errami and Russo ([6]) use a slightly more restrictive assumption, namely: (H_{0+}): $(t,s) \to G(t,s)$ is continuous on $\{(s,t) : 0 \le s \le t \le T\}$.

Note that (H_{0+}), implies (H_3). Actually Γ^2 is continuous and bounded.

If $t \to G(t,s)$ admits a quadratic variation on $(s,T]$ along (D_n), then (H_1) is satisfied.

We shall extend G to the square $[0,T]^2$ by setting $G(t,s) = 0$ if $s > t$.

Theorem 4.1. *If X meets (12) and if G satisfies $(H_0), (H_1), (H_2), (H_3)$, then*

(i) *X is continuous in probability, has finite energy and has an optional modification,*

(ii) *Let us assume that X has a.s. càdlàg trajectories, then X is a weak Dirichlet process with natural decomposition $X = M + A$, such that if M^n is defined as in (3), then for every $t \le T$, $|M_t^n - M_t| \to 0$ in $\mathbf{L}^2$.*

Proof: The proof will be given in several steps.

Lemma 4.1 *The process X is continuous in probability and has finite energy.*

Proof of Lemma 4.1: First of all, from (H_2) and (H_3), for every $t \le T$ X_t is an $\mathcal{F}_t$-measurable square integrable random variable.

Let us write

$$X_{t_{i+1}} - X_{t_i} = \int_0^{t_i} (G(T_{i+1},s) - G(t_i,s))dL_s + \int_{t_i}^{t_{i+1}} G(t_{i+1},s)dL_s.$$

Since L is a square integrable martingale, we get:

$$E(\sum_{t_i \in D_n} (X_{t_{i+1}} - X_{t_i})^2) \le 2E(\sum_{t_i \in D_n} (\int_0^{t_i} (G(t_{i+1}, s) - G(t_i, s))dL_s)^2)$$

$$+2E(\sum_{t_i \in D_n} (\int_{t_i}^{t_{i+1}} G(t_{i+1}, s)dL_s)^2)$$

$$\le 2E(\sum_{t_i \in D_n} \int_0^{t_i} (G(t_{i+1}, s - G(t_i, s))^2 d[L, L]_s)$$

$$+2E(\sum_{t_i \in D_n} \int_{t_i}^{t_{i+1}} (G(t_{i+1}, s))^2 d[L, L]_s = 2E(I_1^n) + 2E(I_2^n).$$

By simple calculations

$$I_1^n = \sum_{t_i \in D_n} \sum_{k=1}^{i} \int_{t_{k-1}}^{t_k} (G(t_{i+1}, s) - G(t_i, s))^2 d[L, L]_s$$

$$= \sum_{t_k \in D_n} \int_{t_{k-1}}^{t_k} \sum_{i>k} (G(t_{i+1}, s) - G(t_i, s))^2 d[L, L]_s$$

$$\le \sum_{t_k \in D_n} \int_{t_{k-1}}^{t_k} V_2^2(G)((s, T], s)d[L, L]_s,$$

and

$$I_2^n \le \sum_{t_i \in D_n} \int_{t_i}^{t_{i+1}} (\Gamma^2(s)d[L, L]_s$$

$$\le \int_0^T \Gamma^2(s)d[L, L]_s.$$

Therefore

$$\sup_n E(\sum_{t_i \in D_n} (X_{t_{i+1}} - X_{t_i})^2 \le 2E \int_0^T V_2^2(G)((s, T], s)d[L, L]_s$$

$$+2E \int_0^T (\Gamma(s))^2 d[L, L]_s$$

$$< \infty.$$

This proves that X has a finite energy.

Now, let us take s, t such that $0 \le s < t \le T$. We get:

$$E[(X_t - X_s)^2] \le 2E \int_0^s (G(t, u) - G(s, u))^2 d[L, L]_u + 2E \int_s^t (G(t, u))^2 d[L, L]_u.$$

Since L is continuous in probability, so is $[L, L]$. Under (H_3),

$$2E \int_s^t (G(t,u))^2 d[L, L]_u \to 0, \quad as\ t \to s,$$

then by continuity of $t \to G(t, s)$ and dominated convergence,

$$E \int_0^s (G(t, u) - G(s, u))^2 d[L, L]_u \to 0.$$

The continuity in probability of the process X follows.

Last, since the process X is $\mathcal{F}_t$-adapted and continuous in probability, it admits an optional modification that we shall denote again by X: see for example [23] pp. 230–231, where Théorème 5 bis is given for a progressively measurable modification, but the sequence of approximating processes introduced in the proof is càdlàg hence optional (see also below the proof of the existence of a predictable modification of the process A).

Therefore Lemma 4.1 is proven.   □

Lemma 4.2 *Consider the decomposition: $X_t = A_t^n + M_t^n$, where as in (3),*

$$M_t^n = \sum_{t_i^n \in D_n, t_i^n \le t} \left[X_{t_i^n} - E[X_{t_i^n} | \mathcal{F}_{t_{i-1}^n}] \right].$$

Then X admits a modification with a decomposition $X_t = A_t + M_t$, where M is the square integrable martingale $M_t = \int_0^t G(s,s) dL_s$, and A a predictable process, such that:

(i) *for every $t \le T$,*

$$|M_t^n - M_t| \to 0\ in\ \mathbf{L}^2,$$

(ii) *for every $t \le T$, $|A_t^n - A_t| \to 0$ in $\mathbf{L}^2$,*

(iii) *for every continuous local martingale N, $[A, N] = 0$.*

Proof of Lemma 4.2: For every t_i, t_{i+1} we have

$$E[X_{t_{i+1}} - X_{t_i} | \mathcal{F}_{t_i}] = \int_0^{t_i} (G(t_{i+1}, s) - G(t_i, s)) dL_s$$

hence for $t \in [0, T]$

$$M_t^n = \sum_{t_{i+1} \le t} \int_{t_i}^{t_{i+1}} G(t_{i+1}, s) dL_s = \int_0^{\rho^n(t)} G^n(s) dL_s,$$

where $G^n(s) = G(t_{i+1}, s)$ for $s \in (t_i, t_{i+1}]$ and $\rho^n(t) = \max\{t_i : t_i \le t\}$. Note that $\rho^n(t) \to t$ as $n \to \infty$.

Define $M_t = \int_0^t G(s,s)dL_s$. By (H_2) and (H_3), for every $\varepsilon > 0$

$$E[(M_\varepsilon)^2] = E \int_0^\varepsilon G^2(s,s)d[L,L]_s \leq E \int_0^\varepsilon \Gamma^2(s)d[L,L]_s$$

and this last expression tends to 0 when ε tends to 0.

Similarly, when $\varepsilon \to 0$

$$\sup_n E[(M_\varepsilon^n)^2] = \sup_n E \int_0^\varepsilon (G^n(s))^2 d[L,L]_s \to 0.$$

Now, let us fix $\varepsilon > 0$ and belonging to D_n. Since $(t,s) \to G(t,s)$ is continuous on $\{(s,t) : \varepsilon \leq s \leq t \leq T\}$, it is also uniformly continuous and therefore:

$$\sup_{\varepsilon \leq s \leq T} |G^n(s) - G(s,s)| \to 0.$$

Hence

$$E[(M_t^n - M_\varepsilon^n - M_t + M_\varepsilon)^2] \leq 2E\Big(\int_\varepsilon^{\rho^n(t)} (G^n(s) - G(s,s))dL_s\Big)^2$$
$$+ 2E\Big(\int_{\rho^n(t)}^t G(s,s)dL_s\Big)^2$$
$$\leq 8E \int_\varepsilon^1 (G^n(s) - G(s,s))^2 d[L,L]_s$$
$$+ 2E \int_{\rho^n(t)}^t G^2(s,s)d[L,L]_s.$$

The first term tends to 0 when $n \to \infty$ because G^n converges uniformly to G on $[\varepsilon, T]$. Now, the continuity in probability of L implies the continuity in probability of M on $[0,T]$, hence the second term tends to 0 when $n \to \infty$.

Note that for every n

$$A_t^n = \sum_{t_i^n \in D_n, t_i^n \leq t} E[X_{t_i^n} - X_{t_{i-1}^n} | \mathcal{F}_{t_{i-1}^n}]$$

and $A = X - M$. A^n is a predictable process and we have $X_{\rho^n(t)} = A_{\rho^n(t)}^n + M_{\rho^n(t)}^n$.

As X is continuous in probability, for every $t \leq T$, $X_{\rho^n(t)} \to X_t$ in $\mathbf{L}^2$; moreover, $M_t^n = M_{\rho^n(t)}^n$ and $A_t^n = A_{\rho^n(t)}^n$: we deduce that $A_t^n \to A_t$ in $\mathbf{L}^2$ for every t.

It follows that A is also adapted and continuous in probability.

Our decomposition $X = M + A$ coincides with the Graversen-Rao decomposition of Theorem 2.1. But in Theorem 2.1, it is assumed that X is càdlàg; here it is not the case, it is necessary to check that A admits a predictable modification.

Actually, since A is continuous in probability on the interval $[0, 1]$, one can find a subsequence $\{n(k)\}_{k \geq 1}$ such that for every $t \in [0, 1]$, $\bar{A}^{n(k)} \to A_t$ a.s. when $k \to \infty$, whith $\bar{A}^n_t = \sum_i 1_{(t^n_i, t^n_{i+1}]}(t) A_{t^n_i}$ and $A^n_0 = A_0$. Since every $\bar{A}^n$ is a step process adapted and left continuous, it is predictable, and the process A' defined by $A'_t = \limsup_k \bar{A}^{n(k)}_t$ is also predictable. But for every t, $A'_t = A_t$ a.s. So, we shall suppose now that $A = A'$, and $X = A' + M$.

Proof of (iii): Let N be a continuous local martingale. Using localization arguments, we will assume that N is a square integrable martingale. For every t we can write:

$$\sum_{t_i \in D_n, t_i \leq t} (N_{t_{i+1}} - N_{t_i})(A_{t_{i+1}} - A_{t_i})$$

$$= \sum_{t_i \in D_n, t_i \leq t} (N_{t_{i+1}} - N_{t_i}) \int_0^{t_i} (G(t_{i+1}, s) - G(t_i, s)) dL_s$$

$$+ \sum_{t_i \in D_n, t_i \leq t} (N_{t_{i+1}} - N_{t_i}) \int_{t_i}^{t_{i+1}} (G(t_{i+1}, s) - G(s, s)) dL_s$$

$$= I^n_1 + I^n_2$$

Since N is a martingale, using the B-D-G inequality and Schwarz' inequality we get:

$$E|I^n_1| \leq cE \left(\sum_{t_i \in D_n} (N_{t_{i+1}} - N_{t_i})^2 \left(\int_0^{t_i} (G(t_{i+1}, s) - G(t_i, s)) dL_s \right)^2 \right)^{1/2}$$

$$\leq cE \left(\max_{t_i \in D_n} |N_{t_{i+1}} - N_{t_i}| \sum_{t_i \in D_n} \left(\int_0^{t_i} (G(t_{i+1}, s) - G(t_i, s)) dL_s \right)^2 \right)^{1/2}$$

$$\leq c(E(\max_{t_i \in D_n} |N_{t_{i+1}} - N_{t_i}|^2))^{1/2}$$

$$\left(E \sum_{t_i \in D_n} \left(\int_0^{t_i} (G(t_{i+1}, s) - G(t_i, s)) dL_s \right)^2 \right)^{1/2}.$$

Because of the continuity of N,

$$E[\max_{t_i \in D_n} |N_{t_{i+1}} - N_{t_i}|^2] \to 0;$$

the second term is estimated as before by

$$(E \int_0^T V_2^2(G)((s, T], s) d[L, L]_s)^{1/2},$$

which is finite.

Now, by Schwarz' inequality

$$E|I_2^n| \leq (E \sum_{t_i \in D_n} (N_{t_{i+1}} - N_{t_i})^2)^{1/2}$$

$$\left(E \sum_{t_i \in D_n} \left(\int_{t_i}^{t_{i+1}} (G(t_{i+1}, s) - G(s, s))dL_s \right)^2 \right)^{1/2}$$

$$\leq (E([N, N]_T))^{1/2} \left(E \int_0^T (G^n(s) - G(s, s))^2 d[L, L]_s \right)^{1/2}$$

where G^n is defined as above.

Since $E \int_0^T (G^n(s) - G(s, s))^2 d[L, L]_s \to 0$, we conclude that for every t

$$\sum_{t_i \in D_n, t_i \leq t} (N_{t_{i+1}} - N_{t_i})(A_{t_{i+1}} - A_{t_i}) \to 0$$

in $\mathbf{L}^1$, and the covariation process $[N, A]$ is null for every continuous martingale N.

The proof of Theorem 4.1 is complete. $\qquad\qquad\square$

Unfortunately we are not able to prove that X admits a modification with càdlàg trajectories. However, in this direction, we have the following lemma.

Lemma 4.3 : *A is continuous (hence X is càdlàg) if the following additional condition is verified:*
(H_c): There exist $\delta > 0, p > 1$ and a function $a(u)$ meeting

$$E\left[(\int_0^T a(u)d[L, L]_u)^p \right] < \infty,$$

and for every s, t, u,

$$\left(G(t, u) - G(s, u) \right)^2 \leq a(u)|t - s|^{\frac{1}{p} + \delta}.$$

Proof : Let us take s, t such that $0 \leq s \leq t \leq T$. We have, with a constant c changing from line to line:

$$E(A_t - A_s)^{2p} \leq E(\int_0^t (G(t, u) - G(u, u))dL_u - \int_0^s (G(s, u) - G(u, u))dL_u)^{2p}$$

$$\leq cE(\int_0^s (G(t, u) - G(s, u))^2 d[L, L]_u)^p$$

$$+ E(\int_s^t (G(t, u) - G(u, u)^2 d[L, L]_u)^p$$

$$\leq c(t - s)^{1+p\delta} E(\int_0^t a(u)d[L, L]_u)^p$$

$$\leq c(t - s)^{1+p\delta}.$$

Hence we get the continuity of A by Kolmogorov's Lemma. $\qquad\qquad\square$

An analogous result under Hölder condition was already given in the paper ([1]), Lemmas 2C and 2D.

We shall suppose by now that the processes given by (12) have a. s. càdlàg trajectories.

We are now interested in investigating conditions on G in order to make X a Dirichlet process or a weak Dirichlet process admitting a quadratic variation. For that let us consider the following hypotheses:

(H_4): For all s, $t \to G(t, s)$ has a bounded variation on $(s, \tau]$, for every $\tau \le T$. (We denote this variation by $|G|((s, \tau], s) < \infty$.)

(H_5) :
$$E \int_0^T |G|((s, T], s)d[L, L]_s < \infty$$

(H_6): For all u, v, $t \to G(t, u)$ and $t \to G(t, v)$ have a finite mutual quadratic covariation with the property (S) on $(\max(u, v), T]$.
(We denote this covariation by $[G(., u), G(., v)]_\tau$.) Moreover we suppose that the convergence involved to define the covariation, is uniform in u, v.

Of course (H_6) implies that $t \to G(t, s)$ admits a quadratic variation on $(s, T]$ for all s with the property (S) and that (H_1) is satisfied.

Theorem 4.2. (i) *Let us assume* $(H_0), (H_3), (H_4), (H_5)$. *Then X is a Dirichlet process (i.e. A is continuous and $[A, A] \equiv 0$).*
(ii) *Let us assume* $(H_0), (H_2), (H_3), (H_6)$. *Then X is a weak Dirichlet process. Moreover if we assume that the process B defined by*

$$B_t = \int_0^t [G(., s), G(., s)]_t d[L, L]_s + 2 \int_0^t \left(\int_0^v [G(., u), G(., v)]_t dL_u \right) dL_v$$
(13)

is a càdlàg process, then X and A admit a quadratic variation such that: $[A, A]_t = B_t$ and

$$[X, X] = [A, A] + [M, M].$$
(14)

Remark 4.2. (i) If $t \to G(t, s)$ is C^1 for every s, denoting by $G_1(t, s)$ its derivative and assuming that $(t, s) \to G_1(t, s)$ is continuous on $[0, 1]^2$, we get $A_t = \int_0^t \left(\int_0^u G_1(u, s)dL_s \right) du$ by applying Fubini's theorem for stochastic integrals, so X is a semimartingale. This result is due to Protter ([15]).
(ii) In the case of a continuous martingale L, part 2) is due to Errami and Russo ([6]) and ([7]).

Proof : (i) First of all, we notice that our hypotheses imply that (H_1) and (H_2) are satisfied for any sequence (D_n) of subdivisions with meshes tending to 0. Since we can write

$$A_{t_{i+1}} - A_{t_i} = \int_0^{t_i} (G(t_{i+1}, s) - G(t_i, s)) dL_s + \int_{t_i}^{t_{i+1}} (G(t_{i+1}, s) - G(s, s)) dL_s,$$

for every $\varepsilon > 0$ we get:

$$E \sum_{t_i \in D_n} (A_{t_{i+1}} - A_{t_i})^2$$

$$\leq E \sum_{t_i \in D_n} \int_0^{t_i} (G(t_{i+1}, s) - G(t_i, s))^2 d[L, L]_s$$

$$+ \sum_{t_i \in D_n} \int_{t_i}^{t_{i+1}} (G(t_{i+1}, s) - G(s, s))^2 d[L, L]_s$$

$$\leq \max_{t_i \in D_n, \varepsilon \leq s \leq T} |G(t_{i+1}, s) - G(t_i, s)| E \sum_{t_i \in D_n} \int_\varepsilon^{t_i} |G(t_{i+1}, s) - G(t_i, s)| d[L, L]_s$$

$$+ E \sum_{t_i \in D_n} \int_0^{t_i \wedge \varepsilon} (G(t_{i+1}, s) - G(t_i, s))^2 d[L, L]_s$$

$$+ E \int_0^T (G^n(s) - G(s, s))^2 d[L, L]_s$$

$$\leq \max_{t_i \in D_n, \varepsilon \leq s \leq T} |G(t_{i+1}, s) - G(t_i, s)| E \int_\varepsilon^1 |G|((s, 1], s) d[L, L]_s$$

$$+ E \int_0^\varepsilon V_2^2(G)((s, T], s) d[L, L]_s + E \int_0^T (G^n(s) - G(s, s))^2 d[L, L]_s$$

Using the following properties:

$$\max_{t_i \in D_n, \varepsilon \leq s \leq T} |G(t_{i+1}, s) - G(t_i, s)| \to 0, \quad \text{when } n \to \infty,$$

$$E \int_\varepsilon^T |G|((s, T], s) d[L, L]_s < \infty,$$

$$E \int_0^\varepsilon V_2^2(G)((s, T], s) d[L, L]_s \to 0, \quad \text{when } \varepsilon \to 0$$

and

$$E \int_0^T (G^n(s) - G(s, s))^2 d[L, L]_s \to 0, \quad \text{when } n \to \infty,$$

we deduce that $[A, A] \equiv 0$. By the inequality

$$\max_{t_i \in D_n} |A_{t_{i+1}} - A_{t_i}|^2 \leq \sum_{t_i \in D_n} (A_{t_{i+1}} - A_{t_i})^2,$$

we get that A is continuous.

(ii) Taking into account Proposition 2.1 we only have to prove property S for the process defined by the right hand side of formula (13). Let us notice first that formula (13) is well-defined, as we take as integrant of dL_s in the last term the predictable projection of the optional process

$$\left(\int_0^s [G(.,u), G(.,s)]_t dL_u \right)_{s \leq t}.$$

See for example Dellacherie-Meyer ([12]), Chap.VI for details.

Taking into account that

$$E \int_0^T (G^n(s) - G(s,s))^2 d[L,L]_s \to 0 \text{ when } n \to \infty,$$

we have:

$$[A,A]_t = \lim_n \sum_{t_i \in D_n, t_i \leq t} (A_{t_{i+1}} - A_{t_i})^2$$

$$= \lim_n \sum_{t_i \in D_n, t_i \leq t} \left(\int_0^{t_i} (G(t_{i+1}, s) - G(t_i, s)) dL_s \right)^2$$

$$= \lim_n 2 \sum_{t_i \in D_n, t_i \leq t} \int_0^{t_i} \left(\int_0^s (G(t_{i+1}, u) - G(t_i, u)) dL_u \right) (G(t_{i+1}, s)$$

$$- G(t_i, s)) dL_s + \lim_n \sum_{t_i \in D_n, t_i \leq t} \int_0^{t_i} (G(t_{i+1}, s) - G(t_i, s))^2 d[L,L]_s$$

$$= 2 \lim_n I_1^n(t) + \lim_n I_2^n(t).$$

Then

$$I_2^n(t) = \sum_k \int_{t_{k-1}}^{t_k} \sum_{i>k, t_i \leq t} (G(t_{i+1}, s) - G(t_i, s))^2 d[L,L]_s$$

This sequence converges to $\int_0^t [G(.,s)]_t d[L,L]_s$ by dominated convergence.

On the other hand, $I_1^n(t)$ can be written

$$I_1^n(t) = \sum_k \int_{t_{k-1}}^{t_k} \left(\int_0^s \sum_{i>k, t_i \leq t} (G(t_{i+1}, u) - G(t_i, u))(G(t_{i+1}, s) - G(t_i, s)) dL_u \right) dL_s$$

$$= \int_0^{\rho^n(t)} \left(\int_0^s [G^n(.,u), G^n(.,s)]_t dL_u \right) dL_s.$$

Write Y^P for the predictable projection of any optional process Y. Then noticing that

$$\left(\int_0^s [G^n(.,u), G^n(.,s)]_t dL_u \right)_- = \left(\int_0^s [G^n(.,u), G^n(.,s)]_t dL_u \right)^P$$

and from classical properties of predictable projections, we get for $t \in D_n$

$$E\left(I_1^n(t) - \int_0^t \left(\int_0^s [G(.,u),G(.,s)]_t dL_u\right) dL_s\right)^2$$

$$= E\left[\int_0^t \left[\left(\int_0^s ([G^n(.,u),G^n(.,s)]_t - [G(.,u),G(.,s)]_t)dL_u\right)^P\right]^2 d<L,L>_s\right]$$

$$\leq E\left[\int_0^t \left[\left(\int_0^s ([G^n(.,u),G^n(.,s)]_t - [G(.,u),G(.,s)]_t)dL_u\right)^2\right]^P d<L,L>_s\right]$$

$$= E\left[\int_0^t \left(\int_0^s ([G^n(.,u),G^n(.,s)]_t - [G(.,u),G(.,s)]_t)dL_u\right)^2 d<L,L>_s\right].$$

We show now that for any $t \in \bigcup_n D_n$

$$\int_0^t \left(\int_0^s ([G^n(.,u),G^n(.,s)]_t - [G(.,u),G(.,s)]_t)dL_u\right)^2 d<L>_s \xrightarrow{P} 0,$$

when $n \to \infty$.

Note that

$$E\left(\int_0^s ([G^n(.,u),G^n(.,s)]_t - [G(.,u),G(.,s)]_t)dL_u\right)^2$$

$$= E\left(\int_0^s ([G^n(.,u),G^n(.,s)]_t - [G(.,u),G(.,s)]_t)^2 d<L,L>_u\right).$$

But for every u,s,t, $[G^n(.,u),G^n(.,s)]_t - [G(.,u),G(.,s)]_t \to 0$ when $n \to \infty$, and we have the estimation

$$|[G^n(.,u),G^n(.,s)]_t - [G(.,u),G(.,s)]_t|$$
$$\leq 1/2([G^n(.,u)]_t + [G^n(.,s)]_t + [G(.,u)]_t + G(.,s)]_t).$$

From (H_6) this last term is bounded, hence by dominated convergence

$$E\left(\int_0^s ([G^n(.,u),G^n(.,s)]_t - [G(.,u),G(.,s)]_t)dL_u\right)^2 \to 0$$

for every s,t. So, we can get easily (18) by localisation of L.

We are finished as soon as we remark that using the continuity of process $<L,L>$, for every t

$$E(I_1^n(t) - I_1^n(\rho^n(t)) - \int_{\rho^n(t)}^t \left(\int_0^s [G(.,u),G(.,s)]dL_u\right) dL_s)^2$$

$$= E\left(\int_{\rho^n(t)}^t \left(\int_0^s [G(.,u),G(.,s)]dL_u\right) dL_s\right)^2$$

$$= E\int_{\rho^n(t)}^t \left(\int_0^s [G(.,u),G(.,s)]dL_u\right)^2 d<L,L>_s$$

converges to 0 when $n \to \infty$. $\qquad\square$

Example 1: Fractional normal processes of index $H > 1/2$

We consider the case where L is a normal martingale (i.e. a square integrable martingale with predictable quadratic variation $< L, L >_t = t$), and $G(t, s)$ given for $t \geq s$ by:

$$G(t, s) = cs^{1/2-H} \int_s^t u^{H-1/2}(u - s)^{H-3/2}du \qquad (15)$$

with c constant. Of course, when L is a standard Brownian motion, X is the classical fractional Brownian motion.

We will now check that $(H_3),(H_4)$ and (H_5) are satisfied (in what follows, c is a constant that may change from line to line).

$$|G|((s, T], s) = cs^{1/2-H} \int_s^T u^{H-1/2}(u - s)^{H-3/2}du$$

$$\leq cs^{1/2-H} \int_s^T (u - s)^{H-3/2}du$$

$$= cs^{1/2-H} \frac{1}{H - 1/2}(T - s)^{H-1/2}du$$

$$\leq \frac{c}{H - 1/2} s^{1/2-H}$$

and

$$\int_0^T |G|((s, T], s)ds \leq \frac{c}{(H - 1/2)(3/2 - H)},$$

hence finally

$$\int_0^T |G^2|((s, T], s)ds \leq \frac{c}{(H - 1/2)(2 - 2H)}.$$

This means that, for every normal martingale L, the process X defined by $X_t = \int_0^t G(t, s)dL_s$ is a Dirichlet process. Since $G(s, s) = 0$, its martingale part is null. In particular X is continuous, even if L is not.

Example 2: Weak Dirichlet process driven by a Brownian motion

For this example we take $L = B$ a standard Brownian motion and $G(t, s) = \beta(t)f(s)$ for $t \geq s$, where $t \to \beta(t)$ is a fixed Brownian trajectory such that its quadratic variation is $[\beta(.)]_t = t$, and f is a real continuous function on $[0, T]$.

Here we can apply part 2) of Theorem 4.2. Actually:

$$[G(; , u), G(., v)]_t = \lim_{t_{i+1} \leq t, t_i \geq \max\{u,v\}} \sum_i (\beta(t_{i+1}) - \beta(t_i))^2 f(u)f(v)$$

and this term converges uniformly in u, v to $(t - \max\{u, v\})f(u)f(v)$.

Therefore we get the decomposition $X = M + A$, with

$$M_t = \int_0^t \beta(s)f(s)dB_s \tag{16}$$

and the formula (13) gives the quadratic variation of A

$$[A, A]_t = \int_0^t (t-s)f^2(s)ds + 2\int_0^t \int_0^s (t-s)f(u)f(s)dB_u dB_s, \tag{17}$$

which can be written

$$[A, A]_t = \int_0^t \left(\int_0^s f(u)dB_u\right)^2 ds.$$

In particular the process $(\int_0^t \int_0^s (t-s)f(u)f(s)dB_u dB_s)_t$ has a finite variation.

Since $E([A, A]_t) = \int_0^t (t-s)f^2(s)ds \neq 0$, X is not a Dirichlet process; however, Theorems 4.1 and 4.2 ensure that X is a weak Dirichlet process admitting quadratic variation.

Actually, this example 2 is a particular case of the following:

Example 3

Let us consider $G(t, s)$ of the form

$$G(t, s) = \int_s^t f(u, s)d\beta_u$$

where for every $s < t$, $u \rightarrow f(u, s)$ has a bounded variation on $(s, t]$, and $(u, s) \rightarrow f(u, s)$ is continuous on $\{(u, s) : 0 \leq u \leq s \leq T\}$. We denote by $df_u(u, s)$ the measure associated to the variation process. We assume also that β is a deterministic real continuous on $[0, T]$ function admitting a quadratic variation along (D_n), and $\beta_0 = 0$.

$X_t = A_t$ and we shall prove the Fubini type formula :

$$A_t = \int_0^t \left(\int_s^t f(u, s)d\beta_u\right) dL_s = \int_0^t \left(\int_0^u f(u, s)dL_s\right) d\beta_u. \tag{18}$$

Indeed, A admits a quadratic variation along (D_n) given by

$$[A, A]_t = \int_0^t \left(\int_0^s f(u, s)dL_u\right)^2 d[\beta, \beta]_s.$$

Actually, taking into account that $[\beta, f(., s)] = 0$, we get

$$\int_s^t f(u, s)d\beta_u = \beta_t f(t, s) - \beta_s f(s, s) - \int_s^t \beta_u df_u(u, s).$$

Then

$$A_t = \beta_t \int_0^t f(t,s)dL_s - \int_0^t \beta_s f(s,s)dL_s - \int_0^t \left(\int_s^t \beta_u df_u(u,s) \right) dL_s.$$

From Theorem 4.2 (i) the process Y defined by $Y_t = \int_0^t f(t,s)dL_s$ is a Dirichlet process and $[\beta, Y] = 0$, then by integration by parts, we get

$$\beta_t Y_t = \int_0^t Y_s d\beta_s + \int_0^t \beta_s dY_s$$

and by using the sequence (D_n)

$$\int_0^t \beta_s dY s = \lim_{D_n} \sum_{t_i \in D_n, t_i \le t} \beta_{t_i} (Y_{t_{i+1}} - Y_{t_i})$$

$$= \lim_{D_n} \sum_{t_i \in D_n, t_i \le t} \beta_{t_i} \int_{t_i}^{t_{i+1}} f(t_{i+1}, s)dL_s$$

$$+ \lim_{D_n} \sum_{t_i \in D_n, t_i \le t} \beta_{t_i} \int_0^{t_i} (f(t_{i+1}, s) - f(t_i, s))dL_s$$

$$= \int_0^t \beta_s f(s,s)dL_s + \int_0^t \left(\int_s^t \beta_u df_u(u,s) \right) dL_s,$$

then we deduce formula (18).

5 Appendix

5.1 A process with finite energy but without quadratic variation along the dyadics.

Let D_k be the k-th dyadic subdivision of $[0,1]$, that is $t_j^k = \dfrac{j}{2^k}, 0 \le j \le 2^k$.

We are willing to build a deterministic function x such that $S_k = 1$ if k is even and greater than 2, and $S_k = 2$ for k odd. Such a funcion has obviously finite energy along the sequence $(D_k)_k$ (and indeed its energy is equal to 2, although is would be equal to 1 along the sequence $(D_{2k})_k$), but has no quadratic variation since the sequence (S_k) has 2 accumulation points.

Let us begin by defining $x_0 = x_1 = 0$ and $x_{1/2} = 1$, so that $S_1 = 2$.

At the second step, we define $x_{1/4} = x_{3/4} = 1/2$, so that $S_2 = 1$

In order to make our construction clear, we go into details for the third step.

We want to define $x_{j/8}$ for odd j in order that $S_3 = 2$. The idea is to compute $x_{j/8}$ such that

$$\left(x_{\frac{j}{8}} - x_{\frac{j-1}{8}}\right)^2 + \left(x_{\frac{j+1}{8}} - x_{\frac{j}{8}}\right)^2 = 2 \times \left(x_{\frac{j+1}{8}} - x_{\frac{j-1}{8}}\right)^2.$$

Actually, this amounts to find a solution y to an equation like

$$(a-y)^2 + (b-y)^2 = 2 \times (a-b)^2. \tag{19}$$

As equation (19) has two solutions, namely $((1+\sqrt{3})a + (1-\sqrt{3})b)/2$ and $((1-\sqrt{3})a + (1+\sqrt{3})b)/2$, we have 2 possible choices for each $x_{j/8}$ with odd j in order that $S_3 = 2$.

This process is then iterated as follows :

1. Assume that we have constructed $x_{\frac{j}{2^{2k-1}}}$, $0 \le j \le 2^{2k-1}$, such that $S_{2k-1} = 2$ for some k. Then we put $x_{\frac{2j+1}{2^{2k}}} = (x_{\frac{2j}{2^{2k}}} + x_{\frac{2j+2}{2^{2k}}})/2$ (so that it is the midpoint of its neighbours). Then it is readily checked that $S_{2k} = 1$.

2. Now we have to choose the $x_{\frac{2j+1}{2^{2k+1}}}$'s. We will proceed as was done above for $k = 1$. Namely, we can always choose $y = x_{\frac{2j+1}{2^{2k+1}}}$ so that it solves equation (19) with $a = x_{\frac{2j}{2^{2k+1}}}$ and $b = x_{\frac{2j+2}{2^{2k+1}}}$, and the result follows the same lines as for $k = 1$.

It remains to check that we can build a real continuous function x on $[0, 1]$ with the specified values on the dyadics.

Let x^n be the piecewise linear function joining the points constructed at rank n. We will show that the sequence (x^n) satisfies a uniform Cauchy criterion, which will give the claim.

First note that it is obvious (again from the solution of equation (19)) that any two neighbours at rank $2k$ or $2k + 1$ have distance from each other at most $(1 + \sqrt{3}/4)^k$. In other words, we always have

$$\left|x^n_{\frac{i+1}{2^n}} - x^n_{\frac{i}{2^n}}\right| \le \left(\frac{1+\sqrt{3}}{4}\right)^{\frac{n}{2}}. \tag{20}$$

Now, fix $\varepsilon > 0$. For positive n and p and for $t \in [0, 1]$, let t^n_i be the point closest to t in D_n, and t^{n+p}_j the point closest to t in D_{n+p}. Without loss of generality we will assume that $t^n_i \le t^{n+p}_j \le t^n_{i+1}$. We have then

$$|x^n_t - x^{n+p}_t| \le |x^n_t - x^n_{t^n_i}| + |x^n_{t^n_i} - x^{n+p}_{t^n_i}|$$
$$+ |x^{n+p}_{t^n_i} - x^{n+p}_{t^{n+p}_j}| + |x^{n+p}_{t^{n+p}_j} - x^{n+p}_t|. \tag{21}$$

From (20) and the definition of x^n it is obvious that the first and last terms in the right-hand side of (5) can be made as small as wanted (say less than $\varepsilon/3$), uniformly in t and p, for n large enough. Moreover, as $D_n \subset D_{n+p}$, the second term is identically zero.

Hence it remains to uniformly estimate $|x^{n+p}_{t^n_i} - x^{n+p}_{t^{n+p}_j}|$ (note that in prin-ciple, at most 2^p points of D_{n+p} may lay between t^n_i and t^{n+p}_j, so that such a sequence has no trivial majorization uniform in p).

Put $t_n^i = k/2^n$

We assume for instance that $l := x_{k/2^n} \le h : x_{(k+1)/2^n}$ and that n is odd. Then clearly $l \le x_{(2k+1)/2^{n+1}} \le h$ (since $x_{(2k+1)/2^{n+1}} = (l+h)/2$), and if we choose $x_{(4k+1)/2^{n+2}} = ((1+\sqrt{3})l + (1-\sqrt{3})(l+h)/2)/2$ and $x_{(4k+3)/2^{n+2}} = ((1+\sqrt{3})(h+l)/2 + (1-\sqrt{3})h)/2$, we can see that both values lay in $[l - (\sqrt{3}-1)(h-l)/4, h]$. Keeping (20) in mind, we conclude that for every point s in $D_{n+2} \cap [k/2^n, (k+1)/2^n]$, and for every $p \ge 2$,

$$l - (\sqrt{3}-1)/4\left(\frac{1+\sqrt{3}}{4}\right)^{\frac{n}{2}} \le x_s^{n+p} \le h.$$

If we iterate this procedure, it is straightforward now that for every point s in $D_{n+2m} \cap [k/2^n, (k+1)/2^n]$, and for every $p \ge 2m$,

$$l - \frac{\sqrt{3}-1}{4}\left(\frac{1+\sqrt{3}}{4}\right)^{\frac{n}{2}} \sum_{i=0}^{m-1}\left(\frac{1+\sqrt{3}}{4}\right)^i \le x_s^{n+p} \le h$$

and eventually, for every point s in $\bigcup_m D_{n+2m} \cap [k/2^n, (k+1)/2^n]$, and for every $p \ge 0$,

$$l - \frac{\sqrt{3}-1}{4}\left(\frac{1+\sqrt{3}}{4}\right)^{\frac{n}{2}}\frac{4}{3-\sqrt{3}} \le x_s^{n+p} \le h.$$

Last, it follows that the third term in the right-hand of (5) can be made as small as wanted, say less than $\varepsilon/3$, for n big enough, uniformly in t and p, and finally we checked the uniform Cauchy criterion for the sequence of functions (x^n). Hence this sequence converges to a continuous function x, such that by construction, for every k

$$\sum_{i \in D_{2k}} (x_{i+1} - x_i)^2 = 1, \text{ and } \sum_{i \in D_{2k+1}} (x_{i+1} - x_i)^2 = 2.$$

This function has finite energy, but no quadratic variation along the dyadics.

5.2 A continuous function with discontinuous pre-quadratic variation

We consider the function introduced in [3], Example 1, that is the piecewise affine function X such that $X_t = 0$ at each $t = 1 - 2^{1-2p}$, $X_t = 1/p^{1/2}$ for $t = 1 - 2^{-2p}$, and X is affine between these points. If we define moreover $X_1 = 0$, X is a continuous function on $[0, 1]$.

It is clear that X is a function of finite variation, hence of zero quadratic variation on every $[0, t]$ with $t < 1$.

On the other hand, it was proven in [3] that X has an infinite quadratic variation on $[0, 1]$ along $S := \{1 - 2^{-2k}, k \ge 1\}$.

For $n > 0$, we define now a subdivision π_n of $[0, 1]$ as follows :

$$\pi_n = \bigcup_{j \leq 2^{2n}-1} \{j2^{-2n}\} \cup \bigcup_{k \geq n} \{1 - 2^{-2k}\}.$$

It is straightforward that for every n X has an infinite quadratic variation along π_n, although its quadratic variation along $\pi_n \cap [0, t]$ goes to zero as $n \to \infty$ for every $t < 1$.

Note that if we modify our example in order that $X_t = 1/p$ for $t = 1 - 2^{-2p}$, $p > 0$, everything else remaining unchanged, the pre-quadratic variation of X on $[0, t]$ is equal to zero if $t < 1$, but finite and non zero for $t = 1$.

5.3 A process continuous in probability may admit no càdlàg modification.

Let X be the piecewise affine function such that $X_t = 0$ at each $t = 1 - 2^{1-2p}$, $X_t = 1$ for $t = 1 - 2^{-2p}$, and X is affine between these points. If we define moreover $X_t = 0$ outside $[0, 1)$, we get a discontinuity of the second kind at 1.

Now define the process Y as follows : $Y_t = X_{t-T} \mathbf{1}_{t \geq T}$, where T is a random variable uniformly distributed on $[0, 1]$, then Y is continuous in probability, but every path of Y has almost surely a discontinuity of the second kind between times 1 and 2.

Note that this result remains true even if we require our process to have a quadratic variation along a sequence (π_n).

References

1. M.A. Berger, V.J. Mizel, Volterra Equations with Itô integrals-I, *J. Integ. Eq.* **2** (1980), 187–245.
2. F. Coquet, J. Mémin and L. Słomiński, On Non-Continuous Dirichlet Processes, *J. Theor. Probab.*, **16** (2003), 197–216.
3. F. Coquet and L. Słomiński, On the convergence of Dirichlet Processes, *Bernoulli*, **5** (1999), 615–639.
4. C. Dellacherie, Capacités et processus stochastiques, 1972, Springer-Verlag, Berlin.
5. C. Dellacherie, P.A. Meyer, Probabilités et potentiel, Chap. V à VIII, 1980, Hermann, Paris.
6. M. Errami, F. Russo, n-covariation, generalized Dirichlet processes and calculus with respect to finite cubic variation processes, *Stochastic Process. Appl.*, **104**, (2003), 259–299.
7. M. Errami, F. Russo, Covariation de convolution de martingales, *C. R. Acad. Sci. Paris*, t. **326**, Série I, p. 601–606, 1998.
8. M. Errami, F. Russo, P. Vallois, Itô formula for $C^{1,\lambda}$- functions of a càdlàg semimartingale. *Probab. Theory Rel. Fields* **122** p. 191–221 (2002).

9. H. Föllmer, Calcul d'Ito sans Probabilités, *in Séminaire de Probabilités XV*, Lecture Notes in Maths 850, Springer-Verlag (1981), 143–150.

10. H. Föllmer, Dirichlet Processes, *in Stochastic Integrals*, Lecture Notes in Maths 851, Springer-Verlag (1981), 476–478.

11. S.E. Graversen and M. Rao, Quadratic variation and Energy, *Nagoya Math. J.*, **100** (1985), 163–180.

12. J. Jacod, Convergence en loi de semimartingales et variation quadratique, *in Séminaire de Probabilités XV*, Lecture Notes in Maths 850, Springer, Berlin, 1981.

13. J. Jacod, A. Shiryaev, Limit Theorems for Stochastic Processes, Springer, Berlin 1987.

14. M. Métivier, Notions fondamentales de la théorie des probabilités, 1968, Dunod, Paris.

15. P. Protter, Volterra equations driven by semimartingales, *The Annals of Probability*, 1985, Vol. **13**, No 2, 519–530.

16. F. Russo, P. Vallois, The generalized covariation process and Ito formula, *Stochastic process. appl*, **59** (1995), 81–104.

A Lost Scroll

John B. Walsh

Department of Mathematics, University of British Columbia, Department of
Mathematics, Vancouver, BC V6T 1Z2, Canada
e-mail: walsh@math.ubc.ca

I recently received a package from a French colleague, together
with a note explaining that the contents had been found while excavating
a dolmen near the coast of Brittany, wrapped around a pre-christian
order of fish and chips. {Yet another evidence of the venerable
origins of the cultural exchange between France and Britain. ed.}
The contents of the package consisted of a single ancient and flaking
scroll of parchment, crowded with hardly-visible runes which were
often obliterated by oily stains, some of which still smel'ed faintly
of druidic vinegar. The note went on to say that the scroll seemed
to concern mathematics and, since I was knowledgeable in the subject,
perhaps I could help to decipher it.

Indeed, the scroll did concern mathematics and, after many long
candle-lit nights of battling with strange ideas and atrocious
handwriting, I was able to aid in some small way with its translation.

Because of the great historical interest of the present paper,
I have been so bold as to circulate a copy among certain close friends.
I beg the readers indulgence, and warn him that the notion of a proof
was less rigorous in the pre-Christian era than it became after the
crucifixion, so that the proofs might seem incomplete to the modern
eye. But even the gap (which the reader will doubtless remark immediately)
in the proof of Theorem 1 is easily bridged, and the resulting
mathematics has a certain timeless quality. I might hesitently
suggest that, even with its admitted imperfections, it has similarities
with some of the most highly regarded of twentieth-century mathematics.

J B W

Université du village gaulois
Séminaire des Sujets Habituels

A NEW RESULT ON THE USUAL SUBJECTS
By Probabilix

Most of our readers are familiar with the usual subjects from articles in previous Seminars, so to keep from boring them with an overlong introduction, we refer to (1) for the usual notation, definitions, and the background to the problem.

The usual way of treating the problem is to first solve it under the usual hypotheses plus hypotheses (B) and (L), and then to take a lexicographic limit as $L \rightarrow Z$. However, in this paper we will assume only the usual hypotheses.

As usual, the first theorem is Theorem one. This is proved as usual, so we will omit it. We can now state the basic

THEOREM 2. Under the usual hypotheses, the usual statement holds.

PROOF. Notice that if we first plunge the problem into a Wry compactification, the fact that càd is dual to làg implies qcg. But this is exactly the situation of (1), so the result follows by the usual methods.

REMARK. Theorem 2 is not new. It was proved in (2), although by a more complicated argument. Notice, however, that our proof never used the fact that the functions were bounded! Thus it is valid uniformly over the political spectrum, so that if we use a "✵" as usual to denote the usual supremum, we have the following much stronger result, which is unusual.

THEOREM 3. Under the usual* hypotheses, the usual* statement holds.

REFERENCES

(1) The usual reference.

(2) Ibid., two pages further on.

Stochastic Integration with Respect to a Sequence of Semimartingales

Marzia De Donno and Maurizio Pratelli

Dipartimento di Matematica, Università di Pisa, via Buonarroti 2, 56127 Pisa, Italy
e-mail: mdedonno@dm.unipi.it, pratelli@dm.unipi.it

1 Introduction

Motivated by a problem in mathematical finance, which, however, will not be discussed in this note, we propose a theory of stochastic integration with respect to a sequence of semimartingales. The case of stochastic integration with respect to a sequence of square integrable martingales, is, in fact, a special case of a theory of cylindrical stochastic integration, developed by Mikulevicius and Rozovskii [15, 16]: indeed, a sequence of martingales can be viewed as a cylindrical martingale with values in the set of all real-valued sequences.

The approach to the general case essentially relies on a paper by Mémin [10], which, in turn, is based on some results due to Dellacherie [4]. The basic idea is the following: by making use of an appropriate change in probability, it is possible to replace the integral with respect to a semimartingale with an integral with respect to the sum of a square integrable martingale and a predictable process with integrable variation; analogously, the integral with respect to a sequence of semimartingales can be replaced with the sum of an integral with respect to a sequence of square integrable martingales and an integral with respect to a sequence of predictable processes with integrable variation. It should be pointed out that the new probability is not "universal", but depends on the particular integral we are calculating.

We show that, with our definition, the stochastic integral keeps some good properties of the integral with respect to a finite-dimensional semimartingale, such as invariance with respect to a change in probability and the so-called "Mémin's theorem", which states that the limit of a sequence of stochastic integrals is still a stochastic integral. Yet, there are also some differences with the finite-dimensional case, and some "bad properties", which will be pointed out by some examples.

2 Definitions and preliminary results

Let be given a filtered probability space $(\Omega, \mathcal{F}, (\mathcal{F}_t)_{t\in[0,T]}, \mathbf{P})$, which fulfills the usual assumptions, and denote by $\mathcal{P}$ the predictable σ-field on $\Omega \times [0,T]$. Let $\mathbf{X} = (X^n)_{n\geqslant 1}$ be a sequence of semimartingales: $\mathbf{X}$ can be viewed as a stochastic process with values in the set of all real sequences $\mathbb{R}^{\mathbb{N}}$.

We call *simple integrand* a finite sequence of predictable bounded processes, that is, a process H of the form

$$H = \sum_{i\leqslant n} h^i e_i, \tag{1}$$

where $\{e_i\}_{i\geqslant 1}$ is the canonical basis in $\mathbb{R}^{\mathbb{N}}$, and h^i are predictable bounded processes. The stochastic integral of a simple integrand with respect to $\mathbf{X}$ is naturally defined: if H has the form (1), then

$$\int H \, d\mathbf{X} = H \cdot \mathbf{X} = \int \sum_{i\leqslant n} h^i \, dX^i.$$

The purpose of this paper is to extend the stochastic integral to an appropriate class of processes, which will be called *generalized integrands*. The construction which we propose keeps some good properties of the classical stochastic integral for finite-dimensional semimartingales: in particular, the stochastic integral is *isometric* in some proper sense and independent of the probability. For the general theory of stochastic integration, we mainly refer to [12]; a good description of this theory can be found also in [8], [9].

We denote by E the set $\mathbb{R}^{\mathbb{N}}$, provided with the product topology: E is a locally convex space. The dual set E' is the space of linear combinations of Dirac measures.

A simple integrand H can be represented as a process with values in E' of the form $H = \sum_{i\leqslant n} h^i \delta_i$, where, as usual, δ_i denotes the Dirac delta at point i (henceforth we will use this notation for H). It is easy to check that a E'-valued process H is *weakly predictable*, that is, He_i is predictable for all i (or, equivalently, all its components are predictable processes) if and only if H is *strongly predictable*, that is, there exists a sequence of simple integrands $(H^n)_{n\geqslant 1}$, such that for all $e \in E$

$$H_{\omega,t}\, e = \lim_{n\to\infty} H^n_{\omega,t}\, e.$$

Métivier [11] constructed an *isometric* integral for the case of a square integrable martingale with values in a Hilbert space: he proved that, in this case, it is necessary to include in the space of integrands some processes with values in the set of not necessarily bounded (or continuous) operators on E, provided with a proper measurability condition with respect to the predictable σ-algebra $\mathcal{P}$.

Following this idea, we denote by $\mathcal{U}$ the set of not necessarily bounded operators on E ($\mathcal{U} \supset E'$) and, for all $h \in \mathcal{U}$, we denote by $\mathcal{D}(h)$ the domain of h ($\mathcal{D}(h) \subset E$). We say that a sequence $(h^n) \in E'$ converges to $h \in \mathcal{U}$ if $\lim_n h^n(x) = h(x)$, for all $x \in \mathcal{D}(h)$.

Analogously to the notion of predictable E'-valued process, we introduce the following definition:

Definition 1. *A process* $\mathbf{H}$ *with values in* $\mathcal{U}$ *is said to be* predictable *if there exists a sequence* (H^n) *of* E'*-valued predictable processes, such that*

$$\mathbf{H} = \lim_{n \to \infty} H^n,$$

in the sense that for all (ω, t)*, and for all* $x \in \mathcal{D}(H_{\omega,t})$*, the sequence* $H^n_{\omega,t}(x)$ *converges to* $H_{\omega,t}(x)$*, as* n *tends to* ∞*.*

Remark 1. For a given sequence (h^n) in E', it always makes sense to define the limit operator $h = \lim_{n \to \infty} h^n$, where $\mathcal{D}(h) = \{x \in E : \lim_{n \to \infty} h^n x \text{ exists}\}$; possibly, $\mathcal{D}(h)$ can be the trivial set $\{0\}$. Hence, for any sequence (H^n) of E'-valued processes, there always exists the limit $\mathbf{H} = \lim_n H^n$, which is a process with values in $\mathcal{U}$.

Besides the usual Banach spaces of semimartingales $\mathbb{H}^p(\mathbf{P})$ (see, for instance, [14]), we will consider the Banach space of special semimartingales $\mathcal{M}^2 \oplus \mathcal{A}(\mathbf{P})$, introduced by Mémin in [10]: a special semimartingale X belongs to $\mathcal{M}^2 \oplus \mathcal{A}(\mathbf{P})$ if its canonical decomposition $X = M + B$ is such that $M \in \mathcal{M}^2(\mathbf{P})$ and $B \in \mathcal{A}(\mathbf{P})$. As usual, $\mathcal{M}^2(\mathbf{P})$ denotes the set of square integrable martingales, while $\mathcal{A}(\mathbf{P})$ denotes the set of predictable processes B with integrable variation, such that $B_0 = 0$. The norm on $\mathcal{M}^2 \oplus \mathcal{A}(\mathbf{P})$ is defined by the formula:

$$\|X\|_{\mathcal{M}^2 \oplus \mathcal{A}} = \|M\|_{\mathcal{M}^2} + \|B\|_{\mathcal{A}}.$$

The importance of this space is evident in the following result, due to Dellacherie. We state it as formulated by Mémin:

Lemma 1 ([4], Theorem 5; [10], Lemma I.3). *Let be given a sequence* $(X^i)_{i \geq 1}$ *of semimartingales. There exists a probability measure* $\mathbf{Q}$*, equivalent to* $\mathbf{P}$*, with* $d\mathbf{Q}/d\mathbf{P} \in L^\infty(\mathbf{P})$*, such that under* $\mathbf{Q}$*,*

(i) X^i *is a special semimartingale, with canonical decomposition*

$$X^i = M^i + B^i;$$

(ii) $M^i \in \mathcal{M}^2(\mathbf{Q})$*;*

(iii) $B^i \in \mathcal{A}(\mathbf{Q})$*.*

In other words: $X^i \in \mathcal{M}^2 \oplus \mathcal{A}(\mathbf{Q})$*, for all* i*.*

The next lemma is an easy consequence of the proof given by Dellacherie to Lemma 1:

Lemma 2. *Let* $X \in \mathcal{M}^2 \oplus \mathcal{A}(\mathbf{P})$. *If* $\mathbf{Q}$ *is a probability measure equivalent to* $\mathbf{P}$, *such that* $d\mathbf{Q}/d\mathbf{P} \in L^\infty(\mathbf{P})$, *then,* $X \in \mathcal{M}^2 \oplus \mathcal{A}(\mathbf{Q})$ *and*

$$\|X\|_{\mathcal{M}^2 \oplus \mathcal{A}(\mathbf{Q})} \leqslant C \|X\|_{\mathcal{M}^2 \oplus \mathcal{A}(\mathbf{P})}, \tag{2}$$

where C *is a constant which depends on* $\|d\mathbf{Q}/d\mathbf{P}\|_{L^\infty}$.

Proof. Let $X = M + B$ be the canonical decomposition of X under $\mathbf{P}$. Denote by Z the density of $\mathbf{Q}$ with respect to $\mathbf{P}$ and by $(Z_t)_{t \leqslant T}$ the right-continuous, bounded, positive martingale $(\mathbb{E}_{\mathbf{P}}[Z \mid \mathcal{F}_t])_{t \leqslant T}$. Then, by the Girsanov theorem (see, for instance, [9], Theorem III.3.11), X is still a semimartingale under $\mathbf{Q}$: the canonical decomposition under the new probability measure is given by $X = N + D$, where

$$N = M - Z_-^{-1} \cdot \langle M, Z \rangle, \qquad D = B + Z_-^{-1} \cdot \langle M, Z \rangle.$$

Clearly, both random variables $V(B)_T$ (where $V(B)$ denotes the variation of B) and $[M, M]_T$ are in $L^1(\mathbf{Q})$: in particular,

$$\mathbb{E}_{\mathbf{Q}}[V(B)_T] \leqslant \|Z\|_{L^\infty} \mathbb{E}_{\mathbf{P}}[V(B)_T] = \|Z\|_{L^\infty} \|B\|_{\mathcal{A}(\mathbf{P})},$$
$$\mathbb{E}_{\mathbf{Q}}\big[[M, M]_T\big] \leqslant \|Z\|_{L^\infty} \mathbb{E}_{\mathbf{P}}\big[[M, M]_T\big] = \|Z\|_{L^\infty} \|M\|_{\mathcal{M}^2(\mathbf{P})}^2.$$

As in the proof of Theorem 5 in [4], one can show that:

$$\mathbb{E}_{\mathbf{Q}}\left[V\left(\frac{1}{Z_-} \cdot \langle M, Z \rangle\right)_T\right] = \mathbb{E}_{\mathbf{Q}}\left[\int_0^T \frac{|d\langle M, Z \rangle_s|}{Z_{s-}}\right]$$
$$= \mathbb{E}_{\mathbf{P}}\left[\int_0^T \frac{Z_{s-} |d\langle M, Z \rangle_s|}{Z_{s-}}\right]$$
$$= \mathbb{E}_{\mathbf{P}}\left[\int_0^T |d\langle M, Z \rangle_s|\right]$$
$$\leqslant \|M\|_{\mathcal{M}^2(\mathbf{P})} \|Z\|_{\mathcal{M}^2(\mathbf{P})}.$$

Then, D is a predictable process such that $V(D)_T \in L^1(\mathbf{Q})$ and

$$\|D\|_{\mathcal{A}(\mathbf{Q})} \leqslant K_1(\|M\|_{\mathcal{M}^2(\mathbf{P})} + \|B\|_{\mathcal{A}(\mathbf{P})}),$$

where K_1 is a proper constant. Moreover, $\mathbb{E}_{\mathbf{Q}}\big[[N, N]_T\big] \leqslant K_2 \mathbb{E}_{\mathbf{Q}}\big[[M, M]_T\big]$ (see, for instance, [10], Lemma I.1). Hence, inequality (2) holds. $\square$

Lemma 1 shows that, possibly by taking an appropriate equivalent probability, we can always assume that X^i is a special semimartingale, which is the sum of a square integrable martingale and a process with integrable variation. In this case, the following result can be proved:

Lemma 3. *Let $(X^i)_{i \geqslant 1}$ be a sequence in $\mathcal{M}^2 \oplus \mathcal{A}(\mathbf{P})$, with canonical decomposition $X^i = M^i + B^i$. Then, there exist:*

(i) *an increasing predictable process A_t, such that $\mathbb{E}[A_T] < \infty$,*

(ii) *a family $Q = (Q^{ij})_{i,j \geqslant 1}$ of predictable processes, such that Q is symmetric and non-negative, in the sense that $Q^{ij} = Q^{ji}$ and $\sum_{i,j \leqslant d} x_i Q^{ij} x_j \geqslant 0$, for all $d \in \mathbb{N}$, for all $x \in \mathbb{R}^d$, $\mathrm{d}\mathbf{P}\, \mathrm{d}A$ a.s.,*

(iii) *a sequence $b = (b^i)_{i \geqslant 1}$ of predictable processes,*

such that

$$\langle M^i, M^j \rangle_t(\omega) = \int_0^t Q^{ij}_{s,\omega}\, \mathrm{d}A_s(\omega), \qquad B^i_t(\omega) = \int_0^t b^i_s(\omega)\, \mathrm{d}A_s(\omega). \qquad (3)$$

Proof. Let $(c_i)_{i \geqslant 1}$ be a sequence of strictly positive numbers, such that

$$\sum_{i \geqslant 1} c_i\, \mathbb{E}\big[\langle M^i, M^i \rangle_T + V(B^i)_T \big] < \infty,$$

and define the process

$$A_t = \sum_{i \geqslant 1} c_i \big(\langle M^i, M^i \rangle_t + V(B^i)_t \big).$$

This process satisfies the condition $\mathbb{E}[A_T] < \infty$; moreover, $\mathrm{d}\langle M^i, M^i \rangle_t$ and $\mathrm{d}V(B^i)$ are absolutely continuous with respect to $\mathrm{d}A_t$ by definition. Finally, for $i \neq j$, the measure $\mathrm{d}\langle M^i, M^j \rangle_t$ is absolutely continuous with respect to $\mathrm{d}\langle M^i, M^i \rangle_t$ and $\mathrm{d}\langle M^j, M^j \rangle_t$ by the Kunita–Watanabe inequality (see, e.g., [12]). Define $(Q^{i,j}_t)_{i,j \geqslant 1}$ and $(b^i)_{i \geqslant 1}$ as follows

$$Q^{ij}_t(\omega) = \begin{cases} \dfrac{\mathrm{d}\langle M^i, M^j \rangle_t(\omega)}{\mathrm{d}A_t(\omega)} & \text{if } \mathrm{d}A_t(\omega) \neq 0 \\ 0 & \text{otherwise.} \end{cases}$$

$$b^i_t = \begin{cases} \dfrac{\mathrm{d}B_t(\omega)}{\mathrm{d}A_t(\omega)} & \text{if } \mathrm{d}A_t(\omega) \neq 0 \\ 0 & \text{otherwise.} \end{cases} \qquad (4)$$

The processes Q^{ij} and b^i are well-defined: they are predictable and fulfill condition (3). Jacod and Shiryaev have proved, in the case of a finite number of martingales, that Q can be chosen so that it is symmetric and non-negative $\mathrm{d}\mathbf{P}\, \mathrm{d}A$-a.s. ([9], Theorem II.2.9): it is rather easy to adapt their proof to the case of a sequence of martingales. $\square$

Remark 2. The process A is minimal in the following sense: if D is a predictable process such that both measures $\mathrm{d}\mathbf{P}\, \mathrm{d}\langle M^i, M^i \rangle_t$ and $\mathrm{d}\mathbf{P}\, \mathrm{d}V(B^i)_t$ are absolutely continuous with respect to $\mathrm{d}\mathbf{P}\, \mathrm{d}D_t$, then the measure $\mathrm{d}\mathbf{P}\, \mathrm{d}A_t$ is also absolutely continuous with respect to $\mathrm{d}\mathbf{P}\, \mathrm{d}D_t$.

We call *negligible* a predictable set C which is $\mathrm{d}\mathbf{P}\,\mathrm{d}A$-negligible, that is,

$$\mathbb{E}\left[\int_0^T \mathbf{1}_C\,\mathrm{d}A_t\right] = 0.$$

This notion does not depend on the probability $\mathbf{P}$: indeed, it is not difficult to prove that C is $\mathrm{d}\mathbf{P}\,\mathrm{d}A$-negligible if and only $\int h\,\mathrm{d}X^i = 0$ for every i and for every bounded predictable process h, which is zero on the complement of C.

We denote by $\mathcal{S}(\mathbf{P})$ the space of real semimartingales, endowed with the semimartingale topology, which was introduced by Émery [6]. We refer to [6] for general definition and main properties of this topology: it is important to recall that $\mathcal{S}(\mathbf{P})$ is a complete metric space. But we will mainly use this result, which is due to Mémin:

Theorem 1 ([10], Theorem II.3). *Let $(X^i)_{i\geqslant 1}$ be a Cauchy sequence in $\mathcal{S}(\mathbf{P})$. Then, there exist a subsequence (which we still denote by X^i) and a probability measure $\mathbf{Q}$, equivalent to $\mathbf{P}$, such that $\mathrm{d}\mathbf{Q}/\mathrm{d}\mathbf{P} \in L^\infty(\mathbf{P})$ and (X^i) is a Cauchy sequence in $\mathcal{M}^2 \oplus \mathcal{A}(\mathbf{Q})$.*

Remark 3. Lemma 1 and Theorem 1 hold only when the time set is a compact interval $[0, T]$: this explains why we work in this framework. However if the stochastic integral can be defined on a finite interval, then it can be defined on the whole set $[0, +\infty[$ by localization.

Finally, we introduce our definition of integrable process:

Definition 2. *Let $\mathbf{H}$ be a predictable $\mathcal{U}$-valued process. We say that $\mathbf{H}$ is integrable with respect to $\mathbf{X}$ if there exists a sequence (H^n) of simple integrands such that:*

(i) H^n *converges to* $\mathbf{H}$*, a.s.;*

(ii) $(H^n \cdot \mathbf{X})$ *converges to a semimartingale Y in $\mathcal{S}(\mathbf{P})$.*

We call $\mathbf{H}$ a generalized integrand and define $\int \mathbf{H}\,\mathrm{d}\mathbf{X} = \mathbf{H} \cdot \mathbf{X} = Y$.

We denote by $L(\mathbf{X}, \mathcal{U})$ the set of generalized integrands.

Remark 4. Of course, Definition 2 makes sense if we prove that Y is uniquely defined, in the sense that $\mathbf{H} \cdot \mathbf{X}$ is independent of the sequence H^n which approximates $\mathbf{H}$. This result, with all its consequences and applications, will be the object of section 5.

We wish also to point out that our definition of integrable process is very similar to the notion of integrable function with respect to a vector-valued measure ([5], section IV.10.7).

3 Stochastic integration with respect to a sequence of square integrable martingales

For the case of a sequence of square integrable martingales, we refer to the theory on cylindrical integration recently developed by Mikulevicius and Rozovskii [15], [16]: indeed, a sequence of martingales can be viewed as a cylindrical martingale with values in E. In this section, we briefly recall Mikulevicius and Rozovskii's main results, adapted to our setting.

We assume that $X^i = M^i \in \mathcal{M}^2(\mathbf{P})$ for all i and denote by $\mathbf{M}$ the sequence $(M^i)_{i \geqslant 1}$. The aim in this section is to define the stochastic integral $\int H \, d\mathbf{M}$ on a proper class of processes, so that the integral is a square integrable martingale. We recall that, even in the one-dimensional case, it may happen that the integral with respect to a square integrable martingale is not even a local martingale (see, for instance, [7]).

For a simple integrand $H_t = \sum_{i \leqslant n} h_t^i \delta_i$, the Ito isometry holds:

$$
\mathbb{E}\left[\left(\int_0^T H_s \, d\mathbf{M}_s\right)^2\right] = \mathbb{E}\left[\int_0^T \sum_{i,j \leqslant n} h_s^i h_s^j \, d\langle M^i, M^j \rangle_s\right]
$$

$$
= \mathbb{E}\left[\int_0^T \sum_{i,j \leqslant n} h_s^i h_s^j Q_s^{ij} \, dA_s\right], \tag{5}
$$

where Q and A are defined as in Lemma 3. The question is how to complete the set of integrands with respect to the norm induced by the Ito isometry. Intuition may suggest that, to this aim, it is sufficient to consider, as value set of the integrands, the space of the sequences $(h^i)_{i \geqslant 1}$ such that $\mathbb{E}\left[\int \sum_{i,j} h^i h^j Q^{ij} \, dA\right] < \infty$. In fact, this is not sufficient, as we will show in Example 1 below.

Consider Q for fixed (ω, t) (which we omit, for simplicity). We can define on E' a linear mapping with values in E, which we still denote by Q, in the following way: for $h = \sum_{i \leqslant n} h^i \delta_i \in E'$, we define Qh as the sequence whose i-th component is $(Qh)_i = \sum_j Q^{ij} h^j$. It is easy to check that Q is symmetric and non-negative, namely $\langle h, Qk \rangle_{E',E} = \langle k, Qh \rangle_{E',E}$ and $\langle h, Qh \rangle_{E',E} \geqslant 0$ for all h, $k \in E'$, where $\langle \ , \ \rangle_{E',E}$ denotes the duality. The mapping Q induces a seminorm on E', by the formula:

$$
|h|_Q^2 = \langle h, Qh \rangle_{E',E} = \sum_{i,j \geqslant 1} h^i Q^{ij} h^j. \tag{6}
$$

Thus, the main problem is to find a completion of E' with respect to this seminorm. Following the approach by Mikulevicius and Rozovskii [15, 16], we consider on the set QE' the scalar product:

$$
(Qh, Qk)_{QE'} = \langle h, Qk \rangle_{E',E} = \langle k, Qh \rangle_{E',E}.
$$

This scalar product induces a norm, with respect to which QE' is a pre-Hilbert space. Its completion K is a Hilbert space and can be continuously embedded in E. Denote by K' the topological dual of K: the set K' contains E' and coincides with the completion of $E'/\ker Q$, with respect to the norm induced by (6). Thus, if $h \in E'$, then $|h|^2_{K'} = \langle h, Qh \rangle_{E',E}$. We recall that K and K' depend on (ω, t): so we have defined a family of Hilbert spaces, depending on Q. Note that, if H is a simple integrand, the isometry (5) can be rewritten in the form:

$$\mathbb{E}\left[\left(\int_0^T H_s \, d\mathbf{M}_s \right)^2 \right] = \mathbb{E}\left[\int_0^T |H_s|^2_{K'_s} \, dA_s \right].$$

Now, it seems natural to take as generalized integrand a predictable process $\mathbf{H}$ with values in $\mathcal{U}$, such that for all (ω, t), the domain of $\mathbf{H}_{\omega,t}$ contains $K_{\omega,t}$, the restriction of $\mathbf{H}_{\omega,t}$ to $K_{t,\omega}$ is an element of $K'_{\omega,t}$, and $\mathbb{E}\left[\int |\mathbf{H}|^2_{K'} \, dA \right] < \infty$. We observe that if $\mathbf{H}$ is predictable, then $|\mathbf{H}|^2_{K'}$ is also predictable: this is a consequence of the argument below.

Consider now the canonical basis in E' and take the sequence $\eta_n = Q\delta_n$. By a standard orthogonalization procedure, we can construct an orthonormal basis $\{k^i\}_{i \geq 1}$ in K (once again, for simplicity of notations, we omit (ω, t)): every k^i is an element of $\text{span}\{\eta_1, \ldots, \eta_i\}$. Hence, $k^i = Qh^i$, where h^i is an element of K', such that $h^i \in \text{span}(\delta_1, \ldots, \delta_i)$ and $\{h^i\}_{i \geq 1}$ is an orthonormal basis in K' (see [15] pag. 141 for details).

If we consider h^i as a function of (ω, t), it follows that $h^i_t = \sum_{j \leq i} \alpha^{ij}_t \delta_j$ where α^{ij}_t are real predictable. If $\mathbf{H}$ is predictable, in the sense of Definition 1, $\mathbf{H}_t(Q_t \delta_j)$ is predictable and so is $\mathbf{H}(k^j_t)$.

Every process $\mathbf{H}$, such that $\mathbf{H}_{\omega,t} \in K'_{\omega,t}$, can be written in the form:

$$\mathbf{H}_t = \sum_{i \geq 1} \lambda^i_t h^i_t,$$

where $\lambda^i_t = (\mathbf{H}_t, h^i_t)_{K'_t} = \mathbf{H}_t(k^i_t)$; then $|\mathbf{H}_t|^2_{K'_t} = \sum_i (\lambda^i_t)^2$ is predictable. Furthermore, the process $\mathbf{H}$ can be approximated by the sequence

$$H^n_t = \sum_{i \leq n} \lambda^i_t h^i_t = \sum_{i \leq n} \sum_{j \leq i} \lambda^i_t \alpha^{ij}_t \delta_j = \sum_{j \leq n} r^{nj}_t \delta_j \qquad (7)$$

(a proof of this fact can be found in [16], Corollaries 2.2 and 2.3).

The following theorem is essentially due to Mikulevicius and Rozovskii: we give a formulation which is slightly different from the original one, but better fits into our context. We also recall the main steps of their proof.

Theorem 2. *Let $\mathbf{H}$ be a $\mathcal{U}$-valued process such that:*

(i) $\mathcal{D}(\mathbf{H}_{\omega,t}) \supset K_{\omega,t}$ *for all (ω, t);*

(ii) $\mathbf{H}_{\omega,t}|_{K_{\omega,t}} \in K'_{\omega,t}$;

(iii) $\mathbf{H}_t(Q_t \delta_n)$ is predictable for all n;

(iv) $\mathbb{E}\left[\int_0^T |\mathbf{H}_t|^2_{K'_t} \, dA_t\right] < \infty$.

Then, there exists a sequence (H^n) of simple integrands, such that $H^n_{\omega,t}$ converges to $\mathbf{H}_{\omega,t}$ in $K'_{\omega,t}$ for all (ω,t) and $(H^n \cdot \mathbf{M})$ is a Cauchy sequence in $\mathcal{M}^2(\mathbf{P})$.

As a consequence, we can define the stochastic integral $\mathbf{H} \cdot \mathbf{M}$ as the limit of the sequence $(H^n \cdot \mathbf{M})$.

Proof. It is easy to check that the approximating sequence H^n defined by (7) converges to $\mathbf{H}$ in K' and it is such that $(H^n \cdot \mathbf{M})$ is a Cauchy sequence in $\mathcal{M}^2(\mathbf{P})$. However, every H^n may not be a simple integrand. Fix n: as we have already observed, the process H^n is of the form $\sum_{i \leqslant n} r^{ni} \delta_i$, where r^{ni} are predictable. We define the sequence

$$H^{n,m}_t = \sum_{i \leqslant n} r^{ni}_t \delta_i \, \mathbf{1}_{\{\max_{i \leqslant n} |r^{ni}_t| \leqslant m\}},$$

for every $m \in \mathbb{N}$. Then, $(H^{n,m})_{m \geqslant 1}$ is a sequence of simple integrands, which converges to H^n in K' (as $m \to \infty$), and it is such that $(H^{n,m} \cdot \mathbf{M})_{m \geqslant 1}$ is a Cauchy sequence in $\mathcal{M}^2(\mathbf{P})$ (see, for instance, [15], Proposition 9). Thus, by a standard diagonalization procedure, we can build a sequence $\tilde{H}^n = H^{n,m_n}$ of simple integrands, which satisfies the required properties.   □

Remark 5. It may seem sufficient, in order to define the integral $\mathbf{H} \cdot \mathbf{M}$, to know the restriction of $\mathbf{H}_{\omega,t}$ to $K_{\omega,t}$. In fact, this is not exactly true: indeed, let $\mathbf{H}$ be a process which fulfills conditions (i)–(iv), and (H^n) be the sequence defined as in the theorem. If we set $\mathbf{J} = \lim_n H^n$, we can say that $\mathbf{J}$, but not $\mathbf{H}$, is M-integrable, according to Definition 2. This will be evident in Example 1. However, an immediate consequence of Theorem 2 is the following: assume that (H^n) and (J^n) are two sequences of simple integrands, such that both sequences $(H^n \cdot \mathbf{M})$, $(J^n \cdot \mathbf{M})$ are Cauchy sequences in $\mathcal{M}^2(\mathbf{P})$. Define $\mathbf{H} = \lim_n H^n$ and $\mathbf{J} = \lim_n J^n$. If $\mathbf{H}|_K = \mathbf{J}|_K$ a.s., then $\lim_n(H^n \cdot \mathbf{M}) = \lim_n(J^n \cdot \mathbf{M})$; hence, $\mathbf{H} \cdot \mathbf{M} = \mathbf{J} \cdot \mathbf{M}$.

Remark 6. The condition of measurability given by Mikulevicius and Rozovskii (condition (iii) in Theorem 2) is strictly related to the probability measure $\mathbf{P}$: under an equivalent measure $\mathbf{Q}$, the processes M^i may not be martingales and, in any case, the process $d\langle M^i, M^j \rangle$ is no longer the same. For this reason, we have decided to give a different notion of predictable process, which does not depend on the probability.

Remark 7. Consider the stable subspace generated by the sequence $\mathbf{M} = (M^i)_{i \geqslant 1}$ of square integrable martingales (for the definition of stable subspace, we refer to [12]). It is natural to expect that a characterization of this space holds in terms of the stochastic integral with respect to $\mathbf{M}$. Indeed, the following result holds: the set of all the stochastic integrals $\mathbf{H} \cdot \mathbf{M}$, with $\mathbf{H}$ fulfilling conditions (i)–(iv) of Theorem 2, is a closed set in $\mathcal{M}^2(\mathbf{P})$ and coincides with the stable subspace generated by $\mathbf{M}$ in $\mathcal{M}^2(\mathbf{P})$. A proof of this can be found in [16]; it is, in fact, a simple extension of the analogous result in the finite-dimensional case.

Remark 8. One could think that the above mentioned construction can be adapted to the case of a sequence of local martingales, just replacing the predictable quadratic covariation $\langle M^i, M^j \rangle$, with the quadratic covariation $[M^i, M^j]$. In fact, this does not work. The reason is that the process $[M^i, M^j]$ is optional, not predictable: a factorization as in Lemma 3 can still be found, but Q and A are optional. So, it is not possible to repeat the construction of the approximating sequence and the proof of Theorem 2.

4 Stochastic integration with respect to a sequence of predictable processes with finite variation

In this section we assume that $X^i = B^i \in \mathcal{A}(\mathbf{P})$ for all i and denote by $\mathbf{B}$ the sequence $(B^i)_{i \geqslant 1}$. The property of finite variation is invariant with respect to a change in probability. However, if $\mathbf{Q}$ is an equivalent probability measure, it may occur that the variation of B is in $L^1(\mathbf{P})$ but not in $L^1(\mathbf{Q})$.

Assume that B^i has integrable variation for all i. Then, as we have proved in Lemma 3, there exists a factorization $dB_t^i = b_t^i \, dA_t$ where A is an increasing predictable process and $(b^i)_{i \geqslant 1}$ is a sequence of predictable processes. In this case, the construction of the stochastic integral $\mathbf{H} \cdot \mathbf{B}$ is easier than in the previous case, since the set of predictable processes with finite variation is closed in $\mathcal{S}(\mathbf{P})$ (see, for instance, [10], Theorem IV.7).

So, if $\mathbf{H}$ is a $\mathcal{U}$-valued process such that $\mathbf{H} = \lim_n H^n$, where H^n are simple integrands, then it is easy to verify that $\mathbf{H}$ is $\mathbf{B}$-integrable if and only if $b_{\omega,t}$ belongs to the domain of $\mathbf{H}_{\omega,t}$ for all (ω, t) and the random variable

$$\int_0^T |\mathbf{H}_t b_t| \, dA_t \tag{8}$$

is finite $\mathbf{P}$-a.s. Note that $\mathbf{H}b = \lim_n H^n b$ is a predictable process. Moreover, assume that the random varable defined in (9) belongs to $L^1(\mathbf{P})$, and H^n is a sequence of simple integrands which converges to $\mathbf{H}$; then, the sequence

$$J^n = H^n \mathbf{1}_{\{|H^n b| \leqslant 2|\mathbf{H}b|\}}$$

is such that $(J^n \cdot \mathbf{B})$ converges to $\mathbf{H} \cdot \mathbf{B}$ in $\mathcal{A}(\mathbf{P})$.

For this case, the result corresponding to Theorem 2 is much simpler: assume that $\mathbf{H}$ is a $\mathcal{U}$-valued process such that:

(i) $b_{\omega,t} \in \mathcal{D}(\mathbf{H}_{\omega,t})$ for all (ω, t);
(ii) the process $\mathbf{H}b$ is predictable;

(iii) $\mathbb{E}\left[\int_0^T |\mathbf{H}_t b_t|\, \mathrm{d}A_t\right] < \infty.$

Then, there exists a sequence (H^n) of simple integrands such that $H^n b$ converges to $\mathbf{H}b$ and $(H^n \cdot \mathbf{B})$ is a Cauchy sequence in $\mathcal{A}(\mathbf{P})$.

As a consequence, we can define the stochastic integral $\mathbf{H} \cdot \mathbf{B}$ as the limit of the sequence $(H^n \cdot \mathbf{B})$.

In fact, it is sufficient to find an E'-valued process H such that $Hb = \mathbf{H}b$: assume, for simplicity that $b^1_{\omega,t} \neq 0$ for all (ω, t); then, we can define $H_{\omega,t} = \mathbf{H}_{\omega,t} \delta_1 / b^1_{\omega,t}$. If the predictable set $\{b^1 = 0\}$ is not negligible, it is clear how to modify such a construction. Finally, H can be easily approximated by a sequence of simple integrands.

5 The general case

The construction of a stochastic integral with respect to a sequence of semimartingales relies on the two previous cases: an appropriate change of probability allows us to reduce to the case of a stochastic integral with respect to the sum of a sequence of square integrable martingales and a sequence of predictable processes with finite variation.

Let $\mathbf{H}$ be a $\mathcal{U}$-valued process and assume, as in Definition 2, that there exists a sequence (H^n) of simple integrands, such that $\mathbf{H} = \lim_n H^n$ and $(H^n \cdot \mathbf{X})$ is a Cauchy sequence in $\mathcal{S}(\mathbf{P})$.

By Lemma 1, there exists a probability measure $\mathbf{Q}_1$, equivalent to $\mathbf{P}$, with $\mathrm{d}\mathbf{Q}_1/\mathrm{d}\mathbf{P} \in L^\infty(\mathbf{P})$ and such that $\mathbf{X} \in \mathcal{M}^2 \oplus \mathcal{A}(\mathbf{Q}_1)$, that is, every X^i belongs to $\mathcal{M}^2 \oplus \mathcal{A}(\mathbf{Q}_1)$. Furthermore, by Theorem 1, we can find a subsequence, which we still denote by (H^n), and a probability measure $\mathbf{Q}_2$, equivalent to $\mathbf{Q}_1$, such that $\mathrm{d}\mathbf{Q}_2/\mathrm{d}\mathbf{Q}_1 \in L^\infty(\mathbf{Q}_1)$ and $(H^n \cdot \mathbf{X})$ is a Cauchy sequence in $\mathcal{M}^2 \oplus \mathcal{A}(\mathbf{Q}_2)$. Lemma 2 makes sure that $\mathbf{X}$ belongs also to $\mathcal{M}^2 \oplus \mathcal{A}(\mathbf{Q}_2)$: if $\mathbf{X} = \mathbf{M} + \mathbf{B}$ is the canonical decomposition of $\mathbf{X}$ under $\mathbf{Q}_2$, then, for all n, $H^n \cdot \mathbf{M} + H^n \cdot \mathbf{B}$ is the canonical decomposition of $H^n \cdot \mathbf{X}$. Moreover, $(H^n \cdot \mathbf{M})$ is a Cauchy sequence in $\mathcal{M}^2(\mathbf{Q}_2)$, whereas $(H^n \cdot \mathbf{B})$ is a Cauchy sequence in $\mathcal{A}(\mathbf{Q}_2)$. Then $\mathbf{H} \cdot \mathbf{M}$ and $\mathbf{H} \cdot \mathbf{B}$ exist, in the sense shown respectively in sections 3 and 4. The integral $\mathbf{H} \cdot \mathbf{X}$, which is defined as the limit in $\mathcal{S}(\mathbf{P})$ of the sequence $(H^n \cdot \mathbf{X})$, coincides with $\mathbf{H} \cdot \mathbf{M} + \mathbf{H} \cdot \mathbf{B}$.

We are now able to prove what we claimed in Remark 4, namely, that Definition 2 is a good definition:

Proposition 1. *Let (H^n) and (J^n) be two sequences of simple integrands such that:*

(i) *there exists a $\mathcal{U}$-valued process $\mathbf{H}$ such that $\mathbf{H} = \lim_n H^n = \lim_n J^n$;*

(ii) *if $Y^n = H^n \cdot \mathbf{X}$ and $Z^n = J^n \cdot \mathbf{X}$, then (Y^n) and (Z^n) are Cauchy sequences in $\mathcal{S}(\mathbf{P})$.*

Then $\lim_n Y^n = \lim_n Z^n$.

Hence, the process $\mathbf{H} \cdot \mathbf{X} = \lim_n Y^n = \lim_n Z^n$ *is well-defined. In other words, the definition of the integral does not depend on the approximating sequence.*

Proof. If we apply twice Lemma 1 and Theorem 1, we can find a probability measure $\mathbf{Q}$, equivalent to $\mathbf{P}$ and such that $d\mathbf{Q}/d\mathbf{P} \in L^\infty(\mathbf{P})$ and a subsequence (which we still denote by n) such that $\mathbf{X} \in \mathcal{M}^2 \oplus \mathcal{A}(\mathbf{Q})$ and $(Y^n), (Z^n)$ are both Cauchy sequences in $\mathcal{M}^2 \oplus \mathcal{A}(\mathbf{Q})$. This means that, if $\mathbf{X} = \mathbf{M} + \mathbf{B}$ is the canonical decomposition under $\mathbf{Q}$, then, $(H^n \cdot \mathbf{M})$ and $(J^n \cdot \mathbf{M})$ are both Cauchy sequences in $\mathcal{M}^2(\mathbf{Q})$, whereas $(H^n \cdot \mathbf{B})$ and $(J^n \cdot \mathbf{B})$ are both Cauchy sequences in $\mathcal{A}(\mathbf{Q})$. From hypothesis (i) and from the results of the previous sections, it follows immediately that it must be

$$\lim_{n \to \infty} H^n \cdot \mathbf{M} = \lim_{n \to \infty} J^n \cdot \mathbf{M}, \qquad \lim_{n \to \infty} H^n \cdot \mathbf{B} = \lim_{n \to \infty} J^n \cdot \mathbf{B}.$$

So, the claim follows.   □

Remark 9. We have just proved that, when the stochastic integral exists, then, by an appropriate change of probability, it can be represented as the sum of an integral with respect to a sequence of square integrable martingales and an integral with respect to a sequence of processes with bounded variation.

The converse does not hold true: it may happen that the two above-mentioned integrals exist, but the process is not integrable in the sense of Definition 2. Let be given a probability measure $\mathbf{Q}$ such that $\mathbf{X} \in \mathcal{M}^2 \oplus \mathcal{A}(\mathbf{Q})$, with canonical decomposition $\mathbf{X} = \mathbf{M} + \mathbf{B}$, and a process $\mathbf{H}$ which is separately integrable with respect to $\mathbf{M}$ and $\mathbf{B}$ in the sense of section 3 and 4; namely, $\mathbf{H}$ fulfills the conditions of Theorem 2 and condition (i), (ii), (iii) of section 4. Then, Theorem 2 suggests the construction of an approximating sequence H^n such that $H^n|_K$ converges to $\mathbf{H}|_K$ and $(H^n \cdot \mathbf{M})$ is a Cauchy sequence in $\mathcal{M}^2(\mathbf{Q})$. Yet, this does not imply that $H^n b$ converges to $\mathbf{H}b$. So, condition (i) of Definition 2 is not satisfied. However, if (H^n) can be chosen so that $H^n b$ converges to $\mathbf{H}b$, then the stochastic integral does exist in the sense of Definition 2.

The following result is the extension to the infinite-dimensional case of Mémin's theorem ([10], Theorem III.4), which states that the set of stochastic integrals with respect to a semimartingale is closed in $\mathcal{S}(\mathbf{P})$.

Theorem 3. *Let be given a sequence of semimartingales* $\mathbf{X} = (X^i)_{i \geqslant 1}$ *and a sequence* $(\mathbf{H}^n)$ *of generalized integrands: assume that* $(\mathbf{H}^n \cdot \mathbf{X})$ *is a Cauchy sequence in* $\mathcal{S}(\mathbf{P})$. *Then, there exists a generalized integrand* $\mathbf{H}$ *such that* $\lim_{n \to \infty} \mathbf{H}^n \cdot \mathbf{X} = \mathbf{H} \cdot \mathbf{X}$.

Proof. Without loss of generality, we can assume that $\mathbf{H}^n = H^n$ are simple integrands. We can choose an equivalent probability measure $\mathbf{Q}$ and a subsequence, which we still denote by H^n, such that $X^i \in \mathcal{M}^2 \oplus \mathcal{A}(\mathbf{Q})$ and $(H^n \cdot \mathbf{X})$ is a Cauchy sequence in $\mathcal{M}^2 \oplus \mathcal{A}(\mathbf{Q})$. Define $\mathbf{H} = \lim_n H^n$. Since

$$\mathbb{E}\left[\int_0^T |H_t^n - H_t^m|_{K_t'}^2 \, \mathrm{d}A_t\right] \longrightarrow 0 \qquad \text{as } n, m \to \infty,$$

it is clear that the domain of $\mathbf{H}_{t,\omega}$ must contain $K_{\omega,t}$ for all (ω, t) and

$$\mathbb{E}\left[\int_0^T |H_t^n - \mathbf{H}_t|_{K_t'}^2 \, \mathrm{d}A_t\right] \longrightarrow 0 \qquad \text{as } n \to \infty.$$

Analogously, since

$$\mathbb{E}\left[\int_0^T |H_t^n b_t - H_t^m b_t| \, \mathrm{d}A_t\right] \longrightarrow 0 \qquad \text{as } n, m \to \infty,$$

it is clear that the domain of $\mathbf{H}_{t,\omega}$ must contain $b_{\omega,t}$ for all (ω, t) and

$$\mathbb{E}\left[\int_0^T |H_t^n b_t - \mathbf{H}_t b_t| \, \mathrm{d}A_t\right] \longrightarrow 0 \qquad \text{as } n \to \infty.$$

So the claimed result is proved. $\qquad\qquad\qquad\qquad\qquad\qquad\qquad\qquad$ $\square$

6 Examples

In this section, we show some examples to point out the main differences between the stochastic integral with respect to a finite-dimensional semimartingale and with respect to a sequence of semimartingales. In the examples we will consider, the spaces $K_{\omega,t}$ do not depend on (ω, t) and the integrands are constants. Nonetheless, even in these simple cases, some differences are evident.

The first example is taken from [3] (Example 2.1): there it is shown that, even in the case of a sequence of square integrable martingales, the set of E-valued processes is not large enough as set of integrands.

Example 1. Consider the sequence of martingales $\mathbf{M} = (M^i)_{i \geqslant 1}$ defined by:

$$M^i = W + N^i,$$

where W is a Wiener process, (N^i) is a sequence of independent compensated Poisson processes all with the same intensity $\lambda = 1$, such that N^i is independent of W, for all i.

According to Lemma 3, we find a factorization of the quadratic variation of $\mathbf{M}$, by setting $A_t = t$, $Q^{ij} = 1 + \delta_{ij}$. In [3], it has been proved that K is the subset of E of all sequences of the form $(\alpha + y_1, \alpha + y_2, \dots)$, with $\alpha \in \mathbb{R}$ and $(y_i)_{i \geqslant 1} \in l^2$, and the norm of such a sequence in K is $\alpha^2 + \sum_{i \geqslant 1} y_i^2$; the dual set K' is isomorphic to $\mathbb{R} \oplus l^2$.

Consider the sequence of simple integrands $H^n = n^{-1} \sum_{i \leqslant n} \delta_i$. It is clear that H^n converges to the constant $\mathcal{U}$-valued process $\mathbf{H}$, defined by

$$\mathbf{H}(x) = \lim_{n \to \infty} \frac{x_1 + \cdots + x_n}{n}.$$

Notice that H^n does not converge componentwise to $\mathbf{H}$: indeed, $\lim_n H^n_i = 0$, for all i. Moreover,

$$H^n \cdot \mathbf{M} = W + \frac{N^1 + \cdots + N^n}{n},$$

which converges to W in $\mathcal{M}^2(\mathbf{P})$. Therefore, we have that $\mathbf{H} \cdot \mathbf{M} = W$.

Consider now the operator $\mathbf{J}(x) = \lim_n x_n$; it is not difficult to check that $\mathbf{H}$ coincides with $\mathbf{J}$ on K. However, we observe that there does not exist $\lim_n M^n = \lim_n (W + N^n)$. This proves what we have claimed in Remark 5. The sequence H^n converges to $\mathbf{H}$ and not to $\mathbf{J}$ on E: the sequence $(H^n \cdot \mathbf{M}) = (H^n(\mathbf{M}))$ converges in $\mathcal{M}^2(\mathbf{P})$ to $\mathbf{H} \cdot \mathbf{M} = \mathbf{H}(\mathbf{M})$, while $\mathbf{J}(\mathbf{M})$ is not defined, although $\mathbf{H}|_K = \mathbf{J}|_K$. So, we can say that $\mathbf{H}$ is $\mathbf{M}$-integrable, whereas $\mathbf{J}$ is not $\mathbf{M}$- integrable.

The following example is obtained by a modification of an example due to Émery [6].

Example 2. Let $(T_n)_{n \geqslant 1}$ be a sequence of independent random variables, such that T_n is exponentially distributed with $\mathbb{E}[T_n] = n^2$. We take as filtration the smallest filtration which satisfies the usual conditions and such that T_n are stopping times. We define a sequence of martingales M^n as follows:

$$M^n_t = \frac{t \wedge T_n}{n^2} - \mathbf{1}_{\{t \geqslant T_n\}} = \frac{t}{n^2} \mathbf{1}_{\{t < T_n\}} + \left(\frac{T_n}{n^2} - 1 \right) \mathbf{1}_{\{t \geqslant T_n\}}. \qquad (9)$$

Consider the simple integrand $H^n = n^{-1} \sum_{i \leqslant n} i^2 \delta_i$. Then, with the usual notation $\mathbf{M} = (M^i)_{i \geqslant 1}$,

$$H^n \cdot \mathbf{M} = \frac{M^1 + 2^2 M^2 + \cdots + n^2 M^n}{n} = \frac{N^1 + \cdots + N^n}{n},$$

where $N^i = t \wedge T_i - i^2 \mathbf{1}_{\{t \geqslant T_i\}}$. The sequence (H^n) converges (as $n \to \infty$) to the operator $\mathbf{H}$ on E, defined by:

$$\mathbf{H}(x) = \lim_{n \to \infty} \frac{1}{n} \sum_{i \leqslant n} i^2 x_i, \qquad (10)$$

for all $x \in E$ such that this limit does exist, while the sequence $(H^n \cdot \mathbf{M})$ converges to the increasing process $A_t = t$. Consider the stopping times $S_n = \inf_{m \geqslant n} T_m$. Using Borel–Cantelli lemma, it can be proved that S_n tends to infinity (as $n \to \infty$). In particular, the sequence $S_n \wedge T$ converges to T

stationarily, namely, $S_n \equiv T$ definitely, **P**-a.s. So, for fixed ε, there exists some n such that $\mathbf{P}(S_n \leqslant T) < \varepsilon$. On the stochastic interval $[\![0, S_n \wedge T]\!]$, the martingale N^m coincides with the process A, for $m \geqslant n$. Then, if we stop the processes $H^k \cdot \mathbf{M}$ at time S^n, we have that, for $k > n$

$$(H^k \cdot \mathbf{M})^{S_n} = \frac{(N^1 + \cdots + N^n)^{S_n}}{k} + \frac{(k-n)}{k}(t \wedge S_n).$$

It is not difficult to check that the sequence $(H^k \cdot \mathbf{M})^{S_n}$ converges to $t \wedge S_n$ as n tends to ∞: as a consequence, $(H^k \cdot \mathbf{M})$ converges to A_t in $\mathcal{S}(\mathbf{P})$ (see [6] for further details).

In the finite-dimensional case, it is well-known that, if $X \in \mathcal{M}^2 \oplus \mathcal{A}(\mathbf{P})$ with canonical decomposition $X = M + B$, and $H \cdot X$ also belongs to $\mathcal{M}^2 \oplus \mathcal{A}(\mathbf{P})$, then, necessarily, the canonical decomposition of the stochastic integral is $H \cdot X = H \cdot M + H \cdot B$. This is no longer the case, when $\mathbf{X}$ is a sequence of semimartingales, as the previous example shows: indeed, $\mathbf{X} \in \mathcal{M}^2 \oplus \mathcal{A}(\mathbf{P})$ and $\mathbf{X} = \mathbf{M} + \mathbf{0}$, the process $\mathbf{H}$ is $\mathbf{X}$-integrable, the integral $\mathbf{H} \cdot \mathbf{X}$ belongs to $\mathcal{M}^2 \oplus \mathcal{A}(\mathbf{P})$, but the canonical decomposition is $\mathbf{H} \cdot \mathbf{X} = 0 + A$. The reason for this different behaviour will be explained in Remark 10.

Ansel and Stricker [1] have proved that if M is a finite-dimensional local martingale and H an integrable process, such that $H \cdot M$ is bounded from below, then $H \cdot M$ is a local martingale (hence it is a supermartingale). Example 2 shows that, as opposed to the finite-dimensional case, this does not necessarily occur when $\mathbf{M}$ is a sequence of local martingales. Indeed, in the mentioned example, $\mathbf{H}$ is a $\mathbf{M}$-integrable process, such that $\mathbf{H} \cdot \mathbf{M}$ is bounded from below, but the stochastic integral $\mathbf{H} \cdot \mathbf{M}$ is not a local martingale.

However if, for a generalized integrand $\mathbf{H}$, an approximating sequence (H^n) of simple integrands can be found, such that $H^n \cdot \mathbf{M}$ converges to $\mathbf{H} \cdot \mathbf{M}$ and there also exists a random variable $W \in L^1(\mathbf{P})$ such that, for all t, $\int_0^t H_s^n \, d\mathbf{M}_s \geqslant W$, then it can be proved, using Fatou's lemma, that $\mathbf{H} \cdot \mathbf{M}$ is a supermartingale.

Moreover, we observe that, if every X^i is continuous, the result by Ansel and Stricker still holds even in the case of a sequence of martingales; indeed, the set of continuous local martingales is closed in $\mathcal{S}(\mathbf{P})$ (see [10], Theorem IV.5). This is also a consequence of the following remark: assume that $\mathbf{X}$ is continuous and $\mathbf{H}$ is an $\mathbf{X}$-integrable process such that $\mathbf{H} \cdot \mathbf{X} \geqslant -C$, where C is a positive constant. Then, for all $\varepsilon \geqslant 0$, there exists a sequence of simple integrands (H^n) such that $H^n \cdot \mathbf{X} \geqslant -C - \varepsilon$: indeed, given an approximating sequence H^n such that $H^n \cdot \mathbf{X}$ converges to $\mathbf{H} \cdot \mathbf{X}$, we can define the stopping time

$$T_n = \inf\left\{t : \int_0^t H_s^n \, d\mathbf{X}_s < -C - \varepsilon\right\}.$$

Since $\lim_n \mathbf{P}(T_n < T) = 0$, we can find a subsequence (which we still denote by T_n) such that $\sum_n \mathbf{P}(T_n < T) < \infty$. We set $S_n = \inf_{m \geqslant n} T_m$ and $\tilde{H}^n =$

$\mathbf{1}_{[0,S_n]}H^n$. Then, clearly $\tilde{H}^n \cdot \mathbf{X} \geqslant -C - \varepsilon$. Moreover $\tilde{H}^n \cdot \mathbf{X} = H^n \cdot \mathbf{X}$ on the set $[0, S_n]$ and S_n converges to T, $\mathbf{P}$-a.s.; hence $\tilde{H}^n \cdot \mathbf{X}$ converges to $\mathbf{H} \cdot \mathbf{X}$.

Remark 10. When X is a $\mathbb{R}^k$-valued semimartingale, a predictable process H, with values in $\mathbb{R}^k$, is X-integrable if the sequence $H^n = H\mathbf{1}_{\{\|H\|\leqslant n\}}$ is such that $(H^n \cdot X)$ is a Cauchy sequence in $\mathcal{S}(\mathbf{P})$ (see [2]). Moreover, if S is a stopping time, and denoting, as usual, by $\Delta_s X$ the jump $X_s - X_{s-}$, we have that $\Delta_S(H \cdot X) = H_S \Delta_S X$; for fixed t, the sequence $\sum_{s \leqslant t} \Delta_s (H^n \cdot X)^2$ is increasing and converges to $\sum_{s \leqslant t} \Delta_s (H \cdot X)^2$. Hence,

$$\mathbb{E}\left[\sum_{s \leqslant t} \Delta_s (H \cdot X)^2\right] = \lim_{n \to \infty} \mathbb{E}\left[\sum_{s \leqslant t} \Delta_s (H^n \cdot X)^2\right], \tag{11}$$

whether the expectation is finite or not.

Given a sequence of semimartingales $\mathbf{X}$, a generalized integrand $\mathbf{H}$ and an approximating sequence H^n, for all stopping times S, one still has $\Delta_S(H^n \cdot \mathbf{X}) = H_S^n \cdot (\Delta_S \mathbf{X})$; this sequence converges in probability to $\Delta_S(\mathbf{H} \cdot \mathbf{X})$: thus, $\Delta_S \mathbf{X}$ belongs to $\mathcal{D}(\mathbf{H}_S)$ and it is equal to $\mathbf{H}_S \cdot (\Delta_S \mathbf{X})$. However, in this case, (11) does not hold. The sequence $\sum_{s \leqslant t} \Delta_s (H^n \cdot \mathbf{X})^2$ does not necessarily converge: we can only say that

$$\mathbb{E}\left[\sum_{s \leqslant t} \Delta_s (\mathbf{H} \cdot \mathbf{X})^2\right] \leqslant \min \lim_{n \to \infty} \mathbb{E}\left[\sum_{s \leqslant t} \Delta_s (H^n \cdot \mathbf{X})^2\right].$$

In particular, in the finite-dimensional case, if M is a purely discontinuous local martingale such that $\mathbb{E}\left[\sum_{s \leqslant t} H_s^2 (\Delta_s M)^2\right] < \infty$, then $H \cdot M$ is a square integrable martingale. This is no longer true for the case of a sequence of local martingales, as we have seen in Example 2.

References

1. J.P. Ansel, C. Stricker: Couverture des actifs contingents et prix maximum. *Annales Inst. H. Poincaré*, 30, 303–315, 1994.
2. C.S. Chou, P.A. Meyer, C. Stricker: Sur les intégrales stochastiques de processus prévisibles non bornés. *Séminaire de Probabilités XIV*, Lecture Notes in Math., 784, 128–139, Springer, 1980.
3. M. De Donno: A note on completeness in large financial markets. *Mathematical Finance*, 14, 295–315, 2004.
4. C. Dellacherie: Quelques applications du lemme de Borel–Cantelli à la théorie des semimartingales. *Séminaire de Probabilités XII*, Lecture Notes in Math., 649, 742–745, Springer, 1978.
5. N. Dunford, J.T. Schwartz: *Linear operators*, 1, Interscience Publishers, Wiley, 1988
6. M. Émery : Une topologie sur l'espace des semi-martingales. *Séminaire de Probabilités XIII*, Lecture Notes in Math., 721, 260–280, Springer, 1979.

7. M. Émery: Compensation de processus à variation finie non localement intégrables. *Séminaire de Probabilités XIV*, Lecture Notes in Math., 784, 152–160, Springer, 1980.

8. J. Jacod: *Calcul stochastique et problèmes de martingales*, Lecture Notes in Math., 714, Springer, 1979.

9. J. Jacod, A.N. Shiryaev: *Limit Theorems for Stochastic Processes*, Springer, 1987.

10. J. Mémin: Espace de semi-martingales et changement de probabilité.   *Z. Wahrscheinlichkeitstheorie verw. Gebiete*, 52, 9–39, 1980.

11. M. Métivier: *Semimartingales (a Course on Stochastic Processes)*, de Gruyter, Berlin New York, 1982.

12. P.A. Meyer: Un cours sur les intégrales stochastiques. *Séminaire de Probabilités X*, Lecture Notes in Math., 511, 246–400, Springer, 1976.

13. P.A. Meyer: Sur un théorème de C. Stricker. *Séminaire de Probabilités XI*, Lecture Notes in Math., 581, 428–489, Springer, 1977.

14. P.A. Meyer: Inégalités de normes pour les intégrales stochastiques. *Séminaire de Probabilités XII*, Lecture Notes in Math., 649, 757–762, Springer, 1978.

15. R. Mikulevicius, B.L. Rozovskii: Normalized stochastic integrals in topological vector spaces. *Séminaire de Probabilités XXXII*, Lecture Notes in Math., 1686, Springer, 137–165, Springer, 1998.

16. R. Mikulevicius, B.L. Rozovskii: Martingale problems for stochastic PDE's. *Stochastic Partial Differential Equations: Six Perspectives*, (Carmona R., Rozovskii B. Editors), *Math. Surveys and Monographs*, 64, 243–326, Amer. Math. Soc., 1999.

On Almost Sure Convergence Results in Stochastic Calculus

Rajeeva L. Karandikar

Indian Statistical Institute, 7 SJS Sansanwal Marg, New Delhi 110016, India
e-mail: rlk@isid.ac.in

Professor P. A. Meyer was a source of inspiration for me. I have learnt a lot from him, via correspondence (in the pre-email era) during the days that I was a doctoral student and subsequently from discussions on the three occasions that I met him in the eighties. I would like to pay my respects to him by dedicating this survey article to his memory.

As I learnt later, Professor Meyer was a referee for my first published paper and also was a referee for my doctoral thesis submitted to the Indian Statistical Institute. I sent my second paper to him and he wrote that it has nice ideas, but I had not written it well. Then he rewrote the paper himself telling me that instead of writing 10 pages of comments on how to write the paper, he is rewriting it! It is this version that appeared in Strasbourg seminar (Karandikar [4]). The same is the story with another paper (Karandikar [5]). I wish Professor Meyer had put his name as a co-author on both these papers. *Then I would have had the honour of having joint papers with him.*

1 Almost sure Convergence of Stochastic integrals

Throughout this article, $(\Omega, \mathcal{F}, P)$ will denote a complete probability space with a filtration $(\mathcal{F}_t)$. We will assume that $(\mathcal{F}_t)$ is right continuous and $\mathcal{F}_0$ contains all P null sets in $\mathcal{F}$. The terms adaped, predictable, semimartingale, martingale will all be with reference to this filtration. For a function ρ from $[0, \infty)$ into $\mathbb{R}^d$,

$$|\rho|_s^* := \sup_{t \leqslant s} |\rho_t|.$$

Usually, convergence results on stochastic integrals are given in terms of convergence in probability or convergence in L^2. Here is a result on almost sure convergence from Bichteler [1]

Theorem 1. *Let X be a semimartingale and let f^n, f be locally bounded predictable processes such that*

$$|f^n - f|_t^* \leqslant \epsilon_n \tag{1}$$

where ϵ_n is a sequence of numbers such that

$$\sum_{n=1}^{\infty} \epsilon_n^2 < \infty. \tag{2}$$

Let $Z_t^n = \int_0^t f^n dX$ and $Z_t = \int_0^t f dX$. Then

$$\left| Z^n - Z \right|_t^* \to 0 \ a.s. \ for \ all \ t < \infty. \tag{3}$$

Bichteler's proof used his development of stochastic integral as an integral with respect to a vector valued measure along with a deep result in functional analysis on factorisation of operators. It is easy to see that the result is true when the semimartingale X is indeed a Wiener Process. Then the inequality

$$\mathbb{E} \left(\sup_{0 \leqslant t \leqslant T} \left| \int_0^t g \, dX \right|^2 \right) \leqslant 4 \mathbb{E} \left(\int_0^T g^2 \, dt \right) \tag{4}$$

(valid since X is now a Wiener process) implies that

$$\mathbb{E} \left(|Z^n - Z|_T^{*2} \right) \leqslant 4T\epsilon_n^2 \tag{5}$$

and hence (2) implies that

$$\mathbb{E} \left(\sum_{n=0}^{\infty} |Z^n - Z|_T^{*2} \right) < \infty. \tag{6}$$

Now (3) follows from (6). The same argument is valid if X is any semimartingale for which

$$\mathbb{E} \left(\sup_{0 \leqslant t \leqslant T} \left| \int_0^t g \, dX \right|^2 \right) \leqslant C \mathbb{E} \left(\int_0^T g^2 \, dt \right) \tag{7}$$

for all predictable processes g, where C is a finite constant. Bichteler showed that given any semimartingale X, one can make a change of measure (to an equivalent probability measure, say Q) in such a way that the given semimartingale X satisfies (7) with respect to the measure Q, thus proving the almost sure convergence result.

In Karandikar [3, 4] it was observed that given any continuous semimartingale Y, we can get a (strict) random time change (σ_t) such that X defined by $X_t = Y_{\sigma_t}$ satisfies (7) (with respect to the filtration $(\mathcal{G}_s)$, where $\mathcal{G}_s = \mathcal{F}_{\sigma_s}$). For example, we can take $B_t = t + \langle M, M \rangle_t + |A|_t$ where $Y = M + A$ is the canonical decomposition of the semimartingale Y with M being the local martingale part and A being the part with bounded variation paths, $\langle M, M \rangle_t$ being the quadratic variation of M on $[0, t]$ and $|A|_t$ being the total variation of A on $[0, t]$. Now σ_t defined by

$$\sigma_t = \inf\{s \geqslant 0 : B_s \geqslant t\}$$

satisfies the requirement: namely the time changed semimartingale

$$Y_s = X_{\sigma_s}$$

satisfies (7). It is clear that given finitely many semimartingales, we can choose one time change σ such that all the semimartingales in the changed time scale satisfy (7). Thus it follows that the Theorem 1 is true if X is a vector or matrix valued continuous semimartingale and f^n, f are vector or matrix valued predictable processes with compatible dimensions.

It was observed in Karandikar [3, 4] that the techniques of changing the time scale (for continuous semimartingales) has an added advantage over changing the measure-namely existence and uniqueness of diffusion type Stochastic differential equations driven by a continuous semimartingale can be proven by first changing the time scale so that the driving semimartingale in the changed time scale Y satisfies (7) and then using the classical arguments-successive approximation and Gronwall's lemma-as in the case of SDE's driven by a Wiener process. As Meyer [10] has noted (see p. 110), this time change technique was used by Schwartz [11] subsequently. Using the time change technique, an almost sure convergence result for solutions of stochastic differential equations was proved in Karandikar [3, 4]. Also, almost sure versions of some results in Emery [2] for continuous semimartingales were proven in Karandikar [7] leading to almost sure convergence results on Multiplicative Stochastic Integrals.

It may be noted that the theory of stochastic integrals with respect to continuous semimartingales can be developed without any reference to predictable σ-field, predictable processes and other notions from the general theory of processes. Just following Ito, one can define stochastic integrals for integrands that are progressively measurable-using the time change technique (see Karandikar [5]).

I would like to quote what Professor Meyer wrote (in a letter to me) about this attempt-to develop stochastic calculus for continuous semimartingales without bringing in predictability and other notions from general theory of processes:

```
Your proofs teach us how to prove partial results without the
modern notions. Now when I was a child ( of course I wasn't
doing probability theory then !) probabilists were trying to
do probability theory <<without measure theory>>. When
I was a student, they tried to do stochastic processes
<<without sample functions>> . My claim is that the same
thing is happening now: you are trying to avoid predictable
fields,projections, decomposition of supermartingales,
because these results are now considered abstract, heavy,
complicated... but really they are just TRIVIAL, and the
future generation will consider them so.
```

Meyer drew our attention to *the stopped Doob's inequality* and the notion of *control process* of a semimartingale introduced by Métivier-Pellaumail and its connection with the time change technique in our work. The stopped Doob's inequality is (see Métivier [10]): Let M be a locally square integrable martingale and τ be a stop time. Then

$$\mathbb{E}|M|^{*2}_{\tau-} \leqslant 4\mathbb{E}\{[M, M]_{\tau-} + <M, M>_{\tau-}\}. \tag{8}$$

If X is a semimartingale with $X = M + A$, with M a locally square integrable martingale, A being a process with bounded variation paths and f is a predictable process such that

$$\int_0^t f_s^2 d\langle M, M\rangle_s < \infty,$$

then for all stopping times σ, one has (using (8))

$$\mathbb{E}\left(\sup_{0\leqslant t<\sigma}\left|\int_0^t f\, dM\right|^2\right) \leqslant 4\mathbb{E}\left(\int_0^{\sigma-} f^2\, d\langle M, M\rangle\right) \tag{9}$$

$$+4\mathbb{E}\left(\int_0^{\sigma-} f^2\, d[M, M]\right)$$

and also

$$\mathbb{E}\left(\sup_{0\leqslant t<\sigma}\left|\int_0^t f\, dA\right|^2\right) \leqslant \mathbb{E}\left(\left(\int_0^{\sigma-} |f|\, d|A|\right)^2\right) \tag{10}$$

The point in using the Métivier-Pellaumail inequality is that one has integral over $[0, \sigma)$ instead of $[0, \sigma]$ and given any M, A as above, one can get a sequence of stopping times σ_n such that for bounded functions f, the right hand sides of (9) and (10) are finite for $\sigma = \sigma_n$.

Métivier-Pellaumail combined the two estimates as follows: defining $V_t = 8(1 + \langle M, M\rangle_t + [M, M]_t + |A|_t)$, one has

$$\mathbb{E}\left(\sup_{0\leqslant t<\sigma}\left|\int_0^t f\, dX\right|^2\right) \leqslant 4\mathbb{E}\left(V_{\sigma-}\left(\int_0^{\sigma-} |f|^2\, dV\right)\right). \tag{11}$$

The process V defined above is called a control process of the semimartingale X. The factorisation theorem used by Bichteler, the time change technique used by Karandikar and the control process used by Métivier-Pellaumail all have one thing in common-they make every semimartingale amenable to L^2 theory.

One disadvantage of working with control process is that even if X is *small*, say in the Emery topology, V is not small. With this in mind, Karandikar [7] introduced the following:

An (adapted) increasing process U is said to be a *dominating process* for a semimartingale X if there exists a decomposition $X = M + A$, with M

a locally square integrable martingale, A a process with bounded variation paths such that

$$B_t = U_t - 2(< M, M >_t + [M, M]_t)^{1/2} - |A|_t \tag{12}$$

is an increasing process. It is easy to see (using (8)) that if U dominates a semimartingale X, then one has

$$\mathbb{E}|X|_{\tau-}^{*^2} \leqslant 2\mathbb{E}U_{\tau-}^2 \tag{13}$$

and for a locally bounded predictable process f,

$$\mathbb{E}\left(\sup_{0 \leqslant t < \sigma} \left| \int_0^t f \, dX \right|^2 \right) \leqslant \mathbb{E}\left(\theta_{\sigma-}^2(f, U)\right) \tag{14}$$

where

$$\theta_t(f, U) = \sqrt{2} \left(\left(\int_0^t |f|^2 dU^2 \right)^{\frac{1}{2}} + \int_0^t |f| dU \right). \tag{15}$$

It was shown in [7] that

(i) Let X, Y be semimartingales with dominating processes U, V respectively. Let $Z = X + Y$. Then Z admits a dominating process W such that

$$W_t \leqslant U_t + V_t \quad \forall \, t. \tag{16}$$

(ii) Let X be a semimartingale with a dominating process U and let f be a locally bounded predictable process. Let $Z_t = \int_0^t f \, dX$. Then Z admits a dominating process D such that

$$D_t \leqslant \theta_t(f, U) \quad \forall \, t. \tag{17}$$

(iii) Let X^n, $n \geqslant 1$ and X be semimartingales. Then X^n converges to X in the Emery topology if and only if the semimartingale $Y^n = X^n - X$ admits a dominating process U^n such that

$$|U^n|_t^* \to 0 \quad \text{in probability } \forall t. \tag{18}$$

For the purpose of proving almost sure convergence results, the following notions play an important role: for locally bounded processes f^n, f we say that $f^n \xrightarrow{o} f$ if

$$\sum_{n=1}^{\infty} |f^n - f|_t^{*^2} < \infty \quad \forall \, t \quad \text{a.s.} \tag{19}$$

and for semimartingales X^n, X, we say that $X^n \xrightarrow{*} X$ if for $n \geqslant 1$ the semimartingale $X^n - X$ admits a dominating process V^n such that $V^n \xrightarrow{o} 0$. Clearly,

$$X^n \xrightarrow{o} X \quad \text{implies} \quad X^n \to X \quad \text{almost surely} \tag{20}$$

As a consequence of (13), it follows that

$$X^n \xrightarrow{*} X \quad \text{implies} \quad X^n \xrightarrow{o} X \tag{21}$$

and hence

$$X^n \xrightarrow{*} X \quad \text{implies} \quad X^n \to X \quad \text{almost surely} \tag{22}$$

It is shown in Karandikar [7] that

$$f^n \xrightarrow{o} f, X^n \xrightarrow{*} X \quad \text{implies} \quad \int f^n dX^n \xrightarrow{*} \int f dX.$$

This result can be restated (without the notation $\xrightarrow{o}$, $\xrightarrow{*}$) as follows:

Theorem 2. *Let $\{X^n : n \geqslant 1\}$ and X be semimartingales admitting decompositions $X = M + A$, $X^n = M^n + A^n$ with M, M^n locally square integrable martingales, A, A^n processes with bounded variation paths such that*

$$\sum_{n=1}^{\infty} < M^n - M, M^n - M >_t \quad < \infty \quad \forall t < \infty, \tag{23}$$

$$\sum_{n=1}^{\infty} [M^n - M, M^n - M]_t \quad < \infty \quad \forall t < \infty \tag{24}$$

and

$$\sum_{n=1}^{\infty} |A^n - A|_t^2 \quad < \infty \quad \forall t < \infty \tag{25}$$

Let f^n, f be locally bounded predictable processes such that

$$\sum_{n=1}^{\infty} |f^n - f|_t^{*^2} \quad < \infty \quad \forall t < \infty. \tag{26}$$

Let $Z_t^n = \int_0^t f^n dX$ and $Z_t = \int_0^t f dX$. Then

$$|Z^n - Z|_t^* \to 0 \quad a.s. \text{ for all } t < \infty. \tag{27}$$

Indeed, the assumptions on X^n, X above are exactly the requirement that $X^n \xrightarrow{*} X$. The conclusion can also be strengthened to : $Z^n \xrightarrow{*} Z$.

2 Almost sure convergence of solutions to SDE

We will consider the SDE

$$Z_t = Y_t + \int_0^t G(Z)_s dX_s \tag{28}$$

where X is an $\mathbb{R}^d$-valued semimartingale, Y is an $\mathbb{R}^m$-valued r.c.l.l. process and where G maps $\mathcal{D}(m)$ (= the class of $\mathbb{R}^m$-valued adapted r.c.l.l. process) into $\mathcal{I}(m,d)$ (= the class of $\mathbb{R}^m \otimes \mathbb{R}^d$ valued locally bounded predictable processes). We assume that G satisfies the following Lipschitz type condition: there exists an adapted increasing r.c.l.l. process A such that

$$|G(Z^1)_t - G(Z^2)_t| \leqslant A_{t-}|Z^1 - Z^2|_{t-}^* \tag{29}$$

for all r.c.l.l. adapted processes Z^1, Z^2 and for all t. Here $|\cdot|$ denotes the Euclidean norm (on $\mathbb{R}^m$ or $\mathbb{R}^m \otimes \mathbb{R}^d$).

Existence and uniqueness of solutions to (28) under the condition (29) are well known (see Métivier [10]). We will now state a result on stability of solutions to (28) in the almost sure sense.

Theorem 3. *Let $\{X^n : n \geqslant 1\}$ and X be $\mathbb{R}^d$−valued semimartingales such that $X^n \xrightarrow{*} X$. Let $\{Y^n : n \geqslant 1\}$ and Y be $\mathbb{R}^m$−valued r.c.l.l. processes such that $Y^n \xrightarrow{o} Y$. Let G^n, G be functionals from $\mathcal{D}(m)$ into $\mathcal{I}(m,d)$ such that for an r.c.l.l. adapted process A, and for all r.c.l.l. adapted processes Z^1, Z^2,*

$$|G(Z^1)_t - G(Z^2)_t| \leqslant A_{t-}|Z^1 - Z^2|_{t-}^* \tag{30}$$

$$|G^n(Z^1)_t - G^n(Z^2)_t| \leqslant A_{t-}|Z^1 - Z^2|_{t-}^* \quad n \geqslant 1. \tag{31}$$

Let Z and Z^n be solutions to

$$Z_t^n = Y_t + \int_0^t G^n(Z^n)_s dX_s^n$$

$$Z_t = Y_t + \int_0^t G(Z)_s dX_s.$$

Suppose that
$$G^n(Z) \xrightarrow{o} G(Z). \tag{32}$$

Then $Z^n \xrightarrow{o} Z$ and as a consequence

$$\sup_{0 \leqslant s \leqslant t} |Z_s^n - Z_s| \to 0 \ a.s.$$

If further, $Y^n \xrightarrow{*} Y$, then $Z^n \xrightarrow{*} Z$.

It should be noted that the assumption (32) involves only Z and does not involve Z^n. Alternatively, the assumption (32) can be replaced by

$$G^n(Z^n) - G(Z^n) \xrightarrow{o} 0. \tag{33}$$

This result is given in Karandikar [7]. The convergence in probability version of this result appears in Emery [2].

Let us now suppose that the functional G is given by

$$G(Z)_t(\omega) = b(t-, U(\omega), Z(\omega))$$

where $b : [0, \infty) \times D([0, \infty), \mathbb{R}^k) \times D([0, \infty), \mathbb{R}^m) \to \mathbb{R}^m \otimes \mathbb{R}^d$ is a functional assumed to satisfy for $\rho, \rho_1, \rho_2 \in D([0, \infty), \mathbb{R}^m)$, $\phi, \phi^1, \phi^2 \in D([0, \infty), \mathbb{R}^k)$,

$$b(s, \phi^1, \rho) = b(s, \phi^2, \rho) \text{ if } |\phi^1 - \phi^2|_s^* = 0, \tag{34}$$

and

$$|b(s, \phi, \rho_1) - b(s, \phi, \rho_2)| \leqslant A_s(\phi)|\rho_1 - \rho_2|_s^*, \tag{35}$$

where again $A_s(\phi^1) = A_s(\phi^2)$ if $|\phi^1 - \phi^2|_s^* = 0$ and for each ϕ, $A_s(\phi)$ is an increasing function of s. In this case one can construct an approximation of the solution to the SDE

$$Z_t = Y_t + \int_0^t b(s-, U, Z)dX_s \tag{36}$$

that converges almost surely. (Here X is a $\mathbb{R}^d$ valued semimartingale and Y, U are adapted $\mathbb{R}^m$, $\mathbb{R}^k$ valued r.c.l.l. processes respectively.) This is achieved by appropriately modifying the Eular-Peano approximation scheme as given below:

Let stop times τ_i^n and processes $W^{n,i}$ be defined inductively by:

$$\tau_0^n = 0 \text{ and } W_t^{n,0} \equiv H_0 \tag{37}$$

and having defined $\tau_j^n, W^{n,j}$ for $j \leqslant i$, let

$$\tau_{i+1}^n = \inf\{t > \tau_i^n : |Y_t - Y_{\tau_i^n} + b(\tau_i^n, U, W^{n,i})(X_t - X_{\tau_i^n})| \geqslant 2^{-n}$$
$$\text{or } |b(t, U, W^{n,i}) - b(\tau_i^n, U, W^{n,i})| \geqslant 2^{-n}\}$$

and $W_t^{n,i+1} = W_t^{n,i}$ for $t < \tau_{i+1}^n$ and for $t \geqslant \tau_{i+1}^n$

$$W_t^{n,i+1} = W_{\tau_i^n}^{n,i} + Y_{\tau_{i+1}^n} - Y_{\tau_i^n} + b(\tau_i^n, U, W^{n,i})(X_{\tau_{i+1}^n} - X_{\tau_i^n})$$

Thus, $W^{n,i+1}$ is a process that has jumps at $\tau_1^n, ..., \tau_{i+1}^n$ and is constant on the intervals $[0, \tau_1^n), \cdots, [\tau_j^n, \tau_{j+1}^n), \cdots [\tau_i^n, \tau_{i+1}^n), [\tau_{i+1}^n, \infty)$. Let us piece together these processes $W^{n,i}, i = 1, 2, \ldots$ to define a step process H^n as follows.

$$H_t^n = W_{\tau_i^n}^{n,i} \text{ for } \tau_i^n \leqslant t < \tau_{i+1}^n.$$

Now define Z^n by $Z_0^n = Y_0$ and

$$Z_t^n = H_{\tau_i^n}^n + Y_t - Y_{\tau_i^n} + b(\tau_i^n, U, W^{n,i})(X_t - X_{\tau_i^n}) \text{ for } \tau_i^n < t \leqslant \tau_{i+1}^n.$$

The following result is proven in Karandikar [8].

Theorem 4. *The approximations constructed above, Z^n as well as H^n converge to the solution Z of (36) almost surely, i.e.*

$$\sup_{0 \leqslant s \leqslant t} |Z_s^n - Z_s| \to 0 \ a.s.$$

$$\sup_{0 \leqslant s \leqslant t} |H_s^n - Z_s| \to 0 \ a.s.$$

Moreover, $Z^n - Z \xrightarrow{*} 0$.

Pathwise formulae

The almost sure convergence results given above can be used to deduce pathwise formulae for stochastic integral and for solution to a stochastic differential equation (see Karandikar [9]). We first give a pathwise formula for the stochastic integral.

Let mappings $\mathcal{I}_n$ and $\mathcal{I}$ from $D([0, \infty, I\!\!R)) \times D([0, \infty, I\!\!R))$ into $D([0, \infty, I\!\!R))$ be defined as follows: Fix a sequence ϵ_n such that $\sum \epsilon_n^2 < \infty$. Fix $\rho, \eta \in D([0, \infty, I\!\!R))$. For $n \geqslant 1$, let $\{a_i^n : i \geqslant 1\}$ and $\mathcal{I}_n(\rho, \eta)$ be defined by, $a_0^n = 0$ and for $i \geqslant 0$

$$a_{i+1}^n = \inf\{t \geqslant a_i^n : |\rho(t) - \rho(a_i)| \geqslant \epsilon_n\}$$

$$\mathcal{I}_n(\rho, \eta)(t) = \rho(0)\eta(0) + \sum_{i=0}^{\infty} \rho(a_i^n)\big(\eta(a_{i+1}^n \wedge t) - \eta(a_i^n \wedge t)\big).$$

(Note that the sum above is a finite sum for fixed ρ, η and $t < \infty$.) Then define $\mathcal{I}$ as

$$\mathcal{I}(\rho, \eta) = \lim \mathcal{I}_n(\rho, \eta)$$

if the limit exists in the topology of uniform convergence on compact subsets of $[0, \infty)$, otherwise define $\mathcal{I}(\rho, \eta) \equiv 0$. $\mathcal{I}$ gives a pathwise integration formula in the following sense:

Theorem 5. *Let X be a semimartingale on a complete probability space $(\Omega^1, \mathcal{F}^1, P^1)$, with respect to a filtration $(\mathcal{G}_t)$ satisfying the usual conditions. Let Y be an $(\mathcal{G}_t)$ adapted r.c.l.l. process. Then*

$$Z_t(\omega) = \mathcal{I}(Y(\omega)., X(\omega).)(t)$$

is a version of the stochastic integral $\int_0^t Y_{s-} dX_s$.

This pathwise integration formula shows that when Y is r.c.l.l., the stochastic integral $\int_0^t Y_{s-} dX_s$ does not depend upon the underlying filtration nor upon the underlying probability measure P.

The modified Euler-Peano scheme given above can be recast to yield a pathwise formula for solution to the SDE (36). Let b be a functional satisfying (34) and (35). Functionals $\mathcal{J}^n$ and $\mathcal{J}$ from $D([0, \infty), I\!\!R^m) \times$

$D([0,\infty),I\!\!R^k) \times D([0,\infty),I\!\!R^d)$ into $D([0,\infty),I\!\!R^m)$ are defined as follows. Fix $\psi \in D([0,\infty),I\!\!R^m)$, $\phi \in D([0,\infty),I\!\!R^k)$ and $\eta \in D([0,\infty),I\!\!R^d)$. Let a_i^n and $\theta^{n,i}$ are defined by

$$a_0^n = 0 \text{ and } \theta_t^{n,0} \equiv \psi_0 \tag{38}$$

and having defined $a_j^n, \theta^{n,j}$ for $j \leqslant i$, let

$$a_{i+1}^n = \inf\{t > a_i^n : |\psi_t - \psi_{a_i^n} + b(a_i^n,\phi,\theta^{n,i})(X_t - X_{a_i^n})| \geqslant 2^{-n}$$
$$\text{or } |b(t,\phi,\theta^{n,i}) - b(a_i^n,\phi,\theta^{n,i})| \geqslant 2^{-n}\}$$

$$\theta_t^{n,i+1} = \theta_t^{n,i} \text{ for } t < a_{i+1}^n$$
$$= \theta_{a_i^n}^{n,i} + \psi_{a_{i+1}^n} - \psi_{a_i^n} + b(a_i^n,\phi,\theta^{n,i})(\eta_{a_{i+1}^n} - \eta_{a_i^n}) \text{ for } t \geqslant a_{i+1}^n.$$

Thus, $\theta^{n,i+1}$ is a function that has jumps at $a_1^n, ..., a_{i+1}^n$ and is constant on the intervals $[0,a_1^n), \cdots, [a_j^n, a_{j+1}^n), \cdots [a_i^n, a_{i+1}^n), [a_{i+1}^n,\infty)$. Now define ρ^n by $\rho_0^n = \psi_0$ and

$$\rho_t^n = \theta_{a_i^n}^{n,i} + \psi_t - \psi_{a_i^n} + b(a_i^n,\phi,\theta^{n,i})(\eta_t - \eta_{a_i^n}) \text{ for } a_i^n < t \leqslant a_{i+1}^n. \tag{39}$$

Now define $\mathcal{J}^n(\psi,\phi,\eta) = \rho^n$ and

$$\mathcal{J}(\psi,\phi,\eta) = \lim \mathcal{J}^n(\psi,\phi,\eta)$$

if $\mathcal{J}^n(\psi,\phi,\eta)$ converges uniformly on compacts, and equal to zero otherwise. As a consequence of Theorem 4, we get the following pathwise formula for solutions to the SDE (36).

Theorem 6. *Let X be a semimartingale on a complete probability space $(\Omega^1,\mathcal{F}^1,P^1)$, with respect to a filtration $(\mathcal{G}_t)$ satisfying the usual conditions. Let Y,U be $(\mathcal{G}_t)$ adapted r.c.l.l. processes. Let b satisfy (34) and (35). Then*

$$Z = \mathcal{J}(Y,U,X)$$

is the unique solution to the SDE

$$Z_t = Y_t + \int_0^t b(s-,U,Z)dX_s.$$

References

1. K. Bichteler. Stochastic integration and L^p-theory of Stochastic integration. Ann. Prob., 9, 1981, 48–89.
2. M. Emery. Une topologie sur l'espace des semimartingales. Séminaire de Probabilités XIII, Lecture Notes in Mathematics 721, p. 260–280, Springer-Verlag, Berlin (1979).
3. R.L. Karandikar. Pathwise solution of stochastic differential equations. Sankhya A, 43, 1981, 121–132.

4. R.L. Karandikar. A.s. approximation results for multiplicative stochastic integrals. Séminaire de Probabilités XVI, Lecture Notes in Mathematics 920, p. 384–391, Springer-Verlag, Berlin (1982).
5. R.L. Karandikar. Stochastic integration w.r.t. continuous local martingales. Stochastic processes and their applications 15 203–209 (1983).
6. R.L. Karandikar. Girsanov type formula for a Lie group valued Brownian motion. Séminaire de Probabilités XVII, Lecture Notes in Mathematics 986, p. 198–204, Springer-Verlag, Berlin (1983).
7. R.L. Karandikar. On Métivier-Pellaumail inequality, Emery topology and Pathwise formulae in Stochastic calculus. Sankhya A, 51, 1989, 121–143.
8. R.L. Karandikar. On a.s. convergence of modified Euler-Peano approximations to the solution of a stochastic differential equation. Séminaire de Probabilités XVII, Lecture Notes in Mathematics 1485, p. 113–120, Springer-Verlag, Berlin (1991).
9. R.L. Karandikar. On pathwise stochastic integration, Stochastic processes and their applications 57, 1995, 11–18.
10. M. Métivier. Semimartingales, Walter de Gruyter, Berlin, New York. (1982).
11. P.A. Meyer. Sur la méthode de L. Schwartz pour les E.D.S. Séminaire de Probabilités XVII, Lecture Notes in Mathematics 1485, p. 108–112, Springer-Verlag, Berlin (1991).
12. L. Schwartz. La convergence de la série de Picard pour les E.D.S. Séminaire de Probabilités XXIII, Lecture Notes in Mathematics 1372, p. 343–354, Springer-Verlag, Berlin (1989).

On a Condition that One-Dimensional Diffusion Processes are Martingales

Shinichi Kotani

Department of Mathematics, Osaka University, Machikaneyamachou 1-1,
Toyonaka, Osaka Japan 560-0043
e-mail: kotani@math.sci.osaka-u.ac.jp

1 Introduction

It is surprising that there are examples of local martingales $\{X_t\}$ which are not martingales in spite of the existence of all moments ([6]). Recently there have been several works linking the tail probability of the quadratic variation $\{\langle X \rangle_t\}$ to that of the maximum process $\{\sup_{s \leqslant t} X_s\}$ ([2], [3], [7]). In this note we give a necessary and sufficient condition for one-dimensional diffusion processes to be martingales.

On an interval (l_-, l_+) with $-\infty \leqslant l_- < l_+ \leqslant +\infty$, let $m(\mathrm{d}x)$ be a non-negative measure satisfying $m(I) > 0$ for any non-empty open interval $I \subset (l_-, l_+)$. Denote the minimal diffusion process with speed measure $m(\mathrm{d}x)$ and scale function $s(x) = x$ by

$$\left\{ \{X_t\}_{t \geqslant 0}, \{\mathbb{P}_x\}_{x \in (l_-, l_+)} \right\}.$$

Let

$$\tau_a = \inf \{t \geqslant 0, X_t = a\}, \text{ for } a \in (l_-, l_+),$$

and define

$$\tau_+ = \lim_{a \uparrow l_+} \tau_a, \quad \tau_- = \lim_{a \downarrow l_-} \tau_a.$$

The equivalences

$$
\begin{cases}
\mathbb{P}_r(\tau_+ = \infty) = 1 \iff \displaystyle\int_{[r,l_+)} m([r,x))\, \mathrm{d}x = \infty, \\[2ex]
\mathbb{P}_r(\tau_- = \infty) = 1 \iff \displaystyle\int_{(l_-,r]} m((x,r])\, \mathrm{d}x = \infty
\end{cases}
\tag{1}
$$

are known ([5]). Now our problem is to find a condition under which the continuous local martingale $\{X_{t \wedge \tau_+ \wedge \tau_-}\}_{t \geqslant 0}$ becomes a martingale.

For convenience, we cite an instructive calculation which was employed in [4]. Suppose $(l_-, l_+) = \mathbb{R}$ and $m(\mathbb{R}) = 1$. Then the diffusion process $\{X_t\}$ starting with the initial distribution $m(\mathrm{d}x)$ becomes a stationary process. However if

$$\mathbb{E}\,|X_t| = \int_{\mathbb{R}} |x|\, m(\mathrm{d}x) < \infty ,$$

then $\{X_t\}$ cannot be a martingale, for, otherwise, $X_t \to X \in \mathbb{R}$ as $t \to \infty$, which contradicts the stationarity. This example shows that for any increasing function ϕ we can create a strict local martingale satisfying $\mathbb{E}\,\phi(X_t) < \infty$.

For a probability measure μ on (l_-, l_+), set $\mathbb{P}_\mu \equiv \int_{(l_-, l_+)} \mu(\mathrm{d}x)\,\mathbb{P}_x$. Now our theorem is

Theorem 1. *Suppose $\int_{(l_-, l_+)} |x|\, \mu(\mathrm{d}x) < \infty$. Then $\mathbb{E}_\mu\,\big|X_{t \wedge \tau_+ \wedge \tau_-}\big| < \infty$ for any $t \geqslant 0$, and $\big\{X_{t \wedge \tau_+ \wedge \tau_-}\big\}$ is a martingale under $\mathbb{P}_\mu$ if and only if one of the following conditions holds.*

(i) Both l_+ and l_- are finite.
(ii) If $-\infty < l_- < l_+ = \infty$, then

$$\int_{[r, \infty)} x\, m(\mathrm{d}x) = \infty .$$

(iii) If $-\infty = l_- < l_+ < \infty$, then

$$\int_{(-\infty, r]} |x|\, m(\mathrm{d}x) = \infty .$$

(iv) If $l_+ = \infty$ and $l_- = -\infty$, then

$$\int_{[r, \infty)} x\, m(\mathrm{d}x) = \infty \quad and \quad \int_{(-\infty, r]} |x|\, m(\mathrm{d}x) = \infty .$$

Conversely, if one of these conditions holds, then for any probability measure μ on (l_-, l_+) satisfying $\int_{(l_-, l_+)} |x|\, \mu(\mathrm{d}x) < \infty$, $\{X_t\}$ becomes a martingale under $\mathbb{P}_\mu$.

Example 1. We can apply this theorem to check the martingale property of Bessel processes. The Bessel process with index $d \in \mathbb{R}$ is defined through the SDE

$$\mathrm{d}r_t = \mathrm{d}B_t + \frac{d-1}{2} r_t^{-1}\, \mathrm{d}t, \quad r_0 > 0 .$$

For $d \neq 2$, set $\alpha = \dfrac{d-1}{d-2}$. Then $X_t = r_{(\alpha-1)^2 t}^{-\frac{1}{\alpha-1}}$ satisfies

$$\mathrm{d}X_t = X_t^\alpha\, \mathrm{d}\tilde{B}_t ,$$

with another Brownian motion $\tilde{B}_t$. In this case $l_- = 0$, $l_+ = \infty$ and $m(\mathrm{d}x) = 2x^{-2\alpha}\, \mathrm{d}x$, therefore condition (ii) is satisfied if $\alpha < 1$ and $\mathbb{P}_x(\tau_0 < \infty) = 1$.

Hence $\{X_{t \wedge \tau_0}\}$ is a martingale iff $\alpha < 1$, namely $2 > d$. If $d = 2$, set $X_t = \log r_t$. Then

$$\mathrm{d}X_t = e^{-X_t}\,\mathrm{d}B_t \;,$$

hence $l_- = -\infty$, $l_+ = \infty$ and $m(\mathrm{d}x) = 2e^{2x}\,\mathrm{d}x$. In this case, condition (1) is satisfied; however condition *(iv)* is not, therefore $\{X_t\}$ cannot be a martingale.

2 Proof

If $-\infty < l_- < l_+ < +\infty$, then $\{X_{t \wedge \tau_+ \wedge \tau_-}\}_{t \geqslant 0}$ is bounded, hence it is trivially a martingale. Therefore we assume that one of l_-, l_+ is infinite, for instance l_+ is $+\infty$. Then from (1) we see that $\mathbb{P}_x(\tau_+ = \infty) = 1$ for any $x \in (l_-, +\infty)$. We prepare two lemmas originating from [1]. Denote the generator of the process by L, that is,

$$Lf(x) = \frac{\mathrm{d}^2 f}{m(\mathrm{d}x)\,\mathrm{d}x} \;.$$

Lemma 1. *There exist positive constants c_1 and c_2 such that for any $x \in (l_-, +\infty)$ and $t \geqslant 0$,*

$$\mathbb{E}_x \left| X_{t \wedge \tau_-} \right| \leqslant c_1 + |x| + c_2 t \;.$$

Proof. Let $f(x)$ be a continuous function satisfying

$$Lf(x) \leqslant c \text{ and } f(x) = |x| \text{ for } |x| \text{ sufficiently large,}$$

with some constant c. Then

$$
\begin{aligned}
f(X_{t \wedge \tau_-}) &= f(x) + \int_0^{t \wedge \tau_-} Lf(X_s)\,\mathrm{d}s + M_t \\
&\leqslant f(x) + c(t \wedge \tau_-) + M_t \;,
\end{aligned}
$$

with a local martingale M_t. Let $[a, b]$ be an interval containing x and τ be the first exiting time from $[a, b]$ of the process. Then we have

$$\mathbb{E}_x f(X_{t \wedge \tau}) \leqslant f(x) + c\,\mathbb{E}_x(t \wedge \tau) \;.$$

Letting $a \to l_-$ and $b \to \infty$, we see from Fatou's lemma that

$$\mathbb{E}_x f(X_{t \wedge \tau_-}) \leqslant f(x) + ct \;,$$

which concludes the lemma. □

Lemma 2. *If $\{X_{t \wedge \tau_-}\}$ is a martingale with respect to $\mathbb{P}_\mu$ for a probability measure μ on $(l_-, +\infty)$, then for every $t \geqslant 0$ and $x \in (l_-, +\infty)$ we have*

$$\mathbb{E}_x X_{t \wedge \tau_-} = x \;. \tag{2}$$

Conversely, if (2) is valid, then $\{X_{t \wedge \tau_-}\}$ is a martingale with respect to $\mathbb{P}_\mu$ for any probability measure μ on $(l_-, +\infty)$ satisfying $\int_{l_-}^{\infty} |x|\,\mu(\mathrm{d}x) < \infty$.

Proof. Let τ be the first hitting time of $a \in (l_-, +\infty)$. Suppose $\{X_{t \wedge \tau_-}\}$ is a martingale with respect to $\mathbb{P}_\mu$. Then for $n = 1, 2, \cdots$, we have

$$\mathbb{E}_\mu \left(X_{(t+\tau \wedge n) \wedge \tau_-} | \mathcal{F}_{\tau \wedge n} \right) = X_{\tau \wedge n \wedge \tau_-} .$$

On the other hand, from the strong Markov property it follows that

$$\mathbb{E}_\mu \left(X_{(t+\tau \wedge n) \wedge \tau_-} | \mathcal{F}_{\tau \wedge n} \right) = \mathbb{E}_{X_{\tau \wedge n \wedge \tau_-}} \left(X_{t \wedge \tau_-} \right) .$$

On the event $A = \{\tau < \tau_-\}$, as $n \to \infty$,

$$X_{\tau \wedge n \wedge \tau_-} \to X_{\tau \wedge \tau_-} = a ,$$

$$\mathbb{E}_{X_{\tau \wedge n \wedge \tau_-}} \left(X_{t \wedge \tau_-} \right) \to \mathbb{E}_{X_{\tau \wedge \tau_-}} \left(X_{t \wedge \tau_-} \right) = \mathbb{E}_a \left(X_{t \wedge \tau_-} \right) ;$$

therefore, if $\mathbb{P}_\mu(A) > 0$, then we have $\mathbb{E}_a \left(X_{t \wedge \tau_-} \right) = a$. However, since

$$\mathbb{P}_x(A) = \begin{cases} 1 & \text{if } x \geqslant a \\ \dfrac{x - l_-}{a - l_-} & \text{if } x < a \end{cases}$$

for every $x \in (l_-, +\infty)$, it follows that $\mathbb{P}_\mu(A) > 0$. The rest of the proof is clear from the Markov property. $\square$

The following lemmas are crucial in the proof of the theorem.

Lemma 3. *For $\lambda > 0$, there exists a positive solution f_+ of $Lf = \lambda f$ which is monotonically decreasing on $(l_-, +\infty)$. Let $\alpha_+ = \lim_{x \to \infty} f_+(x)$. Then*

$$\alpha_+ = 0 \Longleftrightarrow \int_a^\infty y\, m(dy) = \infty \Longleftrightarrow \lambda \int_x^\infty (y - x)\, f_+(y)\, m(dy) = f_+(x) .$$

On the other hand, suppose $l_- > -\infty$. Then there exists a positive solution f_- of $Lf = \lambda f$ which is increasing on $(l_-, +\infty)$ and satisfies $f_-(l_-) = 0$. Any other such solution is a constant multiple of f_-.

Proof. The existence of a positive monotonically decreasing or increasing solution of $Lf = \lambda f$ is known. Let f_+ be a positive monotonically decreasing solution. Since $f'_+(x) \leqslant 0$ and f'_+ is nondecreasing, we have

$$f_+(x) = \alpha_+ - \int_x^\infty f'_+(y)\, dy ,$$

which, in particular, implies that $f'_+(x)$ is integrable on $[a, \infty)$. This combined with the fact that f'_+ is nondecreasing shows $\lim_{x \to \infty} f'_+(x) = 0$, hence, recalling that $df'_+(x) = \lambda f_+(x)\, m(dx)$, we see

$$f_+(x) = \alpha_+ - \int_x^\infty f_+'(y)\, dy$$

$$= \alpha_+ + \lambda \int_x^\infty dy \int_y^\infty f_+(z)\, m(dz)$$

$$= \alpha_+ + \lambda \int_x^\infty (y - x)\, f_+(y)\, m(dy)\,. \tag{3}$$

However, noting that $f_+(x) \geqslant \alpha_+$ for every $x \geqslant a$, if $\alpha_+ > 0$, from (3) we must have $\int_a^\infty y\, m(dy) < \infty$. On the other hand, suppose $\int_a^\infty y\, m(dy) < \infty$. Construct a solution by solving the integral equation

$$\varphi(x) = 1 + \lambda \int_x^\infty (y - x)\, \varphi(y)\, m(dy)\,.$$

This equation is solvable by iteration because $\int_a^\infty y\, m(dy) < \infty$. Clearly φ is positive, decreasing and the derivative goes to 0, therefore the Wronskian vanishes:

$$\varphi'(x) f_+(x) - \varphi(x) f_+'(x) = 0\,,$$

which implies that there exists a positive constant c such that $f_+(x) = c\varphi(x)$. Since $\varphi(\infty) = 1$, we see that $\alpha_+ > 0$.

On the other hand, let f_- be any positive increasing solution of $Lf = \lambda f$. Suppose $l_- > -\infty$. Since we have

$$f_-(x) = f_-(l_-) + f_-'(l_-)(x - l_-) + \lambda \int_{l_-}^x (x - y)\, f_-(y)\, m(dy)\,,$$

if $f_-(l_-) > 0$, then

$$\int_{l_-}^a m(dy) < \infty$$

holds. However in this case we can redefine f_- by solving

$$f_-(x) = x - l_- + \lambda \int_{l_-}^x (x - y)\, f_-(y)\, m(dy)\,.$$

The argument using the Wronskian shows that any other such solution is a constant multiple of f_+ or f_-. □

Let τ_a be the first hitting time of $a \in (l_-, +\infty)$. Using the same notations as in Lemma 3, we have

Lemma 4. $\mathbb{E}_x e^{-\lambda \tau_a} = \dfrac{f_+(x)}{f_+(a)}$ *for* $x > a$ *and* $\mathbb{E}_x(e^{-\lambda \tau_b}, \tau_b < \tau_-) = \dfrac{f_-(x)}{f_-(b)}$ *if* $b > x > l_- > -\infty$.

Proof. Let $\tau = \tau_a \wedge \tau_b$. Since $f_+(X_t)e^{-\lambda t} - f_+(x) - \int_0^t e^{-\lambda s}(L-\lambda) f_+(X_s)\,ds$ is a local martingale, the optional stopping theorem shows

$$f_+(x) = \mathbb{E}_x \left\{ f_+(X_\tau)e^{-\lambda\tau} \right\}$$
$$= \mathbb{E}_x \left\{ e^{-\lambda\tau_a}, \tau_a < \tau_b \right\} f_+(a) + \mathbb{E}_x \left\{ e^{-\lambda\tau_b}, \tau_b < \tau_a \right\} f_+(b) .$$

However, (1) implies $\tau_b \to \infty$, as $b \to \infty$; hence, observing that $f_+(b) \to f_+(l_+) < \infty$, we easily get

$$f_+(x) = f_+(a)\mathbb{E}_x e^{-\lambda\tau_a} .$$

The equality for f_- can be derived in the same manner. □

Proof of the theorem. Let $-\infty \leqslant l_- < a < l_+ = \infty$. First we assume that $\{X_{t \wedge \tau_-}\}$ is a martingale with respect to $\mathbb{P}_\mu$. Then Lemma 2 implies that $\{X_{t \wedge \tau_-}\}$ is a martingale with respect to $\mathbb{P}_x$ for every $x \in (l_-, +\infty)$. Hence the optional stopping theorem shows

$$\mathbb{E}_x X_{t \wedge \tau_- \wedge \tau_a} = x \text{ for any } t \geqslant 0 \text{ and } x \in (l_-, +\infty) .$$

Thus

$$\int_0^\infty e^{-\lambda t} \left(\mathbb{E}_x X_{t \wedge \tau_- \wedge \tau_a} \right) dt = \lambda^{-1} x .$$

To compute the left hand side, we introduce the other positive solution ψ of $Lf = \lambda f$ on (a, ∞) such that $\psi'(a) = 1$, $\psi(a) = 0$. Then it is known that for $x \geqslant a$

$$\int_0^\infty e^{-\lambda t} \mathbb{E}_x f(X_{t \wedge \tau_a})\, dt$$
$$= \mathbb{E}_x \left(\int_0^{\tau_a} e^{-\lambda t} f(X_t)\, dt \right) + \frac{1}{\lambda} f(a) \mathbb{E}_x e^{-\lambda\tau_a}$$
$$= \frac{\psi(x)}{f_+(a)} \int_x^\infty f_+(y)\, f(y)\, m(dy) + \frac{f_+(x)}{f_+(a)} \int_a^x \psi(y)\, f(y)\, m(dy) + \frac{f_+(x)}{\lambda f_+(a)} f(a).$$

Set $f(x) = x - a$. Observing that $\tau_- > \tau_a$ if $x \geqslant a$, dividing both sides by $(x - a)$, and letting $x \downarrow a$, we see that

$$\lambda \int_a^\infty f_+(y)\,(y - a)\, m(dy) = f_+(a) .$$

From Lemma 3 we conclude $\int_a^\infty y\, m(dy) = \infty$. If $l_- = -\infty$, then similarly we have $\int_{-\infty}^a |y|\, m(dy) = \infty$.

Conversely suppose $l_- = -\infty$ and $\int_0^\infty y\, m(dy) = \infty$, $\int_{-\infty}^0 |y|\, m(dy) = \infty$. To compute $\int_0^\infty e^{-\lambda t} (\mathbb{E}_x X_t)\, dt$, we introduce the positive increasing solution f_- of $Lf = \lambda f$. Then

$$\int_0^\infty e^{-\lambda t}\,(\mathbb{E}_x X_t)\,\mathrm{d}t$$

$$= h^{-1}\left\{f_-(x)\int_x^\infty f_+(y)\,y\,m(\mathrm{d}y) + f_+(x)\int_{-\infty}^x f_-(y)\,y\,m(\mathrm{d}y)\right\},$$

where h is the Wronskian of $\{f_+, f_-\}$. However, in this case, Lemma 3 implies

$$\lambda\int_x^\infty f_+(y)\,y\,m(\mathrm{d}y) = f_+(x) + \lambda x\int_x^\infty f_+(y)\,m(\mathrm{d}y)$$

$$= f_+(x) - x f_+'(x)\,,$$

$$-\lambda\int_{-\infty}^x f_-(y)\,y\,m(\mathrm{d}y) = f_-(x) - \lambda x\int_{-\infty}^x f_+(y)\,m(\mathrm{d}y)$$

$$= f_-(x) - x f_-'(x)\,,$$

hence

$$\int_0^\infty e^{-\lambda t}\,(\mathbb{E}_x X_t)\,\mathrm{d}t = (\lambda h)^{-1}\left\{-x f_-(x) f_+'(x) + x f_+(x) f_-'(x)\right\}$$

$$= \lambda^{-1} x\,,$$

for all $\lambda > 0$ and $x \in \mathbb{R}$. By uniqueness of the Laplace transform, $\mathbb{E}_x X_t = x$, from which Lemma 2 implies the theorem in the case when $l_\pm = \pm\infty$. Now suppose $l_- > -\infty$. Then

$$\int_0^\infty e^{-\lambda t}\mathbb{E}_x f\left(X_{t\wedge\tau_-}\right)\mathrm{d}t$$

$$= h^{-1}f_+(x)\int_{l_-}^x f_-(y)\,f(y)\,m(\mathrm{d}y) + h^{-1}f_-(x)\int_x^\infty f_+(y)\,f(y)\,m(\mathrm{d}y)$$

if $f(l_-) = 0$, where h is the Wronskian of $\{f_+, f_-\}$. Setting $f(x) = x - l_-$, we see

$$\int_0^\infty e^{-\lambda t}\mathbb{E}_x X_{t\wedge\tau_-}\,\mathrm{d}t - \frac{l_-}{\lambda} = \frac{x - l_-}{\lambda}\,,$$

which completes the proof. $\square$

References

1. K. Akamatsu: *Conditions for local martingales to be martingales.* Master Thesis, Osaka University, 2002.
2. K.D. Elworthy, X.M. Li, and M. Yor: *On the tails of the supremum and the quadratic variation of strictly local martingales.* Séminaire de Probabilités XXXI, Springer Lecture Notes in Math. 1655, 113–125, 1997.
3. K.D. Elworthy, X.M. Li, and M. Yor: *The importance of strictly local martingales; applications to radial Ornstein-Uhlenbeck processes.* Probab. Theory Related Fields **115**, 325–355, 1999.

4. Hu, Feng-Rung: *On Markov chains induced from stock processes having barriers in finance market.* Osaka J. Math., 39, 487–509, 2002.
5. K. Ito, and H. P. McKean, Jr: Diffusion Processes and their Sample Paths, Springer, 1965.
6. D. Revuz, and M. Yor: Continuous Martingales and Brownian Motion (second edition), Springer, 1994.
7. K. Takaoka: *Some remarks on the uniform integrability of continuous martingales.* Séminaire de Probabilités XXXIII, Springer Lecture Notes in Math. 1709, 327–333, 1999.

Ito's Integrated Formula for Strict Local Martingales

Dedicated to the memory of Paul-André Meyer

Dilip B. Madan* and Marc Yor**

*Robert H. Smith School of Business, Van Munching Hall, University of Maryland, College Park, MD 20742, USA
email: dmadan@rhsmith.umd.edu
**Laboratoire de Probabilités et Modèles Aléatoires, Université Pierre et Marie Curie, 4, Place Jussieu F 75252 Paris Cedex 05, France

Abstract: For $F : \mathbb{R} \to \mathbb{R}$ a C^2 function, and more generally a difference of convex functions, $(S_t, t \geqslant 0)$ a continuous strict local martingale taking values in $\mathbb{R}_+$, we investigate under which condition the stochastic integral appearing in Itô's formula applied to $F(S_t), t \geqslant 0$, is a true martingale and, if not, how it may be corrected to become one.

1 Introduction

Throughout this paper, we consider $\mathcal{S} = (S_t, t \geqslant 0)$ an $\mathbb{R}_+$ − valued *continuous* local martingale, which is strict, i.e. it is not a martingale. We assume S_0 is a fixed positive real. Since $(S_t, t \geqslant 0)$ is a strict local martingale, some care is needed when applying to it, or to martingale functionals of $\mathcal{S}$, the optional stopping theorem. This needed care is illustrated by the "correct" form of the Itô-Tanaka formula which we state now and which will play a central part in this paper.

Theorem 1. *If τ is a $(\mathcal{F}_t)$ stopping time which is a.s. finite, and $K \geqslant 0$ there is the identity:*

$$E\left[(S_\tau - K)^+\right] = (S_0 - K)^+ + \frac{1}{2}E\left[\mathcal{L}_\tau^K\right] - c_\mathcal{S}(\tau) \tag{1}$$

where $(\mathcal{L}_\tau^K)$ denotes the local time of (S_t) at level K and time τ and

$$c_S(\tau) = \lim_{n \to \infty} nP \left(\sup_{u \leqslant \tau} S_u \geqslant n \right) \tag{2}$$

$$= \lim_{m \to \infty} \sqrt{\frac{\pi}{2}} \left(mP \left(\sqrt{\langle S \rangle_\tau} \geqslant m \right) \right)$$

$$= E \left[S_0 - S_\tau \right].$$

Note that the penalty $c_S(\tau)$ which turns out to be equal, in particular to: $E[S_0 - S_\tau]$, measures the lack of martingality of $(S_{t \wedge \tau}, t \geqslant 0)$. The identity between the three quantities in equation (2) was already obtained in Theorem 1 of Elworthy, Li and Yor (1999) completing the results in Elworthy, Li and Yor (1997) and Azema, Gundy and Yor (1980).

As a consequence of formula (1), we obtain a further complement to the different expressions of $c_S(\tau)$, namely

$$c_S(\tau) = \lim_{K \to \infty} \frac{1}{2} E \left[\mathcal{L}_\tau^K \right] \tag{3}$$

which follows upon letting $K \to \infty$ on both sides of (1).

We now explain the "financial motivation" which led us to the identity (1) above and its financial meaning. In applications within mathematical finance the process S could represent a discounted price process or the process for a forward or futures price. It is problematic, as noted in Heston, Loewenstein, and Willard (2004) to take for the price of a European option written on S with maturity T and strike K the value

$$C(K,T) \stackrel{\mathrm{def}}{=} E \left[(S_T - K)^+ \right] \tag{4}$$

which is finite; in fact, assuming that $E[S_T] < S_0$ (which happens if $(S_{t \wedge T})$ is a strict local martingale), then we still get:

$$E \left[(S_T - K)^+ \right] < (S_0 - K)^+ \tag{5}$$

for K small enough. Moreover, the inequality (5) cannot hold for all $K's$ since the right hand side is zero for $K > S_0$, and in general the left hand side will never be zero.

In financial analysis the option to buy is generally viewed as more valuable than the value of a forward contract that represents the obligation to buy at T for K dollars. The obligation to buy is a simple portfolio that is long the stock and short K dollars at T. The current value of the stock being S_0 leads to the value of the forward as S_0 less the present value of the strike and hence market call prices are *larger* than $(S_0 - K)^+$. Hence the problem with modeling the call price using (4) when we have a strict local martingale for the stock price process.

In terms of hedging principles we note that one may replicate the forward stock at the cost of S_0, by buying and holding the stock to the forward date. However, this is not the most efficient way to replicate the forward stock at the precise forward date, in this model. Here there is a cheaper strategy that involves an initial cost of $E[S_T] < S_0$ coupled with a dynamic trading strategy accomplishing this objective. This latter hedge for the forward stock is tailored to the forward date and is not robust to early liquidation considerations. In fact it is exposed to unbounded exposure on a marked to market basis if the short forward stock is marked as a liability with the value of S_t. The valuation we propose considers the limit of values taken at a reducing sequence of stopping times and is thereby robust to random early liquidations.[1]

More precisely, we propose instead as a pricing model, the limiting value attained in a sequence of approximating martingale models. Let $(T_n, n \geqslant 1)$ be any sequence of stopping times, increasing to $+\infty$, and reducing S, i.e. $S^{(n)} = (S_{t \wedge T_n}, t \geqslant 0)$ is a uniformly integrable martingale. As an example for the reducing times one may take the first time the process crosses the level n. Such a choice forces the price process, in the continuous case for example, to live in the rectangle $([0, n] \times [0, T])$ for the practical purpose of price determination. Hence, we propose as the price of the European option written on S :

$$C^{strict}(K, T) \stackrel{def}{=} \lim_{n \to \infty} E\left[(S_{T \wedge T_n} - K)^+\right]. \tag{6}$$

We shall show, however, that the proposed limit is independent of the selection of the reducing stopping times.

First, we note that the sequence $\left\{ E\left[(S_{T \wedge T_n} - K)^+\right], n = 1, 2, \cdots \right\}$ is increasing and bounded above by S_0; hence it converges and the RHS of (6) is well defined. It is also easily shown that this sequence is bounded below by $(S_0 - K)^+$ and hence the intrinsic value remains a lower bound. The same argument holds if we replace T by any stopping time τ (with respect to the filtration $(\mathcal{F}_t)$). Thus, instead of (6) we shall consider more generally

$$C^{strict}(K, \tau) \stackrel{def}{=} \lim_{n \to \infty} E\left[(S_{\tau \wedge T_n} - K)^+\right] \tag{7}$$

where τ always denotes an a.s. finite stopping time. The supremum over all stopping times of (7) would be our candidate for the American option price with respect to a strict local martingale for the underlying asset price.

In order that this definition be sensible, we need to show that the RHS of (7) does not depend on the reducing sequence (T_n). In the next proposition we give two intrinsic expressions of $C^{strict}(K, \tau)$, which proves a fortiori that the definition (7) does not depend on the reducing sequence (T_n).

We in fact have that

$$C^{strict}(K, \tau) = \sup_{\sigma} E\left[(S_{\sigma \wedge \tau} - K)^+\right]. \tag{8}$$

---

[1] We are indebted to Freddy Delbaen for discussions on this point.

Proposition 2. *The value of* $C^{strict}(K, \tau)$ *is equal to:*

$$C^{strict}(K, \tau) = (S_0 - K)^+ + \frac{1}{2} E\left[\mathcal{L}_\tau^K\right] \tag{9}$$

$$= E\left[(S_\tau - K)^+\right] + \lim_{n \to \infty} \left(nP\left(M_\tau \geqslant n\right)\right) \tag{10}$$

where on the RHS, $(\mathcal{L}_t^K, t \geqslant 0)$ *denotes the local time at level K of S, as defined in Meyer (1976) and discussed further in, e.g. Yor (1994), and* $M_t = \sup_{s \leqslant t}(S_s)$.

We observe that the correction term of $\lim_{n \to \infty} (nP(M_\tau \geqslant n))$ in equation (10) for the value $C^{strict}(K, \tau)$ relative to the direct expectation $E\left[(S_\tau - K)^+\right]$ is independent of the strike and hence is absent from all call spreads or other bounded claims. Hence, the computations of put option prices, in for example Emmanuel and MacBeth (1982), is in accord with our computation of

$$P^{strict}(K, \tau) \stackrel{def}{=} \lim_{n \to \infty} E\left[(K - S_{\tau \wedge T_n})^+\right].$$

It follows from the martingale property for the stopped processes that the Emmanuel and Macbeth (1982) call price is in accord with $C^{strict}(K, \tau)$ and above the value for $E\left[(S_\tau - K)^+\right]$ by the shortfall of the correction term $\lim_{n \to \infty} (nP(M_\tau \geqslant n))$ or $c_\mathcal{S}(\tau)$.[2]

Much in the same way as Ito's formula may be recovered from Tanaka's formula, we shall deduce from (6) an integrated and corrected form of Ito's formula. For this purpose, let us associate with any $f \in L^1(\mathbb{R}_+, dK)$ its second antiderivative:

$$F_f(x) = \int_0^\infty dK f(K)(x - K)^+;$$

and more generally, if $\mu(dK)$ is any finite measure on $\mathbb{R}_+$, we define

$$F_\mu(x) = \int_0^\infty \mu(dK)(x - K)^+.$$

We also define by

$$\overline{f} = \int_0^\infty dK \ f(K); \quad \overline{\mu} = \int_0^\infty \mu(dK).$$

Then we present the following form of Ito's formula.

Proposition 3. *With the previous notation and assumptions, we obtain*

$$E\left[F_f(S_\tau)\right] = F_f(S_0) + \frac{1}{2} E\left[\int_0^\tau d\langle S \rangle_u f(S_u)\right] - \overline{f} c_\mathcal{S}(\tau).$$

---

[2] We are indebted to Peter Carr for discussions on this point.

More generally there is the formula

$$E\left[F_\mu(S_\tau)\right] = F_\mu(S_0) + \frac{1}{2}E\left[\int_0^\infty \mu(dK)L_\tau^K\right] - \bar{\mu}c_S(\tau).$$

As a noteworthy consequence of Proposition 3, we get the following

Corollary 4. *The processes*

$$F_f(S_t) - F_f(S_0) - \frac{1}{2}\int_0^t d\langle S\rangle_u f(S_u) - \bar{f}(S_t - S_0), \; t \geqslant 0$$

and

$$F_\mu(S_t) - F_\mu(S_0) - \frac{1}{2}\int_0^\infty \mu(dK)\mathcal{L}_t^K - \bar{\mu}(S_t - S_0), \; t \geqslant 0$$

are martingales.

The rest of the paper is organized as follows. In section 2 we give a proof of Theorem 1, as the identity (1), which differs from that of Proposition 2, as the latter relies upon the two expressions of $C^{strict}(K,\tau)$ given by equations (9) and (10). The proof of Proposition 2 is also provided in section 2. The other results follow easily. Section 3 presents an explicit computation of all the terms $E\left[(S_T - K)^+\right]$, $E\left[\mathcal{L}_T^K\right]$, and $\lim_{n\to\infty}\left(nP\left(M_T \geqslant n\right)\right)$ for the archetypical strict local martingale:

$$S_t = \frac{1}{R_t}$$

where $(R_t, t \geqslant 0)$ denotes a $BES(3)$ process, starting from $r > 0$.

With the help of the $h-process$ relationship between the $BES(3)$ process, and the Brownian motion process, which we write as

$$P_{r|\mathcal{F}_t}^{(3)} = \left(\frac{X_{t\wedge T_0}}{r}\right) \cdot W_{r|\mathcal{F}_t} \tag{11}$$

where $P_r^{(3)}$ denotes the law of the $BES(3)$ process starting from r, and W_r denotes the Wiener measure such that $W_r\left(X_0 = r\right) = 1$. We are able to express the identity (1) in terms of a martingale property involving Brownian motion and its local time at a given level.

The financial relevance of strict local martingales and the "so called" CEV (constant elasticity of variance) processes with diffusions coefficients given by a power of the stock price, with a power greater than one, as studied for example in Heston, Loewenstein and Willard (2004) or Delbaen and Schachermayer (1995) may well arise in the context of modeling defaults. In the current literature (Jarrow and Turnbull (1995), Duffie and Singleton (1999)) the default is modeled by a sudden jump to a default state and the drift in the stock is adjusted upward by the hazard rate exposure to the default event to recover a local martingale by compensating for this sudden loss in value. Alternatively, the strict local martingales offer the possibility of diffusing to default with the compensation arising in terms of a substantial probability of spending time at high levels of the stock price as described by equation (3). Interestingly, one could employ a knowledge of supermartingale exit times as described in Föllmer (1972) to help calibrate the implied default time density to the credit default swap markets. Of course, in this case, we advocate pricing European call options in accordance with equation (6).

2 Proofs of Theorem 1 and Propositions 2, 3 and Corollary 4

We present in this section the proof of Theorem 1 and the Propositions stated in the Introduction.

Proof. of Theorem 1.

a) It is well known that if an $(\mathcal{F}_t)$ adapted process (M_t) satisfies: $E\left(|M_\tau|\right) < \infty$ and $E\left[M_\tau\right] = 0$, for every bounded stopping time τ, then it is a $(\mathcal{F}_t)$ martingale, (and the converse is also true).

Thus, to prove (1), using the expression $c_S(\tau) = E\left[S_0 - S_\tau\right]$, we need to show:

$$N_t^K = \left((S_t - K)^+ - S_t\right) - \left((S_0 - K)^+ - S_0 - \frac{1}{2}\mathcal{L}_t^K\right), \ t \geqslant 0$$

is a martingale.

Remarking that: $\left((S_t - K)^+ - S_t\right) = -\left(S_t \wedge K\right)$, we get,

$$-N_t^K = (S_t \wedge K) - (S_0 \wedge K) + \frac{1}{2}\mathcal{L}_t^K, \ t \geqslant 0.$$

It follows from Tanaka's formula that $(-N_t^K, t \geqslant 0)$ is a local martingale, but, in fact, much more is true: indeed, it is uniformly integrable, and in fact, it is BMO, as we shall now show.

b) Since $(S_t - K)^+ - \frac{1}{2}\mathcal{L}_t^K$, $t \geqslant 0$, is a local martingale, we deduce easily

$$\frac{1}{2}E\left[\mathcal{L}_t^K\right] \leqslant S_0$$

and hence :

$$\frac{1}{2}E\left[\mathcal{L}_\infty^K\right] \leqslant S_0.$$

Hence $(-M_t)$ is a uniformly integrable martingale. We deduce therefrom that

$$\frac{1}{2}E\left[\mathcal{L}_\infty^K - \mathcal{L}_t^K | \mathcal{F}_t\right] = (S_t \wedge K) - E\left[(S_\infty \wedge K)|\mathcal{F}_t\right] \leqslant K$$

which proves the desired result.

c) The preceding implies that, not only is formula (1) true for bounded stopping times; it is in fact true for every stopping time, bounded or not. □

Proof. of Proposition 2.

We first write Tanaka's formula

$$(S_t - K)^+ = (S_0 - K)^+ + \int_0^t \mathbf{1}_{S_u > K} dS_u + \frac{1}{2}\mathcal{L}_t^K$$

We remark that

$$S_t^K \overset{def}{=} \int_0^t \mathbf{1}_{S_u > K} dS_u$$

is also a local martingale.

We now consider $\mathcal{R}$ a stopping time which reduces $(S_u, u \geqslant 0)$, and $(\mathcal{R}_p^K, p = 1, 2, \cdots)$ a sequence of stopping times increasing to $+\infty$ which reduces $(S_t^K, t \geqslant 0)$. Thus we can write

$$E\left[\left(S_{(\tau \wedge \mathcal{R} \wedge \mathcal{R}_p^K)} - K\right)^+\right] = (S_0 - K)^+ + \frac{1}{2}E\left[\mathcal{L}_{(\tau \wedge \mathcal{R} \wedge \mathcal{R}_p^K)}^K\right]. \qquad (12)$$

We then let p increase to $+\infty$ and we get, on the left hand side, the convergence in $\mathcal{L}^1$ of

$$\left\{S_{(\tau \wedge \mathcal{R} \wedge \mathcal{R}_p^K)}, p \to \infty\right\}$$

towards $S_{\tau \wedge \mathcal{R}}$; On the right hand side, it suffices to use Beppo-Levi; thus we deduce from equation (12) that

$$E\left[(S_{\tau \wedge \mathcal{R}} - K)^+\right] = (S_0 - K)^+ + \frac{1}{2}E\left[\mathcal{L}_{\tau \wedge \mathcal{R}}^K\right] \qquad (13)$$

for every reducing stopping time of $(S_t, t \geqslant 0)$.

Finally, using (13), we get (9) by invoking again Beppo-Levi, when using the sequence $(\mathcal{R}_n, n \to \infty)$.

We now assume that $(S_t, t \geqslant 0)$ is continuous, and we take for $\mathcal{R}_n$ the times

$$\mathcal{R}_n = \inf\{t : S_t \geqslant n\}.$$

Next we write:

$$E\left[(S_{\tau \wedge \mathcal{R}_n} - K)^+\right] = E\left[(S_\tau - K)^+ \mathbf{1}_{\tau \leqslant \mathcal{R}_n}\right] + (n - K)^+ P(\mathcal{R}_n \leqslant \tau) \qquad (14)$$

The first expression in (14) converges as $n \to \infty$ towards

$$E\left[(S_\tau - K)^+\right].$$

Hence the second also converges, and is equal to

$$\lim_{n \to \infty}(n - K)^+ P(n \leqslant M_\tau) \qquad (15)$$

where $M_t = \sup_{s \leqslant t}(S_s)$. It is now immediate that the limit in (15) is that of

$$nP(n \leqslant M_\tau)$$

and hence it does not depend on K. $\square$

Proof. of Equation (8).

In order to prove the result (8) we replace the local submartingale $(S_t - K)^+, t \geqslant 0$, by a general *continuous* $\mathbb{R}^+$−valued local submartingale $\Sigma_t, t \geqslant 0$, with canonical Doob-Meyer decomposition:

$$\Sigma_t = \Sigma_0 + N_t + A_t, t \geqslant 0$$

where $(N_t, t \geqslant 0)$ is a local martingale, $(A_t, t \geqslant 0)$ is a continuous increasing process, $N_0 = A_0 = 0$. Then, we shall show:

$$\sup_{\sigma \leqslant \tau, \sigma \text{ stopping time}} E[\Sigma_\sigma] = \Sigma_0 + E[A_\tau]. \tag{16}$$

We consider $(R_n, n \to \infty)$ a sequence of stopping times increasing to $+\infty$ which reduce the local martingale $(M_{t \wedge \tau}, t \geqslant 0)$. Then if $\sigma \leqslant \tau$, σ a stopping time, we obtain

$$E[\Sigma_\sigma] \leqslant \underline{\lim}_{n \to \infty} E\left[\Sigma_{(\sigma \wedge R_n)}\right] , \text{ (from Fatou's Lemma)}$$

$$\leqslant \lim_{n \to \infty} E[\Sigma_0 + A_{\sigma \wedge R_n}]$$

$$= \Sigma_0 + E[A_\sigma]$$

From Beppo-Levi (16) follows by taking finally, $\sigma = \tau$.

The same kind of argument shows that if T is any stopping time which reduces $(S_{t \wedge \tau}, t \geqslant 0)$ then:

$$E\left[(S_T - K)^+\right] = (S_0 - K)^+ + \frac{1}{2} E[\mathcal{L}_T^K]$$

Thus for any reducing sequence $(T_n, n \to \infty)$ of $(S_{t \wedge \tau}, t \geqslant 0)$ we obtain

$$\lim_{n \to \infty} E\left[(S_{T_n} - K)^+\right] = (S_0 - K)^+ + \frac{1}{2} E[\mathcal{L}_T^K]$$

thus proving (9).

In order to prove (10) we take

$$T_n^* = \inf\{t : S_t \geqslant n\}$$

Then

$$C^{strict}(\tau, K) = \lim_{n \to \infty} E\left[(S_{\tau \wedge T_n^*} - K)^+\right] \tag{17}$$

$$= \lim_{n \to \infty} \left\{ \begin{array}{l} E\left[(S_\tau - K)^+\right] \mathbf{1}_{(\tau \leqslant T_n^*)} \\ + (n - K)^+ P(T_n^* \leqslant \tau) \end{array} \right\}$$

It is now easily shown, using expressions for $c_S(\tau)$ (see Theorem 1 above) that (10) follows from (17). □

As noted the equality between (9) and (10) also follows from Theorem 1 but here we gave a direct proof. Conversely, a second proof of Theorem 1 is provided by the equality of the right hand side of (9) and (10).

Proof. Second proof of Theorem 1.

It follows immediately from the equality of the right hand side of (9) and (10).

Proof. of Proposition 3.

It suffices to integrate both sides of (1) with respect to $f(K)dK$, respectively $\mu(dK)$ and use Fubini's theorem, as well as

$$\int_0^\tau d\langle S\rangle_u f(S_u) = \int dK f(K) L_\tau^K.$$

$\square$

Proof. of Corollary 4.

Write

$$N_t^f = F_f(S_t) - F_f(S_0) - \frac{1}{2}\int_0^t (d\langle S\rangle_u)\, f(S_u) - \overline{f}(S_t - S_0)$$

and similarly $(N_t^\mu, t \geqslant 0)$.

Then:

$$E\left[N_\tau^f\right] = 0 \text{ and } E\left[N_\tau^\mu\right] = 0$$

for every bounded stopping time. As recalled at the beginning of the proof of Theorem 1, it is well known that this is equivalent to $\left(N_t^f\right)$, respectively (N_t^μ) being a martingale. $\square$

3 The BES(3) example, and its translation in terms of Brownian motion

We now consider the particular case of the identity (1) when $S_t = \frac{1}{R_t}, t \geqslant 0$, under $P_r^{(3)}$. As we have seen in the proof of (1) given in section 2, it is equivalent to the property that

$$\left(\frac{1}{R_t}\right) \wedge K + \frac{1}{2}\mathcal{L}_t^K\left(\frac{1}{R}\right),\ t \geqslant 0 \tag{18}$$

is a (uniformly integrable) martingale. We would like to translate the property in (18) in terms of Brownian motion, thanks to the absolute continuity relation (11). Thus we deduce from (18) that under W_r :

$$\left(\left(\frac{1}{X_t}\right) \wedge K + \frac{1}{2}\mathcal{L}_t^K\left(\frac{1}{X}\right)\right) X_{t \wedge T_0} \tag{19}$$

is a W_r martingale.

It is easily seen that,

$$\mathcal{L}_t^K \left(\frac{1}{X} \right) = K^2 \mathcal{L}_t^{\frac{1}{K}} (X)$$

therefore we can write (19) as:

$$(1 \wedge (K X_{t \wedge T_0})) + \frac{1}{2} K^2 \mathcal{L}_t^{\frac{1}{K}} (X) X_{t \wedge T_0} \tag{20}$$

is a local martingale under W_r, and it is easily shown in fact that it is a martingale. Equivalently:

$$\left(\frac{1}{K} \right) \wedge (X_{t \wedge T_0}) + \frac{1}{2} K \mathcal{L}_t^{\frac{1}{K}} (X) X_{t \wedge T_0} \tag{21}$$

is a W_r martingale.

We now check this property: We use the notation $U \sim V$ if $U - V$ is a local martingale; then (21) translates as

$$\left(\frac{1}{K} \right) \wedge (X_{t \wedge T_0}) \sim -\frac{1}{2} K \mathcal{L}_t^{\frac{1}{K}} (X) X_{t \wedge T_0} \tag{22}$$

whereas

$$\frac{K}{2} \mathcal{L}_t^{\frac{1}{K}} (X) X_{t \wedge T_0} \sim \frac{1}{2} \mathcal{L}_{t \wedge T_0}^{\frac{1}{K}} (X) . \tag{23}$$

Thus we see that by adding (22) and (23) term by term, we get that the expression (21), hence that in (20) is a martingale under W_r.

A similar discussion may be undertaken for any Bessel process $(R_t, t \geqslant 0)$, with dimension $\delta > 2$, starting with the strict local martingale

$$\left\{ \left(\frac{1}{R_t} \right)^{\delta-2} , t \geqslant 0 \right\}$$

The following relation holds

$$P_{r|\mathcal{F}_t}^{(\delta)} = \left(\frac{R_{t \wedge T_0}}{r} \right)^{\delta-2} \cdot P_{r|\mathcal{F}_t}^{(4-\delta)}$$

(See for example Revuz-Yor (1999)).

The role played by the pair:

$$\left\{ \frac{1}{R_t}, \text{under } P_r^{(3)} \right\} \text{ and } \{ X_{t \wedge T_0}, \text{under } W_r \}$$

is now played by

$$\left\{ \frac{1}{(R_t)^{\delta-2}}, \text{under } P_r^{(\delta)} \right\} \text{ and } \left\{ (R_{t \wedge T_0})^{\delta-2}, \text{under } P_r^{(4-\delta)} \right\} .$$

We note that the strict local martingale (that is the CEV process studied in Heston, Loewenstein and Willard (2004)) :

$$\mathcal{S}_t^{(\delta)} \overset{def}{=} \frac{1}{(R_t)^{\delta-2}}, t \geqslant 0$$

satisfies:

$$\mathcal{S}_t^{(\delta)} = \frac{1}{r^{\delta-2}} + (\delta - 2) \int_0^t \left(\mathcal{S}_u^{(\delta)}\right)^{\left(\frac{\delta-1}{\delta-2}\right)} d\beta(u)$$

for a Brownian motion β ($(-\beta_u)$ is the martingale part in the semimartingale decomposition of R_u, under $P_r^{(\delta)}$).

4 A Further Discussion of the Correction Term

We start again from formula (1) and (2), which yield:

$$E\left[(S_\tau - K)^+\right] = (S_0 - K)^+ + \frac{1}{2}E\left[\mathcal{L}_\tau^K\right] - E\left[S_0 - S_\tau\right].$$

since we have

$$S_\tau \wedge K = S_\tau - (S_\tau - K)^+$$

it follows that

$$E\left[(S_\tau \wedge K)\right] = (S_0 \wedge K) - \frac{1}{2}E\left[\mathcal{L}_\tau^K\right]$$

which, since this is true for all stopping times is equivalent to

$$\left\{(S_t \wedge K) - (S_0 \wedge K) + \frac{1}{2}\mathcal{L}_t^K, t \geqslant 0\right\}$$

is a true uniformly integrable martingale.
The correction is therefore equivalent to

$$\left((S_t \wedge K) + \frac{1}{2}\mathcal{L}_t^K, t \geqslant 0\right) \equiv \left(N_t^K, t \geqslant 0\right) \qquad (24)$$

is a positive martingale. These martingales are in fact in BMO. To observe this we note first that

$$\left(N_t^K, t \geqslant 0\right)$$

is uniformly integrable as

$$E\left[\frac{1}{2}\mathcal{L}_\infty^K\right] \leqslant S_0.$$

Hence we have that

$$E\left[\mathcal{L}_\infty^K\right] < \infty$$

But from (24) we have

$$\frac{1}{2}E\left[\mathcal{L}_{\infty}^{K} - \mathcal{L}_{t}^{K}|\mathcal{F}_t\right] = (S_t \wedge K) - E\left[(S_{\infty} \wedge K)|\mathcal{F}_t\right] \qquad (25)$$

then: $\mathcal{L}_{\infty}^{K}$, as a random variable, or $E\left[\mathcal{L}_{\infty}^{K}|\mathcal{F}_t\right]$ as a process is in BMO. This fact will be vastly generalized in the next section.

Given the previous discussion, the following remark is immediate, but, nonetheless quite noteworthy.

Proposition 5. *For any* $K \geqslant 0$, *and every stopping time* τ, *one has:*

$$\frac{1}{2}E\left[\mathcal{L}_{\tau}^{K}\right] = S_0 \wedge K - E\left[S_{\tau} \wedge K\right] \qquad (26)$$

In particular, if $S_{\tau} = 0$, *then one has:*

$$\frac{1}{2}E\left[\mathcal{L}_{\tau}^{K}\right] = S_0 \wedge K \qquad (27)$$

We illustrate the identity (27) by taking, for $(S_{t\wedge\tau})$, Brownian motion starting from $S_0 > 0$, and considered up to $\tau = \inf\{t, S_t = 0\}$. Then, the Ray-Knight theorem which describes the law of $\{\mathcal{L}_{\tau}^{K}, K \geqslant 0\}$ (see e.g. Revuz and Yor (1999)) asserts that $\left(\mathcal{L}_{\tau}^{K}, K \leqslant S_0\right)$ is the square of a $2-$dimensional BES process, whereas for $K \geqslant S_0$, it is the square of a $0-$dimensional BES process. This implies a fortiori the identity (27).

5 Universal Reducing Processes

In this section we shall exhibit some classes of deterministic functions f and ϕ such that, for any continuous local martingale $(S_t, t \geqslant 0)$ taking values in $\mathbb{R}_+$,

$$\int_0^t f\left(\langle S\rangle_u\right) dS_u$$

$$\int_0^t \phi\left(S_u\right) dS_u$$

are uniformly integrable martingales.

We call such functions, universal reducing functions. In the following Proposition, we exhibit such functions, whose reducing properties are particularly strong.

Proposition 6. *(i) If* f *belongs to* $L^2\left(\mathbb{R}_+, dx\right)$, *then*

$$U^f = \left(\int_0^t f\left(\langle S\rangle_u\right) dS_u, t \geqslant 0\right)$$

belongs to H^∞, *i.e. its bracket is uniformly bounded* $\left(by: \int_0^\infty f^2(x)dx\right)$; *a fortiori,* $\left(U_t^f, t \geqslant 0\right)$ *belongs to BMO.*

(ii) If ϕ belongs to $L^2\left(\mathbb{R}_+, dK\right)$, then

$$V^\phi = \left(\int_0^t \phi(S_u)dS_u, t \geqslant 0\right)$$

is a square integrable martingale;

(iii) If furthermore, $\int_0^\infty dK\phi^2(K)K < \infty$ then

$$\left(V_t^\phi, t \geqslant 0\right)$$

belongs to BMO.

Proof. (i) This property follows immediately from :

$$\langle \int_0^{\cdot} f(\langle S\rangle_u)dS_u \rangle_t = \int_0^t f^2(\langle S\rangle_u)d\langle S\rangle_u$$

$$= \int_0^{\langle S\rangle_t} f^2(x)dx$$

$$\leqslant \int_0^\infty f^2(x)dx.$$

(ii) We may write

$$\int_0^\infty \phi^2(S_u)d\langle S\rangle_u = \int_0^\infty \phi^2(K)\mathcal{L}_\infty^K dK$$

and we use the inequality

$$E\left[\mathcal{L}_\infty^K\right] \leqslant 2S_0$$

which we have already noted.

(iii) The result follows from:

$$E\left[\int_t^\infty \phi^2(S_u)d\langle S\rangle_u|\mathcal{F}_t\right] = \int_0^\infty dK\phi^2(K)E\left[\mathcal{L}_\infty^K - \mathcal{L}_t^K|\mathcal{F}_t\right]$$

$$\leqslant \int_0^\infty dK\phi^2(K)2K,$$

from (25). $\square$

References

1. J. Azéma, R. Gundy and M. Yor (1980), "Sur l'intégrabilité uniforme des martingales locales continues," *Sem. Prob. XIV*, Lecture Notes 784, Springer Verlag.

2. F. Delbaen, and W. Schachermayer (1995), "Arbitrage possibilities in Bessel processes and their relations with local martingales, PTRF 102, 357–366.
3. D. Duffie, and K. Singleton (1999), "Modeling the Term Structure of Defaultable Bonds," *Review of Financial Studies*, 12, 687–720.
4. K.D. Elworthy, X.M. Li, and M. Yor (1997), "On the tails of the supremum and the quadratic variation of strictly local martingales." *Séminaire Prob. XXXI*, Lecture Notes 1655, Springer Verlag.
5. K.D. Elworthy, X.M. Li, and M. Yor (1999), "The importance of strictly local martingales; applications to radial Ornstein-Uhlenbeck processes," PTRF, 115, 325–355.
6. D.C. Emmanuel, and J.D. Macbeth (1982), "Further results on the constant elasticity of variance call option pricing model," *Journal of Financial and Quantitative Analysis*, 4, 533–554.
7. H. Föllmer (1972), "The Exit Measure of a Supermartingale," *Zeitschrift für Wahrscheinlichkeitstheorie und verwandte Gebiete*, 21, 154–166.
8. S. Heston, M. Loewenstein, and G. Willard (2004), "Options and Bubbles," working paper, University of Maryland.
9. R.A. Jarrow, and S. Turnbull (1995), "Pricing Derivatives on Financial Securities Subject to Credit Risk," *Journal of Finance*, 50, 53–85.
10. D. Revuz, and M. Yor (1999), *Continuous Martingales and Brownian Motion*, Third Edition, Springer Verlag.
11. P.A. Meyer (1976), "Un Cours sur les Intégrales Stochastiques", *Sem. Prob. X*, Lecture Notes 511, Springer Verlag.
12. C. Sin (1998), "Complications with stochastic volatility models," Adv. App. Prob. 30, 256–268.
13. M. Yor (1994), "Local Times and Excursion Theory for Brownian Motion: a concise introduction," *Lectures in Mathematics*, University of Caracas, Venezuela.

Martingale-Valued Measures, Ornstein-Uhlenbeck Processes with Jumps and Operator Self-Decomposability in Hilbert Space

Dedicated to the memory of Paul-André Meyer

David Applebaum[**]

[**]Probability and Statistics Department, University of Sheffield, Hicks Building, Hounsfield Road, Sheffield, England, S3 7RH
e-mail: D.Applebaum@sheffield.ac.uk

Summary. We investigate a class of Hilbert space valued martingale-valued measures whose covariance structure is determined by a trace class positive operator valued measure. The paradigm example is the martingale part of a Lévy process. We develop both weak and strong stochastic integration with respect to such martingale-valued measures. As an application, we investigate the stochastic convolution of a C_0-semigroup with a Lévy process and the associated Ornstein-Uhlenbeck process. We give an infinite dimensional generalisation of the concept of operator self-decomposability and find conditions for random variables of this type to be embedded into a stationary Ornstein-Uhlenbeck process.

Key Words and Phrases:- martingale-valued measure, positive operator valued measure, trace class operator, nuclear, decomposable, Lévy process, C_0-semigroup, stochastic convolution, Ornstein-Uhlenbeck process, operator self-decomposability, exponentially stable semigroup.

1 Introduction

The aim of this paper is to introduce some new concepts into stochastic analysis of Hilbert space valued processes with a view to gaining deeper insights into the structure of Lévy processes and other processes which can be built from these.

We begin with an investigation of Hilbert space valued martingale-valued measures. Finite dimensional versions of these (called "martingale measures"

[**] Work carried out at The Nottingham Trent University

therein) were first introduced by Walsh [38] to formulate stochastic partial differential equations (SPDEs) driven by a continuous space-time white noise. They were further developed in [23] and in [3] they were generalised to deal with SPDEs with jumps. In [2], the author found them a convenient tool for si-multaneously dealing with stochastic integration with respect to the Brownian and the compensated small jumps part of a Lévy process, when the integrands depend both on time and the jump-space variable. Here we extend this latter construction to the infinite dimensional context. In particular, we investigate a class of martingale-valued measures whose covariance structure is deter-mined by a trace class positive operator valued measure. This is precisely the covariance structure found in the martingale part of a Lévy process- indeed it is well known that the covariance operator of Brownian motion is trace class (see e.g. [11], proposition 2.15, p.55). Here we show that the covariance of the compensated small jumps is also determined by such operators, which in this case are a continuous superposition of finite rank operators. Our approach exploits the Lévy-Itô decomposition of a Lévy process into drift, Brownian, small jump and large jump parts which has recently been extended to type 2 Banach spaces by Albeverio and Rüdiger [1].

Having established a natural class of martingale-valued measures M, we develop both weak and strong stochastic integrals of suitable predictable processes. In the first of these the integrand $(F(t,x), t \geqslant 0, x \in E)$ (where E is a Lusin space) is vector valued and we generalise the approach of Kunita [24], who dealt with the case where M is a martingale, to construct the scalar val-ued process $\int_0^t \int_E (F(s,x), M(ds, dx))_H$, where $(\cdot, \cdot)_H$ is the inner product in the Hilbert space H. In the second of these, $(G(t,x), t \geqslant 0, x \in E)$ is operator-valued and we generalise the stochastic integral of Métivier [28] who dealt with the case where M is a martingale (see also [29], [11] for the case of Brownian motion), to construct the Hilbert space valued object $\int_0^t \int_E G(s,x) M(ds, dx)$.

As an application of these techniques, we first study the stochastic con-volution $\int_0^t S(r) dX(r)$, of a C_0-semigroup $(S(r), r \geqslant 0)$ with infinitesimal generator J with a Lévy process $X = (X(t), t \geqslant 0)$. We then apply this to investigate the generalised Langevin equation

$$dY(t) = JY(t) + dX(t), \tag{1}$$

whose unique weak solution is the Ornstein-Uhlenbeck process. Equations of this type driven by general Lévy processes, were first considered by S.J. Wolfe [39] in the scalar case where J is a negative constant. Sato and Yamazato [36], [37] generalised this to the multi-dimensional case wherein $-J$ is a matrix all of whose eigenvalues have positive real parts. The generalisation to infinite dimensions was first carried out by A. Chojnowska-Michalik [9], [8] (see also [11], [6] for the Brownian motion case). Using our stochastic integration theory we are able to give an alternative construction of the solution in which the Lévy-Itô decomposition is preserved within its structure. This is useful for later analysis as we see below.

We remark that, in the finite dimensional case, Ornstein-Uhlenbeck processes driven by non-Gaussian Lévy processes have recently been applied to the construction of self-similar processes via the Lamperti transform ([5]) and to models of stochastic volatility in the theory of option pricing [5], [31]. In the latter case, it may be that the infinite dimensional model as considered here, is more appropriate, as it can approximate the very large number of incremental market activities which lead to volatility change.

Finally, we consider an infinite-dimensional generalisation of self-decomposability. We recall that a real-valued random variable X is self-decomposable if for any $0 < c < 1$, there exists a random variable Y_c, which is independent of X such that

$$X \stackrel{d}{=} cX + Y_c. \tag{2}$$

Such random variables were first studied by Paul Lévy and they arise naturally as weak limits of normalised sums of independent (but not necessarily identically distributed) random variables (see e.g. [35], section 3.15). The definition was extended to Banach space valued random variables by Jurek and Vervaat [22] (with c still a scalar) while Jurek and Mason [21] considered the finite-dimensional case of "operator self-decomposability" where c is replaced by a semigroup $(e^{-tJ}, t \geqslant 0)$, with J an invertible matrix. Jurek [19] also investigated the case where J is a bounded operator in a Banach space. It is a consequence of results found in [39], [21], [22] and [37] that X is (operator) self-decomposable if and only if it can be embedded as $X(0)$ in a stationary Ornstein-Uhlenbeck process. Furthermore a necessary and sufficient condition for the required stationarity is that the Lévy measure ν of X has a certain logarithmic moment, more precisely $\int_{|x|\geqslant 1} \log(1 + |x|)\nu(dx) < \infty$, so we see that this is a condition on the large jumps of X (see also [16]).

Here we generalise operator self-decomposability by taking $(e^{-tJ}, t \geqslant 0)$ to be a contraction semigroup acting in a Hilbert space H (see [20] for the case where it is a group acting in a Banach space). We emphasise that, in contrast to the cases discussed in the previous paragraph, J is typically an unbounded operator. We are able to obtain a partial generalisation of the circle of ideas described above which relates self-decomposability, stationary Ornstein-Uhlenbeck processes and logarithmic moments of the Lévy measure. The failure to obtain a full generalisation arises from dropping the condition that J is invertible, which appears to be unnatural in this setting and also from the fact that the operators e^{-tJ} are no longer invertible. We note that the link between stationarity and logarithmic moments has also been established in [9] using different methods, and by a more indirect route than that given here (see also [15]).

The stochastic integration theory developed herein will have extensive further applications. In particular, it can be used to construct solutions to stochastic differential equations driven by Hilbert space valued processes with jumps, generalising the Brownian motion case ([11], [25]). The details will appear elsewhere (see [27] for work in a similar direction).

Notation. $\mathbb{R}^+ = [0, \infty)$. If X is a topological space, then $\mathcal{B}(X)$ denotes its Borel σ-algebra. If H is a real separable Hilbert space, $b(H)$ is the space of bounded Borel measurable real-valued functions on H and $\mathcal{L}(H)$ is the *-algebra of all bounded linear operators on H. The domain of a linear operator T acting in H is denoted as $\mathrm{Dom}(T)$.

Acknowledgement. I am grateful to Zbigniew Jurek for some useful comments and the referee for a number of useful observations. Thanks are also due to Fangjun Xu for valuable suggestions for improvements.

2 Martingale-Valued Measures With Values in a Hilbert Space

2.1 Hilbert Space Valued Martingales

Let $(\Omega, \mathcal{F}, (\mathcal{F}_t, t \geqslant 0), P)$ be a stochastic base wherein the filtration $(\mathcal{F}_t, t \geqslant 0)$ satisfies the usual hypotheses of completeness and right continuity. Let H be a real separable Hilbert space with inner product $(\cdot, \cdot)_H$ and associated norm $||\cdot||$. Throughout this article, unless contra-indicated, all random variables and processes are understood to be H-valued. To any such random variable X, we associate the real-valued random variable $||X||$, where $||X||(\omega) = ||X(\omega)||$, for each $\omega \in \Omega$.

The predictable σ-algebra $\mathcal{P}$ is the smallest sub-σ-algebra of $\mathcal{B}(\mathbb{R}^+) \otimes \mathcal{F}$ with respect to which all mappings $F : \mathbb{R}^+ \times \Omega \to H$ are measurable, wherein $(F(t), t \geqslant 0)$ is adapted and $t \to F(t, \omega)$ is strongly left continuous for each $\omega \in \Omega$.

If $\mathcal{G}$ is a sub-σ-algebra of $\mathcal{F}$ and X is a random variable such that $\mathbb{E}(||X||) < \infty$, the conditional expectation of X given $\mathcal{G}$ is the unique $\mathcal{G}$-measurable random variable $\mathbb{E}_{\mathcal{G}}(X)$ for which

$$\mathbb{E}(1_A X) = \mathbb{E}(1_A \mathbb{E}_{\mathcal{G}}(X)),$$

for all $A \in \mathcal{G}$ (see e.g. [11], section 1.3). Many familiar properties of conditional expectation from the case $H = \mathbb{R}$ carry over to the general case, in particular

$$\mathbb{E}_{\mathcal{G}}((X, Y)_H) = (X, \mathbb{E}_{\mathcal{G}}(Y))_H, \text{ a.s.}$$

if $\mathbb{E}(||X|| \vee ||Y||) < \infty$ and X is $\mathcal{G}$-measurable.

An adapted process $X = (X(t), t \geqslant 0)$ is a *martingale* if $\mathbb{E}(||X(t)||) < \infty$ and $\mathbb{E}(X(t)|\mathcal{F}_s) = X(s)$ (a.s.), for all $0 \leqslant s \leqslant t < \infty$. A martingale is said to be *square-integrable* if $\mathbb{E}(||X(t)||^2) < \infty$, for all $t \geqslant 0$. By proposition 3 of [24], any square-integrable martingale has a strongly càdlàg modification.

If X is a square-integrable martingale, then $(||X(t)||^2, t \geqslant 0)$ is a non-negative uniformly integrable submartingale, hence by the Doob-Meyer decomposition, there is a unique increasing, predictable integrable process

$(\langle X\rangle(t), t \geqslant 0)$ such that $(\|X(t)\|^2 - \langle X\rangle(t), t \geqslant 0)$ is a real-valued martingale (see e.g. [24]). If $Y = (Y(t), t \geqslant 0)$ is another square-integrable martingale, we may, for each $t \geqslant 0$, define $\langle X, Y\rangle(t)$ in the usual way by polarisation, i.e.

$$\langle X, Y\rangle(t) = \frac{1}{4}[\langle X + Y\rangle(t) - \langle X - Y\rangle(t)].$$

Note that for all $0 \leqslant s \leqslant t < \infty$,

$$\mathbb{E}((X(t) - X(s), Y(t) - Y(s))_H | \mathcal{F}_s) = \mathbb{E}(\langle X, Y\rangle(t) - \langle X, Y\rangle(s) | \mathcal{F}(s)).$$

Two square-integrable martingales X and Y are said to be *orthogonal* if $\langle X, Y\rangle(t) = 0$, for all $t \geqslant 0$ (or equivalently, if $((X(t), Y(t))_H, t \geqslant 0)$ is a real-valued martingale).

Note. A different definition of $\langle \cdot \rangle$ for Hilbert space valued martingales is given in [11], section 3.4. We prefer to use that of [24] as it appears to be more general.

2.2 Martingale-Valued Measures

Let (S, Σ) be a Lusin space, so that S is a Hausdorff space which is the image of a Polish space under a continuous bijection and Σ is a Borel subalgebra of $\mathcal{B}(S)$ (see e.g. Chapter 8 of [10]). We assume that there is a ring $\mathcal{A} \subset \Sigma$ and an increasing sequence $(S_n, n \in \mathbb{N})$ in Σ such that

- $S = \bigcup_{n \in \mathbb{N}} S_n$
- $\Sigma_n := \Sigma|_{S_n} \subseteq \mathcal{A}$, for all $n \in \mathbb{N}$.

A *martingale-valued measure* is a set function $M : \mathbb{R}^+ \times \mathcal{A} \times \Omega \to H$ which satisfies the following (c.f [38], [23]):

1. $M(0, A) = M(t, \varnothing) = 0$ (a.s.), for all $A \in \mathcal{A}, t \geqslant 0$.
2. $M(t, A \cup B) = M(t, A) + M(t, B)$ (a.s.), for all $t \geqslant 0$ and all disjoint $A, B \in \mathcal{A}$.
3. $(M(t, A), t \geqslant 0)$ is a square-integrable martingale for each $A \in \mathcal{A}$ and is orthogonal to $(M(t, B), t \geqslant 0)$, whenever $A, B \in \mathcal{A}$ are disjoint.
4. $\sup\{\mathbb{E}(\|M(t, A)\|^2), A \in \Sigma_n\} < \infty$, for all $n \in \mathbb{N}, t > 0$.

Note. In Walsh's terminology [38], M is a "σ-finite L^2-valued orthogonal martingale measure".

Whenever $0 \leqslant s \leqslant t \leqslant \infty, M((s, t], \cdot) := M(t, \cdot) - M(s, \cdot)$. M is said to have *independent increments* if $M((s, t], A)$ is independent of $\mathcal{F}_s$ for all $A \in \mathcal{A}, 0 \leqslant s \leqslant t < \infty$.

Given a martingale valued measure M, for each $t \geqslant 0$, we can define a (random) real-valued set function $\langle M\rangle(t, \cdot)$ on $\mathcal{A}$ and (3) ensures that

$$\langle M\rangle(t, A \cup B) = \langle M\rangle(t, A) + \langle M\rangle(t, B) \text{ a.s.}$$

for all $t \geqslant 0$ and all disjoint $A, B \in \mathcal{A}$. A theorem of Walsh ([38], theorem 2.7, p.299) enables us to "regularise" $\langle M \rangle$ to obtain a (random) predictable σ-finite measure on $\mathcal{B}(\mathbb{R}^+) \otimes \Sigma$, which coincides with $\langle M \rangle$ (a.s.) on sets of the form $[0, t] \times A$, where $t > 0, A \in \mathcal{A}$. In the sequel, we will abuse notation to the extent of also denoting this measure by $\langle M \rangle$.

A *positive-operator valued measure* or (*POV measure* for short) on (S, Σ) is a family $(T_A, A \in \mathcal{A})$ of bounded positive self-adjoint operators in H for which

- $T_\varnothing = 0$,
- $T_{A \cup B} = T_A + T_B$, for all disjoint $A, B \in \mathcal{A}$.

Note. This is a slightly different use of the term POV measure than that employed in the theory of measurement in quantum mechanics (see e.g. [13], section 3.1).

We say that a POV measure is *decomposable* if there exists a strongly measurable family of bounded positive self-adjoint operators in H, $\{T_x, x \in S\}$ and a σ-finite measure λ on (S, Σ) such that

$$T_A \psi = \int_A T_x \psi \lambda(dx),$$

for each $A \in \mathcal{A}, \psi \in H$, where the integral is understood in the Bochner sense.

We recall that a bounded linear operator Z on H is *trace class* if $\mathrm{tr}(|Z|) < \infty$, where $|Z| = (ZZ^*)^{\frac{1}{2}}$. Let $\mathcal{L}_1(H)$ denote the space of all trace class operators on H, then $\mathcal{L}_1(H)$ is a real Banach space with respect to the norm $||Z||_1 = \mathrm{tr}(|Z|)$ (see e.g. [34], section VI.6). A POV measure is said to be trace class if each of its constituent operators is.

Now let M be a martingale-valued measure on $\mathbb{R}^+ \times S$. We say that it is *nuclear* if there exists a pair (T, ρ) where

- $T = (T_A, A \in \mathcal{A})$ is a trace class POV measure in H,
- ρ is a σ-finite measure on $\mathbb{R}^+$,

such that for all $0 \leqslant s \leqslant t < \infty, A \in \mathcal{A}, \psi \in H$,

$$\mathbb{E}(|(M((s, t], A), \psi)_H|^2) = (\psi, T_A \psi)\rho((s, t]). \tag{3}$$

A nuclear martingale-valued measure is *decomposable* if $(T_A, A \in \mathcal{A})$ is decomposable.

Proposition 2.1 *If M is a nuclear martingale-valued measure, then for all $t \geqslant 0, A \in \mathcal{A}$,*

$$\mathbb{E}(\langle M \rangle(t, A)) = ||T_A||_1 \rho((0, t]).$$

Proof. Let $(e_n, n \in \mathbb{N})$ be a maximal orthonormal set in H. We have

$$
\begin{aligned}
\mathbb{E}(\langle M \rangle(t, A)) &= \mathbb{E}(||M(t, A)||^2) \\
&= \sum_{n=1}^{\infty} \mathbb{E}(|(e_n, M(t, A))_H|^2) \\
&= \rho((0, t]) \sum_{n=1}^{\infty} (e_n, T_A e_n) \\
&= \rho((0, t]) \mathrm{tr}(T_A). \qquad \qquad \square
\end{aligned}
$$

2.3 Lévy Processes

Let X be a *Lévy process* taking values in H, so that X has stationary and independent increments, is stochastically continuous and satisfies $X(0) = 0$ (a.s.). If p_t is the law of $X(t)$ for each $t \geqslant 0$, then $(p_t, t \geqslant 0)$ is a weakly continuous convolution semigroup of probability measures on H. We have the *Lévy-Khinchine formula* (see e.g. [32]) which yields for all $t \geqslant 0, \psi \in H$,

$$
\mathbb{E}(\exp(i(\psi, X(t))_H) = e^{ta(\psi)},
$$

where

$$
a(\psi) = i(\zeta, \psi)_H - \frac{1}{2}(\psi, Q\psi)
$$
$$
+ \int_{H - \{0\}} (e^{i(u, \psi)_H} - 1 - i(u, \psi)_H 1_{\{||u|| < 1\}}) \nu(du), \qquad (4)
$$

where $\zeta \in H, Q$ is a positive, self-adjoint trace class operator on H and ν is a *Lévy measure* on $H - \{0\}$, i.e. $\int_{H - \{0\}} (||x||^2 \wedge 1) \nu(dx) < \infty$. We call the triple (ζ, Q, ν) the *characteristics* of the process X and the mapping a, the *characteristic exponent* of X.

Example 1 (*Q*-Brownian motion)

Q-Brownian motion $B_Q = (B_Q(t), t \geqslant 0)$ has characteristics $(0, Q, 0)$. It is a Gaussian process with continuous sample paths and covariance operator Q (see e.g. [11], section 4.1) so that $\mathbb{E}((\psi, B_Q(t))_H^2) = t(\psi, Q\psi)$, for each $\psi \in H, t \geqslant 0$. If $(\lambda_n, n \in \mathbb{N})$ are the eigenvalues of Q and $(e_n, n \in \mathbb{N})$ are the corresponding normalised eigenvectors, we have the useful representation of B_Q as an L^2-convergent series:

$$
B_Q(t) = \sum_{n=1}^{\infty} \sqrt{\lambda_n} \beta_n(t) e_n, \qquad (5)
$$

for each $t \geqslant 0$, where $(\beta_n, n \in \mathbb{N})$ are independent standard real-valued Brownian motions.

In the sequel, a Lévy process with characteristics $(\zeta, Q, 0)$ will be called a *Q-Brownian motion with drift*, while a Lévy process with characteristics $(\zeta, 0, \nu)$ will be said to be *non-Gaussian*.

Example 2 (α-Stable Lévy Processes)

A Lévy process is said to be *stable* if p_t is a stable law for each $t \geqslant 0$, i.e. for all $a, b > 0$, there exists $\phi \in H$ and $c > 0$ such that

$$(\tau_a p_t) * (\tau_b p_t) = \delta_\phi * (\tau_c p_t),$$

where for any measure q on H, $(\tau_a q)(E) = q(a^{-1}E)$, for all $E \in \mathcal{B}(H)$.

By a theorem of Jajte [17], a Lévy process X is stable iff it is a Q-Brownian motion with drift or it is non-Gaussian and there exists $0 < \alpha < 2$ such that $\tau_c \nu = c^\alpha \nu$, for all $c > 0$. We call this latter case an α-*stable Lévy process*. An extensive account of stable distributions in Hilbert and Banach spaces can be found in Chapters 6 and 7 of [26].

From now on we will always assume that Lévy processes have strongly càdlàg paths. We also strengthen the independent increments requirement on X by assuming that $X(t) - X(s)$ is independent of $\mathcal{F}_s$ for all $0 \leqslant s < t < \infty$.

If X is a Lévy process, we write $\Delta X(t) = X(t) - X(t-)$, for all $t > 0$. We obtain a Poisson random measure N on $\mathbb{R}^+ \times (H - \{0\})$ by the prescription:

$$N(t, E) = \#\{0 \leqslant s \leqslant t; \Delta X(s) \in E\},$$

for each $t \geqslant 0, E \in \mathcal{B}(H - \{0\})$. The associated compensated Poisson random measure $\tilde{N}$ is defined by

$$\tilde{N}(dt, dx) = N(dt, dx) - dt\nu(dx).$$

Let $A \in \mathcal{B}(H - \{0\})$ with $0 \notin \overline{A}$. If $f : A \to H$ is measurable, we may define

$$\int_A f(x)N(t, dx) = \sum_{0 \leqslant s \leqslant t} f(\Delta X(s))1_A(\Delta X(s))$$

as a random finite sum. Let ν_A denote the restriction of the measure ν to A, so that ν_A is finite. If $f \in L^2(A, \nu_A; H)$, we define

$$\int_A f(x)\tilde{N}(t, dx) = \int_A f(x)N(t, dx) - t\int_A f(x)\nu(dx),$$

then by standard arguments (see e.g. [2], Chapter 2) we see that $(\int_A f(x)\tilde{N}(t, dx), t \geqslant 0)$ is a centred square-integrable martingale with

$$\mathbb{E}\left(\left\|\int_A f(x)\tilde{N}(t, dx)\right\|^2\right) = t\int_A \|f(x)\|^2\nu(dx), \tag{6}$$

for each $t \geqslant 0$ (see also theorem 3.2.5 in [1]).

The Lévy-Itô decomposition for a càdlàg Lévy process taking values in a separable type 2 Banach space is established in [1]. We only need the Hilbert space version here:

Theorem 1. *[1] If H is a separable Hilbert space and $X = (X(t), t \geqslant 0)$ is a càdlàg H-valued Lévy process with characteristic exponent given by (4), then for each $t \geqslant 0$,*

$$X(t) = t\zeta + B_Q(t) + \int_{||x|| < 1} x\tilde{N}(t, dx) + \int_{||x|| \geqslant 1} xN(t, dx), \qquad (7)$$

where B_Q is a Brownian motion which is independent of N.

In (7),

$$\int_{||x|| < 1} x\tilde{N}(t, dx) = \lim_{n \to \infty} \int_{\frac{1}{n} < ||x|| < 1} x\tilde{N}(t, dx),$$

where the limit is taken in the L^2-sense, and it is a square-integrable martingale.

Let $S = \{x \in H; ||x|| < 1\}$ and take Σ to be its Borel σ-algebra, then it is easy to check that M is a martingale valued measure on $\mathbb{R}^+ \times S$, where

$$M(t, A) = B_Q(t)\delta_0(A) + \int_{A - \{0\}} x\tilde{N}(t, dx), \qquad (8)$$

for each $t \geqslant 0, A \in \mathcal{A}$. We call M a *Lévy martingale-valued measure*.

Here we take $\mathcal{A} = \mathcal{A}_0 \cup \{0\}$, where $\mathcal{A}_0 = \{A \in \Sigma; 0 \notin \overline{A}\}$ and each $S_n = \{x \in S : \frac{1}{n} < ||x|| < 1\}$.

We now aim to show that M is nuclear. To this end, we introduce the family of linear operators $(F_A, A \in \mathcal{A}_0)$ on H given by

$$F_A y = \int_A (x, y)_H x\nu(dx),$$

so that each F_A is a continuous superposition of finite-rank operators (using the Dirac notation employed in physics, we would write "$F_A = \int_A (|x\rangle\langle x|)\nu(dx)$").

It is easy to see that each $||F_A|| \leqslant \int_A ||x||^2 \nu(dx) < \infty$, hence F_A is bounded. Straightforward manipulations show that F_A is positive, self-adjoint. F_A is also trace class. To see this, let $(e_n, n \in \mathbb{N})$ be a maximal orthonormal set in H, then

$$\text{tr}(F_A) = \sum_{n=1}^{\infty} (e_n, F_A e_n)_H$$

$$= \sum_{n=1}^{\infty} \int_A (x, e_n)_H^2 \nu(dx)$$

$$= \int_A ||x||^2 \nu(dx) < \infty.$$

Theorem 2. *If M is a Lévy martingale-valued measure of the form (8), then M is nuclear with ρ being Lebesgue measure on $\mathbb{R}^+$ and*

$$T_A = Q\delta_0(A) + F_{A-\{0\}}, \tag{9}$$

for all $A \in \mathcal{A}$.

The proof follows easily from the above calculations and (6).

It is straightforward to deduce that $(T_A, A \in \mathcal{A})$ is decomposable, wherein $\lambda = \nu + \delta_0$ and

$$T(x) = \begin{cases} Q & \text{if } x = 0 \\ (x, \cdot)_H x & \text{if } x \neq 0. \end{cases}$$

3 Stochastic Integration

3.1 Weak Stochastic Integration

Let M be a martingale-valued measure. Fix $T > 0$. We denote by $\mathcal{H}_2^M(T; S)$ the space of all $\mathcal{P} \otimes \Sigma$-measurable mappings $F : [0, T] \times S \times \Omega \to H$ for which

$$\mathbb{E}\left(\int_0^T \int_S ||F(s, x)||^2 \langle M \rangle (ds, dx) \right) < \infty.$$

Then $\mathcal{H}_2^M(T; S)$ is a real Hilbert space. $\mathcal{S}(T; S)$ is the subspace of all $F \in \mathcal{H}_2^M(T; S)$ for which

$$F = \sum_{i=0}^{N_1} \sum_{j=0}^{N_2} F_{ij} 1_{(t_i, t_{i+1}]} 1_{A_j},$$

where $N_1, N_2 \in \mathbb{N}, 0 = t_0 < t_1 < \cdots < t_{N_1+1} = T, A_0, \ldots, A_{N_2}$ are disjoint sets in $\mathcal{A}$ and each F_{ij} is a bounded $\mathcal{F}_{t_i}$-measurable random variable. $\mathcal{S}(T; S)$ is dense in $\mathcal{H}_2^M(T; S)$ (see e.g. [2], section 4.1). We generalise the construction of stochastic integrals with respect to martingales as developed in [24]. For each $F \in \mathcal{S}(T; S), 0 \leqslant t \leqslant T$, define

$$I_t(F) = \sum_{i=0}^{N_1} \sum_{j=0}^{N_2} (F_{ij}, M((t \wedge t_i, t \wedge t_{i+1}], A_j))_H.$$

Then

$$\mathbb{E}(|I_t(F)|^2)$$
$$= \sum_{i=0}^{N_1} \sum_{j=0}^{N_2} \sum_{k=0}^{N_1} \sum_{l=0}^{N_2} \mathbb{E}[(F_{ij}, M((t \wedge t_i, t \wedge t_{i+1}], A_j))_H$$
$$(F_{kl}, M((t \wedge t_k, t \wedge t_{k+1}], A_l))_H]$$

$$= \sum_{i=0}^{N_1} \sum_{j=0}^{N_2} \mathbb{E}[|(F_{ij}, M((t \wedge t_i, t \wedge t_{i+1}], A_j))_H|^2]$$

$$\leqslant \sum_{i=0}^{N_1} \sum_{j=0}^{N_2} \mathbb{E}[||F_{ij}||^2 . ||M((t \wedge t_i, t \wedge t_{i+1}], A_j)||^2]$$

$$= \mathbb{E}\left[\sum_{i=0}^{N_1} \sum_{j=0}^{N_2} ||F_{ij}||^2 \langle M \rangle ([t \wedge t_i, t \wedge t_{i+1}], A_j)\right]$$

$$= \mathbb{E}\left[\int_0^t \int_S ||F(s,x)||^2 \langle M \rangle (ds, dx)\right].$$

Hence I_t extends to a contraction from $\mathcal{H}_2^M(T; S)$ to $L^2(\Omega, \mathcal{F}, P)$. For each $0 \leqslant t \leqslant T, F \in \mathcal{H}_2^M(T; S)$,

$$\int_0^t \int_S (F(s,x), M(ds, dx))_H := I_t(F).$$

By standard arguments, we see that $(I_t(T); 0 \leqslant t \leqslant T)$ is a centred square-integrable real-valued martingale with

$$\mathbb{E}(|I_t(F)|^2 \leqslant \mathbb{E}\left[\int_0^t \int_S ||F(s,x)||^2 \langle M \rangle (ds, dx)\right],$$

for all $0 \leqslant s \leqslant T, F \in \mathcal{H}_2^M(T; S)$.

3.2 Strong Stochastic Integration

In this section we will take the martingale-valued measure M to be nuclear and decomposable. Let $(R(t,x), t \in [0,T], x \in S)$ be a family of bounded linear operators on H. We say that they are predictable if the mappings $[0,T] \times S \to H$, given by $(t,x) \to R(t,x)\psi$ are $\mathcal{P} \otimes \Sigma$-measurable, for each $\psi \in H$. Our aim is to define $\int_0^t \int_S R(s,x) M(ds, dx)$ as random vectors, for each $t \geqslant 0$. We follow the approach given in section 4.2 of [11] for the case of Brownian motion (see also [28], section 4.22). Let $\mathcal{H}_2(T; \lambda, \rho)$ be the real Hilbert space of all predictable R for which

$$\mathbb{E}\left(\int_0^T \int_S \text{tr}(R(t,x) T_x R(t,x)^*) \lambda(dx) \rho(dt)\right) < \infty. \tag{10}$$

We denote by $S(T; \lambda, \rho)$ the dense linear space of all $R \in \mathcal{H}_2(T; \lambda, \rho)$, which take the form

$$R = \sum_{i=0}^{N_1} \sum_{j=0}^{N_2} R_{ij} 1_{(t_i, t_{i+1}]} 1_{A_j},$$

where $N_1, N_2 \in \mathbb{N}, 0 = t_0 < t_1 < \cdots < t_{N_1+1} = T, A_0, \ldots, A_{N_2}$ are disjoint sets in $\mathcal{A}$ and each R_{ij} is a bounded operator valued $\mathcal{F}_{t_i}$-measurable random variable. For each $R \in S(T; \lambda, \rho), 0 \leqslant t \leqslant T$, define

$$J_t(R) = \sum_{i=0}^{N_1} \sum_{j=0}^{N_2} R_{ij} M((t \wedge t_i, t \wedge t_{i+1}], A_j).$$

Let $(e_n, n \in \mathbb{N})$ be a maximal orthonormal set in H. We compute

$$\mathbb{E}(J_t(R)) = \sum_{n=1}^{\infty} \sum_{i=0}^{N_1} \sum_{j=0}^{N_2} \mathbb{E}((R_{ij}^* e_n, M((t \wedge t_i, t \wedge t_{i+1}], A_j))_H) e_n$$

$$= \sum_{n=1}^{\infty} \sum_{m=1}^{\infty} \sum_{i=0}^{N_1} \sum_{j=0}^{N_2} \mathbb{E}((R_{ij}^* e_n, e_m)_H (e_m, M((t \wedge t_i, t \wedge t_{i+1}], A_j))_H) e_n$$

$$= 0.$$

Similar arguments yield

$$\mathbb{E}(||J_t(R)||^2) = \sum_{i=0}^{N_1} \sum_{j=0}^{N_2} \mathbb{E}(||R_{ij} M((t \wedge t_i, t \wedge t_{i+1}], A_j)||^2)$$

$$= \sum_{n=1}^{\infty} \sum_{i=0}^{N_1} \sum_{j=0}^{N_2} \mathbb{E}(|(R_{ij} M((t \wedge t_i, t \wedge t_{i+1}], A_j), e_n)_H|^2)$$

$$= \sum_{n=1}^{\infty} \sum_{i=0}^{N_1} \sum_{j=0}^{N_2} \mathbb{E}((R_{ij}^* e_n, T_{A_j} R_{ij}^* e_n)_H) \rho((t_i \wedge t, t_{i+1} \wedge t])$$

$$= \sum_{i=0}^{N_1} \sum_{j=0}^{N_2} \mathbb{E}(\mathrm{tr}(R_{ij} T_{A_j} R_{ij}^*)) \rho((t_i \wedge t, t_{i+1} \wedge t])$$

$$= \sum_{i=0}^{N_1} \sum_{j=0}^{N_2} \mathbb{E} \left(\int_{A_j} \mathrm{tr}(R_{ij} T_x R_{ij}^*) \lambda(dx) \right) \rho((t_i \wedge t, t_{i+1} \wedge t]).$$

Hence each J_t extends to an isometry from $\mathcal{H}_2(T; \lambda, \rho)$ into $L^2(\Omega, \mathcal{F}, P; H)$ and we write $\int_0^t \int_S R(s, x) M(ds, dx) := J_t(R)$, for each $0 \leqslant t \leqslant T, R \in \mathcal{H}_2(T; \lambda, \rho)$. The process $(J_t, t \geqslant 0)$ is a square-integrable centred martingale. Henceforth we will always take a strongly càdlàg version.

Notes 1). The condition (10) can be rewritten as

$$\mathbb{E} \left(\int_0^T \int_S ||R(t, x) T_x^{\frac{1}{2}}||_2 \lambda(dx) \rho(dt) \right) < \infty,$$

where $|| \cdot ||_2$ is the Hilbert-Schmidt norm, i.e. $||C||_2^2 = \mathrm{tr}(CC^*)$ for $C \in L(H)$. The set of all $C \in L(H)$ for which $||C||_2 < \infty$ is a Hilbert space with respect to the inner product $(C, D)_2 = \text{tr}(CD^*)$, which we denote as $\mathcal{L}_2(H)$, (see e.g. [34], section VI.6 for further details).

2) $\mathcal{L}_2(H)$ is a two-sided $L(H)$-ideal with $||C_1DC_2||_2 \leqslant ||C_1||.||C_2||.||D||_2$, for all $C_1, C_2 \in L(H), D \in \mathcal{L}_2(H)$. From this we easily deduce that

$$\int_0^T \int_S \mathbb{E}(||R(t,x)||^2)\text{tr}(T_x)\lambda(dx)\rho(dt) < \infty \qquad (11)$$

is a sufficient condition for (10).

3) The construction of this section is easily extended to the conceptually simpler case of deterministic operator-valued families $(R(t,x), t \in [0,T], x \in S)$ satisfying $\int_0^T \int_S \text{tr}(R(t,x)T_xR(t,x)^*)\lambda(dx)\rho(dt) < \infty$.

If $C \in \mathcal{L}(H)$ and $R = (R(t,x), t \in [0,T], x \in S)$, we define $CR = (CR(t,x), t \in [0,T], x \in S)$. We will need the following result in section 4.3 below.

Theorem 3. *If $C \in B(H)$ and $R \in \mathcal{H}_2(T; \lambda, \rho)$ then $CR \in \mathcal{H}_2(T; \lambda, \rho)$ and*

$$C\int_0^t \int_S R(s,x)M(ds, dx) = \int_0^t \int_S CR(s,x)M(ds, dx),$$

for all $t \geqslant 0$.

Proof. $CR \in \mathcal{H}_2(T; \lambda, \rho)$ follows easily from Note 2 above. The identity is immediate if $R \in S(T; \lambda, \rho)$. More generally, let $(R_n, n \in \mathbb{N})$ be a sequence in $S(T; \lambda, \rho)$ converging to $R \in \mathcal{H}_2(T; \lambda, \rho)$, then for all $t \geqslant 0$,

$$\mathbb{E}\left(\left|\left|\int_0^t \int_S CR(s,x)M(ds, dx) - \int_0^t \int_S CR_n(s,x)M(ds, dx)\right|\right|^2\right)$$

$$= \mathbb{E}\left(\left|\left|\int_0^t \int_S C[R(s,x) - R_n(s,x)]M(ds, dx)\right|\right|^2\right)$$

$$= \mathbb{E}\left(\int_0^t \int_S \text{tr}(C[R(s,x) - R_n(s,x)]F_x[R(s,x)^* - R_n(s,x)^*C^*])\lambda(dx)\rho(dt)\right)$$

$$\leqslant ||C||^2\mathbb{E}\left(\int_0^t \int_S \text{tr}([R(s,x) - R_n(s,x)]F_x[R(s,x)^* - R_n(s,x)^*])\lambda(dx)\rho(dt)\right)$$

$$\to 0 \text{ as } n \to \infty,$$

and the result follows. $\qquad\qquad\qquad\qquad\qquad\qquad\qquad\qquad\qquad\qquad\qquad\qquad\square$

3.3 Weak-Strong Compatibility

In this subsection we will assume that the operator-valued family $(R(t,x), t \in [0,T], x \in S)$ is such that the mappings $[0,T] \times S \to H$, given by $(t,x) \to R(t,x)^*\psi$ are $\mathcal{P} \otimes \Sigma$-measurable, for each $\psi \in H$.

Theorem 4. *If M is a decomposable nuclear martingale-valued measure with independent increments and the operator-valued family $(R(t, x), t \in [0, T], x \in S)$ satisfies (11) then for all $0 \leqslant t \leqslant T, \psi \in H$*

$$\left(\psi, \int_0^t \int_S R(s, x) M(ds, dx) \right)_H = \int_0^t \int_S (R(s, x)^* \psi, M(ds, dx))_H. \quad (12)$$

Proof. First note that since (11) holds, the strong integral appearing on the left hand side of (12) exists. The weak integral on the right hand side also exists, since by the independent increments property of M, proposition 2.1 and (11),

$$\mathbb{E} \left(\int_0^T \int_S ||R(s, x)^* \psi||^2 \langle M \rangle (ds, dx) \right) = \int_0^T \int_S \mathbb{E}(||R(s, x)^* \psi||^2) \mathbb{E}(\langle M \rangle (ds, dx))$$

$$\leqslant \int_0^T \int_S \mathbb{E}(||R(s, x)||^2) \mathrm{tr}(T_x) \lambda(dx) \rho(ds) ||\psi||^2$$

$$< \infty.$$

To establish the result, first let $R \in S(T; \lambda, \rho)$, then

$$\left(\psi, \int_0^t \int_S R(s, x) M(ds, dx) \right)_H = \sum_{i=0}^{N_1} \sum_{j=0}^{N_2} (\psi, R_{ij} M((t_i, t_{i+1}], A_j))_H$$

$$= \sum_{i=0}^{N_1} \sum_{j=0}^{N_2} (R_{ij}^* \psi, M((t_i, t_{i+1}], A_j))_H$$

$$= \int_0^t \int_S (R(s, x)^* \psi, M(ds, dx))_H.$$

The general result follows by a straightforward limiting argument. $\qquad \square$

3.4 A Stochastic Fubini Theorem

The result to be established is in some respects quite simple, however it is adequate for our later needs. Let N be a Poisson random measure defined on $\mathbb{R}^+ \times (H - \{0\})$ as in section 1.3 and let ν be its intensity measure, which we will assume to be a Lévy measure. Let $E \in \mathcal{B}(H - \{0\})$. If $F : \mathbb{R}^+ \times H \to \mathbb{R}$ is $\mathcal{P} \otimes E$-measurable and $\int_0^t \int_E \mathbb{E}(|F(s, x)|^2) \nu(dx) ds < \infty$, we can construct the stochastic integral $\int_0^t \int_E F(s, x) \tilde{N}(ds, dx)$. It is a centred square-integrable martingale with

$$\mathbb{E} \left(\left| \int_0^t \int_E F(s, x) \tilde{N}(ds, dx) \right|^2 \right) = \int_0^t \int_E \mathbb{E}(|F(s, x)|^2) \nu(dx) ds,$$

see e.g. Chapter 4 of [2].

Now let $(W, \mathcal{W}, \mu)$ be a finite measure space and let $\mathcal{H}_2(T, E, W)$ be the real Hilbert space of all $\mathcal{P} \otimes \mathcal{B}(E) \otimes \mathcal{W}$-measurable functions G from $[0, T] \times E \times W \to \mathbb{R}$ for which $\int_W \int_0^t \int_E \mathbb{E}(|G(s, x, w)|^2)\nu(dx)ds\mu(dw) < \infty$. The space $S(T, E, W)$ is dense in $\mathcal{H}_2(T, E, W)$, where $G \in S(T, E, W)$ if

$$G = \sum_{i=0}^{N_1} \sum_{j=0}^{N_2} \sum_{k=0}^{N_3} G_{ijk} 1_{(t_i, t_{i+1}]} 1_{A_j} 1_{B_k},$$

where $N_1, N_2, N_3 \in \mathbb{N}, 0 = t_0 < t_1 < \cdots < t_{N_1+1} = T, A_0, \ldots, A_{N_2}$ are disjoint sets in $\mathcal{A}, B_0, \ldots, B_{N_3}$ is a partition of W, wherein each $B_k \in \mathcal{W}$ and each G_{ijk} is a bounded $\mathcal{F}_{t_i}$-measurable random variable.

Theorem 5. *If $G \in \mathcal{H}_2(T, E, W)$, then for each $0 \leqslant t \leqslant T$,*

$$\int_W \left(\int_0^t \int_E G(s, x, w) \tilde{N}(ds, dx) \right) \mu(dy)$$
$$= \int_0^t \int_E \left(\int_W G(s, x, w) \mu(dy) \right) \tilde{N}(ds, dx) \quad a.e. \tag{13}$$

Proof. First note that both integrals in (13) are easily seen to exist in $L^2(\Omega, \mathcal{F}, P)$. If $G \in S(T, E, W)$, then the result holds with both sides of (13) equal to

$$\sum_{i=0}^{N_1} \sum_{j=0}^{N_2} \sum_{k=0}^{N_3} G_{ijk} \tilde{N}((t_i, t_{i+1}], A_j) \mu(B_k).$$

Now suppose that $(G_n, n \in \mathbb{N})$ is a sequence of mappings in $S(T, E, W)$ converging to $G \in \mathcal{H}_2(T, E, W)$, then

$$\mathbb{E}\left(\left| \int_0^t \int_E \left(\int_W [G(s, x, w) - G_n(s, x, w)] \mu(dw) \right) \tilde{N}(ds, dx) \right|^2 \right)$$
$$= \int_0^t \int_E \mathbb{E}\left(\left| \int_W [G(s, x, w) - G_n(s, x, w)] \mu(dw) \right|^2 \right) \nu(dx) ds$$
$$\leqslant \mu(W) \int_0^t \int_E \int_W \mathbb{E}(|G(s, x, w) - G_n(s, x, w)|^2) \mu(dw) \nu(dx) ds$$
$$\to 0 \text{ as } n \to \infty.$$

A similar argument shows that

$$\lim_{n \to \infty} \mathbb{E}\left(\left| \int_W \left(\int_0^t \int_E [G_n(s, x, w) - G(s, x, w)] \tilde{N}(ds, dx) \right) \mu(dy) \right|^2 \right) = 0,$$

and the result follows. $\qquad\qquad\square$

4 Ornstein-Uhlenbeck Processes

4.1 Stochastic Convolution

Let X be a strongly càdlàg Lévy process and let $(S(t), t \geqslant 0)$ be a C_0-semigroup (i.e. a strongly continuous one-parameter semigroup of linear operators) acting in H. Basic facts about such semigroups can be found in e.g. Chapter 1 of [14]. We note in particular that there exists $M > 0, \beta \in \mathbb{R}$ such that

$$||S(t)|| \leqslant Me^{\beta t}, \tag{14}$$

for all $t \geqslant 0$. J will denote the infinitesimal generator of $(S(t), t \geqslant 0)$. It is a closed, densely defined linear operator in H and hence its adjoint J^* is also densely defined.

Let $C \in B(H)$. Our aim in this subsection is to define the *stochastic convolution*

$$X_{J,C}(t) := \int_0^t S(t-s)CdX(s), \tag{15}$$

for all $t \geqslant 0$. We do this by employing the Lévy-Itô decomposition (7) to write each

$$X_{J,C}(t) = \int_0^t S(t-s)C\zeta ds + \int_0^t S(t-s)CdB_Q(s)$$

$$+ \int_0^t \int_{||x||<1} S(t-s)Cx\tilde{N}(ds, dx)$$

$$+ \int_0^t \int_{||x|\geqslant 1} S(t-s)CxN(ds, dx). \tag{16}$$

We need to establish condition under which the process $(X_{J,C}(t), t \geqslant 0)$ exists. To do this we consider each term in (16) in turn. We define $\int_0^t S(t-s)C\zeta ds$ as a standard Bochner integral. Indeed using (14) we obtain

$$\left|\left|\int_0^t S(t-s)C\zeta ds\right|\right| \leqslant \left(\int_0^t ||S(t-s)||ds\right)||C\zeta||$$

$$\leqslant \begin{cases} M\beta^{-1}(e^{\beta t}-1)||C\zeta|| & \text{if } \beta \neq 0 \\ Mt||C\zeta|| & \text{if } \beta = 0. \end{cases}$$

The terms $\int_0^t S(t-s)CdB_Q(s)$ and $\int_0^t \int_{||x||<1} S(t-s)Cx\tilde{N}(ds, dx)$ are dealt with using the (deterministic version) of strong stochastic integration as described in section 3.2. In fact the first of these terms was discussed in [11], section 5.1.2. (see also [7]). Using the estimate (11) we find that $\int_0^t S(t-s)CdB_Q(s)$ exists as a strong integral provided $\left(\int_0^t ||S(t-s)||^2 ds\right) \text{tr}(Q) < \infty$, which by (14) is always satisfied.

Note. [11] impose the weaker condition (10) as they want to explore the degenerate case where $Q = I$. This falls outside the context of the current work as B_Q is not then a Lévy process when H is infinite dimensional.

Again using (11) for the compensated Poisson integral, we must estimate

$$\int_0^t ||S(t-s)||^2 ds \int_{||x||<1} \text{tr}(T_x)\nu(dx) = \int_0^t ||S(t-s)||^2 ds \int_{||x||<1} ||x||^2 \nu(dx)$$
$$< \infty.$$

Hence we see that $\int_0^t \int_{||x||<1} S(t-s)Cx\tilde{N}(ds,dx)$ also always exists.

Finally, we may define the final Poisson integral as a finite (random) sum:

$$\int_0^t \int_{||x||>1} S(t-s)CxN(ds,dx) = \sum_{0 \leqslant s \leqslant t} S(t-s)C\Delta X(s)1_{\{||\Delta X(s)||\geqslant 1\}}(\Delta X(s)).$$

In conclusion, we have established the following:

Theorem 6. *If X is a càdlàg Lévy process, $C \in \mathcal{L}(H)$ and $(S(t), t \geqslant 0)$ is a C_0-semigroup with generator J, the stochastic convolution $X_{J,C}(t) = \int_0^t S(t-s)CdX(s)$ exists in H for all $t \geqslant 0$.*

We note that the process $X_{J,C} = (X_{J,C}(t), t \geqslant 0)$ inherits strongly càdlàg paths from X.

Note. An alternative approach to defining the stochastic convolution is to employ (weak) integration by parts to write, for each $\psi \in \text{Dom}(J^*)$,

$$(\psi, X_{J,C}(t))_H = (\psi, [S(0)CX(t) - S(t)CX(0)])_H + \int_0^t (C^*J^*\psi, S(t-s)X(s-))_H ds.$$

Another approach, using convergence in probability rather than L^2-convergence can be found in [9].

4.2 Existence and Uniqueness for Ornstein-Uhlenbeck Processes

The development of this section closely parallels that of [11], section 5.2. We consider the *generalised Langevin equation* in Hilbert space, i.e.

$$dY(t) = JY(t)dt + CdX(t), \tag{17}$$

with the initial condition $Y(0) = Y_0$ (a.s.), where Y_0 is a given $\mathcal{F}_0$-measurable random variable. We consider (17) as a *weak sense stochastic differential equation*. By this we mean that $Y = (Y(t), t \geqslant 0)$ is a solution to (17) if for all $t \geqslant 0, \psi \in \text{Dom}(J^*)$,

$$(\psi, Y(t) - Y_0)_H = (C^*\psi, X(t))_H + \int_0^t (J^*\psi, Y(s))_H ds. \tag{18}$$

Our candidate solution to (17) is given by the usual stochastic version of the variation of constants formula

$$Z(t) = S(t)Y_0 + \int_0^t S(t-s)CdX(s), \tag{19}$$

for each $t \geqslant 0$. It follows from Theorem 6 that $Z(t)$ exists for all $t \geqslant 0$.

Note. Da Prato and Zabczyk [11] consider (18) in the case where X is a Brownian motion. In their formalism, the operators J and the process X are associated to different Hilbert spaces H_1 and H_2, respectively and C maps H_2 to H_1. Our approach herein is easily extended to this level of generality (in fact one can just take $H = H_1 \oplus H_2$).

Theorem 7. *(19) is the unique weak solution to (18).*

Proof. We extend the argument used to prove theorem 5.4 in [11]. See [9] for an alternative approach.

Existence. First note that if $X_{J,C}$ solves (18) with the initial condition $Y_0 = 0$ (a.s.), then it is clear that Z, as given by (19) solves (18) with the arbitrary initial condition. Hence we may restrict ourselves to the former problem.

For each $t \geqslant 0, \psi \in \text{Dom}(J^*)$, using (12) we obtain

$$
\begin{aligned}
(\psi, X_{J,C}(t))_H - (C^*\psi, X(t))_H &= \left(\psi, \int_0^t [S(t-s) - I]CdX(s)\right)_H \\
&= \int_0^t (C^*[S(t-s)^* - I]\psi, dX(s))_H \\
&= \int_0^t \left(\int_0^{t-s} C^*S(r)^*J^*dr\psi, dX(s)\right)_H \\
&= \int_0^t \left(\int_s^t 1_{[0,r)}(s)C^*S(r-s)^*J^*dr\psi, dX(s)\right)_H.
\end{aligned}
$$

We now need to change the order of integration. Using (16), we employ (13) for the compensated Poisson integral, the stochastic Fubini theorem of [11] (theorem 4.18) for the Brownian integral and the usual Fubini theorem for the Lebesgue integral to deduce that

$$
\begin{aligned}
(\psi, X_{J,C}(t))_H - (C^*\psi, X(t))_H &= \int_0^t \left(J^*\psi, \int_0^t 1_{[0,r)}(s)S(r-s)CdX(s)\right)_H \\
&= \int_0^t (J^*\psi, X_{J,C}(r))_H dr,
\end{aligned}
$$

as was required.

Uniqueness. This is established in exactly the same way as in [11] (pp. 122-3). □

It follows immediately from (19) that Y has strongly càdlàg paths.

Example Let $H = L^2(U)$ where U is a regular domain in $\mathbb{R}^d$. If Δ denotes the usual (Dirichlet) Laplacian acting in H then for each $0 < \alpha < 2$, we can define the fractional power $-(-\Delta^{\frac{\alpha}{2}})$ by e.g. spectral theory, or as a pseudo-differential operator. Indeed when $U = \mathbb{R}^d$, $-(-\Delta^{\frac{\alpha}{2}})$ is a positive self-adjoint operator on the domain $\mathcal{H}_\alpha(\mathbb{R}^d) = \left\{ f \in L^2(\mathbb{R}^d); \int_{\mathbb{R}^d} |v|^{2\alpha} |\hat{f}(v)|^2 dv < \infty \right\}$, where $\hat{f}$ denotes the Fourier transform of f, and $-(-\Delta^{\frac{\alpha}{2}})$ generates a self-adjoint contraction semigroup on H (see e.g. [2], Chapter 3). By the results of this section we know there is a unique weak solution to the equation

$$dY(t) = -(-\Delta^{\frac{\alpha}{2}})Y(t)dt + CdX(t),$$

for any $C \in B(H)$ and any Lévy process X. In particular, one can take X to be α-stable (c.f. [30]).

4.3 Flow and Markov Properties

For each $0 \leqslant s \leqslant t < \infty$, define a two-parameter family of mappings $\Phi_{s,t} : H \times \Omega \to H$ by

$$\Phi_{s,t}(y) = S(t-s)y + \int_s^t S(t-r)CdX(r).$$

The following establishes that $\{\Phi_{s,t}; 0 \leqslant s \leqslant t < \infty\}$ is a *stochastic flow.*

Proposition 4.1 *For all* $0 \leqslant r \leqslant s \leqslant t < \infty$,

$$\Phi_{s,t} \circ \Phi_{r,s} = \Phi_{r,t}.$$

Using the semigroup property and Theorem 3, for each $y \in H$, we obtain

$$\Phi_{s,t}(\Phi_{r,s}(y)) = S(t-s)\Phi_{r,s}(y) + \int_s^t S(t-u)CdX(u)$$

$$= S(t-s)S(s-r)y + S(t-s)\int_r^s S(s-u)CdX(u)$$

$$+ \int_s^t S(t-u)CdX(u)$$

$$= S(t-r)y + \int_r^s S(t-u)CdX(u) + \int_s^t S(t-u)CdX(u)$$

$$= \Phi_{r,t}(y). \qquad \square$$

By the construction of stochastic integrals, we deduce that each $\Phi_{s,t}(y)$ is $\mathcal{G}_{s,t}$-measurable where $\mathcal{G}_{s,t} = \sigma\{X(u) - X(v); s \leqslant u < v \leqslant t\}$, and hence by the independent increment property of X, it follows that $\Phi_{s,t}(y)$ is independent of $\mathcal{F}_s$.

From this fact and Proposition 4.1, we can apply standard arguments (see e.g [2], section 6.4 or [33], section 5.6) to establish the *strong Markov property* for the solution to (18), i.e. if τ is a stopping time with $P(\tau < \infty) = 1$ then for each $f \in b(H), t \geqslant 0$

$$\mathbb{E}(f(Y(\tau + t))|\mathcal{F}_\tau) = \mathbb{E}(f(Y(\tau + t))|Y(\tau)),$$

where $\mathcal{F}_\tau$ is the usual stopped σ-algebra.

By the stationary increments of X it follows that Y is a time-homogeneous Markov process and hence we obtain a contraction semigroup of linear operators $(T_t, t \geqslant 0)$ on $b(H)$ via the prescription

$$(T_t f)(y) = \mathbb{E}(f(Y(t))|Y(0) = y),$$

for each $t \geqslant 0, f \in b(H), y \in H$. We easily verify that $T_t : C_b(H) \subseteq C_b(H)$ for each $t \geqslant 0$, by a routine application of dominated convergence. In fact $(T_t, t \geqslant 0)$ is a *generalised Mehler semigroup* in the sense of [6], [15].

5 Operator Self-decomposability

Generalising ideas developed in [21], we say that a random variable Z is *operator self-decomposable* with respect to a C_0-semigroup $(S(t), t \geqslant 0)$ if for all $t \geqslant 0$, there exists a random variable Z_t which is independent of Z, such that

$$Z \stackrel{d}{=} S(t)Z + Z_t. \tag{20}$$

We aim to show that random variables of the form $Z = \int_0^\infty S(r)dX(r)$, where X is a Lévy process are operator self-decomposable, when the limit makes sense. For each $t \geqslant 0$, we define $\int_0^t S(r)dX(r)$ by employing the Lévy-Itô decomposition, as in (16). We assume throughout this section that the semigroup $(S(t), t \geqslant 0)$ is *exponentially stable*, i.e. (14) holds with $\beta < 0$, e.g. all self-adjoint semigroups whose generator has a spectrum which is bounded away from zero are exponentially stable. In [12], it is shown that a C_0-semigroup $(S(t), t \geqslant 0)$ is exponentially stable if and only if $\int_0^\infty ||S(t)x||^2 dt < \infty$, for all $x \in H$.

Under this assumption, given any sequence $(t_n, n \in \mathbb{N})$ in $[0, \infty)$ with $\lim_{n \to \infty} t_n = \infty$, we can assert the existence of the following limits:

$$\int_0^\infty S(r)\zeta dr = \lim_{n \to \infty} \int_0^{t_n} S(r)\zeta dr$$

$$\int_0^\infty S(r)dB_Q(r) = \lim_{n \to \infty} \int_0^{t_n} S(r)dB_Q(r)$$

$$\int_0^\infty \int_{||x||<1} S(r)x\tilde{N}(dr, dx) = \lim_{n \to \infty} \int_0^{t_n} \int_{||x||<1} S(r)x\tilde{N}(dr, dx),$$

where the first limit is taken in H and the other two in $L^2(\Omega, \mathcal{F}, P; H)$. We need to work harder to consider the limiting behaviour as $t \to \infty$ of $\Pi_{S,N}(t) := \int_0^t \int_{||x|| \geqslant 1} S(r)x N(dr, dx)$.

Lemma 1. *Let $A \in L(H)$ with $||A|| \leqslant 1$ and $(\xi_n, n \in \mathbb{N})$ be a sequence of iid random variables. If $\mathbb{E}(\log(1 + ||\xi_1||)) < \infty$, then $\sum_{n=1}^{\infty} A^n \xi_n$ converges a.s.*

The proof is exactly as in [21], lemma 3.6.5 (p. 121). Note that these authors are able to prove 'if and only if' by assuming that A is invertible. That assumption would be unnatural in our context.

This next result is related to Proposition 1.8.13 in [21], p. 36, although the proof is quite different.

Lemma 2. *Let $f : H \to \mathbb{R}^+$ be measurable and subadditive. If $\int_0^t \int_{||x|| \geqslant 1} f(S(r)x)\nu(dx)dr < \infty$ then $\mathbb{E}(f(\Pi_{S,N}(t))) < \infty$, for each $t \geqslant 0$.*

Proof. By subadditivity of f, for each $t \geqslant 0$ we have

$$f(\Pi_{S,N}(t)) = f \left(\sum_{0 \leqslant r \leqslant t} S(r)\Delta X(r) 1_{\{||\Delta X(s)|| \geqslant 1\}}(\Delta X(r)) \right)$$

$$\leqslant \sum_{0 \leqslant r \leqslant t} f(S(r)\Delta X(r)) 1_{\{||\Delta X(s)|| \geqslant 1\}}(\Delta X(r))$$

$$= \int_0^t \int_{||x|| \geqslant 1} f(S(r)x) N(dr, dx).$$

Hence

$$\mathbb{E}(f(\Pi_{S,N}(t))) \leqslant \mathbb{E} \left(\int_0^t \int_{||x|| \geqslant 1} f(S(r)x) N(dr, dx) \right)$$

$$= \int_0^t \int_{||x|| \geqslant 1} f(S(r)x)\nu(dx)dr < \infty. \qquad \square$$

Theorem 8. *(c.f. [39], [22], [21]) Let $(S(t), t \geqslant 0)$ be an exponentially stable contraction semigroup in H. If $\int_{||x|| \geqslant 1} \log(1 + ||x||)\nu(dx) < \infty$ then $\lim_{t \to \infty} \int_0^t \int_{||x|| < 1} S(r)x N(dr, dx)$ exists in distribution.*

Proof. We follow the approach of [21], theorem 3.6.6 (p.123). By stationary increments of X and the semigroup property, for each $n \in \mathbb{N}$,

$$\int_0^n \int_{||x|| \geqslant 1} S(r)x N(dr, dx) = \sum_{k=0}^{n-1} \int_k^{k+1} \int_{||x|| \geqslant 1} S(r)x N(dr, dx)$$

$$= \sum_{k=0}^{n-1} \int_0^1 \int_{||x|| \geqslant 1} S(r+k)x N(dr+k, dx)$$

$$= \sum_{k=0}^{n-1} S(1)^k \int_0^1 \int_{||x|| \geqslant 1} S(r) x N(dr+k, dx)$$

$$\stackrel{d}{=} \sum_{k=0}^{n-1} S(1)^k \int_0^1 \int_{||x|| \geqslant 1} S(r) x N(dr, dx)$$

$$\stackrel{d}{=} \sum_{k=0}^{n-1} S(1)^k M_k,$$

where each $M_k := \int_k^{k+1} \int_{||x|| \geqslant 1} S(r-k) x N(dr, dx)$. The M_k's are independent by the independent increment property of N. Moreover by the stationary increment property of N, each

$$M_k = \int_0^1 \int_{||x|| \geqslant 1} S(r) x N(dr+k, dx) \stackrel{d}{=} \int_0^1 \int_{||x|| \geqslant 1} S(r) x N(dr, dx).$$

We deduce the convergence in distribution as $n \to \infty$ of $\int_0^n S(r) x N(dr, dx)$ by lemmas 1 and 2 together with the estimate

$$\int_0^t \int_{||x|| \geqslant 1} \log(1 + ||S(r)x||) \nu(dx) dr \leqslant \int_0^t \int_{||x|| \geqslant 1} \log(1 + ||x||) \nu(dx) dr$$

$$= t \int_{||x|| \geqslant 1} \log(1 + ||x||) \nu(dx).$$

Now let $(s_n, n \in \mathbb{N})$ be an arbitrary sequence in $[0, 1]$. By stationary increments of N, for each $n \in \mathbb{N}$,

$$\int_n^{n+s_n} \int_{||x|| \geqslant 1} S(r) x N(dr, dx) \stackrel{d}{=} S(n) \int_0^{s_n} \int_{||x|| \geqslant 1} S(r) x N(dr, dx).$$

Since $t \to \int_0^t S(r) x N(dr, dx)$ is a.s. càdlàg, we deduce that

$$\left\| S(n) \int_0^{s_n} \int_{||x|| \geqslant 1} S(r) x N(dr, dx) \right\|$$

$$\leqslant ||S(n)|| \sup_{t \in [0,1]} \left\| \int_0^t S(r) \int_{||x|| \geqslant 1} x N(dr, dx) \right\|$$

$$\to 0 \text{ as } n \to \infty \text{ a.s.}$$

Hence, given any sequence $(t_n, n \in \mathbb{N})$ diverging to ∞, we can deduce the convergence in distribution as $t_n \to \infty$ of

$$\int_0^{t_n} S(r) \int_{||x|| \geqslant 1} x N(dr, dx) = \int_0^{[t_n]} \int_{||x|| \geqslant 1} S(r) x N(dr, dx)$$

$$+ \int_{[t_n]}^{t_n} \int_{||x|| \geqslant 1} S(r) x N(dr, dx). \qquad \square$$

Note. In [9] it is shown that the following conditions are necessary and sufficient for the existence (in distribution) of $\lim_{t \to \infty} \int_0^t S(r) dK(r)$ where K is the jump part of X, i.e. $K(t) = X(t) - t\zeta - B_Q(t)$, for each $t \geqslant 0$:

$$\int_0^\infty \int_{H-\{0\}} (||S(r)x||^2 \wedge 1) \nu(dx) dr \quad < \quad \infty$$

$$\lim_{t \to \infty} \int_0^t \int_{H-\{0\}} S(r)x[1_{B_1}(S(r)(x)) - 1_{B_1}(x)] \nu(dx) ds \quad \text{exists} . \quad (21)$$

These may be difficult to verify in practice.

The main result of this section is the following:

Theorem 9. *If $(S(t), t \geqslant 0)$ is an exponentially stable contraction semigroup in H and X is a Lévy process with Lévy measure ν for which $\int_{||x|| \geqslant 1} \log(1 + ||x||) \nu(dx) < \infty$, then $\lim_{t \to \infty} \int_0^t S(r) dX(r)$ exists in distribution and is operator self-decomposable with respect to $(S(t), t \geqslant 0)$.*

Proof. It follows from the Lévy-Itô decomposition that $\int_0^t S(r) dX(r) - \Pi_{S,N}(t)$ and $\Pi_{S,N}(t)$ are independent. Since each of these terms converges in distribution as $t \to \infty$, it follows that their sum also does. For the self-decomposability, we define $Z = \int_0^\infty S(r) dX(r)$, then

$$Z = \int_0^t S(r) dX(r) + \int_t^\infty S(r) dX(r),$$

and these terms are independent, by the independent increment property of X. Now

$$\int_t^\infty S(r) dX(r) = \int_0^\infty S(r + t) dX(r + t) \overset{d}{=} S(t) \int_0^\infty S(r) dX(r),$$

by the stationary increment property of X. Hence we have (20) with $Z_t = \int_0^t S(r) dX(r)$. $\qquad \square$

Finally, there is an interesting link between self-decomposability and Ornstein-Uhlenbeck processes (c.f. [39], [4] for the finite-dimensional case).

Suppose that X is a Lévy process with characteristics (ζ, Q, ν), and define the process $\tilde{X} = (-X(t), t \geqslant 0)$, then $\tilde{X}$ is a Lévy process with characteristics $(-\zeta, Q, \tilde{\nu})$, where $\tilde{\nu}(A) = \nu(-A)$, for all $A \in \mathcal{B}(H - \{0\})$. In the following, we define $(X(t), t < 0)$ to be an independent copy of $\tilde{X}$.

We recall the Ornstein-Uhlenbeck process (19)

$$Y(t) = S(t)Y_0 + \int_0^t S(t - s) dX(s),$$

for each $t \geqslant 0$, where we have taken $C = I$.

Theorem 10. *If the Ornstein-Uhlenbeck process $(Y(t), t \geqslant 0)$ is stationary, then Y_0 is self-decomposable. Conversely, if $(S(t), t \geqslant 0)$ is an exponentially stable contraction semigroup in H and $\int_{||x|| \geqslant 1} \log(1 + ||x||)\nu(dx) < \infty$, then there exists a self-decomposable Y_0 such that $Y = (Y(t), t \geqslant 0)$ is stationary.*

Proof. Suppose that $Y = (Y(t), t \geqslant 0)$ is stationary, then for each $t \geqslant 0$,

$$Y_0 \stackrel{d}{=} Y(t) = S(t)Y_0 + \int_0^t S(t-r)dX(r),$$

so Y_0 is self-decomposable. Conversely, define $Y_0 := \int_{-\infty}^0 S(-r)dX(r)$, then Y_0 is self-decomposable by theorem 9. By theorem 3 and the semigroup property, for each $t \geqslant 0$, we have

$$Y(t) = \int_{-\infty}^0 S(t-r)dX(r) + \int_0^t S(t-r)dX(r)$$
$$= \int_{-\infty}^t S(t-r)dX(r).$$

Clearly $Y(t+h) \stackrel{d}{=} Y(t)$, for each $h > 0$. More generally, by stationary increments of Z we can easily deduce (as in [2], theorem 4.3.16) that

$$\mathbb{E}\left(\exp\left\{i\sum_{j=1}^n (u_j, Y(t_j+h))_H\right\}\right) = \mathbb{E}\left(\exp\left\{i\sum_{j=1}^n (u_j, Y(t_j))_H\right\}\right),$$

for each $n \in \mathbb{N}, u_1, \ldots, u_n \in H, t_1, \ldots, t_n \in \mathbb{R}^+$ □.

Note. In [9], it is shown that the conditions (21) are necessary and sufficient for Y to have a stationary solution and the condition $\int_{||x|| \geqslant 1} \log(1 + ||x||)\nu(dx) < \infty$ is demonstrated to be sufficient for these to hold. In [8] an example is constructed which demonstrates that this condition is not necessary when H is infinite dimensional.

References

1. S. Albeverio, B. Rüdiger, Stochastic integrals and the Lévy-Itô decomposition on separable Banach spaces, *Stoch. Anal. and Applns.* **23**, 217-53 (2005)
2. D. Applebaum, *Lévy Processes and Stochastic Calculus*, Cambridge University Press (2004)
3. D. Applebaum, J-L. Wu, Stochastic partial differential equations driven by Lévy space-time white noise, *Random Operators and Stochastic Equations* **8**, 245–61 (2000)
4. O.E. Barndorff-Nielsen, J.L. Jensen, M. Sørensen, Some stationary processes in discrete and continuous time, *Adv. Appl. Prob.* **30**, 989–1007 (1998)

5. O.E. Barndorff-Nielsen, N. Shephard, Non-Gaussian Ornstein-Uhlenbeck-based models and some of their uses in financial economics, *J. R. Statis. Soc. B*, **63**, 167–241 (2001)
6. V.I. Bogachev, M. Röckner, B. Schmuland, Generalised Mehler semigroups and applications, *Prob. Th. Rel. Fields* **105**, 193–225 (1996)
7. Z. Brzeźniak, On stochastic convolution in Banach spaces and applications, *Stoch. and Stoch. Rep.* **61**, 245–95 (1997)
8. A. Chojnowska-Michalik, Stationary distributions for ∞-dimensional linear equations with general noise, in *Lect. Notes Control and Inf. Sci.* **69**, Springer-Verlag, Berlin, 14–25 (1985)
9. A. Chojnowska-Michalik, On processes of Ornstein-Uhlenbeck type in Hilbert space, *Stochastics* **21**, 251–86 (1987)
10. D.L. Cohn, *Measure Theory*, Birkhaüser, Boston (1980)
11. G. Da Prato, J. Zabczyk, *Stochastic Equations in Infinite Dimensions*, Cambridge University Press (1992)
12. R. Datko, Extending a theorem of A.M.Liapunov to Hilbert space, *J. Math. Anal. Appl.* **32**, 610–6 (1970)
13. E.B. Davies, *Quantum Theory of Open Systems*, Academic Press (1976)
14. E.B. Davies, *One-Parameter Semigroups*, Academic Press (1980)
15. M. Fuhrman, M. Röckner, Generalized Mehler semigroups: the non-Gaussian case, *Potential Anal.* **12**, 1–47 (2000)
16. J. Jacod, Grossissement de filtration et processus d'Ornstein-Uhlenbeck généralisé, in Séminaire de Calcul Stochastique, eds Th. Jeulin, M. Yor, *Lecture Notes in Math.* **1118**, Springer, Berlin 36–44 (1985)
17. R. Jajte, On stable distributions in Hilbert space, *Studia Math.* **30**, 63–71 (1968)
18. M. Jeanblanc, J. Pitman, M. Yor, Self-similar processes with independent increments associated with Lévy and Bessel processes, *Stoch. Proc. Appl.* **100**, 223–31 (2002)
19. Z.J. Jurek, An integral representation of operator-self-decomposable random variables, *Bull. Acad. Pol. Sci.* **30**, 385–93 (1982)
20. Z.J. Jurek, Limit distributions and one-parameter groups of linear operators on Banach spaces, *J. Multivar. Analysis* **13**, 578–604 (1983)
21. Z.J. Jurek, J.D. Mason, *Operator-Limit Distributions in Probability Theory*, J.Wiley and Sons Ltd (1993)
22. Z.J. Jurek, W. Vervaat, An integral representation for selfdecomposable Banach space valued random variables, *Z.Wahrscheinlichkeitstheorie verw. Gebiete* **62**, 247–62 (1983)
23. N. El Karoui, S. Méléard, Martingale measures and stochastic calculus, *Probab. Th. Rel. Fields* **84**, 83–101 (1990)
24. H. Kunita, Stochastic integrals based on martingales taking values in Hilbert space, *Nagoya Math. J.* **38**, 41–52 (1970)
25. G. Leha, G. Ritter, On diffusion processes and their semigroups in Hilbert spaces with an application to interacting stochastic systems, *Ann. Prob.* **12**, 1077–1112 (1984)
26. W. Linde, *Probability in Banach Spaces - Stable and Infinitely Divisible Distributions*, Wiley-Interscience (1986)
27. V. Mandrekar, B. Rüdiger, Existence and uniqueness of pathwise solutions for stochastic integral equations driven by non-Gaussian noise on separable Banach spaces, preprint (2003)

28. M. Métivier, *Semimartingales, a Course on Stochastic Processes*, W. de Gruyter and Co., Berlin (1982)
29. M. Métivier, J. Pellaumail, *Stochastic Integration*, Academic Press, New York (1980)
30. L. Mytnik, Stochastic partial differential equation driven by stable noise, *Probab. Th. Rel. Fields* **123**, 157–201 (2002)
31. E. Nicolato, E. Vernados, Option pricing in stochastic volatility models of the Ornstein-Uhlenbeck type, *Math. Finance* **13**, 445–66 (2003)
32. K.R. Parthasarathy, *Probability Measures on Metric Spaces*, Academic Press, New York (1967)
33. P. Protter, *Stochastic Integration and Differential Equations*, Springer-Verlag, Berlin Heidelberg (1992)
34. M. Reed, B. Simon, *Methods of Modern Mathematical Physics, Volume 1 : Functional Analysis* (revised and enlarged edition), Academic Press (1980)
35. K.-I. Sato, *Lévy Processes and Infinite Divisibility*, Cambridge University Press (1999)
36. K.-I. Sato, M. Yamazato, Stationary processes of Ornstein-Uhlenbeck type, in *Probability and Mathematical Statistics*, ed. K.Itô and J.V. Prohorov. Lecture Notes in Mathematics **1021**, 541–51, Springer-Verlag, Berlin (1982)
37. K.-I. Sato, M. Yamazato, Operator-selfdecomposable distributions as limit distributions of processes of Ornstein-Uhlenbeck type, *Stoch. Proc. Appl.* **17**, 73–100 (1984)
38. J.B. Walsh, An introduction to stochastic partial differential equations, in Ecole d'Été de Probabilités de St. Flour XIV, pp. 266–439, Lect. Notes in Math. **1180**, Springer-Verlag, Berlin (1986)
39. S.J. Wolfe, On a continuous analogue of the stochastic difference equation $X_n = \rho X_{n-1} + B_n$, *Stoch. Proc. Appl.* **12**, 301–12 (1982)

Sandwiched Filtrations and Lévy Processes

À la mémoire de Paul André Meyer

Michel Émery

IRMA, 7 rue René Descartes, F-67084 Strasbourg Cedex, France
e-mail: emery@math.u-strasbg.fr

> Les processus à accroissements indépendants, stationnaires et positifs (appelés ci-dessous, pour abréger, processus de LÉVY, bien que LÉVY ait inventé quantité d'autres processus!) sont des processus extraordinaires.
>
> P. A. MEYER. *Séminaire de Probabilités III.*

Summary. A simple example, namely the filtrations of two Lévy processes, shows that a filtration sandwiched (in the sense of immersions) between two filtrations isomorphic to each other, need not be isomorphic to them.

1 Isomorphic probability spaces

This section fixes the vocabulary and contains nothing new, on the contrary: it recalls a few of the classical facts established, at the dawn of abstract measure theory, by such pioneers as Carathéodory, Halmos, Kolmogorov, von Neumann, Maharam, Rohlin, etc.

A.s. equality is an equivalence relation between events; to any probability space $(\Omega, \mathcal{A}, \mathbb{P})$ one can associate the quotient σ-field $\mathcal{A}/\mathbb{P}$, whose elements are the classes of a.s. equal events. It is an abstract σ-field endowed with a probability ('abstract' meaning that it is not a σ-field of subsets of some set).

Definition 1. *Two probability spaces* $(\Omega', \mathcal{A}', \mathbb{P}')$ *and* $(\Omega'', \mathcal{A}'', \mathbb{P}'')$ *will be called* isomorphic *if there exists a bijection between* $\mathcal{A}'/\mathbb{P}'$ *and* $\mathcal{A}''/\mathbb{P}''$ *which preserves the abstract σ-field structure and the probability.*

We choose this definition, instead of the existence of a measure-preserving bijection between full subsets of Ω' and Ω'', because our interest will focus on filtrations. There is not much difference, though, between both definitions: as early as 1932, von Neumann showed in [7] that for Polish spaces with their Borel σ-fields, isomorphism implies the existence of a bimeasurable, measure-preserving bijection between subsets of full measure. But if $\mathcal{F}$ is a filtration on $(\Omega, \mathcal{A}, \mathbb{P})$, quotienting Ω so that $\mathcal{F}_t$ separates points would lead to a whole family (Ω_t) of different quotient spaces; we prefer keeping Ω fixed at the cost of loosening the isomorphisms.

With definition 1, a probability space $(\Omega, \mathcal{A}, \mathbb{P})$ is always isomorphic to its own completion $(\Omega, \bar{\mathcal{A}}, \mathbb{P})$. An isomorphism Ψ between $(\Omega', \mathcal{A}', \mathbb{P}')$ and $(\Omega'', \mathcal{A}'', \mathbb{P}'')$ easily extends to random variables: it yields a bijection (also denoted by Ψ) from $L^0(\Omega', \mathcal{A}', \mathbb{P}')$ to $L^0(\Omega'', \mathcal{A}'', \mathbb{P}'')$ which preserves all Borel relations between countably many random variables and also preserves the laws of all random variables. If $\mathcal{B}'$ is a sub-σ-field of $\mathcal{A}'$, there exists a sub-σ-field $\Psi\mathcal{B}'$ of $\mathcal{A}''$, unique if we require it to contain all negligible events in $\mathcal{A}''$, that corresponds to $\mathcal{B}'$ by the isomorphism; if furthermore X' is any r.v. in $L^1(\Omega', \mathcal{A}', \mathbb{P}')$, then $\mathbb{E}''[\Psi X' | \Psi\mathcal{B}'] = \Psi(\mathbb{E}'[X'|\mathcal{B}'])$. This property and similar ones are trivial and will be admitted without further ado in the sequel.

Recall that a probability space $(\Omega, \mathcal{A}, \mathbb{P})$ is called *essentially separable* if $\mathcal{A}$ is generated by its null sets and countably many events, or equivalently by its null sets and some real r.v., or its null sets and countably many r.v. This is also equivalent to the Banach space $L^1(\Omega, \mathcal{A}, \mathbb{P})$ (or $L^2(\Omega, \mathcal{A}, \mathbb{P})$) being separable; consequently, if $(\Omega, \mathcal{A}, \mathbb{P})$ is essentially separable, so is also $(\Omega, \mathcal{B}, \mathbb{P})$ for any sub-σ-field $\mathcal{B}$ of $\mathcal{A}$.

The classification of essentially separable probability spaces up to isomorphism involves only the masses of their atoms. Recall that an *atom* of $(\Omega, \mathcal{A}, \mathbb{P})$ is an event $A \in \mathcal{A}$ (more precisely, $A \in \mathcal{A}/\mathbb{P}$) such that $\mathbb{P}[A] > 0$ and all r.v. are a.s. constant on A; equivalently, $\mathbb{P}[A] > 0$ and every measurable subset of A has probability 0 or $\mathbb{P}[A]$. To each probability space one can associate the list of measures of its atoms, in decreasing order (the list may be empty, finite or infinite; it contains k times the number p iff there are k atoms with probability p).

Theorem 1. *Let $(\Omega', \mathcal{A}', \mathbb{P}')$ and $(\Omega'', \mathcal{A}'', \mathbb{P}'')$ be two essentially separable probability spaces.*

They are isomorphic if and only if their lists ℓ' and ℓ'' of masses of atoms are the same. (In particular, all essentially separable, diffuse probability spaces are isomorphic to each other.)

The following three conditions are equivalent:

(i) there exists a sub-σ-field $\mathcal{B}'$ of $\mathcal{A}'$ such that $(\Omega', \mathcal{B}', \mathbb{P}')$ is isomorphic to $(\Omega'', \mathcal{A}'', \mathbb{P}'')$;

(ii) there exists a measurable, measure-preserving map from $(\Omega', \bar{\mathcal{A}}', \mathbb{P}')$ to $(\Omega'', \mathcal{A}'', \mathbb{P}'')$;

(iii) there exists a map $f : \ell' \to \ell''$ such that

$$\forall p'' \in \ell'' \qquad p'' \geqslant \sum_{\substack{p' \in \ell' \\ f(p') = p''}} p' .$$

Since essentially separable spaces are generated by real-valued random variables, the first half of the theorem is classically obtained by working with probabilities on the real line; the second half follows from the first one.

Corollary 1 (sandwiched probability spaces). *Let $(\Omega, \mathcal{A}, \mathbb{P})$ be essentially separable and $\mathcal{C} \subset \mathcal{B}$ be two sub-σ-fields of $\mathcal{A}$. If $(\Omega, \mathcal{C}, \mathbb{P})$ is isomorphic to $(\Omega, \mathcal{A}, \mathbb{P})$, then $(\Omega, \mathcal{B}, \mathbb{P})$ is isomorphic to both of them.*

Proof. As $\mathcal{A}$ and $\mathcal{C}$ have the same list $\ell = (p_1 \geqslant p_2 \geqslant \ldots)$ of probabilities of atoms, the probability of the heaviest atom of $\mathcal{B}$ must be at least p_1 since $\mathcal{B} \subset \mathcal{A}$ and at most p_1 since $\mathcal{C} \subset \mathcal{B}$, so it must be p_1; then similarly the second heaviest atom of $\mathcal{B}$ must have mass p_2, etc.   □

Remark. Essential separability can be dispensed of in this corollary: the same conclusion holds when $\mathcal{A}$ is not essentially separable. The proof is similar, but uses the Kolmogorov-Maharam classification of probability spaces instead of Theorem 1.

The question we shall be interested in is, do filtrations verify a sandwich property similar to Corollary 1? Before addressing this question, we recall the analogue of Theorem 1 for σ-finite measures, which will be needed later; its proof is similar to that of Theorem 1.

If $(\Omega, \mathcal{A}, \mu)$ is a measure space, with μ positive and σ-finite, one can consider the masses of its atoms, which form a countable[1] collection $c(\mu)$ of strictly positive numbers, with finite or infinite repetitions allowed. One also considers the total mass $\delta(\mu) \leqslant \infty$ of the diffuse part of μ; naturally,
$$\delta(\mu) + \sum_{m \in c(\mu)} m = \mu(\Omega) \leqslant \infty.$$

Theorem 2. *Let $(\Omega', \mathcal{A}', \mu')$ and $(\Omega'', \mathcal{A}'', \mu'')$ be two separable measurable spaces endowed with positive, σ-finite measures.*
They are isomorphic if and only if $c(\mu') = c(\mu'')$ and $\delta(\mu') = \delta(\mu'')$.
The following three conditions are equivalent:

(i) there exists a sub-σ-field $\mathcal{B}'$ of $\mathcal{A}'$ such that $(\Omega', \mathcal{B}', \mu')$ is isomorphic to $(\Omega'', \mathcal{A}'', \mu'')$;
(ii) there exists a measurable, measure-preserving map from $(\Omega', \bar{\mathcal{A}}', \mu')$ to $(\Omega'', \mathcal{A}'', \mu'')$;
(iii) there exist two maps $f : c(\mu') \to c(\mu'')$ and $u : c(\mu'') \to [0, \infty]$ such that

$$\delta(\mu') = \delta(\mu'') + \sum_{p'' \in c(\mu'')} u(p'') \quad and \quad \forall p'' \in c(\mu'') \quad p'' = u(p'') + \sum_{\substack{p' \in c(\mu') \\ f(p') = p''}} p' .$$

The following two conditions are equivalent:

(iv) there exist an $A' \in \mathcal{A}'$ and a measurable, measure-preserving map from $(\Omega', \bar{\mathcal{A}}', \mathbf{1}_{A'} \cdot \mu')$ to $(\Omega'', \mathcal{A}'', \mu'')$;

---

[1] Empty, finite, or countably infinite.

(v) there exist a sub-collection $c' \subset c(\mu')$, a map $f : c' \to c(\mu'')$ and a map $u : c(\mu'') \to [0, \infty]$ such that

$$\delta(\mu') \geqslant \delta(\mu'') + \sum_{p'' \in c(\mu'')} u(p'') \quad \text{and} \quad \forall p'' \in c(\mu'') \quad p'' = u(p'') + \sum_{\substack{p' \in c' \\ f(p') = p''}} p' \; .$$

In conditions (iii) and (v), $f(p'')$ indicates which atoms of Ω' (or A') coalesce to form an atom of Ω''; $u(p'')$ is the amount of diffuse mass from μ' (or $\mathbf{1}_{A'} \cdot \mu'$) which contributes to that atom.

2 Filtrations: inclusions

Given $(\Omega, \mathcal{A}, \mathbb{P})$, an increasing, right-continuous family $\mathcal{F} = (\mathcal{F}_t)_{t \geqslant 0}$ of sub-σ-fields of $\mathcal{A}$ is called a filtration, and $(\Omega, \mathcal{A}, \mathbb{P}, \mathcal{F})$ is called a filtered probability space. The σ-field $\bigvee_{t \geqslant 0} \mathcal{F}_t$ is denoted by $\mathcal{F}_\infty$.

Definition 2. *Given two filtered probability spaces $(\Omega', \mathcal{A}', \mathbb{P}', \mathcal{F}')$ and $(\Omega'', \mathcal{A}'', \mathbb{P}'', \mathcal{F}'')$, one says that the filtrations $\mathcal{F}'$ and $\mathcal{F}''$ are isomorphic, and one writes $\mathcal{F}' \approx \mathcal{F}''$, if there exists an isomorphism Ψ between $(\Omega', \mathcal{F}'_\infty, \mathbb{P}')$ and $(\Omega'', \mathcal{F}''_\infty, \mathbb{P}'')$ such that $\Psi \mathcal{F}'_t = \mathcal{F}''_t$ for each $t \geqslant 0$.*
This definition is an abuse of language, for such an isomorphism involves not only the filtrations $\mathcal{F}'$ and $\mathcal{F}''$, but also the probabilities $\mathbb{P}'$ and $\mathbb{P}''$.

Definition 3. *Let $\mathcal{F}$ and $\mathcal{G}$ be two filtrations on the same probability space. We say that $\mathcal{F}$ is included in $\mathcal{G}$ (and we write $\mathcal{F} \subset \mathcal{G}$) if $\mathcal{F}_t \subset \mathcal{G}_t$ for all t. (One also says that $\mathcal{F}$ is a sub-filtration of $\mathcal{G}$, or that $\mathcal{G}$ is an enlargement of $\mathcal{F}$.)*
   Let $(\Omega', \mathcal{A}', \mathbb{P}', \mathcal{F}')$ and $(\Omega'', \mathcal{A}'', \mathbb{P}'', \mathcal{F}'')$ be two filtered probability spaces. We say that $\mathcal{F}'$ is *includable* into $\mathcal{F}''$ (and we write $\mathcal{F}' \underset{\sim}{\subset} \mathcal{F}''$) if there exists a probability space $(\Omega, \mathcal{A}, \mathbb{P})$ endowed with two filtrations $\mathcal{F}^*$ and $\mathcal{F}^{**}$, respectively isomorphic to $\mathcal{F}'$ and $\mathcal{F}''$, and such that $\mathcal{F}^* \subset \mathcal{F}^{**}$.
Observe that in the definition of 'includable' it is always possible to choose $(\Omega, \mathcal{A}, \mathbb{P}, \mathcal{F}^{**})$ equal to $(\Omega'', \mathcal{A}'', \mathbb{P}'', \mathcal{F}'')$. But one cannot always take $\Omega = \Omega'$ and $\mathcal{F}^* = \mathcal{F}'$, for there may not be enough room inside $\mathcal{A}'$ (or even inside Ω') to accommodate an enlargement of $\mathcal{F}'$ isomorphic to $\mathcal{F}''$.
   The transitivity property $\mathcal{F}' \underset{\sim}{\subset} \mathcal{F}'' \underset{\sim}{\subset} \mathcal{F}''' \implies \mathcal{F}' \underset{\sim}{\subset} \mathcal{F}'''$ is readily verified.
   In view of Corollary 1, one may be tempted to ask: If three filtrations $\mathcal{F}, \mathcal{G}$ and $\mathcal{H}$ verify $\mathcal{F} \subset \mathcal{G} \subset \mathcal{H}$ and if $\mathcal{F}$ and $\mathcal{H}$ are isomorphic, is $\mathcal{G}$ isomorphic to them? Inclusion of filtrations is so loose a property that this is very far from being true; we shall see it by simply putting together two known facts.
   Call $\mathcal{B}$ "the" filtration of a one-dimensional Brownian motion started at the origin (this filtration is unique up to isomorphisms).

The first known fact coming into play says that $\mathcal{B}$ is includable into many filtrations:

Proposition 1 (Jeulin [6]). *Let $\mathcal{F}$ be some filtration. One has $\mathcal{B} \subseteq \mathcal{F}$ if and only if, for each $t > 0$, $\mathcal{F}_t$ has no atom.*

Saying that $\mathcal{B}$ is includable into $\mathcal{F}$ amounts to saying that there exists an $\mathcal{F}$-adapted Brownian motion (which may be an $\mathcal{F}$-Brownian motion or not!); saying that $\mathcal{F}_t$ has no atom is equivalent to saying that there exists an $\mathcal{F}_t$-measurable r.v. with diffuse law.

The second fact is that many filtrations are includable into $\mathcal{B}$:

Proposition 2 (Vershik). *Let $\mathcal{F}$ be some filtration. One has $\mathcal{F} \subseteq \mathcal{B}$ if and only if $\mathcal{F}_0$ is degenerate and $\mathcal{F}_\infty$ is essentially separable.*

This statement has been know to Vershik for many years; it is a straightforward corollary, in continuous time, of his lacunary isomorphism theorem concerning decreasing sequences of σ-fields (see [9]). This theorem is a deep and powerful result, and apparently the proof of Proposition 2 does not use its full power; so maybe Proposition 2 admits a more elementary proof.

Proof. 'Only if' is trivial. Conversely, suppose $\mathcal{F}_0$ to be degenerate and $\mathcal{F}_\infty$ essentially separable. On some auxiliary sample space Ω', let $(U_k)_{k \geqslant 0}$ be an independent sequence of uniform random variables, and call $\mathcal{S} = (\mathcal{S}_k)_{k \geqslant 0}$ the reverse filtration in discrete time defined by $\mathcal{S}_k = \sigma(U_k, U_{k+1}, \ldots)$. Extract from $\mathcal{F}$ a reverse filtration $\mathcal{G} = (\mathcal{G}_n)_{n \geqslant 0}$ by setting $\mathcal{G}_n = \mathcal{F}_{1/n}$. As $\bigcap_n \mathcal{G}_n = \mathcal{F}_0$ is degenerate and $\mathcal{G}_0 = \mathcal{F}_\infty$ is essentially separable, Vershik's lacunary isomorphism theorem gives the existence of a strictly increasing sequence $(n_k)_{k \geqslant 0}$ with $n_0 = 0$, such that the reverse filtration $\mathcal{G}'$ defined by $\mathcal{G}'_k = \mathcal{G}_{n_k}$ is includable[2] into $\mathcal{S}$. So, putting $\tau_k = 1/n_k$, there exists on Ω' a filtration $\mathcal{F}'$ isomorphic to $\mathcal{F}$ and such that $\mathcal{F}'_{\tau_k} \subset \sigma(U_k, U_{k+1}, \ldots)$, with $\tau_0 = \infty$ and $\tau_k \downarrow 0$. One obtains a Brownian motion B stopped at τ_1 and whose natural filtration contains $\mathcal{F}'$, by defining B in such a way that, for each $k \geqslant 0$, the process $(B_t - B_{\tau_{k+2}}, \; t \in [\tau_{k+2}, \tau_{k+1}])$ generates the same σ-field as the r.v. U_k. It follows that $\mathcal{F} \subseteq \mathcal{B}^{\tau_1]}$, where $\mathcal{B}^{\tau_1]}$ denotes the Brownian filtration stopped at τ_1; and a fortiori, $\mathcal{F} \subseteq \mathcal{B}$. □

Corollary 2. *A filtration $\mathcal{F}$ verifies $\mathcal{B} \subseteq \mathcal{F} \subseteq \mathcal{B}$ if and only if $\mathcal{F}_0$ is degenerate, $\mathcal{F}_\infty$ is essentially separable, and, for each $t > 0$, $\mathcal{F}_t$ has no atom.*

Proof. Trivial from Propositions 1 and 2.

This corollary applies, for instance, if $\mathcal{F}$ is the natural filtration of a Lévy process with infinite Lévy measure. But such a filtration is very far from being isomorphic to $\mathcal{B}$, because it has many purely discontinuous martingales, whereas all $\mathcal{B}$-martingales are continuous.

---

[2] Vershik's theorem actually yields a $\mathcal{G}'$ *immersible* into $\mathcal{S}$. Immersibility is much stronger than mere includability, and will be defined in the next section.

Corollary 3. *Let $\mathcal{F}$ be a filtration such that $\mathcal{F}_0$ is degenerate and $\mathcal{F}_\infty$ is essentially separable; let $\mathcal{G}$ be a filtration such that, for each $t > 0$, $\mathcal{G}_t$ has no atom. Then $\mathcal{F}$ is includable into $\mathcal{G}$.*

Proof. Propositions 1 and 2 give $\mathcal{F} \underset{\sim}{\subseteq} \mathcal{B} \underset{\sim}{\subseteq} \mathcal{G}$, and the result follows by transitivity. □

Remarks. 1) A particular instance of Corollary 3 was established in 1977 by Dudley and Gutmann [2]: given $\mathcal{G}$ as in Corollary 3 and a probability law μ on $]0, \infty]$, there exists a $\mathcal{G}$-stopping time with law μ. (Consider a r.v. T with law μ, and call $\mathcal{F}$ the smallest filtration making T a stopping time.)

2) Let $\mathcal{F}$ and $\mathcal{G}$ be two filtrations. If $\mathcal{F}_\infty$ is essentially separable and $\mathcal{G}_0$ has no atom, $\mathcal{F}$ is includable into $\mathcal{G}$. This statement looks similar to Corollary 3, but it is much simpler: it can be proved by simply observing that $\mathcal{F} \underset{\sim}{\subseteq} \mathcal{F}' \underset{\sim}{\subseteq} \mathcal{G}' \underset{\sim}{\subseteq} \mathcal{G}$, where $\mathcal{F}'$ and $\mathcal{G}'$ are constant filtrations, the former being constantly equal to $\mathcal{F}_\infty$ and the latter to $\mathcal{G}_0$.

3) Applying twice the preceding remark yields the following statement. Let $\mathcal{F}$ and $\mathcal{G}$ be any two filtrations such that $\mathcal{F}_0$ and $\mathcal{G}_0$ have no atom, and $\mathcal{F}_\infty$ and $\mathcal{G}_\infty$ are essentially separable. Then $\mathcal{F} \underset{\sim}{\subseteq} \mathcal{G} \underset{\sim}{\subseteq} \mathcal{F}$.

4) If $\mathcal{F}''$ is includable into $\mathcal{F}'$, for each fixed t the probability space $(\Omega'', \mathcal{F}''_t, \mathbb{P}'')$ is includable into $(\Omega', \mathcal{F}'_t, \mathbb{P}')$ in the sense of conditions (i), (ii) and (iii) of Theorem 1. But the converse does not hold: even if for each t $(\Omega', \mathcal{F}'_t, \mathbb{P}')$ and $(\Omega'', \mathcal{F}''_t, \mathbb{P}'')$ are isomorphic probability spaces, it may happen that neither filtration is includable into the other.

The simplest examples of such situations involve a time-axis consisting of two instants only, 0 and 1. Suppose that $\mathcal{F}_0$ is made of two atoms with masses $\frac{2}{5}$ and $\frac{3}{5}$, and $\mathcal{F}_1$ of four atoms with masses $\frac{1}{5}, \frac{1}{5}, \frac{1}{5}, \frac{2}{5}$. There are two different filtrations with these properties (the $\frac{2}{5}$-atom of $\mathcal{F}_1$ may be the $\frac{2}{5}$-atom of $\mathcal{F}_0$ or a part of the $\frac{3}{5}$-atom of $\mathcal{F}_0$), and neither one is includable into the other.

Corollary 2 and Remark 3) yield a wealth of examples where $\mathcal{F} \underset{\sim}{\subseteq} \mathcal{G} \underset{\sim}{\subseteq} \mathcal{F}$ but $\mathcal{F} \not\approx \mathcal{G}$; this indicates that includability is not a proper tool for comparing filtrations.

3 Filtrations: immersions

Definition 4. *Let $\mathcal{F}$ and $\mathcal{G}$ be two filtrations on the same probability space. We say that $\mathcal{F}$ is immersed in $\mathcal{G}$, and we write $\mathcal{F} \overset{m}{\subset} \mathcal{G}$, if $\mathcal{F}$ is included in $\mathcal{G}$ and every $\mathcal{F}$-martingale is a $\mathcal{G}$-martingale.[3]*

Let $(\Omega', \mathcal{A}', \mathbb{P}', \mathcal{F}')$ and $(\Omega'', \mathcal{A}'', \mathbb{P}'', \mathcal{F}'')$ be two filtered probability spaces. We say that $\mathcal{F}'$ is immersible into $\mathcal{F}''$, and we write $\mathcal{F}' \underset{\sim}{\overset{m}{\subset}} \mathcal{F}''$, if there exists a probability space $(\Omega, \mathcal{A}, \mathbb{P})$ endowed with two filtrations $\mathcal{F}^*$ and $\mathcal{F}^{**}$, respectively isomorphic to $\mathcal{F}'$ and $\mathcal{F}''$, and such that $\mathcal{F}^* \overset{m}{\subset} \mathcal{F}^{**}$.

---

[3] The requirement that $\mathcal{F}$ be included in $\mathcal{G}$ is redundant and can be dropped.

This is again an abuse of language: immersibility is not a property of $\mathcal{F}'$ and $\mathcal{F}''$, but of $(\Omega', \mathcal{F}'_\infty, \mathbb{P}', \mathcal{F}')$ and $(\Omega'', \mathcal{F}''_\infty, \mathbb{P}'', \mathcal{F}'')$.

Immersion of filtrations is much stronger than mere inclusion; it is also much more significant for stochastic calculus. See [1] page 241 and [4] page 268 for a few references concerning this notion.

As was already the case with includability, it is always possible to choose $(\Omega, \mathcal{A}, \mathbb{P}, \mathcal{F}^{**})$ equal to $(\Omega'', \mathcal{F}''_\infty, \mathbb{P}'', \mathcal{F}'')$ in Definition 4.

Immersibility is transitive: $\mathcal{F}' \overset{m}{\underset{\sim}{\subseteq}} \mathcal{F}'' \overset{m}{\underset{\sim}{\subseteq}} \mathcal{F}''' \implies \mathcal{F}' \overset{m}{\underset{\sim}{\subseteq}} \mathcal{F}'''$.

In the case when $\mathcal{F}$ is the natural filtration of a Markov (resp. Lévy, Poisson, Wiener) process X, $\mathcal{F}$ is immersed in $\mathcal{G}$ if and only X is a Markov (resp. Lévy, Poisson, Wiener) process for the filtration $\mathcal{G}$; but $\mathcal{F}$ is included in $\mathcal{G}$ if and only if X is adapted to $\mathcal{G}$.

Immersibility is also much stronger than includability. For instance, if a filtration is immersible into a Brownian filtration, it inherits the property that all martingales are continuous, with absolutely continuous brackets; on the other hand, by Proposition 2, includability into a Brownian filtration is much less stringent.

A natural question is:

$$(\text{Q}) \qquad\qquad ¿ \quad \mathcal{G} \overset{m}{\underset{\sim}{\subseteq}} \mathcal{F} \overset{m}{\underset{\sim}{\subseteq}} \mathcal{G} \implies \mathcal{F} \approx \mathcal{G} \quad ?$$

It would be interesting to answer (Q) when $\mathcal{G}$ is the filtration $\mathcal{B}^{(\infty)}$ of infinite-dimensional Brownian motion. If $\mathcal{B}^{(\infty)}$, which is stable under independent products, is also the only filtration $\mathcal{F}$ verifying $\mathcal{B}^{(\infty)} \overset{m}{\underset{\sim}{\subseteq}} \mathcal{F} \overset{m}{\underset{\sim}{\subseteq}} \mathcal{B}^{(\infty)}$, then it can play in continuous time a rôle similar to that played in Vershik's theory by the standard, non-atomic filtration in discrete, negative time. This is an open problem, closely linked to the question raised p. 219 of Revuz-Yor [8]; a small step in that direction is made in [3].

But it is easy to exhibit examples showing that the answer to (Q) is sometimes negative, as we shall now see with Lévy processes.

Proposition 3 (Jacod). *Let X' and X'' be two Lévy processes (not necessarily on the same probability space); call d' and d'' the dimensions of their Gaussian components and μ' and μ'' their Lévy measures (defined on the state spaces $\mathbb{R}^{n'}$ and $\mathbb{R}^{n''}$ where X' and X'' live). Denote by $\mathcal{F}'$ and $\mathcal{F}''$ the filtrations generated by X' and X''.*

The following three conditions are equivalent:

(i) $\mathcal{F}'$ and $\mathcal{F}''$ are isomorphic filtrations;
(ii) $d' = d''$, and $(\mathbb{R}^{n'}, \mu')$ and $(\mathbb{R}^{n''}, \mu'')$ are isomorphic measure spaces;
(ii') $d' = d''$, $c(\mu') = c(\mu'')$ and $\delta(\mu') = \delta(\mu'')$.

The following three conditions are equivalent:

(iii) $\mathcal{F}''$ is immersible into $\mathcal{F}'$;
(iv) $d'' \leqslant d'$, and there exist a Borel subset B' of $\mathbb{R}^{n'}$ and a measurable map from B' to $\mathbb{R}^{n''}$ transforming $\mu'_{|B'}$ into μ'';

(iv') $d'' \leqslant d'$, and there exist a sub-collection $c' \subset c(\mu')$ and two maps $f : c' \to c(\mu'')$ and $u : c(\mu'') \to [0, \infty]$ such that

$$\delta(\mu') \geqslant \delta(\mu'') + \sum_{p'' \in c(\mu'')} u(p'') \quad and \quad \forall p'' \in c(\mu'') \quad p'' = u(p'') + \sum_{\substack{p' \in c' \\ f(p') = p''}} p' \,.$$

Proof. Both equivalences (ii) $\Longleftrightarrow$ (ii') and (iv) $\Longleftrightarrow$ (iv') are copied from Theorem 2. The real content of the proposition is implications (i) $\Longrightarrow$ (ii) and (iii) $\Longrightarrow$ (iv), which are due to Jacod: implication (i) $\Longrightarrow$ (ii) is Jacod's Theorem (13.8) of [5]; and (iii) $\Longrightarrow$ (iv), although not explicitly stated by him, is implicitly established in his proof of (13.8), on page 411 of [5].

The converse statements (ii) $\Longrightarrow$ (i) and (iv) $\Longrightarrow$ (iii) are straightforward. Assuming for instance (iv), call $f : B' \to \mathbb{R}^{n''}$ a map carrying $\mu'_{|B'}$ onto μ'', U'' the unit ball in $\mathbb{R}^{n''}$, G' (resp. G'') the Gaussian part of X' (resp. X''), γ' (resp. γ'') the covariance matrix of G' (resp. G''), $\Lambda'(\mathrm{d}t, \mathrm{d}x')$ the $\mathcal{F}'$-optional measure counting the jumps of X', and $\tilde{\Lambda}'(\mathrm{d}t, \mathrm{d}x') = \mathrm{d}t\,\mu'(\mathrm{d}x')$ its $\mathcal{F}'$-predictable compensator. As the ranks d' and d'' of γ' and γ'' verify $d'' \leqslant d'$, there exists at least one $(n'' \times n')$-matrix ϕ such that $\gamma'' = \phi\,\gamma'\,{}^t\phi$. By definition of f, one has on the one hand $\mu'\big(B' \cap f^{-1}(U''^c)\big) = \mu''(U''^c) < \infty$, and on the other hand

$$\int_{B' \cap f^{-1}(U'' \setminus \{0\})} |f(x')|^2 \, \mu'(\mathrm{d}x') = \int_{U'' \setminus \{0\}} |x''|^2 \, \mu''(\mathrm{d}x'') < \infty \,.$$

So it is possible to define an $\mathbb{R}^{n''}$-valued $\mathcal{F}'$-Lévy process Y' by

$$Y'_t = \phi\,G'_t + \sum_{\substack{s \leqslant t \\ \Delta X'_s \in B' \setminus f^{-1}(U'')}} f(\Delta X'_s) + \int_0^t \int_{B' \cap f^{-1}(U'' \setminus \{0\})} f(x') \, (\Lambda' - \tilde{\Lambda}')(\mathrm{d}s, \mathrm{d}x') \,;$$

its Gaussian part $\phi\,G'$ has covariance $\phi\,\gamma'\,{}^t\phi = \gamma''$, and its Lévy measure is $f(\mathbf{1}_{B'} \cdot \mu') = \mu''$. This implies the existence of a vector $a'' \in \mathbb{R}^{n''}$ such that the processes Y'_t and $X''_t + a''t$ have the same law. Consequently, the filtration $\mathcal{G}$ generated by Y' is isomorphic to $\mathcal{F}''$, and $\mathcal{G} \overset{m}{\subset} \mathcal{F}'$ implies $\mathcal{F}'' \overset{m}{\underset{\sim}{\subset}} \mathcal{F}'$.

The proof of (ii) $\Longrightarrow$ (i) is similar: observe first that, according to von Neumann [7], there exist two Borel subsets in $\mathbb{R}^{n'}$ and $\mathbb{R}^{n''}$, with full Lévy measures, and a bimeasurable bijection between them, exchanging the Lévy measures; then proceed as above to exhibit a process Y' having the same filtration as X' and the same law as $X''_t + a''t$ for some a''. $\qquad\square$

Corollary 4. *Let $\mathcal{F}'$ and $\mathcal{F}''$ denote the filtrations generated by two Lévy processes X' and X'' with no Gaussian component. If the Lévy measure of X' has an infinite diffuse part, $\mathcal{F}''$ is immersible into $\mathcal{F}'$.*

Proof. As $\delta(\mu') = \infty$, take $c' = \varnothing$ and $u(p'') = p''$ in Proposition 3 (iv'). $\qquad\square$

Corollary 5. *The filtration of any Lévy process is always immersible into the filtration generated by an infinite-dimensional Brownian motion and an independent Cauchy process.*

Proof. The Gaussian part has a Brownian filtration, immersible into $\mathcal{B}^{(\infty)}$; and the jump part is immersible into the Cauchy filtration by Corollary 4. □

By Corollary 4, the filtrations of any two Lévy processes with no Gaussian component and whose Lévy measures have infinite diffuse parts are always immersible into each other. But, by Proposition 3, they are isomorphic to each other only if their Lévy measures also have the same list of masses of atoms. Thus the answer to (Q) is negative. For instance, if S is a $\frac{1}{2}$-stable subordinator and N an independent standard Poisson process, the filtrations of S and $S + N$ are immersible into each other; but they are not isomorphic, because the Lévy measure of S is infinite and diffuse, and that of $S + N$ has an infinite diffuse part and an atom with mass 1.

The reason why these two filtrations are not isomorphic is hidden inside Jacod's proof of (i) $\Longrightarrow$ (ii) in Proposition 3. It is possible to extract it from that proof and to state it as an independent proposition, which explicitly describes how the invariants of the Lévy measure can be seen in the filtration of a Lévy process. This will be done below: Proposition 4 directly deduces (ii′) from (i), without going through (ii), by looking at the structure of Poisson counters in the filtration. All the arguments in the proof are copied, almost verbatim, from page 411 of Jacod [5].

Definition 5. *Fix a filtered probability space* $(\Omega, \mathcal{A}, \mathbb{P}, \mathcal{F})$.

A counter *is an increasing, adapted, right-continuous process* N, *having integer values and unit jumps, such that* $N_0 = 0$.

A counter N *is* indivisible *if, for every decomposition* $N = N' + N''$ *of* N *as the sum of two counters, there exists a predictable set* A *such that* $N'_t = \int_0^t \mathbf{1}_A(s)\, dN_s$ *and* $N''_t = \int_0^t \mathbf{1}_{A^c}(s)\, dN_s$.

A Poisson counter *is a counter* N *such that* $N_t - at$ *is a martingale for a deterministic number* $a \geqslant 0$, *called the* intensity *of this Poisson counter.*

A Poisson counter with intensity a *is* infinitely divisible *if, for every* $k \geqslant 1$, *it is the sum of* k *independent[4] Poisson counters with intensity* a/k.

For instance, a Poisson counter is always indivisible relative to its own filtration (this is a particular instance of Proposition 4 (i) below); but in a larger filtration it can be divisible, or infinitely divisible, or even no longer Poisson.

Proposition 4. *Let* X *be a Lévy process, with Lévy measure* μ. *Fix* $a > 0$ *and call* M_a *the (at most countable) set of all atoms of* μ *having mass* a. *The ambient filtration will be the natural filtration of* X.

---

[4] This is redundant: as their sum is a counter, they have no common jumps, and Poisson counters with no common jumps are always independent.

(i) If ξ is an atom of μ, the Poisson counter $N_t = \sum_{s \leqslant t} \mathbf{1}_{\{\Delta X_s = \xi\}}$ is indivisible.

(ii) Conversely, if N is an indivisible Poisson counter with intensity a, there exists a predictable, M_a-valued process H such that $N_t = \sum_{s \leqslant t} \mathbf{1}_{\{\Delta X_s = H_s\}}$.

(iii) If two indivisible Poisson counters N^1 and N^2 with intensity a are independent, the predictable processes H^1 and H^2 associated to N^1 and N^2 by (ii) verify $H^1_t(\omega) \neq H^2_t(\omega)$ for almost all (t,ω).

(iv) The set M_a has at least k elements if and only if there exist k independent, indivisible Poisson counters with intensity a.

(v) The mass of the diffuse part of μ is at least a if and only if there exists an infinitely divisible Poisson counter with intensity a.

So, in the previous example, the filtrations generated by S and $S + N$ are not isomorphic because, by (i), N is an indivisible Poisson counter in the filtration of $S + N$, whereas, by (iv), the filtration of S admits no indivisible Poisson counter.

Proof. Recall that, in the filtration of our Lévy process X, every martingale with finite variation can be written as a compensated sum

$$(*) \qquad \sum_{0 < s \leqslant t} K(s,\omega,\Delta X_s) - \int_0^t ds \int_{\mathbb{R}^n \setminus \{0\}} \mu(dx)\, K(s,\omega,x) \,,$$

where $K(t,\omega,x)$ is predictable in (t,ω) and null for $x = 0$.

(i) Consider two counters N'_t and N''_t with sum $N_t = \sum_{s \leqslant t} \mathbf{1}_{\{\Delta X_s = \xi\}}$. Their jump times are totally inaccessible, hence their compensators are continuous, and applying $(*)$ gives $N'_t = \sum_{s \leqslant t} K(s,\omega,\Delta X_s)$. Since the jumps of N' are equal to 1 and are also jumps of N, it suffices to introduce the predictable set $A = \{(t,\omega) : K(t,\omega,\xi) \neq 0\}$, and one has $dN'_t = \mathbf{1}_A(t)\, dN_t$.

(ii) If N is a Poisson counter with intensity a, the martingale $N_t - at$ has the form $(*)$. Identifying the jumps shows that K may be chosen with values in $\{0,1\}$; we shall call $J(t,\omega)$ the Borel subset of $\mathbb{R}^n \setminus \{0\}$ such that $K(t,\omega,x) = \mathbf{1}_{J(t,\omega)}(x)$. Identifying the compensators now gives the mass of $J(t,\omega)$:

$$N_t = \sum_{s \leqslant t} \mathbf{1}_{\{\Delta X_s \in J(s,\omega)\}} \quad ; \quad \mu\big(J(t,\omega)\big) = a \text{ for almost all } (t,\omega) \,.$$

On the predictable set $\Gamma = \{(t,\omega) : J(t,\omega) \text{ meets } M_a\}$, the set $J(t,\omega)$ must contain exactly one atom with mass a. Calling $H_t(\omega)$ this atom, H is a predictable process with values in M_a, defined (almost everywhere) on Γ, and such that the difference $J(t,\omega) \setminus \{H_t(\omega)\}$ is μ-negligible. Thus

$$\int_0^t \mathbf{1}_\Gamma\, dN_s = \sum_{s \leqslant t} \mathbf{1}_\Gamma(s)\, \mathbf{1}_{\{\Delta X_s = H_s\}} \,;$$

and to establish (ii) it suffices to check that, if furthermore N is indivisible, the predictable set $\Gamma^c = \{(t,\omega) : J(t,\omega) \cap M_a = \varnothing\}$ is negligible.

For $(t,\omega) \in \Gamma^c$, the set $J(t,\omega)$ has mass a but contains no atom with mass a. So it may be decomposed as the disjoint union of two non-negligible subsets $J'(t,\omega)$ and $J''(t,\omega)$, which depend predictably[5] upon (t,ω). Now observe that the counters

$$N'_t = \sum_{0<s\leqslant t} \mathbf{1}_{\Gamma^c}(s)\,\mathbf{1}_{\{\Delta X_s \in J'(s,\omega)\}}$$

$$N''_t = \sum_{0<s\leqslant t} \left[\mathbf{1}_{\Gamma^c}(s)\,\mathbf{1}_{\{\Delta X_s \in J''(s,\omega)\}} + \mathbf{1}_{\Gamma}(s)\,\mathbf{1}_{\{\Delta X_s \in J(s,\omega)\}}\right]$$

verify $N' + N'' = N$. So if N is supposed to be indivisible, there exists a predictable set A such that N'_t can also be written $\int_0^t \mathbf{1}_A(s)\,dN_s$; expressing the compensator of N' in two different ways gives

$$\int_0^t \mathbf{1}_{\Gamma^c}(s)\,\mu\big(J'(s,\omega)\big)\,ds = a\int_0^t \mathbf{1}_A(s)\,ds \ .$$

But we know that, for $(s,\omega) \in \Gamma^c$, the mass $\mu\big(J'(s,\omega)\big)$ lies strictly between 0 and a; so Γ^c must be negligible for $dt \times \mathbb{P}(d\omega)$, and (ii) is established.

(iii) More generally, let H^1 and H^2 be any two $\mathbb{R}^n$-valued predictable processes such that

$$N^1_t = \sum_{0<s\leqslant t} \mathbf{1}_{\{\Delta X_s = H^1_s\}} \qquad \text{and} \qquad N^2_t = \sum_{0<s\leqslant t} \mathbf{1}_{\{\Delta X_s = H^2_s\}}$$

are independent Poisson counters, N^1 having intensity $a^1 > 0$. The counter $\int_0^t \mathbf{1}_{\{H^1_s=H^2_s\}}\,dN^1_s$ of the common jumps of N^1 and N^2 must vanish identically; its compensator $a^1\int_0^t \mathbf{1}_{\{H^1_s=H^2_s\}}\,ds$ vanishes too, giving $H^1_t(\omega) \neq H^2_t(\omega)$ for almost all (l,ω).

(iv) Given k elements of M_a, (i) yields k independent, indivisible Poisson counters with intensity a.

Conversely, if the filtration contains k independent, indivisible Poisson counters with intensity a, by (ii) and (iii) there exist k processes having values in M_a and almost always avoiding each other; therefore M_a must contain at least k points.

(v) If there exists a Borel subset B of $\mathbb{R}^n \setminus \{0\}$ such that $\mu(B) = a$ and $\mu_{|B}$ is diffuse, B can be partitioned into k subsets with mass a/k; consequently the Poisson counter $\sum_{s\leqslant t} \mathbf{1}_{\{\Delta X_s \in B\}}$ is infinitely divisible.

---

[5] For instance, choose a numbering of the atoms of μ and a bi-Borel bijection ψ between $\mathbb{R}^n \setminus \{0\}$ and $\mathbb{R}_+$, which transforms the diffuse part of μ into the Lebesgue measure on the interval $[0, \delta(\mu)[$. Then take $J'(t,\omega)$ equal to the first (numbering order) atom included in $J(t,\omega)$ if there is such an atom, and else put $J'(t,\omega) = J(t,\omega) \cap \psi^{-1}\big([0,u[\big)$, where u is such that $\mu\big(J'(t,\omega)\big) = a/2$.

Conversely, if N is an infinitely divisible Poisson counter with intensity a, let $M^1 + \ldots + M^k$ be a decomposition of N into k Poisson counters with common intensity a/k. The same argument as in the beginning of (ii) gives

$$N_t = \sum_{s \leqslant t} \mathbf{1}_{\{\Delta X_s \in J(s,\omega)\}} \quad ; \quad \mu\big(J(t,\omega)\big) = a \text{ for almost all } (t,\omega) ;$$

$$M_t^i = \sum_{s \leqslant t} \mathbf{1}_{\{\Delta X_s \in R^i(s,\omega)\}} \quad ; \quad \mu\big(R^i(t,\omega)\big) = \frac{a}{k} \text{ for almost all } (t,\omega) .$$

For $i \neq j$, one has

$$\mathbb{E}\left[\int_0^\infty \mu\big(R^i(t,\omega) \cap R^j(t,\omega)\big) \, \mathrm{d}t\right] = \mathbb{E}\left[\sum_{t \geqslant 0} \mathbf{1}_{\{\Delta X_t \in R^i(t,\omega) \cap R^j(t,\omega)\}}\right] = 0$$

because $M^i + M^j$ can have no jumps of amplitude 2; so, for almost all (t,ω), $R^i(t,\omega) \cap R^j(t,\omega)$ is μ-negligible, showing that $J(t,\omega)$ can be partitioned into k subsets with measure a/k. This holds for all k; therefore almost every $J(t,\omega)$ is a set with measure a and containing no atom. □

References

1. S. Beghdadi-Sakrani, M. Émery (1999): On certain probabilities equivalent to coin-tossing, d'après Schachermayer. Séminaire de Probabilités XXXIII, Springer Lecture Notes in Mathematics 1709, 240–256.
2. R.M. Dudley, S. Gutmann, (1977): Stopping times with given laws. Séminaire de Probabilités XI, Springer Lecture Notes in Mathematics 581, 51–58.
3. M. Émery (2005): On certain almost Brownian filtrations. Ann. Inst. Henri Poincaré section B (Probabilités et Statistiques) 41 (special issue in memory of P. A. Meyer), 285–305.
4. M. Émery, W. Schachermayer (2001): On Vershik's standardness criterion and Tsirelson's notion of cosiness. Séminaire de Probabilités XXXV, Springer Lecture Notes in Mathematics 1755, 265–305.
5. J. Jacod (1979): Calcul Stochastique et Problèmes de Martingales. Lecture Notes in Mathematics 714. Springer, Berlin.
6. T. Jeulin (1996): Filtrations, sous-filtrations : propriétés élémentaires. Hommage à P. A. Meyer et J. Neveu, Astérisque 236, 163–170.
7. J. von Neumann (1932): Einige Sätze über meßbare Abbildungen. Ann. Math. 33, 574–586.
8. D. Revuz, M. Yor (1999): Continuous Martingales and Brownian Motion. Third edition. Springer, Berlin.
9. A.M. Vershik (1995): The theory of decreasing sequences of measurable partitions. St. Petersburg Math. J. 6, 705–761.

The Dalang–Morton–Willinger Theorem Under Delayed and Restricted Information

Yuri Kabanov[1,2] and Christophe Stricker[1]

[1] UMR 6623, Laboratoire de Mathématiques, Université de Franche-Comté, 16 Route de Gray, F-25030 Besançon Cedex, France
e-mail: kabanov@math.univ-fcomte.fr, stricker@math.univ-fcomte.fr
[2] Central Economics and Mathematics Institute, Moscow, Russia

Summary. We extend the classical no-arbitrage criteria to the case of a model where the investor's decisions are based on a partial information (e.g., because of delay or round-off errors), that is the portfolio strategies are predictable with respect to a subfiltration. Our main result is a ramification of the famous Dalang–Morton–Willinger theorem: the model is arbitrage-free if and only if there exists an equivalent probability measure $\tilde{P}$ such that the optional projection of the price process with respect to $\tilde{P}$ is a $\tilde{P}$-martingale.

Key words: no-arbitrage criteria, martingale measure, optional projection.

Mathematics Subject Classification 2000: 60G42

1. Introduction. The Dalang–Morton–Willinger theorem asserts, for the standard discrete-time finite horizon model of security market, that there is no arbitrage if and only if the price process is a martingale with respect to an equivalent probability measure. This remarkable result sometimes is referred to as the Fundamental Theorem of Arbitrage (or Asset) Pricing (FTAP) or simply the First Fundamental Theorem, [5]. Its various aspects have been thoroughly investigated, the theorem has been augmented by additional equivalent conditions and extended in many directions, see, e.g. [2].

In this note we deal with the same model modified in only one aspect: the agent's decision are based on a restricted information flow described by a filtration which can be smaller than a filtration generated by the price process. Apparently, such situations may arise if the information arrives with a delay or based on quantified prices and so on. We show that there is no-arbitrage under partial information iff there exists an equivalent probability $\tilde{P}$ such that the optional projection with respect to $\tilde{P}$ is a $\tilde{P}$-martingale. Surprisingly, this natural generalization was not studied previously and there is a certain explanation for this. Almost all available proofs use a reduction to the one-period model. This reduction is possible because in the standard setting, as was observed already in [1], the NA property is equivalent to the absence of arbitrage on each step. Due to this, all efforts are concentrated to construct a martingale density for the one-step model; proofs are accomplished by assembling the required density process from the one-step densities using the procedure suggested in [1]. Unfortunately, attempts to follow the same strategy of proof for the partial information case cannot be fruitful: as we show below, the equivalence between the "global" NA and the collection of one-step NA properties fails in general. However, the proof of NA criteria given in [3] and reproduced in [2] (to our knowledge, the unique one which does not rely upon the reduction to the one-step model) works well and requires only minor changes.

2. No-arbitrage criteria under delayed information. Let $(\Omega, \mathcal{G}, P)$ be a probability space equipped with a filtration $\mathbf{F} = (\mathcal{F}_t)$, $t = 0, 1, \ldots, T$, with $\mathcal{F}_T \subseteq \mathcal{G}$. We are given a d-dimensional process $S = (S_t)$ which is not necessarily adapted. Let

$$R_T := \{\xi : \ \xi = H \cdot S_T, \ H \in \mathcal{P}\}$$

where $\mathcal{P}$ is the set of all predictable d-dimensional processes with respect to $\mathbf{F}$ (i.e. H_t is $\mathcal{F}_{t-1}$-measurable) and

$$H \cdot S_T := \sum_{t=1}^{T} H_t \Delta S_t, \qquad \Delta S_t := S_t - S_{t-1}.$$

Put $A_T := R_T - L_+^0$; $\bar{A}_T$ is the closure of A_T in probability, L_+^0 is the set of non-negative random variables. In the context of mathematical finance the process S describes the discounted price process of d assets. The assumption that the strategy H is predictable with respect to a filtration $\mathbf{F}$ to which S may not be adapted, covers, in particular, the situation where the investor has no access to the full information contained in the price process (for instance, he may observe the price process after some delay).

We formulate our main result in the same manner as in [3].

Theorem 1. *The following conditions are equivalent:*

(a) $A_T \cap L_+^0 = \{0\}$;
(b) $A_T \cap L_+^0 = \{0\}$ *and* $A_T = \bar{A}_T$;
(c) $\bar{A}_T \cap L_+^0 = \{0\}$;
(d) there is a probability $\tilde{P} \sim P$ with $d\tilde{P}/dP \in L^\infty$ such that all S_t are $\tilde{P}$-integrable and $\tilde{E}(S_{t+1}|\mathcal{F}_t) = \tilde{E}(S_t|\mathcal{F}_t)$ for $t = 0, \ldots, T-1$.

The last condition means that the $(\mathbf{F}, \tilde{P})$-optional projection $\tilde{S}$ is an $(\mathcal{F}, \tilde{P})$-martingale (in discrete time $\tilde{S}_n = \tilde{E}(S_n|\mathcal{F}_n)$ by definition).

Condition (a) is interpreted as the absence of arbitrage; it can be written in the obviously equivalent form $R_T \cap L_+^0 = \{0\}$ (or $H \cdot S_T \geqslant 0 \Rightarrow H \cdot S_T = 0$). When S is adapted to $\mathbf{F}$, (a) is equivalent to condition:

(a') $H_t \Delta S_t \geqslant 0 \Rightarrow H_t \Delta S_t = 0$ for all $t = 1, \ldots, T$.

This is no longer true when S is not adapted. Consider the following simple example where $T = 2$, $\mathcal{F}_0 = \mathcal{F}_1 = \{\varnothing, \Omega\}$ but there is $A \in \mathcal{G}$ such that $0 < P(A) < 1$. Put

$$\Delta S_1 := I_A - \frac{1}{2} I_{A^c}, \qquad \Delta S_2 := -\frac{1}{2} I_A + I_{A^c}.$$

There is no arbitrage on each of two steps but the constant process with $H_1 = H_2 = 1$ is an arbitrage strategy for the two-step model.

3. Proof of Theorem 1. $(a) \Rightarrow (b)$ For the sake of completeness and the reader's convenience we repeat the arguments from [3] which are based on the following observation due to H.-J. Engelbert and H. von Weizsäcker (see [3] for a proof).

Lemma 1. *Let* $\eta^n \in L^0(\mathbf{R}^d)$ *be such that* $\liminf |\eta^n| < \infty$. *Then there are* $\tilde{\eta}^k \in L^0(\mathbf{R}^d)$ *such that for all* ω *the sequence of* $\tilde{\eta}^k(\omega)$ *is a convergent subsequence of the sequence of* $\eta^n(\omega)$.

To show that A_T is closed we proceed by induction. Let $T = 1$. Suppose that $H_1^n \Delta S_1 - r^n \to \zeta$ a.s. where H_1^n is $\mathcal{F}_0$-measurable and $r^n \in L_+^0$. It is sufficient to find $\mathcal{F}_0$-measurable random variables $\tilde{H}_1^k$ which are a.s. convergent and $\tilde{r}^k \in L_+^0$ such that $\tilde{H}_1^k \Delta S_1 - \tilde{r}^k \to \zeta$ a.s.

Suppose that certain sets $\Omega_i \in \mathcal{F}_0$ form a finite partition of Ω. Obviously, we may argue on each Ω_i separately as on an autonomous measure space (considering the restrictions of random variables and traces of σ-algebras).

Let $\underline{H}_1 := \liminf |H_1^n|$. On the set $\Omega_1 := \{\underline{H}_1 < \infty\}$ we can take, using Lemma 1, $\mathcal{F}_0$-measurable $\tilde{H}_1^k$ such that $\tilde{H}_1^k(\omega)$ is a convergent subsequence of $H_1^n(\omega)$ for every ω; $\tilde{r}^k$ are defined correspondingly. Thus, if Ω_1 is of full measure, the goal is achieved.

On $\Omega_2 := \{\underline{H}_1 = \infty\}$ we put $G_1^n := H_1^n / |H_1^n|$ and $h_1^n := r^n / |H_1^n|$ and observe that $G_1^n \Delta S_1 - h_1^n \to 0$ a.s. By Lemma 1 we find $\mathcal{F}_0$-measurable $\tilde{G}_1^k$ such that $\tilde{G}_1^k(\omega)$ is a convergent subsequence of $G_1^n(\omega)$ for every ω. Denoting the limit by $\tilde{G}_1$, we obtain that $\tilde{G}_1 \Delta S_1 = \tilde{h}_1$ where $\tilde{h}_1$ is non-negative, hence, in virtue of (a), $\tilde{G}_1 \Delta S_1 = 0$.

As $\tilde{G}_1(\omega) \neq 0$, there exists a partition of Ω_2 into d disjoint subsets $\Omega_2^i \in \mathcal{F}_0$ such that the ith coordinate $\tilde{G}_1^i \neq 0$ on Ω_2^i. Define $\bar{H}_1^n := H_1^n - \beta^n \tilde{G}_1$ where $\beta^n := H_1^{ni} / \tilde{G}_1^i$ on Ω_2^i. Then $\bar{H}_1^n \Delta S_1 = H_1^n \Delta S_1$ on Ω_2. As it was mentioned above we may consider as isolated the set $\Omega_2^i \in \mathcal{F}_0$. The replacement of the sequence (H_1^n) by the $(\bar{H}_1^n)$ does not change the limits. Represent these sequences as infinite matrices with infinitely many columns, H_1^n and $\bar{H}_1^n$, respectively. The difference is that in the second matrix the ith row is zero and if the first matrix already has null rows, they remain null in the second one. We restart the entire procedure on Ω_2^i with the sequence $\bar{H}_1^n$ such that $\bar{H}_1^{ni} = 0$ for all n. Since at each step the number of zero lines increases, the process stops after a finite number of steps. The induction step from $T - 1$

to T can be done exactly in the same way, considering always the sequence of (H_1^n) to construct a partition of Ω and making the needed operations also with (H_t^n) for $t \geqslant 2$: this is legitimate since they do not destroy measurability.

$(b) \Rightarrow (c)$ Trivial.

$(c) \Rightarrow (d)$ Notice that for any random variable η there is an equivalent probability P' with bounded density such that $\eta \in L^1(P')$ (e.g., one can take $P' = CE^{-|\eta|}P$). Property (c) (as well as (a) and (b)) is invariant under equivalent change of probability. This consideration allows us to assume from the very beginning that all S_t are integrable. The convex set $A_T^1 := \bar{A}_T \cap L^1$ is closed in L^1 and hence satisfies the hypotheses of the well-known result due to Kreps and Yan, [4], [6] (its proof can also be found in [3] or [2]).

Lemma 2. *Let $K \supseteq -L_+^1$ be a closed convex cone in L^1 with $K \cap L_+^1 = \{0\}$. Then there is a probability $\tilde{P} \sim P$ with $d\tilde{P}/dP \in L^\infty$ such that $\tilde{E}\,\xi \leqslant 0$ for all $\xi \in K$.*

This lemma ensures the existence of $\tilde{P} \sim P$ with bounded density and such that $\tilde{E}\,\xi \leqslant 0$ for all $\xi \in A_T^1$, in particular, for $\xi = \pm H_t \Delta S_t$ where H_t is bounded and $\mathcal{F}_{t-1}$-measurable. Thus, $\tilde{E}(\Delta S_t | \mathcal{F}_{t-1}) = 0$.

$(d) \Rightarrow (a)$ Let $\xi \in A_T \cap L_+^0$, i.e. $0 \leqslant \xi \leqslant H \cdot S_T$. As $\tilde{E}(H_t \Delta S_t | \mathcal{F}_{t-1}) = 0$, we obtain by conditioning that $\tilde{E}\,H \cdot S_T = 0$. Thus, $\xi = 0$.

4. Optional projection.

It may happen that the $(\mathbf{F}, P)$-optional projection of S does not satisfy our NA condition although S does. Indeed, consider again the two-step model with $\mathcal{F}_0 = \mathcal{F}_1 = \{\varnothing, \Omega\}$. Let $\Delta S_1, \Delta S_2$ be independent random variables uniformly distributed on $[-1, 3]$. Then $E(\Delta S_i | \mathcal{F}_i) = 1$ for $i = 1, 2$ but for any point $(H_1, H_2) \in \mathbf{R}^2$ different from the origin the distribution of the random variable $H_1 \Delta S_1 + H_2 \Delta S_2$ charges both $]-\infty, 0[$ and $]0, \infty[$ and, therefore, S has the NA property with restricted information.

5. Comment on continuous-time models.

The following example illustrates that in the continuous-time setting where $t \in \mathbf{R}_+$, the question of absence of arbitrage under partial information can be posed even if the price process is not a semimartingale.

Let $B = (B_s)$ be a standard Brownian motion with respect to a filtration $(\mathcal{H}_t)$ satisfying the usual conditions. Let $\mathbf{F}$ be the trivial filtration formed by the σ-algebras $\mathcal{F}_t$ generated by the null sets from $\mathcal{H}_\infty$.

Put $\phi(x) := \pi + \arctan x$, $\theta(t) := \mathbf{E}^t$, and $S_t := \int_0^t \phi(B_s)\,dB_s + \int_0^t B_{\theta(s)}\,ds$. Let $\mathbf{G}$ be any filtration satisfying the usual conditions for which the process S is $\mathbf{G}$-adapted. If the diameters of the partitions $\pi_n := \{0 = t_0 \leqslant \dots \leqslant t_n = t\}$ converge to zero, $\sum_{\pi_n} (S_{t_{i+1}} - S_{t_i})^2 \to [S, S]_t$ in probability. So the process $[S, S]_t = \int_0^t \phi^2(B_s)\,ds$ is $\mathbf{G}$-adapted. It follows that (B_t) is $\mathbf{G}$-adapted and hence also $(B_{\theta(t)})$. Therefore, (B_t) cannot be a $\mathbf{G}$-semimartingale. Thus, (S_t) is not a $\mathbf{G}$-semimartingale. Nevertheless, we can define the stochastic integral $h \cdot S_t$ for any bounded Borel function and, moreover, for this integral $E\,h \cdot S_t = 0$ and, hence, the $\mathbf{F}$-optional projection of S is a martingale.

References

1. R.C. Dalang, A. Morton, W. Willinger: Equivalent martingale measures and no-arbitrage in stochastic securities market model. Stochastics and Stochastic Reports, **29**, 185–201, (1990)
2. Yu.M. Kabanov: Arbitrage theory. In: J. Cvitanić, E. Joini, M. Musiela (eds.) Handbooks in Mathematical Finance. Option Pricing: Theory and Practice. Cambridge University Press, Cambridge (2001)
3. Yu.M. Kabanov, C. Stricker: A teacher's note on no arbitrage criteria. Séminaire de Probabilités XXXV. Lect. Notes Math., **1755**, 149–152 (2001)
4. D.M. Kreps: Arbitrage and equilibrium in economies with infinitely many commodities. *J. Math. Economics*, **8**, 15–35 (1981)
5. A.N. Shiryaev: Essentials of Stochastic Mathematical Finance. World Scientific, Singapour (1999)
6. J.A. Yan: Caractérisation d'une classe d'ensembles convexes de L^1 et H^1. Séminaire de Probabilités XIV. Lect. Notes Math., **784**, 220–222 (1980)

The Structure of m–Stable Sets and in Particular of the Set of Risk Neutral Measures

Freddy Delbaen

Department of Mathematics Eidgenössische Technische Hochschule, Zürich,
CH 8092 Zurich, Switzerland
e-mail: delbaen@math.ethz.ch

Abstract. The study of dynamic coherent risk measures and risk adjusted values as introduced by Artzner, Delbaen, Eber, Heath and Ku, leads to a property called fork convexity, rectangularity or m–stability. We give necessary and sufficient conditions for a closed convex set of measures to be fork convex. Since the set of martingale measures for price processes is m–stable, this leads to a characterisation of closed convex sets that can be obtained as the set of risk neutral measures in an arbitrage free model of security prices. We also relate the property of m–stability with the validity of Bellman's principle. It turns out that the stability property investigated in this paper is equivalent to properties known as time-consistency and rectangularity as used in multiprior Bayesian decision theory. Finally we characterise the set of risk neutral measures for continuous price processes.

Keywords: arbitrage theory, capital requirement, coherent risk measure, coherent utility functions, capacity theory, dynamic risk measures, martingale measures, multiple priors, rectangularity, risk neutral measures, Snell envelope, shortfall, submartingale method, time consistency, risk neutral measures.

1 Introduction and Notation

The concept of coherent risk measures together with its axiomatic characterization was introduced in the paper [2] and further developed in [3] and [10]. The idea of dynamic coherent risk measures or parallel to it, dynamic risk adjusted values was introduced in [4]. A characterisation of the risk measures defined on the space of bounded càdlàg processes is given by Cheridito-Delbaen-Kupper [7]. The relation between their theory and the present paper is the subject of ongoing research, [22]. Some of the examples given in [4] require the use of sets satisfying a property that is called multiplicative stability. The name is given since it is the multiplicative equivalent of a stability property introduced in stochastic integration, see [12], page 370. After taking the stochastic logarithm, the property becomes equivalent to the property "stable par bifurcation" of stochastic control theory (definition 1.6 in the Saint-Flour lecture notes of El Karoui, [13]). This property already plays a role in non-stochastic control theory. Another name for this concept is fork convexity (a good translation of "stable par bifurcation"), a terminology that was introduced by Zitkovic [29]. Earlier the same concept was introduced under the name "predictably convex set" by Kramkov, [19] and by Föllmer and Kramkov, [17]. These authors showed that in the case of financial markets "predictably convex sets" play a fundamental role. In decision theory the property of multiplicative stability is known as rectangularity, see Epstein and Schneider, Wang ([16], [28]). These papers deal with the case of finite Ω. It turns out that there are natural examples of multiplicatively stable convex sets. One of these examples is the set of absolutely continuous risk neutral measures for an arbitrage free (or better NFLVR) price process, see below for precise statements. In this paper we give necessary and sufficient conditions for a closed convex set of measures to satisfy this stability property. The conditions are related to concepts such as "price of risk" and fit well in economic theory. Applying this characterisation to the situation of arbitrage free price processes, we will give a characterisation of those sets that can arise as sets of risk neutral measures. Especially in the case of filtrations where all martingales are continuous, we will solve the problem completely.

There is also a relation with $g-$expectations introduced via Backward Stochastic Differential Equations. These were introduced by Peng, [24]. Basic papers explaining the relation with time consistent risk measures are [6], [8]. For an introduction to BSDE as well as their use in mathematical finance we refer to [14] as well as [15]. The relation between these approaches and the concept of m–stable sets is not the topic of this presentation. We refer the reader to the already cited literature as well as to [26] and the overview of the literature given there. Throughout the paper, we will work with a fixed, filtered probability space, denoted as $\left(\Omega, \mathcal{F}_\infty, (\mathcal{F}_t)_{t \geqslant 0}, \mathbb{P}\right)$. The filtration $\mathcal{F}$ is supposed to satisfy the usual assumptions, i.e. the filtration is right continuous and $\mathcal{F}_0$ contains all the null sets of the complete $\sigma-$algebra $\mathcal{F}_\infty$. The time set is supposed to be $\mathbb{R}_+$. The reader can check that this is the most general case. By using suitable imbeddings it covers the case of discrete, finite as well as infinite time sets. With $\mathbf{L}^\infty(\Omega, \mathcal{F}, \mathbb{P})$ (or $\mathbf{L}^\infty(\mathbb{P})$ or even $\mathbf{L}^\infty$ if no confusion is possible), we mean the space of all equivalence classes of bounded real valued random variables. The space $\mathbf{L}^0(\Omega, \mathcal{F}, \mathbb{P})$ (or $\mathbf{L}^0(\mathbb{P})$ or simply $\mathbf{L}^0$) denotes the space of all equivalence classes of real valued random variables. The space $\mathbf{L}^0$ is equipped with the topology of convergence in probability. The space $\mathbf{L}^\infty(\mathbb{P})$, equipped with the usual $\mathbf{L}^\infty$ norm, is the dual space of the space of integrable (equivalence classes of) random variables, $\mathbf{L}^1(\Omega, \mathcal{F}, \mathbb{P})$ (also denoted by $\mathbf{L}^1(\mathbb{P})$ or $\mathbf{L}^1$ if no confusion is possible). The spaces $\mathbf{L}^p$ for $0 < p < \infty$ are defined in the usual way. A useful result in integration theory is the so-called Scheffé's lemma. It says that if a sequence of nonnegative random variables f_n tends in probability to a random variable f, if moreover $\mathbb{E}[f_n]$ tends to $\mathbb{E}[f] < \infty$, then necessarily the convergence takes place in $\mathbf{L}^1$ and the sequence is therefore uniformly integrable. For notions from the general theory of stochastic processes, we refer the reader to Dellacherie-Meyer [12]. In particular, if X is a stochastic process and T is a stopping time, the process X^T is defined through $X_t^T = X_{T \wedge t}$. Since stochastic intervals play a special role, let us recall from [12] some of these notions. If $T \leqslant S$ are two stopping times, then the stochastic intervals are defined as follows

$$[\![T, S]\!] = \{(t, \omega) \mid t \in \mathbb{R}_+ \text{ and } T(\omega) \leqslant t \leqslant S(\omega)\}.$$

The other intervals are defined in a similar way. In case $T = S$ we simply write $[\![T, S]\!] = [\![T]\!] = \{(t, \omega) \mid T(\omega) < \infty\}$. If T is a stopping time and if $A \in \mathcal{F}_T$, then T_A denotes the stopping time defined as $T_A = T$ on the set A and $T_A = \infty$ on the set $A^c = \Omega \setminus A$. In particular for $t \in \mathbb{R}_+$ and $A \in \mathcal{F}_t$ we have $[\![t_A]\!] = \{t\} \times A$. With the given filtration we will construct the σ−algebras of predictable and optional sets. The predictable σ−algebra, denoted by $\mathcal{P}$, is the smallest σ−algebra on $\mathbb{R}_+ \times \Omega$ that contains sets of the form $[\![0_A]\!] = \{0\} \times A$ with $A \in \mathcal{F}_0$, as well as for each stopping time T, the stochastic interval

$$[\![0, T]\!] = \{(t, \omega) \mid t \leqslant T(\omega) \text{ and } t < \infty\}.$$

The optional σ−algebra, denoted by $\mathcal{O}$, is the smallest σ−algebra on $\mathbb{R}_+ \times \Omega$ that contains sets of the form $\{0\} \times A$ with $A \in \mathcal{F}_0$, as well as for each stopping time T, the stochastic interval

$$[\![0, T[\![= \{(t, \omega) \mid t < T(\omega)\}.$$

We remark that the indicator functions of elements of the generating set of $\mathcal{P}$ are left continuous adapted processes and that the indicator functions of elements of the generating sets of $\mathcal{O}$ are right continuous adapted processes. It can easily be checked that $\mathcal{P} \subset \mathcal{O}$. Let us recall that the class of predictable sets

$$\{[\![0_A]\!] \mid A \in \mathcal{F}_0\} \cup \{[\![T, S]\!] \mid T \leqslant S \text{ stopping times}\},$$

forms a semi–algebra that generates $\mathcal{P}$. The Boolean algebra generated by this class is simply

$$\mathcal{A} = \Big\{ [\![0_A]\!] \cup]\!]T_0, T_1]\!] \cup]\!]T_1, T_2]\!] \ldots \cup]\!]T_{n-1}, T_n]\!] \mid n \geqslant 1; A \in \mathcal{F}_0 \text{ and }$$

$$0 \leqslant T_0 \leqslant T_1 \leqslant \ldots T_n \leqslant +\infty \text{ are all stopping times} \Big\}.$$

The importance of this class lies in the following density result from general measure theory. The proof of the lemma is included in the proof of the Carathéodory extension theorem.

Lemma 1. *Let μ be a nonnegative finite $\sigma-$additive measure on $\mathcal{P}$, then for each $\varepsilon > 0$ and for each set $B \in \mathcal{P}$, there is a set $A \in \mathcal{A}$ such that $\mu(A \Delta B) \leqslant \varepsilon$.*

If $\mathbb{Q}$ is a probability defined on the $\sigma-$algebra $\mathcal{F}_\infty$, we will use the notation $\mathbb{E}_\mathbb{Q}$ or $\mathbb{Q}$, to denote the expected value operator defined by the probability $\mathbb{Q}$. So we will write $\mathbb{E}_\mathbb{Q}[f]$ or $\mathbb{Q}[f]$ to denote the expected value of f. Since the filtration satisfies the usual assumptions, we will suppose that all the (sub–, super–) martingales are càdlàg, meaning they are right continuous and have left limits. When we deal with the construction of the Snell envelope, we will pay attention to this continuity property and the reader will notice similar difficulties as in the work of Mertens see [23] and [12], appendix, see also [13]. Although we treat the case of supermartingales with respect to a family of measures, there is no essential difference with the case of a fixed probability measure. The proof we present here is standard and is only added for completeness to make the text more self contained. We will identify, through the Radon–Nikodým theorem, finite measures ν on $\mathcal{F}_\infty$, that are absolutely continuous with respect to $\mathbb{P}$, with their densities $\frac{d\nu}{d\mathbb{P}}$, i.e. with functions in $\mathbf{L}^1$. Furthermore we will sometimes identify this measure with the càdlàg martingale $Z_t = \mathbb{E}_\mathbb{P}\left[\frac{d\nu}{d\mathbb{P}} \mid \mathcal{F}_t\right]$. We hope that these identifications will not cause too many problems. We can now state the definition of multiplicatively stable sets. The definition is related to the concept of stable sets as in [12]. To simplify the writing of the definition we suppose that $\mathcal{S}$ is a set of probability measures, all elements of which are absolutely continuous with respect to $\mathbb{P}$. The elements $\mathbb{Q}$ of the set $\mathcal{S}$ will (as already said above) be identified with their Radon–Nikodým derivative $\frac{d\mathbb{Q}}{d\mathbb{P}}$ and therefore we see $\mathcal{S}$ as a subset of $\mathbf{L}^1$ Most of the time, the set $\mathcal{S}$ will be supposed to be convex, also we will always have that $\mathbb{P} \in \mathcal{S}$. In that case we have for all $1 > \varepsilon > 0$ and all $\mathbb{Q} \in \mathcal{S}$ that $(1 - \varepsilon)\mathbb{Q} + \varepsilon\mathbb{P} \in \mathcal{S}$. This means that every element in $\mathcal{S}$ can be approximated (in $\mathbf{L}^1$-norm) by elements in $\mathcal{S}$ that are also equivalent to $\mathbb{P}$. The set of elements in $\mathcal{S}$ that are also equivalent to $\mathbb{P}$ is denoted by $\mathcal{S}^e$. If $\mathbb{Q} \in \mathcal{S}^e$ then the martingale $Z_t = \mathbb{E}\left[\frac{d\mathbb{Q}}{d\mathbb{P}} \mid \mathcal{F}_t\right]$ has the property $\inf_{t \in \mathbb{R}_+} Z_t > 0$, $\mathbb{P}$ a.s. (see [12] page 85). If $\mathbb{Q} \sim \mathbb{P}$, Bayes' rule implies that $\mathbb{E}_\mathbb{Q}[f \mid \mathcal{F}_T] = \mathbb{E}_\mathbb{P}[f \frac{Z_\infty}{Z_T} \mid \mathcal{F}_T]$, here T denotes a stopping time.

Remark 1. Standing assumption and notation We will always assume that $\mathbb{P} \in \mathcal{S}$ and if Z is a nonnegative (local) martingale, the expression that Z is positive (we will rather say strictly positive to avoid linguistic difficulties) means that $Z_\infty > 0$ a.s. . As a consequence we have that if Z is strictly positive then we have that a.s. : $\inf_t Z_t > 0$. The latter is of course stronger than $Z_t > 0$ a.s. for every $t \geqslant 0$.

Definition 1 *We say that a set of probability measures $\mathcal{S} \subset \mathbf{L}^1$, is multiplicatively stable, (m–stable for short) if for elements $\mathbb{Q}^0 \in \mathcal{S}, \mathbb{Q} \in \mathcal{S}^e$ with associate martingales $Z_t^0 = \mathbb{E}\left[\frac{d\mathbb{Q}^0}{d\mathbb{P}} \mid \mathcal{F}_t\right]$ and $Z_t = \mathbb{E}\left[\frac{d\mathbb{Q}}{d\mathbb{P}} \mid \mathcal{F}_t\right]$, and for each stopping* time T, the element L defined as $L_t = Z_t^0$ for $t \leqslant T$ and $L_t = Z_T^0 Z_t / Z_T$ for $t \geqslant T$ is a martingale that defines an element in $\mathcal{S}$. We also assume that every $\mathcal{F}_0$–measurable nonnegative function Z_0 such that $\mathbb{E}_\mathbb{P}[Z_0] = 1$, defines an element $d\mathbb{Q} = Z_0 \, d\mathbb{P}$ that is in $\mathcal{S}$.

Remark 2. The reader can check that indeed $\mathbb{E}[L_\infty] = 1$.

Remark 3. The second part of the definition is required to be sure that when $\mathcal{F}_0$ is not trivial, the set $\mathcal{S}$ is big enough. That part of the definition does not follow from the concatenation property. In most of the cases the σ–algebra $\mathcal{F}_0$ will be trivial and then the assumption only implies that $\mathbb{P} \in \mathcal{S}$.

Remark 4. If the set $\mathcal{S}$ is m–stable and closed in $\mathbf{L}^1$, it also satisfies the property: for elements $\mathbb{Q}^0 \in \mathcal{S}, \mathbb{Q} \in \mathcal{S}^e$ with associated martingales $Z_t^0 = \mathbb{E}\left[\frac{d\mathbb{Q}^0}{d\mathbb{P}} \mid \mathcal{F}_t\right]$ and $Z_t = \mathbb{E}\left[\frac{d\mathbb{Q}}{d\mathbb{P}} \mid \mathcal{F}_t\right]$, and for each *predictable* stopping time T, the element L defined as $L_t = Z_t^0$ for $t < T$ and $L_t = Z_{T-}^0 Z_t / Z_{T-}$ for $t \geqslant T$, is a martingale that defines an element in $\mathcal{S}$. On the set $\{T = 0\}$, Z_{T-} is, as usual, taken to be equal to Z_0. The proof that $L \in \mathcal{S}$ is quite straightforward. If T_n is a sequence that announces T then $L_t^n = Z_t^0$ for $t \leqslant T_n$ and $L_t^n = Z_{T_n}^0 Z_t / Z_{T_n}$ for $t \geqslant T_n$ defines elements in $\mathcal{S}$. Because T is predictable, we have that on the set $\{T > 0\}$, $Z_{T_n} = \mathbb{E}_\mathbb{P}[Z_\infty \mid \mathcal{F}_{T_n}]$ tends to $\mathbb{E}_\mathbb{P}[Z_\infty \mid \mathcal{F}_{T-}] = Z_{T-}$, whereas on the set $\{T = 0\}$, we always have that $Z_{T_n} = Z_0 = Z_{T-}$. It is now clear that L_∞^n tends to L_∞ a.s. and therefore also in $\mathbf{L}^1$ by Scheffé's lemma. Since $\mathcal{S}$ is closed this implies $L \in \mathcal{S}$.

Remark 5. The reader familiar with the concept of stable sets of martingales, can see the resemblance between the concept of being m–stable and the usual concept of stable spaces. As already mentioned in the introduction the definition of m–stable sets is the multiplicative version of the stability property, encountered in stochastic analysis. Here a set (most of the time a subspace) E of processes is called stable if for each stopping time T and each pair of processes $X, Y \in E$, the process $X^T + (Y - Y_T)\mathbf{1}_{\llbracket T, +\infty \llbracket} \in E$. The references given in the introduction, [13], [19], [17], [29], [16], [28] introduced in some way the concept of stability (called rectangularity) but there was no separate study or the probability space was supposed to be finite.

Remark 6. Let us now analyse how to concatenate two elements in $\mathbb{Q}^0, \mathbb{Q} \in \mathcal{S}$ that are only absolutely continuous (and not necessarily equivalent) to $\mathbb{P}$. We suppose that the set $\mathcal{S}$ is convex. For each $1 > \varepsilon > 0$, let us define the probability $\mathbb{Q}^\varepsilon = \varepsilon \mathbb{P} + (1 - \varepsilon)\mathbb{Q} \in \mathcal{S}$. The associated martingales are denoted by Z^0, Z and Z^ε. If T is a stopping time we define L_t^ε as above, namely for $t < T$ we put $L_t^\varepsilon = Z_t^0$ and for $t \geqslant T$ we put $L_t^\varepsilon = Z_T^0 Z_t^\varepsilon / Z_T^\varepsilon$. On the set $\{Z_T > 0\}$ we have that L_∞^ε tends to $L_\infty = Z_T^0 Z_\infty / Z_T$ and on the set $\{Z_T = 0\}$, we must have that $Z_t = 0$ for all $t \geqslant T$ and hence we have that

L_∞^ε tends to $L_\infty = Z_T^0$. We still have that $\mathbb{E}_\mathbb{P}[L_\infty] = 1$. Indeed

$$\mathbb{E}_\mathbb{P}[L_\infty] = \mathbb{E}_\mathbb{P}[Z_T^0 Z_\infty / Z_T \mathbf{1}_{\{Z_T>0\}}] + \mathbb{E}_\mathbb{P}[Z_T^0 \mathbf{1}_{\{Z_T=0\}}]$$

$$= \mathbb{E}_\mathbb{P}\left[Z_T^0 \mathbf{1}_{\{Z_T>0\}} \mathbb{E}_\mathbb{P}[Z_\infty / Z_T \mid \mathcal{F}_T]\right] + \mathbb{E}_\mathbb{P}[Z_T^0 \mathbf{1}_{\{Z_T=0\}}]$$

$$= \mathbb{E}_\mathbb{P}[Z_T^0 \mathbf{1}_{\{Z_T>0\}}] + \mathbb{E}_\mathbb{P}[Z_T^0 \mathbf{1}_{\{Z_T=0\}}]$$

$$= \mathbb{E}_\mathbb{P}[Z_T^0] = 1.$$

It seems that the calculations are done as in the case where $\mathbb{Q} \in \mathcal{S}^e$ but with the extra notation that on the set $\{Z_T = 0\}$ we put, in a naive way, $Z_\infty/Z_T = 1$.

We will frequently use stochastic exponentials. For strictly positive martingales Z, with $Z_0 = 1$ — such as density processes of measures $\mathbb{Q}$ that are equivalent to $\mathbb{P}$ — we can take the stochastic logarithm defined as $N = \frac{1}{Z_-} \cdot Z$. This stochastic integral is always defined and we have that $Z = \mathcal{E}(N)$ where $\mathcal{E}$ is the stochastic exponential or Doléans-Dade exponential (see [25] for precise definitions). The main theorem of this paper deals with the structure of m–stable convex closed sets $\mathcal{S} \subset \mathbf{L}^1$ of probability measures. Before we state the theorem, let us give an example of such a set (the proof that such sets are indeed m–stable is deferred). First let us recall what is usually called a multivalued mapping. For each $(t, \omega) \in \mathbb{R}_+ \times \Omega$ we give a nonempty closed convex set $C(t, \omega)$ of $\mathbb{R}^d$. The graph of C is then the set $\{(t, \omega, x) \mid x \in C(t, \omega)\}$. Set-theoretically we can identify the graph of C with C itself. In case the sets $C(t, \omega)$ are one–point sets, the object C simply defines a mapping from $\mathbb{R}_+ \times \Omega$ into $\mathbb{R}^d$. In our, more general, case we say that C is a *multivalued mapping* from $\mathbb{R}_+ \times \Omega$ into $\mathbb{R}^d$. We realise that from the set–theoretic viewpoint this terminology is horrible. However it is quite standard and it is widely used in the literature. Other names are multi-function, in French "multi-application" and set-valued function. We will make use of the integration theory for multivalued mappings later on. The multivalued mapping C is called predictable if the graph of C belongs to the product σ−algebra $\mathcal{P} \otimes \mathcal{B}(\mathbb{R}^d)$, where $\mathcal{B}(\mathbb{R}^d)$ is the Borel σ−algebra of $\mathbb{R}^d$. The following result gives a method to construct m–stable sets. The statement uses the technical assumption of what we call the predictable range of a σ−martingale. This concept is explained in the appendix. The concept is needed to deal with predictable processes q that are not identically zero but are such that the stochastic integral $q \cdot M$ is zero. The reader can now see that the concept of m–stable set is indeed closely related to the property "stable par bifurcation" introduced in [13].

Theorem 1 *Let C be a predictable convex closed multivalued mapping from $\mathbb{R}_+ \times \Omega$ into $\mathbb{R}^d$. Let an $\mathbb{R}^d$−valued martingale M be given. Suppose that for each (t, ω), $0 \in C(t, \omega)$ and suppose that the projection of C on the predictable range of the process M is closed. Then the $\mathbf{L}^1$−closure $\mathcal{S}$ of the set*

$$\mathcal{S}^e = \left\{ \mathcal{E}(q \cdot M)_\infty \left| \begin{array}{l} q \text{ is predictable} \\ q(t, \omega) \in C(t, \omega) \\ \mathcal{E}(q \cdot M) \text{ is a positive uniformly integrable martingale} \end{array} \right. \right\}$$

is an m–stable convex closed set such that $\mathbb{P} \in \mathcal{S}$. Furthermore the set $\{\mathbb{Q} \in \mathcal{S} \mid \mathbb{Q} \sim \mathbb{P}\}$ is precisely the set $\mathcal{S}^e$ defined above.

Remark 7. There are two extreme cases that deserve attention. The first case is when $C(t, \omega) = \{0\}$ in which case we have that $\mathcal{S} = \{\mathbb{P}\}$. The second case is when $C(t, \omega) = \mathbb{R}^d$ in which case we have that $\mathcal{S}$ is the set of all absolutely continuous probability measures $\mathbb{Q}$, whose density process $Z_t = \mathbb{E}\left[\frac{d\mathbb{Q}}{d\mathbb{P}} \mid \mathcal{F}_t\right]$, is a stochastic integral with respect to the martingale M.

In case the m–stable set $\mathcal{S}$ has only elements of the form $\mathcal{E}(q \cdot M)$, where M is a *continuous* martingale, we can also prove a converse to the preceding result.

Theorem 2 *Let $\mathcal{S} \subset \mathbf{L}^1$ be an m–stable convex closed set of probability measures such that $\mathbb{P} \in \mathcal{S}$. Suppose that there is an $\mathbb{R}^d$–valued continuous martingale M such that for each $\mathbb{Q} \in \mathcal{S}^e$ there is a predictable, $\mathbb{R}^d$–valued process q such that $\mathbb{E}\left[\frac{d\mathbb{Q}}{d\mathbb{P}} \mid \mathcal{F}_t\right] = \mathcal{E}(q \cdot M)_t$. Then there is a predictable, convex closed multivalued mapping C from $\mathbb{R}_+ \times \Omega$ into $\mathbb{R}^d$ such that $0 \in C(t, \omega)$ and such that*

$$\mathcal{S}^e = \left\{ \mathcal{E}(q \cdot M)_\infty \left| \begin{array}{l} q \text{ is predictable} \\ q(t, \omega) \in C(t, \omega) \\ \mathcal{E}(q \cdot M) \text{ is a positive uniformly integrable martingale} \end{array} \right. \right\}$$

Of course the latter theorem is more difficult since we have to find the multivalued mapping C. This will be done through the theory of multivalued measures and their corresponding Radon-Nikodým theorems. This theory was developed during the end of the sixties and is fundamental in control theory and in mathematical economics. Let us briefly describe what we will need from this theory. Let us suppose that $(G, \mathcal{G}, \mu)$ is a probability space. For a multivalued measurable mapping, C from G into $\mathbb{R}^d$, we define the integral of C as follows

$$\int_G C(g)\mu(dg) = \left\{ \int_G q(g)\mu(dg) \left| \begin{array}{l} q(g) \in C(g), \mu \text{ a.e.} \\ q \text{ is integrable} \end{array} \right. \right\}.$$

It turns out that if μ is atomless, then, by the Lyapunov theorem, the integral is automatically a convex set. The existence of elements q that are measurable follows from the measurable selections theorems, see [5] for the integration theory of set–valued mappings. The completeness of the probability space is not really needed. However in case the measure space is not complete, there are not necessarily measurable selections. The best one can obtain is an almost everywhere selection that is measurable. The existence of *integrable* selections has to be dealt with through boundedness conditions on the sets $C(g)$. A mapping Φ that assigns with each element A from $\mathcal{G}$, a set $\Phi(A) \subset \mathbb{R}^d$, is called a set–valued measure if whenever $A = \cup_n A_n$ is a union of pairwise disjoint sets in $\mathcal{G}$, we can write that

$$\Phi(A) = \sum_n \Phi(A_n)$$

$$= \left\{ \sum x_n \mid x_n \in \Phi(A_n) \text{ the sum being absolutely convergent} \right\}.$$

The set-valued measure Φ is called μ absolutely continuous if $\mu(A) = 0$ implies that $\Phi(A) = \{0\}$. We say that C is the Radon-Nikodým derivative of Φ if for each A we have $\int_A C(g)\mu(dg) = \Phi(A)$. In case the set–valued measure is bounded, convex and closed valued, the absolute continuity of Φ with respect to μ guarantees the existence of a Radon-Nikodým derivative. This is a consequence of the theorem of Debreu–Schmeidler ([9]). In case the set–valued measure is not convex compact valued, the situation is different. The reader can consult Debreu–Schmeidler ([9]) and Artstein's paper ([4]) to have an idea of the difficulties that arise.

2 Elementary Stability Properties of m–Stable Sets

The definition of m–stable sets allows for immediate extensions. More precisely we have the following property, that may explain why m–stable sets are also called fork convex.

Proposition 1. *Let $\mathcal{S} \subset \mathbf{L}^1$ be an m–stable set. Let $Z^0, Z^1, \ldots Z^n$ be density processes that are elements of $\mathcal{S}^e$. Suppose that T is a stopping time and suppose that $A_1, \ldots, A_n$ are elements of $\mathcal{F}_T$ that form a partition of Ω. The element*

$$L_t = Z_t^0 \quad \text{if} \quad t \leqslant T$$

$$= \sum_{k=1}^{n} \mathbf{1}_{A_k} Z_T^0 \frac{Z_t^k}{Z_T^k} \quad \text{if} \quad t \geqslant T,$$

defines an element of $\mathcal{S}$.

Proof The proof is by induction on n. Let us put $L^0 = Z^0$ and for $k \geqslant 1$ let us define $L_t^k = Z_t^0$ if $t \leqslant T$ and if $t \geqslant T$ let us put

$$L_t^k = \mathbf{1}_{A_k^c} L_t^{k-1} + \mathbf{1}_{A_k} L_T^{k-1} \frac{Z_t^k}{Z_T^k}$$

$$= \sum_{j=1}^{k} \mathbf{1}_{A_j} Z_T^0 \frac{Z_t^k}{Z_T^k} + \mathbf{1}_{\cup_{j>k} A_j} Z_t^0.$$

From the definition of m–stable sets, applied to the stopping time

$$T_{A_k} = T \mathbf{1}_{A_k} + \infty \mathbf{1}_{A_k^c},$$

it follows that if $L^{k-1} \in \mathcal{S}$, then also $L^k \in \mathcal{S}$. An induction argument now shows that $L = L^n \in \mathcal{S}$.   □

Corollary 1 *Let $\mathcal{S} \subset \mathbf{L}^1$ be an $(\mathbf{L}^1-)$closed m–stable set. Let $Z^0 \in \mathcal{S}$ and let $Z^n, n \geqslant 1$ be density processes that are elements of $\mathcal{S}^e$. Suppose that T is a stopping time and suppose that $A_n, n \geqslant 1$ are elements of $\mathcal{F}_T$ that form a partition of Ω. The element*

$$L_t = Z_t^0 \quad \text{if} \quad t \leqslant T$$

$$= \sum_{k \geqslant 1} \mathbf{1}_{A_k} Z_T^0 \frac{Z_t^k}{Z_T^k} \quad \text{if} \quad t \geqslant T,$$

defines an element of $\mathcal{S}$.

Proof This is easily seen. We define exactly as in the proof of the proposition, the sequence L^k. It is clear that $\mathbb{E}[L_\infty^k] = 1$ and by conditioning on the σ–algebra $\mathcal{F}_T$ it also follows that $\mathbb{E}[L_\infty] = \mathbb{E}[\mathbb{E}[L_\infty \mid \mathcal{F}_T]] = \mathbb{E}[L_T^0] = 1$. Furthermore we have that $L_\infty^k \to L_\infty$ a.s. . From this and Scheffé's lemma it follows that $L_\infty^k \to L_\infty$ in the $\mathbf{L}^1$–norm, implying that $L \in \mathcal{S}$.   □

We will now prove the

Theorem 3 *Suppose that $\mathcal{F}_0$ is trivial. Let C be a predictable convex closed multivalued mapping from $\mathbb{R}_+ \times \Omega$ into $\mathbb{R}^d$. Let an $\mathbb{R}^d$–valued martingale M be given. Suppose that for each (t, ω), $0 \in C(t, \omega)$ and suppose that the projection of C on the predictable range of the process M is closed. Then the $\mathbf{L}^1$ closure $\mathcal{S}$ of the set*

$$\mathcal{S}^e = \left\{ \mathcal{E}(q \cdot M)_\infty \,\middle|\, \begin{array}{l} q \text{ is predictable} \\ q(t, \omega) \in C(t, \omega) \\ \mathcal{E}(q \cdot M) \text{ a positive uniformly integrable martingale} \end{array} \right\}$$

is an m–stable convex closed set such that $\mathbb{P} \in \mathcal{S}$. Moreover $\{\mathbb{Q} \in \mathcal{S} \mid \mathbb{Q} \sim \mathbb{P}\} = \mathcal{S}^e$ (as the notation suggests).

Proof The concatenation property is trivial. The proof of the convexity mainly follows from Itô's lemma. First of all it is trivial that $\mathbb{P} \in \mathcal{S}$. Let $Z^1 = \mathcal{E}(q^1 \cdot M)$ and $Z^2 = \mathcal{E}(q^2 \cdot M)$ be two strictly positive uniformly integrable martingales coming from elements in $\mathcal{S}^e$. Let $0 < \alpha < 1$ be fixed. Put $Z = \alpha Z^1 + (1 - \alpha)Z^2$, which is a strictly positive uniformly integrable martingale. We have to show that $Z_\infty \in \mathcal{S}^e$. Itô's lemma gives

$$dZ_t = \alpha Z_{t-}^1 q_t^1 \, dM_t + (1 - \alpha)Z_{t-}^2 q_t^2 \, dM_t$$

We can proceed as follows

$$dZ_t = Z_{t-} \left(\frac{\alpha Z_{t-}^1}{Z_{t-}} q_t^1 + \frac{(1-\alpha)Z_{t-}^2}{Z_{t-}} q_t^2 \right) dM_t$$

$$= Z_{t-} q_t \, dM_t,$$

where

$$q_t = \left(\frac{\alpha Z_{t-}^1}{Z_{t-}} q_t^1 + \frac{(1-\alpha)Z_{t-}^2}{Z_{t-}} q_t^2 \right)$$

is in the set C since it is a convex combination of two elements in C. This proves convexity. Since $0 \in C$ we have that $\mathbb{P} \in \mathcal{S}^e$. We still have to show that elements of the closure of $\mathcal{S}^e$ and that are equivalent to $\mathbb{P}$ are of the form stated in the description of $\mathcal{S}^e$. To do this, consider a sequence $Z^n \in \mathcal{S}$ and suppose that Z^n converges to the strictly positive martingale Z. Each Z^n can be written as

$$Z^n = \mathcal{E}\left(q^n \cdot M \right),$$

where $q^n \in C$. The sequence q^n does not have to converge but its projection onto the predictable range of M does. Indeed we have that a.s. the brackets

$$[(q^n - q^m) \cdot M, (q^n - q^m) \cdot M]_\infty$$

tend to zero in the space of nonnegative definite matrices. It follows that there is a vector valued predictable process q' such that $q^n \cdot M$ tends to $q' \cdot M$ in the space of local martingales. But the hypothesis that the projection of C onto the predictable range of M is closed, implies (together with the measurable selection theorem) that q' is the projection of a predictable selection q of the set valued mapping C. This implies that $Z = \mathcal{E}(q \cdot M)$ as desired. □

Remark 8. In case $\mathcal{F}_0$ is not trivial, we have to include the factors of the form $\exp(Z_0)$ where Z_0 is $\mathcal{F}_0$ measurable. The author thanks the referee for pointing out this and other imprecisions.

3 The Characterisation of m-Stable Sets in the Continuous Case

In this section we suppose that the set $\mathcal{S}$ is a closed convex $m-$stable set. Furthermore we suppose that there is a *continuous* $\mathbb{R}^d-$valued martingale M so that each element $\mathbb{Q}$ of $\mathcal{S}^e$ can be written as

$$\frac{d\mathbb{Q}}{d\mathbb{P}} = \mathcal{E}(q \cdot M),$$

where q is an $\mathbb{R}^d-$valued predictable process. Such a situation occurs when there is a finite dimensional martingale that has the predictable representation property. But for the moment we do not need this more restrictive assumption. The main object of this section is to prove theorem 2 of the introduction. As the reader can verify, it does not harm to suppose that the bracket of the martingale M is bounded by 1, i.e. $Trace\langle M, M\rangle_\infty < 1$. If this is not the case we may replace M by the martingale defined by the stochastic integral

$$\int \frac{1}{1 + \exp\left(200\, Trace\langle M, M\rangle\right)}\, dM.$$

This assumption simplifies the notation of the proof considerably. Before we prove the theorem let us make the precise statement (including the simplifications we introduced).

Theorem 4 *Let $\mathcal{S} \subset \mathbf{L}^1$ be an m–stable convex closed set of probability measures. Suppose that there is an $\mathbb{R}^d$–valued continuous martingale M, verifying $M_0 = 0$ and such that for each $\mathbb{Q} \in \mathcal{S}^e$ there is a predictable, $\mathbb{R}^d$–valued process q such that $\mathbb{E}\left[\frac{d\mathbb{Q}}{d\mathbb{P}} \mid \mathcal{F}_t\right] = \mathcal{E}(q \cdot M)_t$. Suppose further that $Trace\langle M, M\rangle_\infty < 1$. Then there is a predictable, convex closed set-valued mapping C from $\mathbb{R}_+ \times \Omega$ into $\mathbb{R}^d$ such that $0 \in C(t, \omega)$ and such that*

$$\mathcal{S}^e = \left\{ \mathcal{E}(q \cdot M)_\infty \;\middle|\; \begin{array}{l} q \text{ is predictable} \\[4pt] q(t, \omega) \in C(t, \omega) \\[4pt] \mathcal{E}(q \cdot M) \text{ is a positive uniformly integrable martingale} \end{array} \right\}.$$

The proof is divided into several steps. Because of this we will introduce an extra notation. The finite measure μ on $\mathcal{P}$ is defined through the formula

$$\mu(A) = \mathbb{E}\left[\int_{\mathbb{R}_+} \mathbf{1}_A \, d\, Trace\langle M, M\rangle\right].$$

This measure will serve as a control measure. The first step of the proof is to generalise the stability property.

Lemma 2. *Let $\mathbb{Q} \in \mathcal{S}^e$, suppose that $d\mathbb{Q}/d\mathbb{P} = \mathcal{E}(q \cdot M)$ and let $A \in \mathcal{P}$ be such that $\mathbb{E}\left[\mathcal{E}\left(q\mathbf{1}_A \cdot M\right)_\infty\right] = 1$, then we have that*

$$\mathcal{E}\left(q\mathbf{1}_A \cdot M\right) \in \mathcal{S}^e.$$

Proof The proof follows from Proposition 1 as soon as the predictable set $A \in \mathcal{A}$. For the general case we take a sequence of predictable sets $A_n \in \mathcal{A}$ so that $\mu(A_n \Delta A) \to 0$. Of course we have that each element $\mathcal{E}(q\mathbf{1}_{A_n} \cdot M) \in \mathcal{S}^e$. The sequence $\mathcal{E}(q\mathbf{1}_{A_n} \cdot M)_\infty$ tends in probability to the element $\mathcal{E}(q\mathbf{1}_A \cdot M)_\infty$ and Scheffé's lemma implies that the convergence takes place in $\mathbf{L}^1$. Since $\mathcal{S}$ is closed we have that $\mathcal{E}(q\mathbf{1}_A \cdot M) \in \mathcal{S}$. But $\langle q\mathbf{1}_A \cdot M, q\mathbf{1}_A \cdot M\rangle_\infty \leqslant \langle q \cdot M, q \cdot M\rangle_\infty < \infty$ a.s. and therefore $\mathcal{E}(q\mathbf{1}_A \cdot M)_\infty > 0$ a.s. . Therefore $\mathcal{E}(q\mathbf{1}_A \cdot M) \in \mathcal{S}^e$. The proof of the lemma is complete. $\qquad\square$

Remark 9. It is not true that the stochastic exponential $\mathcal{E}(q\mathbf{1}_A \cdot M)$ is uniformly integrable as soon as the stochastic exponential $\mathcal{E}(q \cdot M)$ is uniformly integrable. The assumption $\mathbb{E}\left[\mathcal{E}\left(q\mathbf{1}_A \cdot M\right)_\infty\right] = 1$ cannot be omitted.

Lemma 3. *Let* $\mathbb{Q}^1, \mathbb{Q}^2 \in \mathcal{S}^e$, *suppose that* $d\mathbb{Q}^i/d\mathbb{P} = \mathcal{E}(q^i \cdot M)$ *and let* $A \in \mathcal{P}$ *be such that* $\mathbb{E}\left[\mathcal{E}\left((q^1\mathbf{1}_A + q^2\mathbf{1}_{A^c}) \cdot M\right)_\infty\right] = 1$, *then we have*

$$\mathcal{E}\left((q^1\mathbf{1}_A + q^2\mathbf{1}_{A^c}) \cdot M\right) \in \mathcal{S}^e.$$

Proof We omit the proof since it is almost a copy of the proof of the previous lemma. In fact we could have proved this lemma first. The previous lemma is then a special case. $\square$

The next step is to reduce our attention to elements of $\mathcal{S}$ that come from bounded integrands. More precisely, for each $\lambda > 0$ we introduce

$$\mathcal{S}^\lambda = \left\{\mathbb{Q} \in \mathcal{S} \mid \frac{d\mathbb{Q}}{d\mathbb{P}} = \mathcal{E}(q \cdot M) \text{ and } \|q\| \leqslant \lambda\right\} = \left(\mathcal{S}^\lambda\right)^e.$$

The previous lemma allows us to prove the following density result

Lemma 4. *The sets* $\mathcal{S}^\lambda$ *are m–stable, form an increasing family and the union* $\cup_{\lambda>0}\mathcal{S}^\lambda$ *is* $\mathbf{L}^1-$*dense in* $\mathcal{S}$.

Proof The stability of the sets $\mathcal{S}^\lambda$ is obvious from the definition of $m-$stability. That they are increasing in λ is also obvious. That the sets are subsets of $\mathcal{S}$ follows from the previous lemma. Indeed if $\mathcal{E}(q \cdot M) \in \mathcal{S}$ then necessarily we must have that $\mathcal{E}(q\mathbf{1}_{\|q\|\leqslant\lambda} \cdot M) \in \mathcal{S}$. Indeed by Novikov's criterion (see [27]), the stochastic exponential $\mathcal{E}(q\mathbf{1}_{\|q\|\leqslant\lambda} \cdot M)$ is uniformly integrable (remember that $Trace\langle M, M\rangle \leqslant 1$). For $\lambda \to \infty$ we also have that $\mathcal{E}(q\mathbf{1}_{\|q\|\leqslant\lambda} \cdot M)_\infty$ converges in probability to $\mathcal{E}(q\cdot M)_\infty$. Scheffé's lemma transforms the convergence into $\mathbf{L}^1-$convergence, proving the density. $\square$

Lemma 5. *The sets* $\mathcal{S}^\lambda$ *are convex and closed.*

Proof The convexity will be checked using Itô's formula. So let us take

$$Z^1 = \mathcal{E}(q^1 \cdot M) \in \mathcal{S}^\lambda, \|q^1\| \leqslant \lambda \text{ and}$$
$$Z^2 = \mathcal{E}(q^2 \cdot M) \in \mathcal{S}^\lambda, \|q^2\| \leqslant \lambda.$$

Take $0 < \alpha < 1$. Itô's formula now gives

$$
\begin{aligned}
d(\alpha Z^1 &+ (1-\alpha)Z^2)_t \\
&= \alpha Z_t^1 q_t^1 \, dM_t + (1-\alpha)Z_t^2 q_t^2 \, dM_t \\
&= \left(\alpha Z_t^1 + (1-\alpha)Z_t^2\right)\left(\frac{\alpha Z_t^1 q_t^1}{\alpha Z_t^1 + (1-\alpha)Z_t^2} + \frac{(1-\alpha)Z_t^2 q_t^2}{\alpha Z_t^1 + (1-\alpha)Z_t^2}\right) dM_t \\
&= q_t \, dM_t \qquad \text{where} \\
q_t &= \left(\frac{\alpha Z_t^1 q_t^1}{\alpha Z_t^1 + (1-\alpha)Z_t^2} + \frac{(1-\alpha)Z_t^2 q_t^2}{\alpha Z_t^1 + (1-\alpha)Z_t^2}\right).
\end{aligned}
$$

Obviously $\|q_t\| \leqslant \lambda$ since both $\|q_t^1\| \leqslant \lambda$ and $\|q_t^2\| \leqslant \lambda$. Because $\mathcal{S}$ is convex we have that $\mathcal{E}(q \cdot M)$ is already in $\mathcal{S}$. The boundedness on q then implies $\mathcal{E}(q \cdot M) \in \mathcal{S}^\lambda$. We still have to show that the set $\mathcal{S}^\lambda$ is closed. Let us take a sequence $\mathcal{E}(q^n \cdot M) \in \mathcal{S}^\lambda$ that converges in $\mathbf{L}^1$ to a martingale Z. Of course we suppose that $\|q^n\| \leqslant \lambda$ for all n. First observe that by taking the stochastic logarithm, we can easily see that the sequence $q^n \cdot M$ converges in the semi-martingale topology to a martingale N and that $Z = \mathcal{E}(N)$. But by the uniform boundedness of the sequence q^n, we must then also have that the convergence takes place in all spaces $\mathbf{L}^p$. The sequence q^n forms a bounded sequence in the space $\mathbf{L}^\infty(\mu)$ and therefore there is a sequence of convex combinations

$$k_n \in \ conv\{q^n, q^{n+1}, \ldots\},$$

so that $k_n \to k$ in μ–measure. Of course k is predictable and $\|k\| \leqslant \lambda$. We also have that $k_n \cdot M$ converges in probability to $k \cdot M$ (even in all $\mathbf{L}^p$). Therefore we also have that $N = k \cdot M$. This shows that Z is of the form $Z = \mathcal{E}(k \cdot M)$ where k remains bounded by λ. This completes the proof of the lemma.   □

We now introduce the set

$$\mathcal{C}^\lambda = \left\{q \colon \mathbb{R}_+ \times \Omega \to \mathbb{R}^d \mid \|q\| \leqslant \lambda \text{ predictable and } \mathcal{E}(q \cdot M) \in \mathcal{S}^\lambda\right\}.$$

The following lemma seems an obvious consequence of the closedness of the sets $\mathcal{S}^\lambda$, so we omit the proof.

Lemma 6. $\mathcal{C}^\lambda \subset \mathbf{L}^\infty(\mu)$ *is closed for the topology of convergence in* μ-*measure.*

The difficult part of the proof of the main result is to show that the sets $\mathcal{C}^\lambda$ are convex. The rather technical proof of this convexity result is based on the following BMO-style inequality.

Lemma 7. *Let* $(\mathcal{G}_t)_t$ *be a filtration satisfying the usual assumptions. Suppose that V is a continuous martingale adapted to $\mathcal{G}$ and such that $V_0 = 0$. Suppose that $\langle V, V \rangle_\infty \leqslant K$ for some constant K. Then a.s.*

$$\mathbb{E}\left[\left(\frac{\mathcal{E}(V)_\infty + \mathcal{E}(-V)_\infty}{2}\right)^2 \,\middle|\, \mathcal{G}_0\right] \leqslant \cosh(K).$$

Proof The proof uses the DDS-time change theorem, see [27]. This theorem allows us to reduce the problem to the Brownian Motion case. Here are the details. First of all, remark that the martingale V converges at ∞ and therefore we can close the interval $\mathbb{R}_+$ by adding the point $+\infty$. We then transform the interval $[0, +\infty]$ to the interval $[0, 1]$. After time one we continue the process by adding an independent Brownian Motion. The new process is still denoted by V and the filtration is still denoted by $\mathcal{G}$, no confusion is possible. For this new process we define the finite stopping time

$$\tau = \inf\{t \mid \langle V, V \rangle_t > K\}.$$

By the assumption on the bracket of V, $\tau \geqslant 1$. From the DDS theorem it follows that V_τ is a random variable that has a gaussian distribution with mean 0 and variance K. However this random variable is independent of $\mathcal{G}_0$. By Jensen's inequality for conditional expectations, applied to the martingale $\frac{\mathcal{E}(V)+\mathcal{E}(-V)}{2}$, we get that

$$
\mathbb{E}\left[\left(\frac{\mathcal{E}(V)_1 + \mathcal{E}(-V)_1}{2}\right)^2 \middle| \mathcal{G}_0\right] \leqslant
$$
$$
\mathbb{E}\left[\left(\frac{\mathcal{E}(V)_\tau + \mathcal{E}(-V)_\tau}{2}\right)^2 \middle| \mathcal{G}_0\right] =
$$
$$
\mathbb{E}\left[\left(\frac{\mathcal{E}(V)_\tau + \mathcal{E}(-V)_\tau}{2}\right)^2\right].
$$

The latter quantity can easily be calculated and gives

$$
\mathbb{E}\left[\left(\frac{\mathcal{E}(V)_\tau + \mathcal{E}(-V)_\tau}{2}\right)^2\right] = \frac{1}{4}e^{-K}\mathbb{E}\left[\exp(2V_\tau) + \exp(-2V_\tau) + 2\right]
$$
$$
= \frac{1}{4}e^{-K}\left(e^{2K} + e^{2K} + 2\right)
$$
$$
= \frac{1}{2}(e^K + e^{-K}) = \cosh(K).
$$

$\square$

Lemma 8. *Let the sequence of stopping times* $(T_n^k)_{0 \leqslant k \leqslant 2^n}$ *be defined as follows. For each n and $0 \leqslant k \leqslant 2^n$, we define:*

$$
T_n^k = \inf\left\{t \mid \langle M, M\rangle_t \geqslant \frac{k}{2^n}\right\}.
$$

Obviously $T_n^0 = 0$ *and* $T_n^{2^n} = \infty$ *since* $\langle M, M\rangle_\infty < 1$. *Let q^1 and q^2 be predictable $\mathbb{R}^d$ valued processes bounded by λ. For each n we define*

$$
f_n = \prod_{k=0}^{2^n - 1}\left(\frac{1}{2}\mathcal{E}\left(\mathbf{1}_{\rrbracket T_n^k, T_n^{k+1}\rrbracket}q^1 \cdot M\right)_\infty + \frac{1}{2}\mathcal{E}\left(\mathbf{1}_{\rrbracket T_n^k, T_n^{k+1}\rrbracket}q^2 \cdot M\right)_\infty\right).
$$

Let

$$
f = \mathcal{E}\left(\frac{q^1 + q^2}{2} \cdot M\right)_\infty
$$

Then f_n tends to f in $\mathbf{L}^1(\mathbb{P})$.

Proof Clearly $f > 0$ and $\mathbb{E}_\mathbb{P}[f] = 1$. Define the measure $\mathbb{Q}$ as $d\mathbb{Q} = f\,d\mathbb{P}$. We will show that

$$
\|f_n - f\|_{\mathbf{L}^1(\mathbb{P})} = \left\|\frac{f_n}{f} - 1\right\|_{\mathbf{L}^1(\mathbb{Q})}
$$

tends to zero. Obviously $\mathbb{E}_{\mathbb{Q}}\left[\frac{f_n}{f}\right] = 1$. The statement therefore follows as soon as we can prove that $\mathbb{E}_{\mathbb{Q}}\left[\left(\frac{f_n}{f}\right)^2\right] \to 1$. Indeed this convergence immediately implies that $\|\frac{f_n}{f} - 1\|^2_{\mathbf{L}^2(\mathbb{Q})} \to 0$.

Under the measure $\mathbb{Q}$, the martingale M can be decomposed into a martingale N and a process of finite variation. The continuous martingale N has the same bracket as M. Moreover a straightforward calculation shows that each factor, say g_n^k in the expression of $\frac{f_n}{f}$ can be written as

$$g_n^k = \frac{1}{2}\mathcal{E}\left(\mathbf{1}_{]T_n^k, T_n^{k+1}]}\frac{q^1 - q^2}{2} \cdot N\right)_\infty + \frac{1}{2}\mathcal{E}\left(\mathbf{1}_{]T_n^k, T_n^{k+1}]}\frac{q^2 - q^1}{2} \cdot N\right)_\infty.$$

We now repeatedly will use the lemma 8. The bracket we must control is

$$\langle \mathbf{1}_{]T_n^k, T_n^{k+1}]}\frac{q^1 - q^2}{2} \cdot N, \mathbf{1}_{]T_n^k, T_n^{k+1}]}\frac{q^1 - q^2}{2} \cdot N \rangle_\infty \leqslant \lambda^2 2^{-n}.$$

By telescoping and repeated application of lemma 8, we now find

$$\mathbb{E}_{\mathbb{Q}}\left[\prod_{k=0}^{2^n - 1}(g_n^k)^2\right] \leqslant \mathbb{E}_{\mathbb{Q}}\left[\prod_{k=0}^{2^n - 2}(g_n^k)^2 \, \mathbb{E}_{\mathbb{Q}}\left[\left(g_n^{2^n-1}\right)^2 \mid \mathcal{F}_{T_n^{2^n-1}}\right]\right]$$

$$\leqslant \mathbb{E}_{\mathbb{Q}}\left[\prod_{k=0}^{2^n - 2}(g_n^k)^2 \, \cosh(\lambda^2 2^{-n})\right]$$

$$\leqslant \ldots$$

$$\leqslant \left(\cosh(\lambda^2 2^{-n})\right)^{2^n}$$

Since for small x we have $\cosh(x) \approx 1 + x^2/2$, we get that $\left(\cosh(\lambda^2 2^{-n})\right)^{2^n}$ tends to 1 as n tends to infinity. □

Lemma 9. *The set $\mathcal{C}^\lambda$ is convex.*

Proof We use the notation of the previous lemma. Obviously we have that $f_n \in \mathcal{S}^\lambda$, therefore also $f \in \mathcal{S}^\lambda$ and therefore $\frac{q^1 + q^2}{2} \in \mathcal{C}^\lambda$. Since $\mathcal{C}^\lambda$ is already closed for convergence in probability, we get that convex combinations with other coefficients than $1/2$ remain in $\mathcal{C}^\lambda$ as well. □

The next step in the proof is to analyse the structure of the set $\mathcal{C}^\lambda$. The following lemma is obvious in the sense that either the statements were proved before or are trivial

Lemma 10. *The set $\mathcal{C}^\lambda$ satisfies the following properties*

1. $\mathcal{C}^\lambda \subset \mathbf{L}^\infty(\mathcal{P}, \mu; \mathbb{R}^d)$
2. $\mathcal{C}^\lambda$ *is contained in the ball of radius λ*
3. $\mathcal{C}^\lambda$ *is closed for μ convergence*

4. $\mathcal{C}^\lambda$ is convex, therefore it is also weak*, i.e. $\sigma(\mathbf{L}^\infty(\mathcal{P},\mu;\mathbb{R}^d),\mathbf{L}^1(\mathcal{P},\mu;\mathbb{R}^d))$ compact.
5. if $q^1,q^2 \in \mathcal{C}^\lambda$, if $A \in \mathcal{P}$, then $q^1\mathbf{1}_A + q^2\mathbf{1}_{A^c} \in \mathcal{C}^\lambda$

The following lemma can be seen as an extension of the convexity property. This property is sometimes called "predictably convex".

Lemma 11. If $q^1,q^2 \in \mathcal{C}^\lambda$, if h is a real valued predictable process such that $0 \leqslant h \leqslant 1$, then also $h\,q^1 + (1-h)\,q^2 \in \mathcal{C}^\lambda$.

Proof If $h = \sum_{i=1}^n \alpha_i\mathbf{1}_{A_i}$ where $A_i \in \mathcal{P}$ and where the nonnegative numbers α_i sum up to 1, the property follows from convexity. However the set of such convex combinations is dense (for the topology in μ−convergence) in the set of all functions between 0 and 1. The closedness property completes the argument. □

We are now ready to prove the theorem for the set $\mathcal{S}^\lambda$.

Theorem 5 With the notation above, there exists a compact convex set-valued function $\Phi^\lambda : \mathbb{R}_+ \times \Omega \to \mathbb{R}^d$ so that

1. The graph of Φ^λ is in $\mathcal{P} \otimes \mathcal{B}(\mathbb{R}^d)$
2. $0 \in \Phi^\lambda(t,\omega)$ for each (t,ω).
3. $\mathcal{C}^\lambda = \{q : \mathbb{R}_+ \times \Omega \to \mathbb{R}^d \mid q \text{ is predictable and } \mu \text{ a.s. } q(t,\omega) \in \Phi^\lambda(t,\omega)\}$

Proof For each $A \in \mathcal{P}$ we define the set $C(A)$ as follows

$$C(A) = \left\{ \int_A q\,d\mu \mid q \in \mathcal{C}^\lambda \right\}.$$

The set-valued mapping $C : \mathcal{P} \to \mathbb{R}^d$ satisfies

1. $0 \in C(A)$ since $0 \in \mathcal{C}^\lambda$.
2. If $x \in C(A)$ then $\|x\| \leqslant \lambda\mu(A)$. Indeed $\|\int_A q\,d\mu\| \leqslant \int_A \|q\|\,d\mu \leqslant \lambda\mu(A)$.
3. $C(A)$ is convex since $\mathcal{C}^\lambda$ is convex.
4. $C(A)$ is compact since $\mathcal{C}^\lambda$ is weak* compact.
5. If $(A_n)_n$ is a sequence of pairwise disjoint predictable sets with $A = \cup_n A_n$, if $x_n = \int_{A_n} q^n\,d\mu \in C(A_n)$ then the sum $x = \sum_n x_n$ converges and $x \in C(A)$. Indeed the convergence follows from the bound under item 2 and $x = \int_A q\,d\mu$ where $q = \sum_n \mathbf{1}_{A_n}q^n$. The latter follows from Lebesgue's dominated convergence theorem and the closedness of $\mathcal{C}^\lambda$.

The theorem of Debreu-Schmeidler, [9] now gives the existence of a compact convex set-valued, $\mathcal{P} \otimes \mathcal{B}(\mathbb{R}^d)$ measurable mapping Φ^λ so that for all A we have

$$C(A) = \left\{ \int_A q\,d\mu \mid q \in \Phi^\lambda \ \mu \text{ a.s. } \right\}.$$

We still have to show that this allows to find the set $\mathcal{C}^\lambda$. Let us first suppose that $q \in \mathcal{C}^\lambda$. We have to show that μ a.s. we have that $q \in \Phi^\lambda$. In case this were false let us look at the set

$$A = \{(t, \omega) \mid q(t, \omega) \notin \Phi^\lambda(t, \omega)\}.$$

This set is measurable and it has a positive measure $\mu(A) > 0$. Because of the density of the points with rational coordinates and the convexity of the sets Φ^λ, this means that there is a vector $p \in \mathbb{R}^d$ (with rational coordinates) such that the set

$$A_p = \{(t, \omega) \mid \langle p, q(t, \omega) \rangle > \sup \langle p, \Phi^\lambda(t, \omega) \rangle\}$$

also has a positive measure $\mu(A_p) > 0$. Indeed by the separation theorem we can write that

$$A = \cup_{p \in \mathbb{R}^d, p \text{ rational}} A_p.$$

But then we must have that $\int_{A_p} q \, d\mu \notin C(A_p)$, since obviously we have that $\langle p, \int_{A_p} q \, d\mu \rangle = \int_{A_p} \langle p, q \rangle \, d\mu > \int_{A_p} \sup \langle p, \Phi^\lambda(t, \omega) \rangle \, d\mu \geqslant \sup_{g \in \Phi^\lambda} \int_{A_p} \langle p, g \rangle \, d\mu = \sup \langle p, C(A_p) \rangle$. The converse is proved in a similar way. So let q_0 be a predictable selector of Φ^λ, we have to show that $q_0 \in C^\lambda$. If this is not the case, we separate the point q_0 from the compact convex set C^λ. This we can do by the Hahn-Banach theorem. We obtain a function $f \in \mathbf{L}^1(\mu; \mathbb{R}^d)$ so that

$$\int \langle f, q_0 \rangle \, d\mu > \sup_{q \in C^\lambda} \int \langle f, q \rangle \, d\mu.$$

The sup is actually attained because of compactness of C^λ. Let a maximising element be q_1. Necessarily we then must have that for every $q \in C^\lambda$ the inequality $\langle f, q_1 \rangle \geqslant \langle f, q \rangle$ holds μ a.s. This follows from the property 5 of lemma 10. Indeed if there would be an element $q \in C^\lambda$ so that the set $B = \{\langle f, q_1 \rangle < \langle f, q \rangle\}$ is not negligible, we could replace q_1 by $q_1 \mathbf{1}_{B^c} + q \mathbf{1}_B$ yielding a greater expression than the one for q_1. The set $\{\langle f, q_0 \rangle > \langle f, q_1 \rangle\}$ must have a strictly positive μ measure. Since the simple functions are dense in $\mathbf{L}^1$ we can find a vector $p \in \mathbb{R}^d$ as well as an $\epsilon > 0$ so that the set $A_p = \{\langle f, q_0 \rangle > \epsilon + \langle f, q_1 \rangle\} \cap \{\|f - p\|_{\mathbb{R}^d} \leqslant \frac{\epsilon}{4\lambda}\}$ has a nonzero measure $\mu(A_p) > 0$. But this inequality and the fact that all the functions in C^λ are pointwise bounded by λ, implies that

$$\langle p, \int_{A_p} q_0 \, d\mu \rangle \geqslant \int_{A_p} \langle f, q_0 \rangle \, d\mu - \frac{\epsilon}{4} \mathbb{P}[A_p]$$

$$> \int_{A_p} \langle f, q_1 \rangle \, d\mu + \frac{3\epsilon}{4} \mathbb{P}[A_p]$$

$$\geqslant \sup_{q \in C^\lambda} \int_{A_p} \langle f, q \rangle \, d\mu + \frac{3\epsilon}{4} \mathbb{P}[A_p]$$

$$\geqslant \sup_{q \in C^\lambda} \int_{A_p} \langle p, q \rangle \, d\mu + \frac{\epsilon}{2} \mathbb{P}[A_p].$$

And therefore we must have that $\int_{A_p} q_0 \, d\mu \notin C(A_p)$. But this is a contradiction to $q_0 \in \Phi^\lambda$ and the definition of the Radon-Nikodým derivative for set-valued measures. $\qquad\square$

Remark 10. The proof was based on the Radon-Nikodým theorem for set-valued measures. However the proof of the version we need can be given using the support functionals. Since we also needed the support functionals in a later stage of the proof there is a shortcut in the sense of merging the proof of the RN-theorem together with the arguments on the support functionals. However this would have obscured the idea of the proof.

The next step consists in getting rid of the truncation.

Lemma 12. *If $\lambda \leqslant \nu$ then $\Phi^\lambda = \Phi^\nu \cap B_\lambda$, where B_λ denotes the ball of radius λ in the Euclidean space $\mathbb{R}^d$. As a consequence we have $\Phi^\lambda \subset \Phi^\nu$ μ a.s.*

Proof If this is not the case we will make a measurable selection, say q, of the set-valued function $\Phi^\lambda \setminus (\Phi^\nu \cap B_\lambda)$, at least on the predictable set A, where the set $\Phi^\lambda \setminus (\Phi^\nu \cap B_\lambda)$ is nonempty. Let us put q equal to 0 where the set is empty. Since q is in $\mathcal{C}^\lambda$ it has to be in $\mathcal{C}^\nu$ as well. Since obviously the element $q = q\mathbf{1}_A$ is also in $\mathcal{C}^\nu$ it has to be a selector of Φ^ν. But this is a contradiction to the construction of q. The converse inclusion is proved in the same way. $\square$

For each (t, ω) we now define

$$\Phi(t, \omega) = \cup_{\lambda > 0} \Phi^\lambda(t, \omega) = \cup_{n \geqslant 1} \Phi^n(t, \omega).$$

Because Φ is the union of an increasing sequence of convex sets, it is convex. The closedness of Φ is something that needs a proof, since the countable union of closed sets does not have to be closed. But since we have the equality $\Phi^\lambda = \Phi^\nu \cap B_\lambda$ for $\lambda \leqslant \nu$, the union is indeed closed. The last part consists in showing that we get the m–stable set $\mathcal{S}$ back.

Lemma 13. *Let $q \in \Phi$, μ a.s. . Suppose that $\mathbb{E}\left[(\mathcal{E}(q \cdot M))_\infty\right] = 1$, meaning that $\mathcal{E}(q \cdot M)$ is a uniformly integrable martingale. Then $\mathcal{E}(q \cdot M) \in \mathcal{S}$*

Proof By construction we have that $q_n = q\mathbf{1}_{\|q\| \leqslant n}$ is a selector of Φ^n. Therefore it is in $\mathcal{C}^n$. But then we have that $\mathcal{E}(q_n \cdot M)$ is in $\mathcal{S}$. Since $\mathcal{E}(q_n \cdot M)$ tends to $\mathcal{E}(q \cdot M)$ in $\mathbf{L}^1$ (by Scheffé's lemma), we must have that $\mathcal{E}(q \cdot M) \in \mathcal{S}$. $\square$

The following statement concludes the proof of theorem 4 or 2.

Theorem 6 *We have the following equality*

$$\mathcal{S} = \{\mathcal{E}(q \cdot M) \mid q \in \Phi \text{ and } \mathbb{E}\left[(\mathcal{E}(q \cdot M))_\infty\right] = 1\}$$

Proof One inclusion is in the previous lemma. The other inclusion is quite obvious. Take q so that $\mathcal{E}(q \cdot M)$ is in $\mathcal{S}$. Obviously lemma 2 implies that for each n we have that $q\mathbf{1}_{\|q\| \leqslant n}$ is in $\mathcal{C}^n$. Therefore we have that $q\mathbf{1}_{\|q\| \leqslant n} \in \Phi^n \subset \Phi$, μ a.s. . This of course implies that $q \in \Phi$, μ a.s. . $\qquad\square$

Remark 11. Kabanov pointed out that the proof of the above theorem can be simplified when the following special case of his result on extreme points is used, see [18].

Theorem 7 *We use the notation of theorem 3 and we suppose that for each (t, ω) the set $C(t, \omega)$ is closed, convex and bounded by a constant independent of (t, ω). If ∂ denotes the operator that associates with a set S the set of its extreme points ∂S, we have that*

$$\partial S = \{\mathcal{E}(q \cdot M)_\infty \mid q \text{ predictable}, q \in \partial C \text{ and } \mathbb{E}_\mathbb{P} [\mathcal{E}(q \cdot M)_\infty] = 1\}.$$

Remark 12. In case the set S is weakly compact, the m–stability implies extra properties.

Lemma 14. *If Z is a continuous, nonnegative martingale with $Z_0 = 1$, if $\mathbb{P}[Z_\infty = 0] > 0$, if τ_ϵ denotes the stopping time $\tau_\epsilon = \inf\{t \mid Z_t \leqslant \epsilon\}$, then the family $\{Z_\infty/Z_{\tau_\epsilon} \mid 1 > \epsilon > 0\}$ is not relatively weakly compact.*

Proof Let $A_\epsilon = \{\tau_\epsilon < \infty\}$. Clearly $\mathbb{P}[A_\epsilon] \geqslant \mathbb{P}[Z_\infty = 0]$ and for $\epsilon \to 0$ we get $\mathbb{P}[A_\epsilon] \to \mathbb{P}[Z_\infty = 0]$. By the stopping time theorem we get $\int_{A_\epsilon} Z_\infty = \epsilon \mathbb{P}[A_\epsilon]$. Hence we get

$$\int_{A_\epsilon} Z_\infty/Z_{\tau_\epsilon} = \mathbb{P}[A_\epsilon].$$

Since $\mathbf{1}_{A_\epsilon} Z_\infty/Z_{\tau_\epsilon} \to 0$ when $\epsilon \to 0$, we get a contradiction with uniform integrability.   □

Theorem 8 *If S is m–stable and weakly compact, then all continuous elements in S are strictly positive (i.e. satisfy $Z_\infty > 0$ a.s.).*

Proof This immediately follows from the previous lemma and from the fact that the elements $Z_\infty/Z_{\tau_\epsilon} \in S$ by m–stability.   □

4 The m–stable hull and the relation with some risk measures

In this section we will investigate if the classical examples of risk measures ([10]) come from m–stable sets. We start with the obvious

Lemma 15. *If S is a set of probability measures $S \subset \mathbf{L}^1$, then the intersection of all m–stable, convex, closed sets containing S is still an m–stable, closed, convex set. It is the smallest closed, convex, m–stable set containing S and it is called the m–stable hull of S.*

Proof This follows immediately from the definition of m–stable sets.   □

The following theorem deals with the case of Tailvar or CV@R. For the definition see [10]. Just for the information of the reader, let us recall that the σ−algebra $\mathcal{F}_0$ is trivial and therefore every element $Z \in S$ satisfies $Z_0 = 1$.

Theorem 9 *Suppose that all martingales for the filtration $\mathcal{F}$ are continuous (as well known, this is equivalent to the property that all stopping times are predictable). Suppose that $K > 1$ and let*

$$\mathcal{S}_K = \left\{ \mathbb{Q} \mid \frac{d\mathbb{Q}}{d\mathbb{P}} \leqslant K \right\}.$$

Then the m–stable hull of $\mathcal{S}_K$ is the set of all probability measures absolutely continuous with respect to $\mathbb{P}$.

Proof We will show that every probability measure $\mathbb{Q}$ can be approximated by probability measures that are concatenations of elements of $\mathcal{S}_K$. Since the measures $\mathbb{Q}$ with densities that are bounded away from zero form a dense set, it does not harm to suppose that $\mathbb{Q}$ has a density process

$$Z_t = \mathbb{E}\left[\frac{d\mathbb{Q}}{d\mathbb{P}} \mid \mathcal{F}_t \right]$$

satisfying $0 < \epsilon \leqslant Z$. By hypothesis the martingale Z is continuous. Let us now define a sequence of stopping times starting with $T_0 = 0$ and inductively defined as

$$T_k = \inf\left\{ u \mid u > T_{k-1} \text{ and } \frac{Z_u}{Z_{T_{k-1}}} \geqslant K \right\}.$$

Because of the continuity we have that the random variables

$$f_k = \frac{Z_{T_k}}{Z_{T_{k-1}}}$$

are bounded by K and are therefore densities of elements in $\mathcal{S}_K$. Furthermore, because the intervals $]\!]T_{k-1}, T_k]\!]$ are disjoint, the products $Z_{T_N} = \prod_{k=1}^{N} f_k$ are concatenations of elements of $\mathcal{S}_K$ and therefore they are densities of probability measures which, by Proposition 1, are necessarily in the m–stable hull of $\mathcal{S}_K$. Because the martingale Z is continuous and converges at $t = \infty$ (at this point we only need that it has left limits for each $t \leqslant \infty$), we must have that $\mathbb{P}[T_N = \infty]$ tends to 1. Since Z is a uniformly integrable martingale, the convergence of Z_{T_N} to Z_∞ is both a.s. and in $\mathbf{L}^1$. This shows that $\mathbb{Q}$ is in the m–stable hull of $\mathcal{S}_K$.   □

The case of law invariant risk measures deserves special attention. We refer to Kusuoka's paper, [20] for details and for a nice characterisation of law invariant risk measures. For our purposes it suffices to recall that the set $\mathcal{S}$ is called *law invariant* if $Z_\infty \in \mathcal{S}$ and if the distribution of Z'_∞ and Z_∞ are the same, then also $Z'_\infty \in \mathcal{S}$. For law-invariant risk measures we can prove the following

Theorem 10 *Suppose that the filtration $\mathcal{F}$ is generated by a d–dimensional Brownian Motion W. If $\mathcal{S}$ is an m–stable convex closed set, if $\mathcal{S}$ is law-invariant, then either $\mathcal{S} = \{\mathbb{P}\}$ or $\mathcal{S}$ equals the set of all probability measures that are absolutely continuous with respect to $\mathbb{P}$.*

Proof The proof is based on the following

Lemma 16. *(Skorohod embedding theorem) If B is a Brownian Motion (with respect to a filtration $\mathcal{G}$). If ν is a probability on $\mathbb{R}_+$ such that $\int_{\mathbb{R}_+} x\, \nu(dx) = 1$, then there is a $(\mathcal{G})$-stopping time τ such that $\mathcal{E}(B)^\tau$ is uniformly integrable and $\mathcal{E}(B)_\tau$ has the probability ν as its law.*

Proof of the lemma The proof is almost identical to the proof of the usual stopping time problem, see [27] and the references given there. For completeness we present an easy proof. Let $\mathcal{R}^n$ be an increasing sequence of finite σ−algebras on $\mathbb{R}_+$, chosen such that they generate the Borel σ−algebra and such that each atom of $\mathcal{R}^n$ is split into exactly two atoms of $\mathcal{R}^{n+1}$. For convenience we take $\mathcal{R}^0 = \{\varnothing, \mathbb{R}_+\}$. Define the $\mathcal{R}^n$ measurable, conditional expectation $y_n = \mathbb{E}_\nu\left[id_{\mathbb{R}_+} \mid \mathcal{R}^n\right]$. Inductively we define an increasing sequence of stopping times σ_n so that $\mathcal{E}(B)_{\sigma_n}$ has the same law as y_n. For $n = 0$ we take $\sigma_0 = 0$ and $y_0 = 1$. Then y_1 takes two values (at most) and we take σ_1 as the first time that $\mathcal{E}(B)_t$ takes one of these values. Clearly $\mathcal{E}(B)_{\sigma_1}$ takes the same two values (at most). On each of the atoms, generated by the random variable $\mathcal{E}(B)_{\sigma_1}$, we define σ_2 as the first time after σ_1, where $\mathcal{E}(B)$ takes one of the (at most two) corresponding values of y_2. We continue this procedure and we obtain a martingale (not only a local martingale) $(\mathcal{E}(B)_{\sigma_n})_n$. The sequence σ_n can clearly be defined since the only difficulty is when $\sigma_n = \infty$ with some probability, in which case we have $\mathcal{E}(B)_{\sigma_n}$ takes the value zero. When this is the case, the martingale $y_n = 0$ on the corresponding set and consequently the further values of y_k are all zero as well. From the construction it follows that for each n, $\mathcal{E}(B)_{\sigma_n}$ has the same law as y_n. Indeed since there are at most two values, the probabilities of these values are determined by the fact that their average is the preceding value of the martingale. These are the same for the sequence $(y_n)_n$ as for the sequence $(\mathcal{E}(B)_{\sigma_n})_n$. Let us now define $\tau = \lim \sigma_n$. Since the law of $\mathcal{E}(B)_\tau$ is the limit of the laws of $\mathcal{E}(B)_{\sigma_n}$, this law is precisely the limit law of y_n, hence equal to ν. This implies that the process $\mathcal{E}(B)^\tau$ is a uniformly integrable martingale since obviously it is a nonnegative local martingale, starting at 0 and $\mathbb{E}[\mathcal{E}(B)_\tau] = \int_{\mathbb{R}_+} x\, \nu(dx) = 1$. □

We can now continue the proof of the theorem. We suppose that $\mathcal{S}$ contains an element $f > 0$, a.s. , that is different from 1. The law of f will be denoted by ν. We have to show that $\mathcal{S}$ equals the set of all probability measures that are absolutely continuous with respect to $\mathbb{P}$. What we will show is that the set C, from the representation theorem 4, is equal to $\mathbb{R}^d$. Take a vector $x \in \mathbb{R}^d$ of unit length. Let us consider the process $B = x.W$. This process is clearly a 1−dimensional Brownian Motion for the filtration $\mathcal{F}$. Let τ_1 be a stopping time such that $\mathcal{E}(B)_{\tau_1}$ has the law ν. By Skorohod's theorem (see the lemma above), this is possible. Since $f > 0$ a.s. , we must have that $\tau_1 < +\infty$. The stopping time τ_1 can be taken to be a stopping time with respect to the filtration generated by B. Because of Blumenthal's zero-one law (see [27]) we must have that $\tau_1 > 0$ a.s. . The process $q = x\mathbf{1}_{[\![0,\tau_1]\!]}$ is therefore a selector of the set C. If we restart the Brownian Motion B at time τ_1, i.e. if we look at the process $B'_s = 0$

for $s \leqslant \tau_1$ and $B'_s = B_s - B_{\tau_1}$ we can again apply Skorohod's theorem and we get a second stopping time $\tau_2 > \tau_1$ such that $\mathcal{E}(B')_{\tau_2}$ has the same law ν. Moreover the random variable $\mathcal{E}(B')_{\tau_2}$ is independent of $\mathcal{E}(B)_{\tau_1}$. This means that also the process $q = x\mathbf{1}_{]\!]\tau_1,\tau_2]\!]}$ is a selector of C. If we continue in the same way, we get a strictly increasing sequence of stopping times τ_n such that for each k we have that $x\mathbf{1}_{]\!]\tau_{k-1},\tau_k]\!]}$ is a selector of C and such that the random variables $f_k = \mathcal{E}(\mathbf{1}_{]\!]\tau_{k-1},\tau_k]\!]} \cdot B)_{\tau_k}$ are iid with law ν. Since as is easily seen, the product $\Pi_0^\infty f_k$ diverges to zero a.s. , we must have that the sequence τ_k tends to ∞. It follows that $x \in C$ on $\mathbb{R}_+ \times \Omega$. We now apply the same reasoning to the process $B" = nB = nx.W$. Since $B"$ is, up to scaling by $\sqrt{n}$, a Brownian Motion we can find a stopping time $\tau"$ such that $\mathcal{E}(B")_{\tau"}$ has the law ν and $0 < \tau" < +\infty$. This means that $q = nx\mathbf{1}_{]\!]0,\tau"]\!]}$ is a selector of C. But the same reasoning as above, meaning that we restart at time $\tau"$, then gives that the constant process nx is a selector of C. Since C is convex and since for each vector $z \in \mathbb{R}^d$ we now have that $z \in C$ a.s. on $\mathbb{R}_+ \times \Omega$, we must have that $C = \mathbb{R}^d$, a.s. □

Remark 13. The proof can easily be adapted to the case where M is a $d-$dimensional continuous local martingale with the predictable representation property and with the condition that for each coordinate k we have that $\langle M^k, M^k \rangle_\infty = +\infty$. Although we have some generalisations of this situation, we do not know whether the theorem holds in the general case of a filtration where all martingales are continuous.

Remark 14. The previous theorem is not the end of law invariant dynamic coherent risk measures. There is a way out as shown in [21].

5 The construction of the Snell envelope

In this section we will assume without further notice that the set $\mathcal{S}$ is m–stable. The properties of m–stable sets allow us to define a risk adjusted value as a process. The construction goes as follows. For an m–stable set of probability measures $\mathcal{S}$ and for a bounded random variable $f \in \mathbf{L}^\infty$ and under the assumption that $\mathcal{F}_0$ is degenerate, we define the risk adjusted value at time zero, $\Phi_0(X)$ as

$$\Phi_0(f) = \inf \{\mathbb{E}_\mathbb{Q}[f] \mid \mathbb{Q} \in \mathcal{S}\}.$$

At intermediate times $0 \leqslant t \leqslant \infty$, we could try to define the random variable

$$\Phi_t(f) = \text{ess.inf} \{\mathbb{E}_\mathbb{Q}[f \mid \mathcal{F}_t] \mid \mathbb{Q} \in \mathcal{S}\}.$$

For $t = 0$ there is no ambiguity in the definitions since the $\sigma-$algebra $\mathcal{F}_0$ is trivial. The infimum is an infimum of random variables and therefore it has to be seen as an essential infimum. But the measures $\mathbb{Q} \in \mathcal{S}$ are not all equivalent to $\mathbb{P}$ and hence the conditional expectations are not defined $\mathbb{P}$ a.s. . However the random variable $\Phi_t(f)$ can also be defined in another way, thereby avoiding this difficulty. One way is to observe that the set $\mathcal{S}^e$ of the measures in $\mathcal{S}$, equivalent to $\mathbb{P}$, are dense in $\mathcal{S}$. Therefore we can define

$$\Phi_t(f) = \text{ess.inf} \{\mathbb{E}_\mathbb{Q}[f \mid \mathcal{F}_t] \mid \mathbb{Q} \in \mathcal{S}^e\}.$$

By the density argument we have that for *every* $\mathbb{Q}_0 \in \mathcal{S}$:

$$\text{ess.inf} \left\{ \mathbb{E}_{\mathbb{Q}}[f \mid \mathcal{F}_t] \mid \mathbb{Q} \in \mathcal{S}^e \right\} \leqslant \mathbb{E}_{\mathbb{Q}_0}[f \mid \mathcal{F}_t] \qquad \mathbb{Q}_0 \text{ a.s. .}$$

Another solution is to take

$$\Phi_t(f) = \text{ess.sup} \left\{ g \mid \forall \mathbb{Q} \in \mathcal{S} : g \leqslant \mathbb{E}_{\mathbb{Q}}[f \mid \mathcal{F}_t], \mathbb{Q} \text{ a.s. } \right\}.$$

The following theorem deals with risk adjusted values of a stochastic process. We will use a different but similar notation as the one introduced in the beginning of this section.

Theorem 11 *If $\mathcal{S}$ is an m–stable set and if X is a bounded càdlàg adapted stochastic process, then there is a càdlàg process, denoted by $\Psi_t(X)$, so that for every stopping time $0 \leqslant T < \infty$ we have*

$$\Psi_T(X) = \text{ess.inf} \left\{ \mathbb{E}_{\mathbb{Q}}[X_\tau \mid \mathcal{F}_T] \mid \tau \geqslant T \text{ is a stopping time and } \mathbb{Q} \in \mathcal{S}^e \right\}.$$

We call the process $\Psi_t(X)$, the risk adjusted process corresponding to the process X. The process $\Psi(X)$ is a $\mathbb{Q}$–submartingale for every $\mathbb{Q} \in \mathcal{S}$.

Remark 15. The proof follows the same lines as the proof of the existence of the Snell envelope, see [12] pages 431 up to 436 or [13] pages 110–113. Since we have the "extra" difficulty that we have to deal with all the measures in $\mathcal{S}$, we prefer to give the details. The reader can skip the proof.

Remark 16. That the process $\Psi(X)$ is a submartingale for every "test-probability" has a direct interpretation. It shows that as time evolves, the uncertainty on the remaining part decreases. The risk adjusted value therefore increases in expected value.

Proof We start with the definition of a family of random variables, indexed by the set of all stopping times $0 \leqslant T \leqslant \infty$:

$$Y_T = \text{ess.inf} \left\{ \mathbb{E}_{\mathbb{Q}}[X_\tau \mid \mathcal{F}_T] \mid \tau \geqslant T \text{ is a stopping time and } \mathbb{Q} \in \mathcal{S}^e \right\}.$$

We emphasize that this is only a family of random variables and that for the moment there is no process Y that gives the values Y_T at times T. The construction of such a process involves the selection of representatives of the a.s. equivalence classes Y_T.

Lemma 17. *For a fixed stopping time T, the set*

$$\left\{ \mathbb{E}_{\mathbb{Q}}[X_\tau \mid \mathcal{F}_T] \mid \tau \geqslant T \text{ is a stopping time and } \mathbb{Q} \in \mathcal{S} \right\}$$

is a lattice.

Proof Indeed for $\tau_1, \tau_2 \geqslant T$ and $\mathbb{Q}^1, \mathbb{Q}^2 \in \mathcal{S}^e$, (with density processes Z^1, Z^2 resp), we have that

$$\min \left(\mathbb{E}_{\mathbb{Q}^1}[X_{\tau_1} \mid \mathcal{F}_T], \mathbb{E}_{\mathbb{Q}^2}[X_{\tau_2} \mid \mathcal{F}_T] \right) = \mathbb{E}_{\mathbb{Q}}[X_\tau \mid \mathcal{F}_T],$$

where the measure $\mathbb{Q}$ is defined as

$$\frac{d\mathbb{Q}}{d\mathbb{P}} = \frac{Z^1_\infty}{Z^1_T} \quad \text{on the set } \{\mathbb{E}_{\mathbb{Q}^1}[X_{\tau_1} \mid \mathcal{F}_T] < \mathbb{E}_{\mathbb{Q}^2}[X_{\tau_2} \mid \mathcal{F}_T]\}$$

$$= \frac{Z^2_\infty}{Z^2_T} \quad \text{on the set } \{\mathbb{E}_{\mathbb{Q}^1}[X_{\tau_1} \mid \mathcal{F}_T] \geqslant \mathbb{E}_{\mathbb{Q}^2}[X_{\tau_2} \mid \mathcal{F}_T]\}.$$

That the measure $\mathbb{Q}$ is still in $\mathcal{S}$ follows from the m-stability of the set $\mathcal{S}$. In the same way we define

$$\tau = \tau_1 \quad \text{on the set } \{\mathbb{E}_{\mathbb{Q}^1}[X_{\tau_1} \mid \mathcal{F}_T] < \mathbb{E}_{\mathbb{Q}^2}[X_{\tau_2} \mid \mathcal{F}_T]\}$$

$$= \tau_2 \quad \text{on the set } \{\mathbb{E}_{\mathbb{Q}^1}[X_{\tau_1} \mid \mathcal{F}_T] \geqslant \mathbb{E}_{\mathbb{Q}^2}[X_{\tau_2} \mid \mathcal{F}_T]\}.$$

□

Corollary 2 *Because of this lattice property we also have that for every stopping time $0 \leqslant T < \infty$ and for every probability measure $\mu \ll \mathbb{P}$ (not necessarily in $\mathcal{S}$):*

$$\mathbb{E}_\mu[Y_T] = \inf \{\mathbb{E}_\mu [\mathbb{E}_{\mathbb{Q}}[X_\tau \mid \mathcal{F}_T]] \mid \mathbb{Q} \in \mathcal{S}^e; T \leqslant \tau \text{ stopping time}\}.$$

Proof Obvious. □

Lemma 18. *The m-stability of the set $\mathcal{S}$ implies the equality of the following two sets: (here $\infty > \nu \geqslant \sigma$ are two stopping times)*

$$\left\{ \left(\frac{Z_\nu}{Z_\sigma}, \frac{Z_\sigma}{Z_\tau}\right) \mid Z \in \mathcal{S}^e \right\} = \left\{ \left(\frac{Z'_\nu}{Z'_\sigma}, \frac{Z_\sigma}{Z_\tau}\right) \mid Z \in \mathcal{S}^e, Z' \in \mathcal{S}^e \right\}.$$

Proof Obvious since it follows from the definition of m–stability. □

Lemma 19. *For every pair of stopping times $0 \leqslant \tau \leqslant \sigma < \infty$, we have*

$$Y_\tau \leqslant \text{ess.inf} \{\mathbb{E}_{\mathbb{Q}}[Y_\sigma \mid \mathcal{F}_\tau] \mid \mathbb{Q} \in \mathcal{S}^e\}.$$

Proof From the lattice property and the previous lemma, we can easily justify the following calculations

$$Y_\tau \leqslant \mathbb{E}\left[X_\nu \frac{Z_\nu}{Z_\tau} \mid \mathcal{F}_\tau \right] \text{ all } Z \in \mathcal{S}^e, \text{ all } \nu \geqslant \tau$$

$$\leqslant \mathbb{E}\left[\mathbb{E}\left[X_\nu \frac{Z_\nu}{Z_\sigma} \mid \mathcal{F}_\sigma \right] \frac{Z_\sigma}{Z_\tau} \mid \mathcal{F}_\tau \right] \text{ all } Z \in \mathcal{S}^e, \text{ all } \nu \geqslant \sigma$$

$$\leqslant \mathbb{E}\left[\mathbb{E}\left[X_\nu \frac{Z'_\nu}{Z'_\sigma} \mid \mathcal{F}_\sigma \right] \frac{Z_\sigma}{Z_\tau} \mid \mathcal{F}_\tau \right] \text{ all } Z \in \mathcal{S}^e, \text{ all } Z' \in \mathcal{S}^e, \text{ all } \nu \geqslant \sigma$$

$$\leqslant \mathbb{E}\left[Y_\sigma \frac{Z_\sigma}{Z_\tau} \mid \mathcal{F}_\tau \right] \text{ all } Z \in \mathcal{S}^e, \text{ all } \nu \geqslant \sigma.$$

Taking the ess.inf over all $Z \in \mathcal{S}^e$ implies the desired inequality. □

Lemma 20. *The family Y_T constructed above satisfies the following sub-martingale property: for every pair of stopping times $0 \leqslant \tau \leqslant \sigma < \infty$ and every $\mathbb{Q} \in \mathcal{S}$ we have*

$$Y_\tau \leqslant \mathbb{E}_\mathbb{Q}[Y_\sigma \mid \mathcal{F}_\tau].$$

In particular it satisfies this property for $\mathbb{P}$.

Proof Follows directly from the previous lemma. □

Lemma 21. *The family Y_T constructed above satisfies the following right continuity property: if $\infty > T_n$ is a nonincreasing sequence of stopping times converging to T, then $\lim_n \mathbb{E}[Y_{T_n}] = \mathbb{E}[Y_T]$; consequently Y_{T_n} tends to Y_T in $\mathbf{L}^1$.*

Proof Because of the corollary 2, we have for each $\epsilon > 0$ the existence of a stopping time $\sigma \geqslant T$ and $Z \in \mathcal{S}^e$, so that $\mathbb{E}\left[X_\sigma \frac{Z_\sigma}{Z_T}\right] \leqslant \mathbb{E}[Y_T] + \epsilon$. Because of the right continuity of the process X we can also suppose that $\sigma > T$. Indeed on the set $\{\sigma = T\}$ we can replace σ by $\sigma + \delta$ where δ is small enough. This not only uses the right continuity of the process X, it also uses that for $\delta \to 0$, we have that $\frac{Z_{T+\delta}}{Z_T} \to 1$ in $\mathbf{L}^1$. So we may and do suppose that $\sigma > T$. The following sequence of inequalities is now clear

$$\epsilon + \mathbb{E}[Y_T] \geqslant \mathbb{E}\left[X_\sigma \frac{Z_\sigma}{Z_T}\right]$$

$$= \mathbb{E}\left[X_\sigma \frac{Z_\sigma}{Z_T} 1_{\sigma \geqslant T_n}\right] + \mathbb{E}\left[X_\sigma \frac{Z_\sigma}{Z_T} 1_{\sigma < T_n}\right]$$

$$= \mathbb{E}\left[X_\sigma \frac{Z_\sigma}{Z_{T_n}} \frac{Z_{T_n}}{Z_T} 1_{\sigma \geqslant T_n}\right] + \mathbb{E}\left[X_\sigma \frac{Z_\sigma}{Z_T} 1_{\sigma < T_n}\right]$$

$$= \mathbb{E}\left[\mathbb{E}\left[X_\sigma \frac{Z_\sigma}{Z_{T_n}} 1_{\sigma \geqslant T_n} \mid \mathcal{F}_{T_n}\right] \frac{Z_{T_n}}{Z_T}\right] + \mathbb{E}\left[X_\sigma \frac{Z_\sigma}{Z_T} 1_{\sigma < T_n}\right]$$

$$\geqslant \mathbb{E}\left[Y_{T_n} 1_{\sigma \geqslant T_n} \frac{Z_{T_n}}{Z_T}\right] + \mathbb{E}\left[X_\sigma \frac{Z_\sigma}{Z_T} 1_{\sigma < T_n}\right]$$

$$\geqslant \mathbb{E}[Y_{T_n}] + \mathbb{E}\left[Y_{T_n}\left(1_{\sigma \geqslant T_n} \frac{Z_{T_n}}{Z_T} - 1\right)\right] + \mathbb{E}\left[X_\sigma \frac{Z_\sigma}{Z_T} 1_{\sigma < T_n}\right].$$

The third term tends to zero since the set $\{\sigma < T_n\}$ decreases to the empty set. Now the second term tends to zero as $n \to \infty$. Indeed $1_{\sigma \geqslant T_n} \frac{Z_{T_n}}{Z_T} \to 1$ in $\mathbf{L}^1$ and the variables $(Y_{T_n})_n$ are uniformly bounded by the bound on X. As a result we get that for all $\epsilon > 0$:

$$\epsilon + \mathbb{E}[Y_T] \geqslant \lim_n \mathbb{E}[Y_{T_n}].$$

The last statement is easy since the sequence $Y_T, (Y_{T_n})_n$ forms a submartingale that has a right continuous modification (see [12]). This completes the proof of the lemma. □

Lemma 22. *There is a càdlàg process V so that for all stopping times $T < \infty$ we have that $V_T = Y_T$ a.s. .*

Proof This follows from the modification theorem for submartingales. For the appropriate version (stated for supermartingales) see [12] page 73, Théorème 1.   □

    The proof of the theorem is now complete. It is sufficient to take the process V constructed above as a version for the "process" $\Psi(X)$ and to apply the lemma 20.   □

Remark 17. The notation Ψ is reserved for processes, whereas the notation Φ was reserved for random variables. The construction is also different in the sense that in the construction of Ψ, we use an infimum over stopping times as well as an infimum over all elements of $\mathcal{S}$. If with a random variable $f \in \mathbf{L}^\infty$, we associate the càdlàg process $X_t = \mathbb{E}_\mathbb{P}[f \mid \mathcal{F}_t]$, we could associate with any random variable a process $\Psi_t(f)$ (defined as $\Psi(f) = \Psi(X)$). In the next section, we will give conditions under which, for random variables, both families $\Phi_T(f)$ and $\Psi_T(X)$ are the same. For the moment let us show the following property

Lemma 23. *Let $\mathcal{S}$ be m–stable. With the notation introduced above, i.e. $X_t = \mathbb{E}_\mathbb{P}[f \mid \mathcal{F}_t]$, we have for every bounded random variable f*

$$\Psi_t(X) = \Phi_t(f).$$

Proof We obviously have

$$\Psi_\sigma(X) = \text{ess.inf}_{\sigma \leqslant \tau, \mathbb{Q} \in \mathcal{S}^e} \, \mathbb{E}_\mathbb{Q} \left[\mathbb{E}_\mathbb{P}[f \mid \mathcal{F}_\tau] \mid \mathcal{F}_\sigma \right] \leqslant \text{ess.inf}_{\mathbb{Q} \in \mathcal{S}^e} \, \mathbb{E}_\mathbb{Q}[f \mid \mathcal{F}_\sigma]$$
$$= \Phi_\sigma(f).$$

Indeed take $\tau = \infty$ as a special case in the left hand side.

    Conversely, we have that

$$\Psi_\sigma(X) = \text{ess.inf}_{\sigma \leqslant \tau, \mathbb{Q} \in \mathcal{S}^e} \, \mathbb{E}_\mathbb{Q} \left[\mathbb{E}_\mathbb{P}[f \mid \mathcal{F}_\tau] \mid \mathcal{F}_\sigma \right]$$
$$\geqslant \text{ess.inf}_{\tau \geqslant \sigma} \, \text{ess.inf}_{\mathbb{Q} \in \mathcal{S}^e} \, \mathbb{E}_\mathbb{Q} \left[\text{ess.inf}_{\mathbb{Q}' \in \mathcal{S}} \, \mathbb{E}_{\mathbb{Q}'}[f \mid \mathcal{F}_\tau] \mid \mathcal{F}_\sigma \right]$$
$$\geqslant \text{ess.inf}_{\tau \geqslant \sigma} \, \text{ess.inf}_{\mathbb{Q} \in \mathcal{S}^e} \, \mathbb{E}_\mathbb{Q} \left[\mathbb{E}_\mathbb{Q}[f \mid \mathcal{F}_\tau] \mid \mathcal{F}_\sigma \right]$$
$$\geqslant \text{ess.inf}_{\tau \geqslant \sigma} \, \text{ess.inf}_{\mathbb{Q} \in \mathcal{S}^e} \, \mathbb{E}_\mathbb{Q} \left[f \mid \mathcal{F}_\sigma \right]$$
$$\geqslant \text{ess.inf}_{\mathbb{Q} \in \mathcal{S}^e} \, \mathbb{E}_\mathbb{Q} \left[f \mid \mathcal{F}_\sigma \right] = \Phi_\sigma(\Phi_\tau(f)).$$

The third line follows from the m–stability of the set $\mathcal{S}$.   □

Example 1 If $\mathcal{S} = \{\mathbb{P}\}$ the process $\Psi(X)$ coincides with the Snell envelope (up to sign changes, since we take a lower envelope and not an upper envelope). As well known this "upside down" Snell envelope can be calculated as $\Psi_0(X) = \text{ess.inf}_{\sigma < \infty} \, \mathbb{E}_\mathbb{P}[X_\sigma]$. The other extreme example is when $\mathcal{S}$ is the set of all probability measures, absolutely continuous with respect to $\mathbb{P}$. In this case we have

$$\Psi_0(X) = \text{ess.inf} \left(\inf_{t \geqslant 0} X_t \right),$$

the worst possible loss over time and over all "states of nature". This requires some extra proof, since the time where a process attains its minimum (if the minimum is attained) is not a stopping time. (In finance this would have spectacular consequences as we would be able to buy at the lowest price and sell at the highest price, a strategy opening a lot of perspectives). The proof goes as follows. We suppose that the càdlàg process X is bounded by 1. Let $a = \text{ess.inf} \, (\inf_{t \geqslant 0} X_t)$. The set, defined for $\varepsilon > 0$:

$$\pi(A) = \{\omega \mid \text{ there is } t \text{ such that } X_t(\omega) < a + \varepsilon\}$$

is the projection of the optional set

$$A = \{(t, \omega) \mid X_t(\omega) < a + \varepsilon\}.$$

It is therefore measurable, i.e in $\mathcal{F}_\infty$. Moreover the stopping time

$$T(\omega) = \inf \{t \mid (t, \omega) \in A\},$$

is well defined and satisfies $\{T < \infty\} = \pi(A)$. Moreover on $\{T < \infty\}$ we have that $X_T \leqslant a + \varepsilon$, since the process X is càdlàg. We now take n so that $\mathbb{P}[T \leqslant n] \geqslant \mathbb{P}[\pi(A)](1 - \varepsilon) > 0$. The probability measure $\mathbb{Q}$ is defined as the conditional probability measure with respect to $\{T \leqslant n\}$, i.e. it has density $Z_\infty = \mathbf{1}_{\{T \leqslant n\}}/\mathbb{P}[T \leqslant n]$. As easily seen $\mathbb{E}_\mathbb{Q}[X_{T \wedge n}] \leqslant a + \varepsilon$. Since ε was arbitrary we have $\Psi_0(X) = a$. The identification of $\Psi_\tau(X)$ is more or less the same. Roughly speaking we can say that $\Psi_\tau(X)$ is, given the information $\mathcal{F}_\tau$, the infimum of $\{X_t \mid \tau \leqslant t < \infty\}$. This requires the use of conditional distributions but it can also be defined as

$$\text{ess.inf} \, (\inf_{\tau \leqslant t < \infty} X_t) =$$

$$\text{ess.sup} \, \{h \mid h \text{ is } \mathcal{F}_\tau \text{ measurable and } h \leqslant X_t \text{ for all } t \geqslant \tau\}$$

As we will show, this random variable is equal to the random variable $\Psi_\tau(X)$. Since obviously $\text{ess.inf} \, (\inf_{\tau \leqslant t < \infty} X_t) \leqslant \Psi_\tau(X)$, it is sufficient to prove that for every stopping time $\infty > \sigma \geqslant \tau$ we have that $\Psi_\tau(X) \leqslant X_\sigma$ a.s. If this would not be true then the set $C = \{\Psi_\tau(X) > X_\sigma\}$ would not be negligible and we could then take the probability measure $\mathbb{Q}$ with density $\mathbf{1}_C/\mathbb{P}[C]$. For this probability we would get

$$\mathbb{E}_\mathbb{Q}[X_\sigma \mid \mathcal{F}_\tau] < \mathbb{E}_\mathbb{Q}[\Psi_\tau(X) \mid \mathcal{F}_\tau] = \Psi_\tau(X),$$

a contradiction to the definition of $\Psi_\tau(X)$.

Remark 18. (on the motivation) The reader could ask why we took the infimum over all probabilities in $\mathcal{S}$ and over all stopping times. It can be argued that if the economic agent can choose the stopping time, e.g. to stop a project, it would be more realistic to take the supremum over all stopping times. This gives a mathematical problem that is related to a maximin/minimax strategy. The mathematics involved are more complicated as the outcome is not the result of a concave problem but rather of a concave-convex optimisation problem. This approach clearly makes sense if the economic agent can choose the stopping time. If however the economic agent cannot choose the stopping time or if the stopping time is selected by the "enemy", then the infimum makes more sense. In case the economic agent can choose the stopping time, it becomes interesting to have a look at the following generalisation. For convenience and to facilitate the mathematics, let us suppose that we are working on a finite horizon interval $[0, T]$. The stopping time $\tau \leqslant T$ can be identified with the nondecreasing process $A_t = \mathbf{1}_{\tau \leqslant t}$ and has the property that $A_T = 1$. The convex closed envelope of these processes brings us to the set of all càdlàg, adapted, nondecreasing processes such that $A_T = 1$. The value A_t could describe the amount of the process that is already stopped — or closed down — at time t. If the set $\mathcal{S}$ is weakly compact in $\mathbf{L}^1$, we can apply the minimax theorem and we get that

$$\inf_{Q \in \mathcal{S}} \sup_A \mathbb{E}_Q \left[\int_0^T X_u \, dA_u \right] = \sup_A \inf_{Q \in \mathcal{S}} \mathbb{E}_Q \left[\int_0^T X_u \, dA_u \right].$$

We do not develop this theory any further as it does not fit in the approach we present here.

6 The Equivalence between the m-stability property, recursivity and time-consistency

The setup of this section is a little bit more general as we will deal with random variables instead of dealing with stochastic processes. Let us first recall some notations. Let $\mathcal{S}$ be a closed convex set of probability measures $\mathcal{S} \subset \mathbf{L}^1(\Omega, \mathcal{F}, \mathbb{P})$. No stability on $\mathcal{S}$ is assumed. If f is a bounded random variable then for each stopping time $T \leqslant \infty$ we denote by $\Phi_T(f)$ the random variable

$$\Phi_T(f) = \mathrm{ess.inf}_{Q \in \mathcal{S}} \mathbb{E}_Q \left[f \mid \mathcal{F}_T \right].$$

The nonlinear functional Φ_T satisfies the following coherence properties. We omit the straightforward proofs.

1. if $f \geqslant g$ are bounded random variables then $\Phi_T(f) \geqslant \Phi_T(g)$
2. for $\lambda \geqslant 0$ and $f \in \mathbf{L}^\infty$, we have that $\Phi_T(\lambda f) = \lambda \Phi_T(f)$, (the same holds for λ a nonnegative bounded $\mathcal{F}_T$ measurable random variable)
3. for f, g bounded random variables we have $\Phi_T(f + g) \geqslant \Phi_T(f) + \Phi_T(g)$
4. if g is a bounded $\mathcal{F}_T$ measurable random variable, then for any bounded random variable f we have $\Phi_T(f + g) = \Phi_T(f) + g$
5. if f_n is a sequence of random variables $1 \geqslant f_n \geqslant -1$ then $\Phi_T(\limsup_n f_n) \geqslant \limsup_n \Phi_T(f_n)$.

Definition 2 *The set $\mathcal{S}$ is called time consistent if the following holds. For any pair of stopping times $\sigma \leqslant \tau$ and any pair of random variables $f, g \in \mathbf{L}^{\infty}$, we have that $\Phi_{\tau}(f) \leqslant \Phi_{\tau}(g)$ implies that $\Phi_{\sigma}(f) \leqslant \Phi_{\sigma}(g)$.*

The following theorem characterises the convex closed sets $\mathcal{S}$ that are also m–stable. This theorem is related to decision theory with multipriors, where the m-stability is referred to as rectangularity. However the technicalities are different from the ones addressed here. See [16], [28]. The first paper uses as basic concepts, the so called lotteries on a set of actions. Some of the problems here can be translated into the language of [16]. How the paper [16] fits in our approach is not yet clear and subject of ongoing research. The paper [28] is difficult to read since it does not use the structures introduced in general stochastic process theory. Both cited papers only deal with discrete time or/and discrete event spaces.

Theorem 12 *The following are equivalent*

1. *The set $\mathcal{S}$ is m–stable.*
2. *For every bounded random variable f, the family $\Phi_T(f)$ satisfies: for every two stopping times $\sigma \leqslant \tau$ we have $\Phi_{\sigma}(f) = \Phi_{\sigma}(\Phi_{\tau}(f))$.*
3. *For every bounded random variable f, for every stopping time σ we have $\Phi_0(f) \leqslant \Phi_0(\Phi_{\sigma}(f))$.*
4. *The set $\mathcal{S}$ is time consistent.*
5. *The family $\Phi_T(f)$ satisfies the submartingale property: for all $\mathbb{Q} \in \mathcal{S}$ and all pairs of stopping times $\sigma \leqslant \tau$ we have that $\Phi_{\sigma}(f) \leqslant \mathbb{E}_{\mathbb{Q}}[\Phi_{\tau}(f) \mid \mathcal{F}_{\sigma}]$.*

Proof First of all let us show that 1) implies 2). So let $\sigma \leqslant \tau$ be two stopping times and let f be a bounded random variable. By lemma 18 and the lattice property of lemma 17, we have that

$$
\begin{aligned}
\text{ess.inf}_{\mathbb{Q}} \, \mathbb{E}_{\mathbb{Q}}[f \mid \mathcal{F}_{\sigma}] &= \text{ess.inf}_{\mathbb{Q}} \, \mathbb{E}_{\mathbb{Q}} \left[\mathbb{E}_{\mathbb{Q}}[f \mid \mathcal{F}_{\tau}] \mid \mathcal{F}_{\sigma} \right] \\
&= \text{ess.inf}_{\mathbb{Q}} \, \text{ess.inf}_{\mathbb{Q}^1} \, \mathbb{E}_{\mathbb{Q}} \left[\mathbb{E}_{\mathbb{Q}^1}[f \mid \mathcal{F}_{\tau}] \mid \mathcal{F}_{\sigma} \right] \\
&= \text{ess.inf}_{\mathbb{Q}} \, \mathbb{E}_{\mathbb{Q}} \left[\text{ess.inf}_{\mathbb{Q}^1} \, \mathbb{E}_{\mathbb{Q}^1}[f \mid \mathcal{F}_{\tau}] \mid \mathcal{F}_{\sigma} \right] \\
&= \text{ess.inf}_{\mathbb{Q}} \, \mathbb{E}_{\mathbb{Q}} \left[\Phi_{\tau}(f) \mid \mathcal{F}_{\sigma} \right] \\
&= \Phi_{\sigma}(f).
\end{aligned}
$$

Let us now show the equivalence between 2 and 4. Suppose that for two bounded random variables f, g and two stopping times $\sigma \leqslant \tau$, we have $\Phi_{\tau}(f) \leqslant \Phi_{\tau}(g)$. Then we have, because of 4, $\Phi_{\sigma}(f) = \Phi_{\sigma}(\Phi_{\tau}(f)) \leqslant \Phi_{\sigma}(\Phi_{\tau}(g)) = \Phi_{\sigma}(f)$. Conversely to prove 4 out of 2, we take for the two random variables, the functions f and $g = \Phi_{\tau}(f)$. We have equality $\Phi_{\tau}(\Phi_{\tau}(f)) = \Phi_{\tau}(f)$ and therefore (applying item 2 twice) that $\Phi_{\sigma}(\Phi_{\tau}(f)) = \Phi_{\sigma}(f)$.

Obviously we have that 2 implies 3.

We now come to the proof that 3 implies 1, this is the most serious part. So let us suppose that Z^1 and Z^2 are two elements in $\mathcal{S}^e$ – coming from the measures $\mathbb{Q}^1, \mathbb{Q}^2$ – and let σ be a stopping time. Also suppose that the element

$Z_\sigma^1 \frac{Z_\infty^2}{Z_\sigma^2}$ is not in the closed convex set $\mathcal{S}$. By the Hahn–Banach theorem, there is a random variable $f \in \mathbf{L}^\infty$, so that

$$\mathbb{E}_\mathbb{P}\left[Z_\sigma^1 \frac{Z_\infty^2}{Z_\sigma^2} f \right] < \inf_{\mathbb{Q} \in \mathcal{S}} \mathbb{E}_\mathbb{Q}[f].$$

We can write the left hand side as $\mathbb{E}_{\mathbb{Q}^1}[\mathbb{E}_{\mathbb{Q}^2}[f \mid \mathcal{F}_\sigma]]$. This is clearly at least equal to $\mathbb{E}_{\mathbb{Q}^1}[\Phi_\sigma(f)]$, a quantity at least equal to $\Phi_0[\Phi_\sigma(f)]$, hence by property 3, at least equal to $\Phi_0(f)$. This is a contradiction since the right hand side is precisely $\Phi_0(f)$.

We still have to show the equivalence with property 5. That 1 implies 5 follows from Theorem 11 (see lemma 20) of the previous section and lemma 23. Suppose now that 5 holds. We have to show property 2. This means for every $f \in \mathbf{L}^\infty$ and every stopping time σ, we have the inequality $\Phi_0(f) \leqslant \Phi_0(\Phi_\sigma(f))$. Now this inequality follows from the submartingale property for the family $\Phi_T(f)$, T a stopping time. Indeed, for every $\mathbb{Q} \in \mathcal{S}$ we have, by the submartingale property, that $\Phi_0(f) \leqslant \mathbb{E}_\mathbb{Q}[\Phi_\sigma(f)]$. Taking the infimum over all elements $\mathbb{Q} \in \mathcal{S}^e$ then gives $\Phi_0(f) \leqslant \Phi_0(\Phi_\sigma(f))$, as desired. This completes the proof. □

Again, we can give a version of the above theorem when risk adjusted processes are used. Before we can give the precise statement, we need the following proposition.

Proposition 2. *Let $\mathcal{S}$ be a convex closed set of probability measures, all of them absolutely continuous with respect to $\mathbb{P}$, i.e. $\mathcal{S} \subset \mathbf{L}^1$. Let X be a bounded optional, càdlàg process. Then there is a càdlàg process $Y \leqslant X$ that is a $\mathbb{Q}$-submartingale for each $\mathbb{Q} \in \mathcal{S}$ and such that every other càdlàg process $V \leqslant X$, that is a $\mathbb{Q}$-submartingale for each $\mathbb{Q} \in \mathcal{S}$, is necessarily smaller than Y, i.e. satisfies $V \leqslant Y$.*

Proof The proof being a standard construction we only give a sketch. We look at the set

$$\mathcal{V} = \{V \mid V \text{ is a càdlàg submartingale for each } \mathbb{Q} \in \mathcal{S}, V \leqslant X\}.$$

This set is nonempty since the process X is bounded from below by a constant. If V^1 and V^2 are both elements in $\mathcal{V}$, then $V^1 \vee V^2$ is still in $\mathcal{V}$. It follows that there is a sequence $V^n \in \mathcal{V}$ such that for each rational t we have that $\mathbb{E}_\mathbb{P}[V_t^n]$ tends to $\sup\{\mathbb{E}_\mathbb{P}[V_t] \mid V \in \mathcal{V}\}$. Let for t rational: $Y_t' = \sup_n V_t^n$. The family $(Y_t')_{t \text{ rational}}$ is still a $\mathbb{Q}$- submartingale for each $\mathbb{Q} \in \mathcal{S}$. For every t we now define the a.s. limit $Y_t = \lim_{s \downarrow t, s > t} Y_s'$. The process Y is right continuous and is a $\mathbb{Q}$ submartingale for each $\mathbb{Q} \in \mathcal{S}$. It is therefore càdlàg . The process Y is smaller than the process X and every càdlàg process $V \in \mathcal{V}$ necessarily satisfies $V \leqslant Y$. □

Remark 19. In the same way as in the proposition, we can construct a smallest process, that is bigger than X and that is a $\mathbb{Q}$-supermartingale for each

$\mathbb{Q} \in \mathcal{S}$. In particular we can make the following construction. For $f \in \mathbf{L}^\infty$ we define M to be the càdlàg version of the $\mathbb{P}$–martingale $M_t = \mathbb{E}_\mathbb{P}[f \mid \mathcal{F}_t]$. Then we construct a càdlàg process F such that for each $\mathbb{Q} \in \mathcal{S}$, the process F is a $\mathbb{Q}$–supermartingale, $F \geqslant M$ and F is minimal for these properties. The reader can check that the construction yields that $F_\infty = f$, but it might happen that the process F has a jump at time ∞. The reader can also check that F is the smallest process such that F is a $\mathbb{Q}$–supermartingale for each $\mathbb{Q} \in \mathcal{S}$ and $F_\infty \geqslant f$. We will make use of this construction in the next theorem.

Theorem 13 *The following are equivalent*

1. *The set $\mathcal{S}$ is m–stable.*
2. *The family of random variables indexed by the family of stopping times and defined as*

$$\Psi_T(X) = \mathrm{ess.inf}\left\{ \mathbb{E}_\mathbb{Q}[X_\tau \mid \mathcal{F}_T] \mid \tau \geqslant T \text{ is a stopping time and } \mathbb{Q} \in \mathcal{S}^e \right\}$$

satisfies $\Psi_T(X) = Y_T$, where Y is the process introduced in the previous proposition, i.e. Y is the biggest càdlàg process, smaller than X, that is a $\mathbb{Q}$–submartingale for each $\mathbb{Q} \in \mathcal{S}$.

Proof We first prove that 1 implies 2. Since by construction we always have that $\Psi_T(X) \geqslant Y_T$, it suffices to observe that by theorem 11, $\Psi(X)$ defines a càdlàg process that is a $\mathbb{Q}$–submartingale for each $\mathbb{Q} \in \mathcal{S}$.

The converse implication goes as follows. We take a bounded random variable f and define F as in the preceding remark. This means that F is a $\mathbb{Q}$–supermartingale for each $\mathbb{Q} \in \mathcal{S}$, $F_\infty \geqslant f$ and F is minimal for these properties. We now construct the process $\Psi(F)$ and observe that by hypothesis, $\Psi(F)$ is a $\mathbb{Q}$–submartingale for each $\mathbb{Q} \in \mathcal{S}$. However we have that for each stopping time σ:

$$\Psi_\sigma(F) = \mathrm{ess.inf}_{\mathbb{Q} \in \mathcal{S}^e; \tau \geqslant \sigma} \mathbb{E}_\mathbb{Q}[F_\tau \mid \mathcal{F}_\sigma] = \mathrm{ess.inf}_{\mathbb{Q} \in \mathcal{S}^e} \mathbb{E}_\mathbb{Q}[f \mid \mathcal{F}_\sigma] = \Phi_\sigma(f),$$

where τ is a stopping time and where the second inequality follows from the fact that for each $\mathbb{Q} \in \mathcal{S}$, the process F is a $\mathbb{Q}$–supermartingale. This means that the family $\Phi_\sigma(f)$, where σ runs through the set of stopping times, satisfies the submartingale inequality and hence by theorem 12, the set $\mathcal{S}$ is m–stable. □

In case the set $\mathcal{S}$ is weakly compact we can be more precise. For a random variable $f \in \mathbf{L}^\infty$ we saw that there is a càdlàg process $\Phi_t(f)$ which for each $\mathbb{Q} \in \mathcal{S}$ is a submartingale.

Theorem 14 *If $\mathcal{S}$ is weakly compact and m–stable, if $f \in \mathbf{L}^\infty$, then there is a $\mathbb{Q}_0 \in \mathcal{S}$ so that the process $\Phi(f)$ is a $\mathbb{Q}_0$–martingale. Consequently we have for each stopping time τ, $\Phi_\tau(f) = \mathbb{E}_{\mathbb{Q}_0}[f \mid \mathcal{F}_\tau]$, $\mathbb{Q}_0$ a.s. . If moreover all elements of $\mathcal{S}$ are continuous, we can say that $\Phi_\tau(f) = \mathbb{E}_{\mathbb{Q}_0}[f \mid \mathcal{F}_\tau]$, $\mathbb{P}$ a.s. .*

Proof Let us suppose for simplicity that $\mathcal{F}_0$ is trivial, leaving the more general case to the reader. Since $\mathcal{S}$ is weakly compact we can find an element

$\mathbb{Q}_0 \in \mathcal{S}$ so that $\mathbb{E}_{\mathbb{Q}_0}[f] = \inf\{\mathbb{E}_{\mathbb{Q}}[f] \mid \mathbb{Q} \in \mathcal{S}\}$. For an arbitrary stopping time we then have

$$\Phi_0(f) = \mathbb{E}_{\mathbb{Q}_0}[f] = \mathbb{E}_{\mathbb{Q}_0}[\mathbb{E}_{\mathbb{Q}_0}[f \mid \mathcal{F}_\tau]] \geqslant \mathbb{E}_{\mathbb{Q}_0}[\Phi_\tau(f)] \geqslant \Phi_0(\Phi_\tau(f)) = \Phi_0(f).$$

Consequently all inequalities are equalities and we must have that $\mathbb{Q}_0$−a.s. $\Phi_\tau(f) = \mathbb{E}_{\mathbb{Q}_0}[f \mid \mathcal{F}_\tau]$. If all elements of $\mathcal{S}$ are continuous we can apply Theorem 8 to get that $\mathbb{Q}_0 \sim \mathbb{P}$. □

Remark 20. The measure $\mathbb{Q}_0$ can be chosen to be an extremal point of $\mathcal{S}$. In case the elements of $\mathcal{S}$ are all continuous, the extremal points of $\mathcal{S}$ are characterised by Kabanov's theorem (see the remarks in section 3 and see [18]). From there we get a better description of the measure $\mathbb{Q}_0$ and in the case of Brownian Motion we can make a link to the so-called g–expectations as mentioned in the introduction. We do not give details since this is beyond the scope of this paper.

Remark 21. In case $\mathcal{S}$ is not weakly compact there is (by James' theorem on weak compactness) $f \in \mathbf{L}_\infty$ so that $\inf\{\mathbb{E}_{\mathbb{Q}}[f] \mid \mathbb{Q} \in \mathcal{S}\}$ is an infimum which is not attained. So the existence of $\mathbb{Q}_0$ can be seen as a necessary condition for weak compactness.

7 The Characterisation using the cone of Acceptable Elements

From [10] we recall that there is a one to one correspondence between risk measures and weak* closed cones of $\mathbf{L}^\infty$, $\mathcal{A}$ such that $\mathcal{A} \supset L_+^\infty$. The question arises how we can characterise m–stable sets using the cone of acceptable elements. This question was addressed in [4] for the discrete time case. The proofs can be copied without big changes. Let us start with some definitions and notations.

Definition 3 *If $\mathcal{S} \subset \mathbf{L}^1(\mathbb{P})$ is a closed, convex set of probability measures, $\mathbb{P} \in \mathcal{S}$, then with $\mathcal{A}$ we denote the set of acceptable elements:*

$$\mathcal{A} = \{f \mid f \in \mathbf{L}^\infty, \ \text{for all } \mathbb{Q} \in \mathcal{S} : \mathbb{E}_{\mathbb{Q}}[f] \geqslant 0\}.$$

Remark 22. Of course, in the case where $\mathcal{F}_0$ is trivial, we could also have required that $\Phi_0(f) \geqslant 0$.
As observed in [10] we have that $\mathcal{A}$ is weak* closed in $\mathbf{L}^\infty$ and we can recover $\mathcal{S}$ as

$$\mathcal{S} = \{\mathbb{Q} \mid \mathbb{Q} \ll \mathbb{P}; \ \mathbb{Q} \text{ is a probability measure and for all } f \in \mathcal{A} : \mathbb{E}_{\mathbb{Q}}[f] \geqslant 0\}.$$

Definition 4 *and Notation Let τ be a stopping time. A random variable $f \in \mathcal{A}_\tau = \mathbf{L}^\infty(\mathcal{F}_\tau) \cap \mathcal{A}$ is called $\tau-$acceptable. The set $\mathcal{A}'_\tau$ is defined as*

$$\mathcal{A}'_\tau = \{f + g \mid f \in \mathcal{A}_\tau; g \in \mathbf{L}^\infty_+(\mathcal{F}_\infty)\}.$$

Definition 5 *Let τ be a stopping time. An element $f \in \mathbf{L}^\infty(\mathcal{F}_\infty)$ is called acceptable at time τ if for every event $A \in \mathcal{F}_\tau$ we have that $f\mathbf{1}_A \in \mathcal{A}$. By $\mathcal{A}^\tau$ we denote the set of all elements that are acceptable at time τ:*

$$\mathcal{A}^\tau = \{f \mid \text{ for all } A \in \mathcal{F}_\tau, f\mathbf{1}_\tau \in \mathcal{A}\}.$$

The interpretation of both definitions is straightforward. An element f is acceptable at time τ if given the information at time τ, the element f is still acceptable. It could happen that an element f is acceptable at time 0, i.e. $f \in \mathcal{A}$, but as uncertainty is revealed and $A \in \mathcal{F}_\tau$ is realised, we see that the "bad" part of f is realised and hence at time τ the random variable f or better $f\mathbf{1}_A$, should be considered as unacceptable. The following characterisation is straightforward

Proposition 3. *Let τ be a stopping time. We have that*

$$\mathcal{A}^\tau = \{f \mid \Phi_\tau(f) \geqslant 0\}.$$

Proof If $\Phi_\tau(f) \geqslant 0$, then we have for all $\mathbb{Q} \in \mathcal{S}$ and every event $A \in \mathcal{F}_\tau$ that $\mathbb{E}_\mathbb{Q}[\Phi_\tau(f)\mathbf{1}_A] \geqslant 0$. By the definition of Φ this implies that for all $\mathbb{Q} \in \mathcal{S}$ and all events $A \in \mathcal{F}_\tau$ we have $\mathbb{E}_\mathbb{Q}[\mathbb{E}_\mathbb{Q}[f \mid \mathcal{F}_\tau]\mathbf{1}_A] \geqslant 0$. This implies that for all $\mathbb{Q} \in \mathcal{S}$ we have $\mathbb{E}_\mathbb{Q}[f\mathbf{1}_A] \geqslant 0$. Conversely if $f \in \mathcal{A}^\tau$, we must have that for each $\mathbb{Q} \in \mathcal{S}^e$ that $\mathbb{E}_\mathbb{Q}[f\mathbf{1}_A] \geqslant 0$ and this for each $A \in \mathcal{F}_\tau$. This implies that $\mathbb{P}$ a.s we have that $\mathbb{E}_\mathbb{Q}[f \mid \mathcal{F}_\tau] \geqslant 0$. The definition of Φ then implies that $\Phi_\tau(f) \geqslant 0$. $\qquad\square$

Corollary 3 *Let τ be a stopping time. For $f \in \mathbf{L}^\infty$ we have $f - \Phi_\tau(f) \in \mathcal{A}^\tau$.*

Definition 6 *We say that the cone $\mathcal{A}$ of acceptable elements satisfies the decomposition property if for every stopping time τ we have $\mathcal{A} \subset \mathcal{A}_\tau + \mathcal{A}^\tau$.*

The interpretation is clear. Every acceptable element can, for every stopping time τ, be decomposed into two elements. The first element is acceptable when the observation period is stopped at τ. The second element is acceptable when the observation starts at time τ. Of course, since $\mathcal{A}^\tau, \mathcal{A}_\tau \subset \mathcal{A}$, the definition is equivalent to the statement that $\mathcal{A} = \mathcal{A}^\tau + \mathcal{A}_\tau$ for every stopping time τ. Since trivially $\mathbf{L}^\infty_+(\mathcal{F}_\infty) \subset \mathcal{A}^\tau$ we have that $\mathcal{A}^\tau + \mathcal{A}_\tau = \mathcal{A}^\tau + \mathcal{A}'_\tau = \mathcal{A}^\tau + \mathcal{A}_\tau + \mathbf{L}^\infty_+(\mathcal{F}_\infty)$.

Theorem 15 *Let τ be a stopping time, then $f \in \mathcal{A}_\tau + \mathcal{A}^\tau$ if and only if $\Phi_\tau(f) \in \mathcal{A}_\tau$.*

Proof One direction of the proof follows from Corollary 3. Indeed if $\Phi_\tau(f) \in \mathcal{A}_\tau$, the equality $f = \Phi_\tau(f) + (f - \Phi_\tau(f))$ implies that $f \in \mathcal{A}_\tau + \mathcal{A}^\tau$. The other direction is proved as follows. Let $f = g + h$ where $g \in \mathcal{A}_\tau$ and $h \in \mathcal{A}^\tau$. Because of the superadditivity of the functions Φ we have $\Phi_\tau(f) \geqslant \Phi_\tau(g) + \Phi_\tau(h) \geqslant g$, since $\Phi_\tau(g) = g$ and $\Phi_\tau(h) \geqslant 0$. Because $g \in \mathcal{A}_\tau$ and because $\Phi_\tau(f)$ is $\mathcal{F}_\tau$−measurable we get $\Phi_\tau(f) \in \mathcal{A}_\tau$.   □

Corollary 4 *The cone $\mathcal{A}_\tau + \mathcal{A}^\tau$ is always weak* closed.*

Proof We apply the criterion of Krein-Smulian (exactly as in [10]). So let $f_n \in \mathcal{A}_\tau + \mathcal{A}^\tau, \|f_n\|_\infty \leqslant 1$, be a sequence of functions, tending a.s to a function f. Since $\limsup \Phi_\tau(f_n) \leqslant \Phi_\tau(f)$ we deduce from $\|\Phi_\tau(f_n)\|_\infty \leqslant \|f\|_\infty \leqslant 1$ and $\Phi_\tau(f_n) \in \mathcal{A}_\tau$ that also $\Phi_\tau(f) \in \mathcal{A}_\tau$.   □

Theorem 16 *The set $\mathcal{S}$ is m–stable if and only if for each stopping time τ we have $\mathcal{A} = \mathcal{A}_\tau + \mathcal{A}^\tau$. Or in other words, if and only if $\mathcal{A}$ satisfies the decomposition property.*

Proof Let τ be a stopping time. We will use the equivalence (1) and (3) of theorem 12. If $\mathcal{S}$ is m–stable we have for each $f \in \mathbf{L}^\infty$ that $\Phi_0(\Phi_\tau(f)) = \Phi_0(f)$. Consequently we have that $f \in \mathcal{A}$ implies that $\Phi_\tau(f) \in \mathcal{A}$ and hence $\Phi_\tau(f) \in \mathcal{A}_\tau$. Conversely we deduce from the decomposition property that $\Phi_0(f) \geqslant 0$ implies that $\Phi_0(\Phi_\tau(f)) \geqslant 0$. But the translation property then implies that $\Phi_0(f) \leqslant \Phi_0(\Phi_\tau(f))$ for every $f \in \mathbf{L}^\infty$.   □

Remark 23. In [10] the theory of general risk measures was developed using finitely additive measures instead of using σ−additive probability measures. It is not clear how to develop a theory of stable sets in this context. The equivalence of the decomposition property with the m–stability gives us an answer. Since the definitions of $\mathcal{A}^\tau$ and $\mathcal{A}_\tau$ are purely algebraic, they apply to every cone. So these concepts could be used in Definition 6 and Theorem 16 to give an alternative definition of m–stability in the case of risk measures that do not necessarily satisfy the Fatou property. We do not pursue this analysis further.

8 The Relation with Bellman's Principle

In this paragraph we prove that the m–stability is equivalent to the validity of Bellman's principle. The proof is the same as in [4]. Especially in the case of Markov processes such a result can be of great importance. In order not to overload the notation we systematically suppose that $\mathcal{S}$ is a closed convex set of probability measures, $\mathcal{S} \subset \mathbf{L}^1$ and (as always) $\mathbb{P} \in \mathcal{S}$. We also suppose that $\mathcal{F}_0$ is trivial. For a bounded process X and a stopping time σ we defined

$$\Psi_\sigma(X) = \text{ess.inf}_{\mathbb{Q} \in \mathcal{S}^e; \tau \geqslant \sigma} \mathbb{E}_\mathbb{Q}[X_\tau \mid \mathcal{F}_\sigma].$$

We also recall that if σ is a stopping time, the process ${}^\sigma X$ is defined as ${}^\sigma X_s = 0$ if $s \leqslant \sigma$ and ${}^\sigma X_s = X_s - X_\sigma$ if $s \geqslant \sigma$. The process X^σ is defined as $X_s^\sigma = X_s$

if $s \leqslant \sigma$ and $X_s^\sigma = X_\sigma$ if $s \geqslant \sigma$. The proof requires the time interval to be closed from the right, i.e. of the form $[0, t]$ where $0 \leqslant t < +\infty$.

Theorem 17 *In case the time interval is closed from the right, say $[0, t]$, with $0 \leqslant t < +\infty$, the following two properties are equivalent*

1. *$\mathcal{S}$ is m–stable*
2. *(Bellman's principle) For every bounded càdlàg adapted process X and every finite stopping time $\tau \leqslant t$, we have that*

$$\Psi_0(X) = \Psi_0(X^\tau + \Psi_\tau({}^\tau X)\mathbf{1}_{[\![\tau,t]\!]}).$$

Proof   We first show that Bellman's principle implies stability. For $f \in \mathbf{L}^\infty(\mathcal{F}_t)$ we introduce the process X defined as $X_u = \|f\|_\infty$ for $u < t$ and $X_u = f$ for $u \geqslant t$. The value $\Psi_\tau(X)$ then coincides with the value $\Phi_\tau(f)$ and the Bellman principle gives the recursivity for Φ. According to theorem 12 this implies that $\mathcal{S}$ is m–stable.

Conversely let us suppose that $\mathcal{S}$ is m–stable and let us show the Bellman principle. To simplify the notation we will suppose that the measures $\mathbb{Q}, \mathbb{Q}', \mathbb{Q}''$ are taken in $\mathcal{S}^e$, $\sigma \leqslant \tau \leqslant t$ are given stopping times and ν runs through the set of all stopping times $\sigma \leqslant \nu \leqslant t$.

$$\begin{aligned}
\Psi_\sigma(X) &= \mathrm{ess.inf}_{\mathbb{Q},\nu \geqslant \sigma}\, \mathbb{E}_\mathbb{Q}[X_\nu \mid \mathcal{F}_\sigma] \\
&= \mathrm{ess.inf}_{\mathbb{Q},\nu \geqslant \sigma}\, \mathbb{E}_\mathbb{Q}[\mathbb{E}_\mathbb{Q}[X_\nu \mid \mathcal{F}_\tau] \mid \mathcal{F}_\sigma] \\
&= \mathrm{ess.inf}_{\mathbb{Q},\nu \geqslant \sigma}\, \mathbb{E}_\mathbb{Q}[X_\nu \mathbf{1}_{\nu \leqslant \tau} + \mathbb{E}_\mathbb{Q}[X_\nu \mathbf{1}_{\nu > \tau} \mid \mathcal{F}_\tau] \mid \mathcal{F}_\sigma] \\
&= \mathrm{ess.inf}_{\mathbb{Q},\nu \geqslant \sigma}\, \mathbb{E}_\mathbb{Q}[X_\nu \mathbf{1}_{\nu \leqslant \tau} + \mathbf{1}_{\nu > \tau}(X_\tau + \mathbb{E}_\mathbb{Q}[X_\nu - X_\tau \mid \mathcal{F}_\tau]) \mid \mathcal{F}_\sigma] \\
&= \mathrm{ess.inf}_{\mathbb{Q},\nu \geqslant \sigma}\, \mathbb{E}_\mathbb{Q}[X_{\nu \wedge \tau} + \mathbf{1}_{\nu > \tau}\mathbb{E}_\mathbb{Q}[X_\nu - X_\tau \mid \mathcal{F}_\tau] \mid \mathcal{F}_\sigma]
\end{aligned}$$

The Lemma 18 allows us to rewrite the result of the simple ess.inf as a compounded expression:

$$\begin{aligned}
\Psi_\sigma(X) &= \mathrm{ess.inf}_{\mathbb{Q}',\nu \geqslant \sigma}\, \mathbb{E}_{\mathbb{Q}'}[X_{\nu \wedge \tau} \\
&\qquad + \mathbf{1}_{\nu > \tau}\, \mathrm{ess.inf}_{\mathbb{Q}'',\nu' \geqslant \tau}\, \mathbf{1}_{\nu > \tau}\mathbb{E}_{\mathbb{Q}''}[X_{\nu'} - X_\tau \mid \mathcal{F}_\tau] \mid \mathcal{F}_\sigma] \\
&= \mathrm{ess.inf}_{\mathbb{Q}',\nu \geqslant \sigma}\, \mathbb{E}_{\mathbb{Q}'}[X_\nu \mathbf{1}_{\nu < \tau} + \mathbf{1}_{\nu \geqslant \tau}(X_\tau + \Psi_\tau({}^\tau X)) \mid \mathcal{F}_\sigma] \\
&= \mathrm{ess.inf}_{\mathbb{Q}',\nu \geqslant \sigma}\, \mathbb{E}_{\mathbb{Q}'}[(X^\tau + \Psi_\tau({}^\tau X))_\nu \mid \mathcal{F}_\sigma] \\
&= \Psi_\sigma(X^\tau + \Psi_\tau({}^\tau X))
\end{aligned}$$

$\square$

Remark 24. (and Counter-example) An analysis of the proof shows that if the time interval is not closed on the right, the Bellman principle still follows from the fact that the set $\mathcal{S}$ is stable. Whether the Bellman principle implies the stability property is a much more delicate problem. We will give two answers. In case the set $\mathcal{S}$ is weakly compact in $\mathbf{L}^1$, the answer is yes. Afterwards we will give a counter-example in the case where $\mathcal{S}$ is not weakly compact.

Proposition 4. *Suppose that the time interval is $\mathbb{R}_+$, suppose that the Bellman principle holds and suppose that the set $\mathcal{S}$ is weakly compact in $\mathbf{L}^1$, then the set $\mathcal{S}$ is m–stable.*

Proof We will adapt the proof of theorem 17 above. The idea is to show that $\Phi_0(f) = \Phi_0(\Phi_\tau(f))$ for every finite stopping time τ and for every bounded function f that is $\mathcal{F}_\infty$–measurable. Since $\mathcal{S}$ is weakly compact the set

$$\{Z_\sigma \mid \sigma \text{ a finite stopping time}, \ Z \in \mathcal{S}\}$$

is still relatively weakly compact. If we replace f by the sequence $f_n = \mathbb{E}_{\mathbb{P}}[f \mid \mathcal{F}_n]$ then weak-compactness implies that *uniformly* for $\mathbb{Q} \in \mathcal{S}$, f_n approximates f in $\mathbf{L}^1(\mathbb{Q})$. It follows that $\Phi_0(f_n), \Phi_\tau(f_n), \Phi_0(\Phi_\tau(f_n))$ tend to $\Phi_0(f), \Phi_\tau(f), \Phi_0(\Phi_\tau(f))$. It is therefore sufficient to prove the statement for functions that are $\mathcal{F}_n$–measurable. This is done exactly in the same way as in the proof of the theorem. $\qquad\square$

It is clear that a counter-example will have to use the fact that the set $\mathcal{S}$ is big. For notational ease we will work on the time interval $[0, 1[$. This is equivalent to the time interval $\mathbb{R}_+$ (simply use a time transform $u = t/(t+1)$). The use of the time interval $[0, 1[$ allows us to use a Brownian Motion W defined for all times $t < \infty$ even if we only need the part before time 1. Finite stopping times will now be replaced by stopping times $\nu < 1$. The filtration we will use is the usual filtration coming from the process W. The set $\mathcal{S}$ we will use, is defined as

$$\{Z_1 \mid \mathbb{E}_{\mathbb{P}}[Z_1] = 1, Z_1 \geqslant 0, \mathbb{E}_{\mathbb{P}}[Z_1 \operatorname{sign}(W_1)] = 0\}.$$

It is clear that this set is not m–stable. It can be seen using the definition of m–stability but it will also follow from the results below. We first give the sequence of lemma's used to prove the Bellman principle and then we will give the details of the proofs of these lemma's. Since the Bellman principle will be valid, Ψ_0 will in fact be equivalent to the risk adjusted value

$$\Psi_0(X) = \text{ess.inf}\{\inf_{0 \leqslant t < 1} X_t\}.$$

Hence we cannot have m-stability. Indeed $\Phi_0(f) = 0$ for $f = \operatorname{sign}(W_1)$.

Lemma 24. *Let $\nu < 1$ be a stopping time, then the set*

$$\{Z_\nu \mid Z \in \mathcal{S}\}$$

is dense in the set of all $\mathcal{F}_\nu$ measurable densities of probabilities absolutely continuous with respect to $\mathbb{P}$.

Lemma 25. *Bellman's principle is valid.*

Lemma 26. *Let $\mathcal{Q}$ be the set of all density processes Z such that*

1. $Z_1 = \mathcal{E}(q \cdot W)_1 > 0$, $\mathbb{E}_{\mathbb{P}}[Z_1] = 1$
2. $\int_0^1 q_u \, du = 0$ a.s.

We then have that $Q \subset S$. For $\tau < 1$ a stopping time, the set

$$\{Z_\tau \mid Z \in Q\}$$

is dense in the set of all probability densities on the σ–algebra $\mathcal{F}_\tau$.

Lemma 27. *Let $\nu < 1$ be a stopping time and let q be a predictable process, defined on $[0, 1] \times \Omega$ so that*

1. $q_u = 0$ *for* $u \leqslant \nu$
2. q *is measurable for the σ–algebra $\mathcal{R} \times \mathcal{F}_\nu$ where $\mathcal{R}$ is the Borel σ–algebra on $[0, 1]$,*
3. *a.s.* $\int_\nu^1 q_u^2 \, du < \infty$,

then $\mathbb{E}_{\mathbb{P}} \left[\mathcal{E}(q \cdot W)_1 \right] = 1$ and therefore $\mathcal{E}(q \cdot W)_1$ is the density of a probability measure, equivalent to $\mathbb{P}$. Moreover we have

$$\mathbb{E}_{\mathbb{P}} \left[\mathcal{E}(q \cdot W)_1 \mid \mathcal{F}_\nu \right] = 1.$$

Proof of Lemma 27 This is almost trivial. Seen from time ν the process q is deterministic. Here are the details. For each n we put

$$A_n = \left\{ \int_\nu^1 q_u^2 \, du \leqslant n \right\}.$$

Clearly $A_n \in \mathcal{F}_\nu$ and the stochastic exponential $\mathcal{E}(1_{A_n} q \cdot W)$ satisfies Novikov's condition. Therefore we have

$$\mathbb{E}_{\mathbb{P}} \left[1_{A_n} \mathcal{E}(q \cdot W)_1 \right] = \mathbb{E}_{\mathbb{P}} \left[1_{A_n} \mathcal{E}(1_{A_n} q \cdot W)_1 \right] = \mathbb{P}[A_n].$$

We now apply Beppo Levi's theorem to conclude that $\mathbb{E}_{\mathbb{P}} \left[\mathcal{E}(q \cdot W)_1 \right] = 1$ as desired. The statement on the conditional expectation follows from the fact that since $\mathbb{E}_{\mathbb{P}} \left[\mathcal{E}(q \cdot W)_1 \right] = 1$, $\mathcal{E}(q \cdot W)$ must be a uniformly integrable martingale. □

Proof of Lemma 26 and 24 Let Z_τ be the density of a probability measure equivalent to $\mathbb{P}$ on $\mathcal{F}_\tau$. The process Z is supposed to be defined up to time τ. We will now extend it in such a way that it defines an element $Z \in Q$. The process Z is a stochastic exponential and therefore Z_τ can be written as $Z_\tau = \mathcal{E}(q \cdot W)_\tau$. The predictable process q is defined up to time τ. Since $Z_\tau > 0$ we must have that $\int_0^\tau q_u^2 \, du < \infty$ and therefore we also have that $r = \int_0^\tau q_u \, du$ is defined. If we now put for $u > \tau$

$$q_u = \frac{-r}{1 - \tau}$$

we have that $q 1_{]\tau, 1[}$ satisfies the assumptions of lemma 27. We therefore have that

$$\mathbb{E}_{\mathbb{P}} \left[\mathcal{E}(q \cdot W)_1 \right] = \mathbb{E}_{\mathbb{P}} \left[\mathbb{E}_{\mathbb{P}} \left[\mathcal{E}(q \cdot W)_1 \mid \mathcal{F}_\tau \right] \right] = \mathbb{E}_{\mathbb{P}}[1] = 1.$$

Moreover $\int_0^1 q_u\,du = \int_0^\tau q_u\,du + \int_\tau^1 q_u\,du = r + (-r) = 0$. Also, we have that $\int_0^1 q_u^2\,du = \int_0^\tau q_u^2\,du + \int_\tau^1 q_u^2\,du = \int_0^\tau q_u^2\,du + r^2/(1-\tau) < \infty$. Therefore $Z_1 > 0$ and $Z \in \mathcal{Q}$. This proves the density part of the lemma. We now prove that $\mathcal{Q} \subset \mathcal{S}$. For an element $\mathbb{Q} \in \mathcal{Q}$ we have that W is a Brownian motion with drift $q_u\,du$. Therefore the variable W_t is, under the measure $\mathbb{Q}$, equal to a gaussian random variable $+ \int_0^t q_u\,du$. For $t = 1$ this simply means that under $\mathbb{Q}$, the random variable W_1 is still a symmetric gaussian random variable with $\mathbf{L}^2(\mathbb{Q})$ norm 1. In particular we have that $\mathbb{E}_\mathbb{Q}\left[\text{sign}(W_1)\right] = 0$, i.e. $\mathbb{Q} \in \mathcal{S}$. Lemma 24 immediately follows from Lemma 26.   □

Proof of lemma 25 Let us suppose that X is càdlàg , bounded adapted. Furthermore let us fix a stopping time $\nu < 1$. It is clear that

$$\Psi_\nu(X) = X_\nu + \Psi_\nu({}^\nu X).$$

So we have to calculate $\Psi_\nu({}^\nu X)$. By definition we have

$$\Psi_\nu({}^\nu X) = \text{ess.inf}_{\nu \leqslant \sigma < 1}\,\text{ess.inf}_{\mathbb{Q} \in \mathcal{S}^e}\left\{\mathbb{E}_\mathbb{Q}[{}^\nu X_\sigma \mid \mathcal{F}_\nu]\right\}.$$

Because of lemma 24 this can also be written as

$$\Psi_\nu({}^\nu X) = \text{ess.inf}_{\nu \leqslant \sigma < 1}\,\text{ess.inf}_{\mathbb{Q} \sim \mathbb{P}}\left\{\mathbb{E}_\mathbb{Q}[{}^\nu X_\sigma \mid \mathcal{F}_\nu]\right\}.$$

Indeed the set

$$\{Z_\sigma \mid Z \in \mathcal{S}^e\}$$

is dense in the set

$$\{Z_\sigma \mid Z \text{ a nonnegative uniformly integrable martingale with } \mathbb{E}_\mathbb{P}[Z_1] = 1\}.$$

This means that the Ψ−operator is the same when calculated with the set $\mathcal{S}$ as with the set of all probability measures that are absolutely continuous with respect to $\mathbb{P}$. The latter set is stable and therefore the Ψ−operator satisfies Bellman's inequality.   □

We end this analysis with the following

Corollary 5 *The m−stable hull of the set $\mathcal{Q}$ is the set of all probability measures that are absolutely continuous with respect to $\mathbb{P}$.*

Remark 25. That the set $\mathcal{S}$ is not m−stable can also be seen from the criteria in section 7. The calculations are of course similar to the ones above but it might be of pedagogical interest to give the details. Let us have a look at the variable $f = \text{sign}(W_1)$. Because of the definition of $\mathcal{S}$, we have that $f \in \mathcal{A}$. Let τ be a stopping time $0 \leqslant \tau < 1$. We will show that $\Phi_\tau(f) \notin \mathcal{A}_\tau$. According to theorem 16, this is a contradiction to the m−stability of $\mathcal{S}$. The set $\mathcal{A}_\tau$ is the set of all $\mathcal{F}_\tau$−measurable elements g such that for all $Z \in \mathcal{S}$ we have $\mathbb{E}_\mathbb{P}[fZ] = \mathbb{E}_\mathbb{P}[fZ_\tau] \geqslant 0$. But lemma 24 then implies that necessarily $g \geqslant 0$. Hence $\mathcal{A}_\tau = \mathbf{L}_+^\infty(\mathcal{F}_\tau)$. Now let us calculate $\Phi_\tau(f)$. To do so let us define the function $p : \mathbb{R} \to [-1, +1]$ by the relation $p(W_\tau) = \mathbb{E}_\mathbb{P}[f \mid \mathcal{F}_\tau]$. Take now $\frac{d\mathbb{Q}}{d\mathbb{P}} = Z_1 = \mathcal{E}(q \cdot W)_1 \in \mathcal{S}$, implying that W_1 has the same distribution under $\mathbb{P}$ as under $\mathbb{Q}$. Clearly we have that $W_t - \int_0^t q_u \, du$ is a $\mathbb{Q}$–Brownian motion and therefore $\mathbb{E}_\mathbb{Q}[f \mid \mathcal{F}_\tau] = p\left(W_\tau - \int_0^\tau q_u \, du\right)$. Therefore $\Phi_\tau(f) \leqslant \inf_n p\left(W_\tau - \int_0^\tau n \, du\right) = \inf_n p\left(W_\tau - n\tau\right) = -1$. This is sufficient to guarantee that $\Phi_\tau(f) \notin \mathcal{A}_\tau$.

9 The Set of Local Martingale Measures for a finite dimensional locally bounded Price Process

In this section we will prove that for locally bounded processes, the set of martingale measures forms an m–stable set. This allows us to apply our previous results to situations occurring in finance. We will also see which m–stable sets can occur as sets of martingale measures for finite dimensional processes. The latter characterisation is not fully complete since it will only be done in the context of continuous filtrations. Throughout this section we will use the following notation, see [11] for more information.

On the filtered probability space $\left(\Omega, \mathcal{F}_\infty, (\mathcal{F}_t)_{t \geqslant 0}, \mathbb{P}\right)$, let $S : \mathbb{R}_+ \times \Omega \to \mathbb{R}^d$ be an adapted càdlàg process that takes values in the d–dimensional space $\mathbb{R}^d$. We suppose that the process is locally bounded and that the original measure is a local martingale measure for the process S. This is a simplification when compared to the situation in finance, but it simplifies notation without destroying its generality. Since the process S is locally bounded, the set

$$\mathcal{S} = \{\mathbb{Q} \ll \mathbb{P} \mid \text{ the process } S \text{ is a local martingale for } \mathbb{Q}\}$$

is a closed convex set. As the following shows, it is also m–stable..

Proposition 5. *The set $\mathcal{S}$ is m–stable.*

Proof We can suppose that the process S is bounded (in the same way as in [11]). That the set $\mathcal{S}$ is convex and closed is then obvious. The m–stability is also quite obvious. Let us take $\mathbb{Q}^1, \mathbb{Q}^2 \in \mathcal{S}^e$. Let Z^1, Z^2 be the associated density processes. If σ is a stopping time, we have to show that the density process defined as $Z_t = Z_t^1$ for $t \geqslant \sigma$ and $Z_t = Z_\sigma^1 \frac{Z_t^2}{Z_\sigma^2}$, is still in $\mathcal{S}$. To show this, it is sufficient to show that the process ZS is a $\mathbb{P}$–martingale. This is easy. Indeed first observe that the process $Z^1 S$ is a $\mathbb{P}$–martingale (since $\mathbb{Q}^1 \in \mathcal{S}$). The same applies to Z^2 and hence the process $\mathbf{1}_{t \geqslant \sigma}\left(Z_t^2 S_t - Z_\sigma^2 S_\sigma\right)$ is also a $\mathbb{P}$–martingale. It follows that the process:

$$Z_t S_t = Z_{t \wedge \sigma}^1 S_{t \wedge \sigma} + \frac{Z_\sigma^1}{Z_\sigma^2} \mathbf{1}_{t \geqslant \sigma}\left(Z_t^2 S_t - Z_\sigma^2 S_\sigma\right)$$

is also a $\mathbb{P}$–martingale. $\square$

To avoid complicated notation we first introduce some extra notions. We restrict ourselves to the case of a continuous price process S. As above we may and do suppose that S is bounded. If X is a local martingale then there is a decomposition of X with respect to S. This decomposition, called the Kunita-Watanabe-Galtchouk decomposition, allows to write X as a sum of two local martingales. One is a stochastic integral with respect to S, the other part M is strongly orthogonal to S. So let us write $X = H \cdot S + M$. Saying that X is strongly orthogonal to S means that $H \cdot S$ is strongly orthogonal to S. This means that the vector H is orthogonal to the predictable range of S. In other words it means that the measure $H'\, d\langle S, S\rangle H = 0$ and this implies that $H \cdot S = 0$. This can only happen when the price process has some redundance.

Theorem 18 *With the notation of the preceding paragraphs and under the assumption that S is continuous we have that*

$$S^e = \left\{ \mathcal{E}(X) \,\middle|\, \begin{array}{l} \mathcal{E}(X)_\infty > 0, \\ X \text{ is strongly orthogonal to } S,\ \mathcal{E}(X) \text{ is unif. integrable} \end{array} \right\}.$$

Proof The proof is very easy. If $\mathcal{E}(X)$ is a uniformly integrable, nonnegative martingale, where $X = H \cdot S + M$ is the Kunita-Watanabe-Galtchouk decomposition, then $\mathcal{E}(X)S$ is a martingale if and only if X is strongly orthogonal to S. This is equivalent to $H \cdot S$ being strongly orthogonal to S. The latter is equivalent to the fact that every coordinate of S is strongly orthogonal to $H \cdot S$ and hence to the fact that $H'\, d\langle S, S\rangle H = 0$. This in turn is equivalent to the property $PH = H$.   □

There is also a converse to this theorem. The interpretation of such a converse theorem is the following. Given a convex closed set of probabilities, when does there exist a finite dimensional process, say S, such that the given set is the set of absolutely continuous martingale measures for the process S? A necessary condition is certainly that the set is m–stable. In the continuous case the answer is given by the following theorem.

Theorem 19 *Let S be a stable set of probability measures. Let the filtration be so that every local martingale is the stochastic integral with respect to the $d-$dimensional local martingale M. Let S be given by the closure of*

$$S^e = \{ \mathcal{E}(q \cdot M)_\infty \mid q \in \Phi \text{ and } \mathbb{E}\left[(\mathcal{E}(q \cdot M))_\infty \right] = 1 \},$$

where the set-valued predictable process Φ is convex and closed valued. Then the set S is a set of equivalent local martingale measures for a price process if and only if each $\Phi(t, \omega)$ is a subspace. If the predictable projection valued process P is the orthogonal projection on the space $\Phi(t, \omega)$, then the price process S can be chosen as $S = (Id_{\mathbb{R}^d} - P) \cdot M$.

Proof The proof is a reformulation of the above theorem 17 and theorem 4. The details are left to the reader.   □

Remark 26. The situation can be generalised to the setting of theorem 18, in the sense that we may suppose that M only generates the continuous local martingales. This means that every local martingale is given by a decomposition of the form $H \cdot M + N$, where N is purely discontinuous. In that case we get the following theorem

Theorem 20 *With the above notation we have that the closure $\mathcal{S}$ of the set*

$$
\mathcal{S}^e = \left\{ \mathcal{E}(q \cdot M + N) \,\middle|\, \begin{array}{l} q \in \Phi \\ \mathcal{E}(q \cdot M + N) \text{ uniformly integrable and} \\ \text{strictly positive} \\ N \text{ is purely discontinuous} \end{array} \right\},
$$

is a set of risk neutral measures if and only if each $\Phi(t,\omega)$ is a subspace. If the predictable projection valued process P is the orthogonal projection on the space $\Phi(t,\omega)$, then the price process S can be chosen as $S = (Id_{\mathbb{R}^d} - P) \cdot M$.

10 Appendix on the predictable range

If M is a d–dimensional martingale then it may happen that on some time intervals — or on some predictable sets — coordinates are linearly dependent. To avoid difficulties coming from this redundancy we will introduce the predictable range of M. We will only need the concept for continuous martingales. We will not give full details of the proofs, most of them being straightforward. Readers familiar with the theory of stochastic processes can skip this section that indeed does not contain anything new. Only the presentation is (maybe) of some interest.

Lemma 28. *The set*

$$
\mathcal{K} = \{q \mid q \text{ predictable } d - \text{dimensional and } q \cdot [M, M] = 0\}
$$

is a vector space of predictable processes satisfying

$$
\text{if } q \in \mathcal{K}, \text{ if } h \text{ is predictable and real-valued, then } h\,q \in \mathcal{K}.
$$

Lemma 29. *The space*

$$
\mathcal{K}^\perp = \{q \mid q \text{ predictable } d - \text{dimensional and for all } k \in \mathcal{K} : q.k = 0\}
$$

is a vector space of predictable processes satisfying

$$
\text{if } q \in \mathcal{K}^\perp, \text{ if } h \text{ is predictable and real-valued, then } h\,q \in \mathcal{K}^\perp.
$$

Lemma 30. *and Definition There exist predictable processes $e_j : \mathbb{R}_+ \times \Omega \longrightarrow \mathbb{R}^d$, $j = 1,\ldots,d$ such that μ almost everywhere*

1. *for $j \leqslant d - 1$ we have $\{e_{j+1} \neq 0\} \subset \{e_j \neq 0\}$.*
2. *Either $\|e_j(\omega)\| = 1$ or $e_j(\omega) = 0$.*

3. For $j \neq k$ we have $e_j.e_k = 0$.
4. $q \in \mathcal{K}^\perp$ if and only if there are real-valued predictable processes $h_1, \ldots, h_d$ such that $q = \sum_{j \leqslant d} h_j\, e_j$.
5. The orthogonal projection (depending on t and ω), $P = \sum_j e_j \otimes e_j$ satisfies $q \in \mathcal{K}^\perp$ if and only if on $\mathbb{R}_+ \times \Omega$ we have $Pq = q$.
6. We call the range of P the predictable range for the process M.
7. P satisfies: $q \cdot [M, M] = 0$ if and only if $Pq = 0$.

Proof We will not prove all the statements, the reader can easily fill in the details. The only tricky point is how to get the predictable processes e_k. We start by introducing a measure μ defined on the predictable sets. The measure is defined as follows

$$\mu(A) = \mathbb{E}\left[\int_0^\infty e^{-Trace[M,M]_t}(\mathbf{1}_A)_t\, d(Trace[M, M])_t\right].$$

This measure is easily seen to be finite. Furthermore if $q = 0$, μ almost everywhere, we have $q \cdot M = 0$. This will allow us to replace predictable processes q by processes that are equal to q, μ a.e. We now put $\mathcal{K}_1 = \mathcal{K}^\perp$ and we look at the class

$$\mathcal{C} = \{\{q \neq 0\} \mid q \in \mathcal{K}_1\}.$$

It is easily seen that this class of predictable sets is stable for countable unions. Indeed let q^n be a sequence in $\mathcal{K}_1$. Without loss of generality we may suppose that each q^n has a norm equal to either 0 or 3^{-n}, eventually we replace q^n by $\frac{q^n}{3^n\|q^n\|}\mathbf{1}_{\{q^n \neq 0\}}$. We can now verify that $q = \sum_n q^n$ satisfies $\{q \neq 0\} = \cup_n\{q^n \neq 0\}$. Since the class $\mathcal{C}$ is stable for countable unions, it has up to $\mu-$negligible sets a biggest element, coming from say an element q. Of course we may and do suppose that $\|q\|$ is either 0 or 1. Let us put $e_1 = q$. Now we look at the class

$$\mathcal{K}_2 = \{q \in \mathcal{K}_1 \mid q.e_1 = 0\},$$

and we continue with $\mathcal{K}_1$ replaced by $\mathcal{K}_2$. We again find an element e_2 with maximal support. Of course the maximality of the support of e_1 implies that $\{e_2 \neq 0\} \subset \{e_1 \neq e_2\}$. At least μ a.e., but it is easy to adapt the processes in such a way that the inclusion holds as sets. We continue by induction and observe that the procedure stops after d steps, i.e. $\mathcal{K}_{d+1} = \{0\}$. We now prove item 4. Let the space obtained using the procedure of item 4 be $\mathcal{L}$. We claim that $\mathcal{L} = \mathcal{K}^\perp$, up to equality μ a.e.. If not then we take an element $q \in \mathcal{L}\setminus\mathcal{K}^\perp$, q not equal to zero μ a.e.. Replacing q by $q - Pq$ then gives an element $q \in \mathcal{K}^\perp$ such that $q.e_j = 0$ for all $j \leqslant d$. This, by induction, implies that $q \in \mathcal{K}_j$ for each $j \leqslant d$. Consequently we must have $\{q \neq 0\} \subset \{e_j \neq 0\}$ for each j. In the points where $q(t, \omega)$ is not zero this means that the vectors $e_j(t, \omega), j = 1 \ldots d$ are all nonzero and orthogonal. But then $q(t, \omega)$ is perpendicular to a basis of $\mathbb{R}^d$, a contradiction to q not equal to zero μ a.e. $\qquad\square$

References

1. Z. Artstein: Set-valued measures, *Trans. Amer. Math. Soc.* **165**, 103–125, (1972)
2. Ph. Artzner, F. Delbaen, J.-M. Eber, and D. Heath: Thinking Coherently, *RISK*, November 97, 68–71, (1997)
3. Ph. Artzner, F. Delbaen, J.-M. Eber, and D. Heath: Characterisation of Coherent Risk Measures, *Mathematical Finance* **9**, 145–175, (1999)
4. P. Artzner, F. Delbaen, J.-M. Eber, D. Heath, and H. Ku: Coherent Multiperiod Risk Adjusted Values and Bellman's Principle, to appear in Annals of Operations Research, http://www.math.ethz.ch/~delbaen, (2003)
5. R.J. Aumann, Integrals of Set Valued Functions, *Journ. Math. Anal. and Appl.*, **12**, 1–12, (1965)
6. P. Briand, F. Coquet, Y. Hu, J. Mémin, and S. Peng: A converse comparison theorem for BSDE's and related properties of g−expectation, *Elect. Comm. in Probab.*, **5**, 101–117, (2000)
7. P. Cheridito, F. Delbaen, and M. Kupper: Convex measures of risk for càdlàg processes, Stoch Proc. and Appli., **112**, 1–22, (2004)
8. F. Coquet, Y. Hu, J. Mémin, and S. Peng, Filtration consistent nonlinear expectations and related g−expectations, *Probability Theory and Related Fields*, **123**, 1–27, (2002)
9. G. Debreu and D. Schmeidler: The Radon-Nikodým derivative of a correspondence. *Proceedings of the Sixth Berkeley Symposium on Mathematical Statistics and Probability Vol. II: Probability theory*, 41–56, (1972)
10. F. Delbaen: Coherent Risk Measures, *Lectures given at the Cattedra Galileiana at the Scuola Normale di Pisa, March 2000*, Published by the *Scuola Normale di Pisa*, (2002)
11. F. Delbaen and W. Schachermayer: A general version of the fundamental theorem of asset pricing, *Math. Ann.*, **300**, 463–520, (1994)
12. C. Dellacherie and P.-A. Meyer: *Probabilités et potentiel, Chapitres V à VIII, Hermann, Paris*, (1980)
13. N. El Karoui: Les aspects probabilistes du contrôle stochastique *Ecole d'Eté de Probabilités de Saint-Flour IX, 1979, Lecture Notes in Mathematics 876, Springer, Berlin, 1981*, 73–238
14. N. El Karoui: Backward Stochastic Differential Equations: a general introduction in *N. El Karoui and L. Mazliak eds, Pitman Res. Notes Math. Ser.*, **364**, 7–26, (1997)
15. N. El Karoui, S. Peng and M.C. Quenez: Backward Stochastic Differential Equations in Finance, *Mathematical Finance*, **7**, 1–71, (1997)
16. L. Epstein and M. Schneider, Recursive Multiple Priors, *working paper, University of Rochester*, (2002)
17. H. Föllmer and D.O. Kramkov: Optional decompositions under constraints, *Probab. Theory Related Fields*, **109**, 1–25, (1997)
18. Y. Kabanov: On an existence of the optimal solution in a control problem for a counting process, *Mat. Sbornik*, **119**, 431–445, (1982)
19. D.O. Kramkov: Optional decomposition of supermartingales and hedging contingent claims in incomplete security markets, *Probab. Theory Related Fields*, **105**, 459–479, (1996)

20. S. Kusuoka: On Law Invariant Coherent Risk Measures, *Advances in Mathematical Economics*, **3**, 83–95, (2001)
21. S. Kusuoka and J. Morimoto: On Dynamic Law Invariant Coherent Risk Values, *Working Paper, Tokyo University*, (2004)
22. M. Kupper: Ph.D. thesis, Department of Mathematics, ETH Zurich, (2005)
23. J.-F. Mertens: Processus stochastiques généraux et surmartingales, *Zeitschr. für Wahrscheinlichkeitstheorie und verwandte Gebiete*, **22**, 45–68, (1972)
24. S. Peng: Backward SDE and related g−expectations, *in N. El Karoui and L. Mazliak eds, Pitman Res. Notes Math. Ser.*, **364**, 141–159, (1997)
25. P. Protter: Stochastic Integration and Differential Equations: a new approach, Springer-Verlag, Berlin, 1990
26. E. Rosazza Gianin: Some examples of risk measures via g−expectations, *Working paper, University of Naples Federico II*, 1–29, (2004)
27. D. Revuz and M. Yor: Continuous Martingales and Brownian Motion, second edition, Springer-Verlag, Berlin, 1994
28. T. Wang: A Class of Multi-Prior Preferences, *working paper*, (2002)
29. G.A. Zitkovic, A filtered version of the Bipolar Theorem of Brannath and Schachermayer, *Journal of Theoretical Probability*, **15**, 41–61, (2002)

A Path Transformation of Brownian Motion

Bhaskaran Rajeev

Indian Statistical Institute, 8th Mile Mysore Road, R.V. College P.O.,
Bangalore - 560 059, India
e-mail: brajeev@isibang.ac.in

Abstract: In this article we consider a transformation on the paths of Brownian motion induced by a random closed (time) set and prove the Markov property of the resulting process.

1 Introduction and statement of results

In this article we consider a transformation on the paths of a Brownian motion induced by a random closed set H, and discuss the Markov property of the resulting process.

Let $\Omega = C[0, \infty)$ and $\mathcal{F}$ its Borel σ-field. Let for all $t \geqslant 0$, and $\omega \in \Omega, X_t(\omega) - \omega(t)$. Let Γ_0 be the Weiner measure on $(\Omega, \mathcal{F})$ and for each $x \in \mathbb{R}$, P_x be the measure on $(\Omega, \mathcal{F})$ defined by $P_x(A) = P_0(A - x), A \in \mathcal{F}$. Let $\theta_t : \Omega \to \Omega$ be shift operator i.e. $\theta_t(\omega)(s) = \omega(t + s)$ for all $t \geqslant 0, s \geqslant 0$. Let $\mathcal{F}_t$ be the σ-field generated by $(X_s)_{s \geqslant 0}$ up to time t.

The transformation that we consider is the following: Let $a < b$ and $H = \{t : X_t \leqslant a \text{ or } X_t \geqslant b\}$. Define $\sigma_t = \sup\{s \leqslant t : s \in H\}(\sup \phi = 0)$. Define the (discontinuous), $\mathcal{F}_t$-adapted, $\mathbb{R}^2$-valued process $(Z_t)_{t \geqslant 0}$ as follows: $Z_t =: (X_t - X_{\sigma_t}, X_{\sigma_t})$. The component processes $(X_t - X_{\sigma_t})$ and (X_{σ_t}) are known to be $((\mathcal{F}_t)_{t \geqslant 0}, P_x)$ semi-martingales ([5]) for each $x \in \mathbb{R}$. Using the strong Markov property of (X_t) we prove in Theorem 1 the strong Markov property of $(Z_t)_{t \geqslant 0}$ under the measures $\{P_x, x \in \mathbb{R}\}$, with respect to the filtration $(\mathcal{F}_t, t \geqslant 0)$. Note that $P_x\{Z_0 = (0, x)\} = 1$ and that $\{(0, x), x \in \mathbb{R}\}$ is a proper subset of the state space S of (Z_t) (see below). We define (below) 'transition functions' $P(t, z, G)$ with respect to which the Markov property in Theorem 1 is proved. In Theorem 2, we prove the semigroup property of these transition functions (the Chapman-Kolmogorov equations).

The techniques we use suggest that our results are true for more general sets H and for more general processes (X_t). We plan to make a more detailed study later. For a general Markov process (X_t), and H a random closed homogeneous set, it is known that (under suitable assumptions) $(t - \sigma_t, X_{\sigma_t})$ is a strong Markov process (see [1], [2]). The random times (σ_t) are so called 'co-terminal' times and have been long known in the Markov process literature (see [3], [4]). Our results are also related to P. Lévy's classic results on $(M_t - X_t, M_t)$ where $M_t = \sup_{s \leqslant t} X_s$ (see remarks after statement of Theorem 1.2).

To state the main results, we need to describe the state space and transition function of (Z_t). The state space S of the process can be described as follows:

$$S = S_1 \cup S_2 \cup S_3$$
$$\text{where } S_1 = \{(0, y) : y \leqslant a \text{ or } y \geqslant b\}$$
$$S_2 = S_2(a) \cup S_b(b)$$
$$\text{where } S_2(a) = \{(x, a) : 0 < x < b - a\}$$
$$S_2(b) = \{(x, b) : -(b - a) < x < 0\}.$$

Finally, $S_3 = \bigcup_{a < y < b} S_3(y)$ where $S_3(y) = \{(x, y) : a - y < x < b - y\}$.

Clearly S is a Borel subset of $\mathbb{R}^2$. The transition function $P(t, z, G)$ of (Z_t) can be described as follows: Let G be a Borel set, $G \subseteq S$. Let $t \geqslant 0$ and $z = (x, y) \in S$. Define

$$\begin{aligned} P(t, z, G) =: &\ I_{S_1}(z) P_{x+y}\{Z_t \in G\} \\ &+ I_{S_2 \cup S_3}(z)\{P_{x+y}\{\sigma_t = 0, (X_t - y, y) \in G\} \\ &+ P_{x+y}\{\sigma_t > 0, Z_t \in G\}\}. \end{aligned} \tag{1}$$

We now state the main results.

Proposition 1.1 *Let τ be an $(\mathcal{F}_t, t \geqslant 0)$ stopping time such that a.s. $P_x, \tau(\omega) \in H(\omega) = \{t : X_t(\omega) \leqslant a \text{ or } X_t(\omega) \geqslant b\}$. Then a.s. P_x,*

$$P_x\{Z_{t+\tau} \in G | \mathcal{F}_{\tau+}\} = P_{X_\tau}\{Z_t \in G\} \tag{2}$$

for all Borel sets $G \subseteq S$.

Remark Proposition 2 is also a consequence of the fact that $(\Omega, \mathcal{F}, \mathcal{F}_t, (X_t), \theta_t, P_x, R)$ is a regenerative system where

$$R = \inf\{t > 0 : t \in H\}.$$

See [1]. We have stated it in a form for which Theorem 1.2 below is a generalisation.

Theorem 1.2 *Let τ be an $(\mathcal{F}_t)$ stopping time. Let $t \geqslant 0$. Then a.s. P_y,*

$$P_y\{Z_{t+\tau} \in G | \mathcal{F}_{\tau+}\} = P(t, Z_\tau, G) \tag{3}$$

on the set $\{\tau < \infty\}$, for all $G \in \mathcal{B}(S)$.

Remark Let $M_t = \sup\limits_{s \leqslant t} X_s$. Let $H = \{t : M_t = X_t\}$. Let $\sigma_t = \sup\{s \leqslant t : s \in H\}$. Then it is easy to see that $X_{\sigma_t} = M_t$ and hence $(M_t - X_t, M_t) = (-(X_t - X_{\sigma_t}), X_{\sigma_t})$. The (strong) Markov property of the process $(M_t - X_t, M_t)$ is well known from a result due to P. Lévy (see [6], Chap. VI).

Remark We note that $P_y\{Z_0 = (0, y)\} = 1$, for all $y \in \mathbb{R}$. In other words, the Markov property (3) is proved only for 'initial conditions' corresponding to $Z_0 = (0, y), y \in \mathbb{R}$. However for each $y \in \mathbb{R}$, we have the following 'entrance laws' for the process (Z_t) : Define $\lambda_t^y(G) =: P_y\{Z_t \in G\}, G \in \mathcal{B}(S)$. Taking expected values in eqn. (3) we get with $\tau \equiv s > 0$,

$$\lambda_{t+s}^y(G) = \int_S P(t, z', G)\lambda_s^y(dz') \tag{4}$$

From Theorem 1.2, eqn (4) and the Markov property of Brownian motion killed on exiting (a, b), we will deduce the 'Chapman-Kolmogorov' equations or the semi-group property of $P(t, z, G)$. Note that $P(0, z, G) = \delta_z(G), \delta_z$ being the Dirac mass at $z \in S$.

Theorem 1.3 $\forall\, t, s \geqslant 0, z \in S$, we have

$$P(t + s, z, G) = \int_S P(s, z', G)\ P(t, z, dz') \tag{5}$$

for all $G \in \mathcal{B}(S)$.

2 The Proofs

Proof of Proposition 1.1 Since $\tau(\omega) \in H(\omega), P_x$ a.s., we have a.s. P_x,

$$\sigma_{t+\tau}(\omega) = \tau(\omega) + \sigma_t(\theta_\tau(\omega)).$$

Hence,

$$Z_{t+\tau} = Z_t \circ \theta_\tau.$$

Hence if $A \in \mathcal{F}_{\tau+}$,

$$\int_A I_G(Z_{t+\tau})dP_x = \int_A I_G(Z_t \circ \theta_\tau)dP_x$$

$$= \int_A P_{X_\tau}\{Z_t \in G\}dP_x$$

where the last equality follows from the strong Markov property for (X_t). The result follows. $\qquad\square$

Proof of Theorem 1.2 We prove the theorem first when $\tau \equiv s > 0$. To apply the Markov property of $(X_t, t \geqslant 0, \mathcal{F}_t, t \geqslant 0, P_x, x \in \mathbb{R})$ we split the sample space as follows: $\Omega = \Omega_1 \cup \Omega_2$ where $\Omega_2 = \{\sigma_s = s\}, \Omega_1 = \Omega_{11} \cup \Omega_{12}, \Omega_{11} = \{\sigma_s < s < \sigma_{s+t}\}$ and $\Omega_{12} = \{\sigma_s < s, \sigma_{s+t} = \sigma_s\}$. Note that the sets $\{\sigma_s < s\}$ and $\{\sigma_s = s\}$ belong to $\mathcal{F}_s$. Further $\omega \in \Omega_{11}$ iff $\sigma_s(\omega) < s$ and $\sigma_t \circ \theta_s(\omega) > 0$ and $\omega \in \Omega_{12}$ iff $\sigma_s(\omega) < s$ and $\sigma_t \circ \theta_s(\omega) = 0$. On the set $\Omega_2 \cup \Omega_{11}$ we can write $\sigma_{s+t} = s + \sigma_t \circ \theta_s$. Hence

$$
\begin{aligned}
P_y\{\Omega_2 \cap \{Z_{t+s} \in G\}|\mathcal{F}_{s+}\} &= P_y\{\Omega_2 \cap \{Z_t \circ \theta_s \in G\}|\mathcal{F}_s\} \\
&= I_{\Omega_2} P_{X_s}\{Z_t \in G\} \\
&= I_{S_1}(Z_s) P(t, Z_s, G).
\end{aligned}
\tag{6}
$$

Further,

$$
\begin{aligned}
P_y\{\Omega_{11} \cap \{Z_{t+s} \in G\}|\mathcal{F}_{s+}\} &= P_y\{\{\sigma_s < s\} \\
&\quad \cap \{\sigma_t \circ \theta_s > 0\} \cap \{Z_t \circ \theta_s \in G\}|\mathcal{F}_{s+}\} \\
&= I_{\{\sigma_s < s\}} P_{X_s}\{\sigma_t > 0, Z_t \in G\} \\
&= I_{S_2 \cup S_3}(Z_s) P_{X_s}\{\sigma_t > 0, Z_t \in G\}.
\end{aligned}
\tag{7}
$$

If $\omega \in \Omega_{12}$, then we can no longer write, $\sigma_{t+s}(\omega) = s + \sigma_t \circ \theta_s(\omega)$. In this case we write

$$
Z_{t+s} = (X_t \circ \theta_s - X_{\sigma_s}, X_{\sigma_s}).
$$

Hence,

$$
\begin{aligned}
P_y\{\Omega_{12} \cap \{Z_{t+s} \in G\}|\mathcal{F}_{s+}\} &= P_y\{\{\sigma_s < s\} \cap \{\sigma_t \circ \theta_s = 0\} \\
&\quad \cap \{X_t \circ \theta_s - X_{\sigma_s}, X_{\sigma_s}) \in\}|\mathcal{F}_{s+}\} \\
&= I_{\{\sigma_s < s\}} P_y\{\{\sigma_t \circ \theta_s = 0\} \\
&\quad \cap \{(X_t \circ \theta_s - X_{\sigma_s}, X_{\sigma_s}) \in G\}|\mathcal{F}_{s+}\} \\
&= I_{S_2 \cup S_3}(Z_s) P_{X_s}\{\sigma_t = 0, (X_t - y, y) \in G\}_{y=X_{\sigma_s}}
\end{aligned}
\tag{8}
$$

where the last equality follows from Proposition 2.1 below. Adding (6), (7) and (8) we get eqn. (3) and hence Theorem 1 is proved when $\tau = s$. For τ an arbitrary $(\mathcal{F}_t)$ stopping time, note that the sets $\{\sigma_\tau < \tau\}$ and $\{\sigma_\tau = \tau\}$ are in $\mathcal{F}_\tau$ - this follows from the fact that $X_{s \wedge \tau}$ is $\mathcal{F}_\tau$ measurable for all $s \geqslant 0$. Eqn (3) is obvious on the set $\tau = 0$. On the set $\tau > 0$, one can use the strong Markov property to prove the analogs of eqns (6),(7), and (8) when 's' is replaced by the random time 'τ'. Adding the three equations, it is easily seen that eqn (3) holds on the set $\tau > 0$. □

Proposition 2.1 *Let* $G : \Omega \times \Omega \to \mathbb{R}$ *be bounded and* $\mathcal{F} \times \mathcal{F}_\tau$ *measurable. Then a.s.* P_y,

$$
E_y\{G(\theta_\tau, \cdot)|\mathcal{F}_\tau\}(\omega) = E_{X_\tau(\omega)}[G(\cdot, \omega)].
$$

Proof This is immediate from the strong Markov property and the monotone class theorem. □

Proof of Theorem 1.3 The proof makes essential use of eqn (4). Since $S = S_1 \cup S_2 \cup S_3$ and the S_i's are mutually disjoint, to prove (5) we can consider separately the different cases corresponding to the value of the initial variable z in the different sets $S_i, i = 1, 2, 3$, and corresponding to G in different sets $S_i, i = 1, 2, 3$. Since $P(0, z, G) = \delta_z(G)$, we will only consider the case $t > 0, s > 0$.

Case (i): Let $z = (0, y) \in S_1$. Then since $P(t, z, G) = \lambda_t^y(G)$, eqn (5) follows from eqn (4).

Before we consider the other cases we introduce some notation. Let $z = (x, y) \in S_2 \cup S_3$. Define

$$\lambda_{t,1}^{x+y}(G) =: P_{x+y}\{\sigma_t = 0, (X_t - (x+y), x+y) \in G\}$$
$$\lambda_{t,2}^{x+y}(G) =: P_{x+y}\{\sigma_t > 0, (X_t - X_{\sigma_t}, X_{\sigma_t}) \in G\}$$
$$P_1(t, z, G) =: P_{x+y}\{\sigma_t = 0, (X_t - y, y) \in G\}$$
$$P_2(t, z, G) =: P_{x+y}\{\sigma_t > 0, (X_t - X_{\sigma_t}, X_{\sigma_t}) \in G\}$$

Note that $\lambda_t^{x+y}(G) = \lambda_{t,1}^{x+y}(G) + \lambda_{t,2}^{x+y}(G)$ and similarly $P(t, z, G) = P_1(t, z, G) + P_2(t, z, G)$. Clearly, $\lambda_{t,2}^{x+y}(G) = P_2(t, z, G)$. Note that the measure $\lambda_{t,1}^{x+y}(\cdot)$ is supported in $S_3(x+y) \subseteq S_3$ for $z \in S_3 \cup S_2$ whereas $P_1(t, z, \cdot)$ is supported in $S_3(y)$ if $z = (x, y) \in S_3$; $S_2(a)$ if $z = (x, a) \in S_2(a)$; and $S_2(b)$ if $z = (x, b) \in S_2(b)$.

Case (ii): Let $z = (x, y) \in S_3$ and $G \subseteq S_3$. Note that $P(t, z, G) = P_1(t, z, G)$ if $G \subseteq S_3(y)$. Hence it suffices to show that

$$P_1(t + s, z, G) = \int_{S_3(y)} P_1(s, z', G) P_1(t, z, dz').$$

Let $T : S_3(y) \to (a, b), T(z_1, y) = z_1 + y$. For $r \in (a, b), A$ a Borel subset of (a, b) and $\tau = \inf\{s > 0 : X_s \notin (a, b)\}$, let

$$\tilde{P}(t, r, A) = P_r\{t < \tau, X_t \in A\}.$$

Note that

$$\tilde{P}(t, x + y, A) = P_{x+y}\{\sigma_t = 0, (X_t - y, y) \in T^{-1}(A)\}$$
$$= P_1(t, z, \cdot) \circ T^{-1}(A).$$

Further, from the Markov property of Brownian motion killed at time τ, we have

$$\tilde{P}(t + s, r, A) = \int_{(a,b)} \tilde{P}(s, q, A) \tilde{P}(t, r, dq).$$

Hence,

$$P_1(t+s,(x,y),G) = \tilde{P}(t+s,x+y,T(G))$$

$$= \int_{(a,b)} \tilde{P}(s,q,T(G))\tilde{P}(t,x+y,dq)$$

$$= \int_{(a,b)} \tilde{P}(s,q,T(G))P_1(t,(x,y),\cdot)\circ T^{-1}(dq)$$

$$= \int_{S_3} \tilde{P}(s,Tz',TG)P_1(t,z,dz')$$

$$= \int_{S_3} P_1(s,z',G)P_1(t,z,dz')$$

which proves Case (ii).

Case (iii): Let $z=(x,y)\in S_3$ and $G\subseteq S_1\cup S_2$. With the notation as in Case (ii), note that for any $t\geqslant 0$,

$$P(t,z,G) = P_2(t,z,G) = \lambda_{t,2}^{x+y}(G) = \lambda_t^{x+y}(G)$$

$$P(t+s,z,G) = \lambda_{t+s}^{x+y}(G)$$

$$= \int P(s,z',G)\lambda_t^{x+y}(dz')$$

$$= \int_{S_1\cup S_2} P(s,z',G)P(t,z,dz')$$

$$+ \int_{S_3(x+y)} P(s,z',G)\lambda_{t,1}^{(x+y)}(dz'). \qquad (9)$$

We analyse the last integral in the RHS (see Section 1 for the notation $S_3(x+y)$. Let $T:S_3(y)\to S_3(x+y)$ be $T(z_1,y)=(z_1-x,y+x)$. Then it is easy to see that

$$\lambda_{t,1}^{x+y}(\cdot) = P_1(t,z,\cdot)\circ T^{-1}(\cdot).$$

Hence

$$\int_{S_3(x+y)} P(s,z',G)\lambda_{t,1}^{x+y}(dz') = \int_{S_3(x+y)} P(s,z',G)P_1(t,z,\cdot)\circ T^{-1}(dz')$$

$$= \int_{S_3(y)} P(s,Tz',G)P_1(t,z,dz').$$

Note that

$$P(s, Tz', G) = P_2(s, Tz', G)$$
$$= P_2(s, z', G)$$
$$= P(s, z', G)$$

because $P_1(t, z, G)$ does not contribute to transitions from S_3 to $G \subseteq S_1 \cup S_2$. Hence,

$$\int_{S_3(x+y)} P(s, z', G)\lambda_{t,1}^{x+y}(dz') = \int_{S_3(y)} P(s, z', G)P_1(t, z, dz')$$
$$= \int_{S_3} P(s, z', G)P(t, z, dz')$$

where the last equality holds because $P_2(t, z, \cdot)$ is supported in $S_1 \cup S_2$. Substituting back into (9) proves Case (iii).

Case (iv): Let $z = (x, a) \in S_2(a) \subset S_2$. Let $G \subseteq S_1$. Then it is easy to see that

$$P(t, z, G) = P_2(t, z, G) = \lambda_{t,2}^{x+a}(G)$$
$$= \lambda_t^{x+a}(G).$$
Therefore $P(t + s, z, G) = \lambda_{t+s}^{x+a}(G)$
$$= \int P(s, z', G)\lambda_t^{x+a}(dz')$$
$$= \int_{S_1} P(s, z', G)P(t, z, dz')$$
$$+ \int_{S_2 \cup S_3} P(s, z', G)\lambda_t^{x+a}(dz'). \qquad (10)$$

Note that $\lambda_{t,1}^{x+a}(dz')$ is concentrated on $S_3(x+a)$ and $\lambda_{t,2}^{x+a}(dz')$ is concentrated on S_2. Hence

$$\int_{S_2 \cup S_3} P(s, z', G)\lambda_{t,2}^{x+a}(dz') = \int_{S_2} P(s, z', G)P_2(t, z, dz')$$
$$= \int_{S_2 \cup S_3} P(s, z', G)P_2(t, z, dz') \qquad (11)$$

since $P_2(t, z, dz')$ is also concentrated on S_2. Consider $T : S_2(a) \rightarrow S_3(x + a), T(z_1, a) = (z_1 - x, a + x)$. Easy to see that

$$\lambda_{t,1}^{x+a}(\cdot) = P_1(t, z, \cdot) \circ T^{-1}(\cdot).$$

Hence,

$$\int_{S_2 \cup S_3} P(s, z', G)\lambda_{t,1}^{x+a}(dz') = \int_{S_3(x+a)} P(s, z', G)P_1(t, z, \cdot) \circ T^{-1}(dz')$$

$$= \int_{S_2(a)} P(s, Tz', G)P_1(t, z, dz').$$

But since $G \subseteq S_1$ we have

$$P(s, Tz', G) = P_2(s, Tz', G) = P_2(s, z', G) = P(s, z', G).$$

Hence,

$$\int_{S_2 \cup S_3} P(s, z', G)\lambda_{t,1}^{x+a}(dz') = \int_{S_2(a)} P(s, z', G)P_1(t, z, dz')$$

$$= \int_{S_2 \cup S_3} P(s, z', G)P_1(t, z, dz'). \qquad (12)$$

The proof when $z = (x, a)$ is completed using (11), (12) in (10). The proof for $z = (x, b) \in S_2(b)$ is similar.

Case (v): Let $z = (x, a) \in S_2(a)$ and $G \subseteq S_2$. It suffices to show that

$$P_1(t + s, z, G) = \int P_1(s, z', G)P_1(t, z, dz') \qquad (13)$$

and

$$P_2(t + s, z, G) = \int P_2(s, z', G)P_1(t, z, dz') + \int P(s, z', G)P_2(t, z, dz'). (14)$$

Eqn (13) follows from the semi-group property for the Brownian motion killed on exiting (a, b) (see case (ii)). To prove (14), we observe that

$$P_2(t + s, z, G) = \lambda_{s+t,2}^{x+a}(G) = \lambda_{t+s}^{x+a}(G)$$

$$= \int P(s, z', G)\lambda_t^{x+a}(dz')$$

$$= \int P(s, z', G)\lambda_{t,1}^{x+a}(dz')$$

$$+ \int P(s, z', G)P_2(t, z, dz').$$

Hence to prove (14), it is enough to show

$$\int P(s, z', G)\lambda_{t,1}^{x+a}(dz') = \int P_2(s, z', G)P_1(t, z, dz'). \qquad (15)$$

Let $T : S_2(a) \to S_2(x+a), T(z_1, a) = (z_1 - x, x+a)$. Then as before, $\lambda_{t,1}^{x+a}(\cdot) = P_1(t, z, \cdot) \circ T^{-1}(\cdot)$. Hence,

$$
\int P(s, z', G)\lambda_{t,1}^{x+a}(dz') = \int\limits_{S_2(x+a)} P(s, z', G)\lambda_{t,1}^{x+a}(dz')
$$

$$
= \int\limits_{S_2(x+a)} P(s, z', G)P_1(t, z, \cdot) \circ T^{-1}(dz')
$$

$$
= \int\limits_{S_2(a)} P(s, Tz', G)P_1(t, z, dz').
$$

Finally note that since $G \subseteq S_2$,

$$
P(s, Tz', G) = P_2(s, Tz', G) = P_2(s, z', G)
$$

and eqn (15) follows. The proof for the case $z = (x, b)$ is similar.

Case (vi): Let $z = (x, y) \in S_2$ and $G \subseteq S_3$. Then $P(t, z, G) = 0$ for all t and further

$$
\int P(s, z', G)P(t, z, dz') = \int\limits_{S_3} P(s, z', G)P(t, z, dz') = 0.
$$

This completes the proof of Theorem 2. $\qquad\qquad\qquad\qquad\qquad\qquad$ □

References

1. B. Maisonneuve: Systèmes Régénératifs. Astérisque, 15, 1974.
2. B. Maisonneuve: Exit Systems. The Annals of Probability, Vol.3, No.3, 1975.
3. P.A. Meyer, R.T. Smythe and J.L. Walsh: Birth and death of Markov processes, Proc. Sixth Berkely Sympos., Vol.3, Univ. of California press, Berkeley, California, 1973.
4. A.O. Pittenger and C.T. Shih: Coterminal families and the strong Markov property, Transactions of AMS, Vol.182, August 1973.
5. B. Rajeev: First Order Calculus and Last Entrance times, Sém. de Prob. XXX, LNM1626, Springer.
6. D. Revuz and M. Yor: Continuous Martingales and Brownian motion, Springer Verlag, Berlin, 1991.

Two Recursive Decompositions of Brownian Bridge Related to the Asymptotics of Random Mappings

David Aldous and Jim Pitman [*]

Department of Statistics, University of California, 367 Evans Hall # 3860,
Berkeley, CA 94720-3860, Canada
e-mail: aldous@stat.berkeley.edu, pitman@stat.berkeley.edu

Summary. Aldous and Pitman (1994) studied asymptotic distributions as $n \to \infty$, of various functionals of a uniform random mapping of the set $\{1, \ldots, n\}$, by constructing a *mapping-walk* and showing these random walks converge weakly to a reflecting Brownian bridge. Two different ways to encode a mapping as a walk lead to two different decompositions of the Brownian bridge, each defined by cutting the path of the bridge at an increasing sequence of recursively defined random times in the zero set of the bridge. The random mapping asymptotics entail some remarkable identities involving the random occupation measures of the bridge fragments defined by these decompositions. We derive various extensions of these identities for Brownian and Bessel bridges, and characterize the distributions of various path fragments involved, using the Lévy–Itô theory of Poisson processes of excursions for a self-similar Markov process whose zero set is the range of a stable subordinator of index $\alpha \in (0, 1)$.

Key words: Brownian bridge, Brownian excursion, local time, occupation measure, stable subordinator, self-similar Markov process, Bessel process, path decomposition, Poisson–Dirichlet distribution, pseudo-bridge, random mapping, size-biased sampling, weak convergence, exchangeable interval partition.

1 Introduction

In a previous paper [1] we showed how features of a uniformly distributed random mapping M_n, from $[n] := \{1, 2, \ldots, n\}$ to itself, could be encoded as functionals of a particular non-Markovian random walk on the non-negative integers. This *mapping-walk*, suitably rescaled, converges weakly in $C[0, 1]$ as $n \to \infty$ to the distribution of the reflecting Brownian bridge defined by the absolute value of a *standard Brownian bridge* B^{br} with $B_0^{\mathrm{br}} = B_1^{\mathrm{br}} = 0$ obtained by conditioning a standard Brownian motion B on $B_1 = 0$. Two important

---

[*] Research supported in part by N.S.F. Grants DMS-9970901 and DMS-0071448

features of a mapping are the vector of sizes of connected components of its digraph, and the vector of sizes of cycles in its digraph. Results of [1] imply that for a uniform random mapping, as $n \to \infty$, the component sizes rescaled by n, jointly with corresponding cycle sizes rescaled by $\sqrt{n}$, converge in distribution to a limiting bivariate sequence of random variables $(\lambda_{I_j}, L^0_{I_j})_{j=1,2,\dots}$ where $(I_j)_{j=1,2,\dots}$ is a random interval partition of $[0, 1]$, with λ_{I_j} the length of I_j and $L^0_{I_j}$ the increment of local time of B^{br} at 0 over the interval I_j. With the convention for ordering connected components of the mapping digraph used in [1], the limiting interval partition is $(I_j) = (I^D_j)$, according to the following definition. Here, and throughout the paper, U, U_1, U_2, ... denotes a sequence of independent uniform $(0, 1)$ variables, independent of B^{br}.

Definition 1 (the D-partition [1]). *Let $I^D_j := [D_{V_{j-1}}, D_{V_j}]$ where $V_0 = D_{V_0} = 0$ and V_j is defined inductively along with the D_{V_j} for $j \geqslant 1$ as follows: given that D_{V_i} and V_i have been defined for $0 \leqslant i < j$, let*

$$V_j := D_{V_{j-1}} + U_j(1 - D_{V_{j-1}}),$$

so V_j is uniform on $[D_{V_{j-1}}, 1]$ given B^{br} and (V_i, D_{V_i}) for $0 \leqslant i < j$, and let

$$D_{V_j} := \inf\{t \geqslant V_j : B^{\mathrm{br}}_t = 0\}.$$

On the other hand, a variation of the main result of [1] shows that with a different ordering convention, the mapping component sizes rescaled by n, jointly with their cycle sizes rescaled by $\sqrt{n}$, have a limit distribution specified by the sequence of lengths and Brownian local times $(\lambda_{I_j}, L^0_{I_j})_{j=1,2,\dots}$ a differently defined limiting interval partition. This is the partition $(I_j) = (I^T_j)$ defined as follows using the local time process $(L^0_u, 0 \leqslant u \leqslant 1)$ of B^{br} at 0:

Definition 2 (the T-partition). *Let $I^T_j := [T_{j-1}, T_j]$ where $T_0 := 0$, $\widehat{V}_0 := 0$, and for $j \geqslant 1$*

$$\widehat{V}_j := 1 - \prod_{i=1}^{j}(1 - U_i), \tag{1}$$

so $\widehat{V}_j$ is uniform on $[\widehat{V}_{j-1}, 1]$ given B^{br} and $(\widehat{V}_i, T_i)$ for $0 \leqslant i < j$, and

$$T_j := \inf\{u : L^0_u/L^0_1 > \widehat{V}_j\}.$$

For each of these two random interval partitions (I_j) we are interested in the distribution of the bivariate sequence of lengths and local times $(\lambda_{I_j}, L^0_{I_j})_{j=1,2,\dots}$ and the distribution of the associated path fragments $B^{\mathrm{br}}[I_j]$ and standardized fragments $B^{\mathrm{br}}_*[I_j]$. Here for a process $X := (X_t, t \in J)$ parameterized by an interval J, and $I = [G_I, D_I]$ a subinterval of J with length

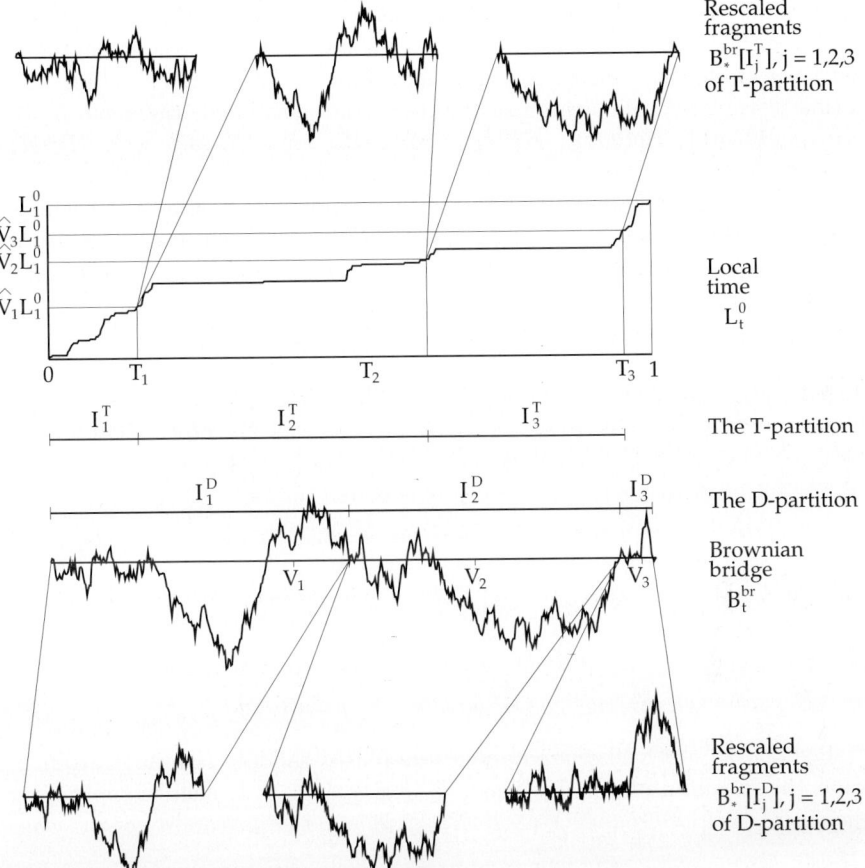

Fig. 1. The two interval partitions.

$\lambda_I := D_I - G_I > 0$, we denote by $X[I]$ or $X[G_I, D_I]$ the *fragment of X on I*, that is the process

$$X[I]_u := X_{G_I+u} \qquad (0 \leqslant u \leqslant \lambda_I). \tag{2}$$

We denote by $X_*[I]$ or $X_*[G_I, D_I]$ the *standardized fragment of X on I*, defined by the *Brownian scaling operation*

$$X_*[I]_u := \frac{X[I]_{u\lambda_I}}{\sqrt{\lambda_I}} := \frac{X_{G_I+u\lambda_I}}{\sqrt{\lambda_I}} \qquad (0 \leqslant u \leqslant 1). \tag{3}$$

Figure 1 illustrates these definitions for a typical path of $X = B^{\mathrm{br}}$. Note that the first interval I_1^D of the D-partition ends at the time D_{U_1}

of the first zero of B^{br} after a uniform$(0,1)$-distributed time U_1, whereas the first interval I_1^T of the T-partition ends at the time T_1 when the local time of B^{br} at 0 has reached a uniform$(0,1)$-distributed fraction of its ultimate value. As illustrated in Figure 1, the associated fragments of B^{br} are qualitatively different: $B_*^{\mathrm{br}}[I_1^D]$ ends with an excursion while $B_*^{\mathrm{br}}[I_1^T]$ does not.

Despite this difference between the fragments of B^{br} over the D- and T-partitions, the random mapping asymptotics have the following corollary. Let $(I_{(j)}^D)$ and $(I_{(j)}^T)$ denote the length-ranked D-partition and the length-ranked T-partition respectively, meaning $I_{(j)}^D$ is the jth longest interval in the D-partition, and $I_{(j)}^T$ is the jth longest interval in the T-partition.

Theorem 3. *Considering the four bivariate sequences* $(\lambda_{I_j}, L_{I_j}^0)_{j=1,2,\dots}$ *of lengths and bridge local times at 0, for* (I_j) *one of the four random interval partitions of* $[0,1]$ *defined by* (I_j^D), $(I_{(j)}^D)$, (I_j^T) *or* $(I_{(j)}^T)$,

(i) *the bivariate sequence has the same distribution for* $(I_{(j)}^D)$ *as for* $(I_{(j)}^T)$;

(ii) *the bivariate sequence for* (I_j^D) *is the bivariate sequence for* $(I_{(j)}^D)$ *in a length-biased order;*

(iii) *the bivariate sequence for* (I_j^T) *is the bivariate sequence for* $(I_{(j)}^T)$ *in an* L^0*-biased order;*

(iv) *the sequence of local times* $(L_{I_j}^0)$ *has the same distribution for* (I_j^D) *as for* (I_j^T), *whereas the sequence of lengths* (λ_{I_j}) *does not.*

See [25, 26] for background about size-biased random orderings. To illustrate the meaning of (iii) for instance, for each $k \geq 1$, conditionally given the entire bivariate sequence $(\lambda_{I_{(j)}^T}, L_{I_{(j)}^T}^0)_{j=1,2,\dots}$, the probability of the event $(I_1^T = I_{(k)}^T)$ is $L_{I_{(k)}^T}^0 / L_1^0$, where $L_1^0 = \sum_j L_{I_{(j)}^T}^0$ almost surely. And given also $(I_1^T = I_{(k)}^T)$, for each $m \geq 1$ with $m \neq k$ the probability of the event $(I_2^T = I_{(m)}^T)$ is $L_{I_{(m)}^T}^0 / (L_1^0 - L_{I_1^T}^0)$ and so on. Put another way, parts (i)-(iii) of the corollary state that the bivariate sequence $(\lambda_{I_j}, L_{I_j}^0)_{j=1,2,\dots}$ for $(I_j) = (I_j^D)$ is distributed like a length-biased rearrangement of the bivariate sequence for $(I_j) = (I_j^T)$, which is in turn distributed like an L^0-biased rearrangement of the bivariate sequence for $(I_j) = (I_j^D)$. Consequently, the distribution of any one of the four bivariate sequences determines the distribution of each of the others.

The rest of this paper is organized as follows. Section 2 explains how we discovered Theorem 3 by consideration of random mapping asymptotics. We recall the theorem from [1] which describes the asymptotics of mapping-walks in terms of the fragments of B^{br} defined by the D-partition, and present the companion result, for a different ordering of components, where the limit involves the fragments of B^{br} defined by the T-partition. Section 3 lays out our results regarding the decomposition of B^{br} into path fragments associated with the D- and T-partitions, in a way which does not depend on the random mapping asymptotics. In particular, we describe the three different distributions of bivariate sequences featuring in the three parts of Theorem 3. We formulate and prove these results more generally, for B^{br} the standardized bridge of a recurrent self-similar Markov process B whose inverse local time process at 0 is a stable subordinator of index α for some $\alpha \in (0,1)$. So $\alpha = 1/2$ for B a standard Brownian motion as supposed in previous paragraphs, and $\alpha = 1 - \delta/2$ for B a Bessel process of dimension $\delta \in (0,2)$. Some of the results in Section 3, like Theorem 3, can be viewed in the Brownian case as asymptotic counterparts (under weak convergence of mapping-walks) of some combinatorial symmetries of random mappings, discussed in Section 2.2. Other results in the Brownian case, especially those involving the method of Poissonization by random scaling [34, 35], are not obvious from the combinatorial perspective, but provide explicit limit distributions for functionals of uniform random mappings. See also [3] where we apply this method to characterize the asymptotic distribution of the diameter of the digraph of a uniform mapping. Sections 4 and 5 provide some proofs and further details of the main results in Section 3, while Section 6 contains various complements. In particular, we show in Section 6.2 that Theorem 3 holds even more generally for interval partitions (I_j^D) and (I_j^T) defined as before, but with the random zero set of B^{br} replaced by the complement of $\bigcup_j I_j^{ex}$, where I_j^{ex} is any exchangeable random partition of $[0,1]$ into an infinite number of intervals, and $(L_u^0, 0 \leqslant u \leqslant 1)$ is the associated local time process, as defined by Kallenberg [16]. This is the limiting case of a corresponding result for a finite exchangeable interval partition of $[0,1]$, which we prove by a combinatorial argument.

In companion papers [2] and [5] we show that Brownian bridge asymptotics apply for models of random mappings more general than the uniform model, in particular for the *p-mapping* model [24, 29], and that proofs can be simplified by use of Joyal's bijection between mappings and trees. See also [30] for a recent review of the applications of Brownian motion and Poisson processes to the asymptotics of various kinds of large combinatorial objects, including partitions, trees, graphs, permutations, and mappings.

2 Random Mappings

In this section we explain how study of random mappings led us to consideration of the two interval partitions of Brownian bridge, and show how the distributions of path fragments of the bridge defined by these partitions encode various asymptotic distributions for mappings.

2.1 Mapping-walks and the two orderings

A mapping $M_n : [n] \to [n]$ can be identified with its digraph of edges $\{(i, M_n(i)), i \in [n]\}$. The connection between random mappings and Brownian bridge developed in [1] can be summarized as follows.

- A mapping digraph can be decomposed as a collection of *rooted trees* together with extra structure (*cycles, basins of attraction*).
- A rooted tree can be coded as a discrete *tree-walk*, a walk excursion starting and ending at 0.
- Given some ordering of tree-components, one can concatenate walk-excursions to define a discrete *mapping-walk* which codes M_n.
- For a uniform random mapping, the induced distribution on tree-components is such that the tree-walks, suitably normalized, converge to Brownian excursion as the tree size increases to infinity.
- So for a uniform random mapping, we expect the mapping-walks, suitably normalized, to converge to a limit process defined by some concatenation of Brownian excursions.
- With appropriate choice of ordering, the limit process is in fact reflecting Brownian bridge.

We now amplify this summary, emphasizing the only subtle issue – the choice of ordering. Fix a mapping M_n. It has a set of *cyclic points*

$$\mathcal{C}_n := \big\{i \in [n] : M_n^k(i) = i \text{ for some } k \geqslant 1\big\},$$

where M_n^k is the kth iterate of M_n. Let $\mathcal{T}_{n,c}$ be the set of vertices of the (perhaps trivial) tree component of the digraph with root $c \in \mathcal{C}_n$. The tree components are bundled by the disjoint cycles $\mathcal{C}_{n,j} \subseteq \mathcal{C}_n$ to form the *basins of attraction* (connected components) of the mapping digraph, say

$$\mathcal{B}_{n,j} := \bigcup_{c \in \mathcal{C}_{n,j}} \mathcal{T}_{n,c} \supseteq \mathcal{C}_{n,j} \quad \text{with} \quad \bigcup_j \mathcal{B}_{n,j} = [n] \quad \text{and} \quad \bigcup_j \mathcal{C}_{n,j} = \mathcal{C}_n \quad (4)$$

where all three unions are disjoint unions, and the $\mathcal{B}_{n,j}$ and $\mathcal{C}_{n,j}$ are indexed in some way by $j = 1, \ldots, K_n$ say. The construction in [1] encodes the restriction of the digraph of M_n to each tree component $\mathcal{T}_{n,c}$ of size k (that is, with k vertices) by $2k$ steps of a *tree-walk* with increments ± 1 on the non-negative integers. The tree-walk proceeds by a suitable search of the set $\mathcal{T}_{n,c}$, making an excursion which starts at 0 and returns to 0 for the first time after $2k$ steps, after reaching a maximum level $1 + h_n(c)$, where $h_n(c)$ is the maximal height above c of all vertices of the tree $\mathcal{T}_{n,c}$ with root c, that is

$$h_n(c) = \max\{h : \exists i \in [n] \text{ with } M_n^h(i) = c \text{ and } M_n^j(i) \notin \mathcal{C}_n \text{ for } 0 \leqslant j < h\}. \quad (5)$$

It was shown in [4] that as $k \to \infty$, the distribution of the tree-walk for a k-vertex random tree, of the kind contained in the digraph of the uniform random mapping M_n for $k \leqslant n$, when scaled to have $2k$ steps of $\pm 1/\sqrt{k}$ per unit time, converges to the distribution $2B^{\mathrm{ex}}$ for B^{ex} a standard Brownian excursion. Subsequent work [22] shows that the same result holds for a variety of codings of trees as walks. Consequently, any of these codings would serve our purpose in the following definitions.

We now define a mapping-walk (to code M_n) as a concatenation of its tree-walks, to make a walk of $2n$ steps starting and ending at 0 with exactly $|\mathcal{C}_n|$ returns to 0, one for each tree component of the mapping digraph. To retain useful information about M_n in the mapping-walk, we want the definition of the walk to respect the cycle and basin structure of the mapping. Here are two orderings that do so.

Definition 4 (cycles-first ordering). *Fix a mapping M_n from $[n]$ to $[n]$. If M_n has K_n cycles, first put the cycles in increasing order of their least elements, say $c_{n,1} < c_{n,2} < \ldots < c_{n,K_n}$. Let $\mathcal{C}_{n,j}$ be the cycle containing $c_{n,j}$, and let $\mathcal{B}_{n,j}$ be the basin containing $\mathcal{C}_{n,j}$. Within cycles, list the trees around the cycles, as follows. If the action of M_n takes $c_{n,j} \to c_{n,j,1} \to \cdots \to c_{n,j}$ for each $1 \leqslant j \leqslant K_n$, the tree components $\mathcal{T}_{n,c}$ are listed with c in the order*

$$
(\overbrace{c_{n,1,1}, \ldots, c_{n,1}}^{\mathcal{C}_{n,1}}, \overbrace{c_{n,2,1}, \ldots, c_{n,2}}^{\mathcal{C}_{n,2}}, \ldots, \overbrace{c_{n,K_n,1}, \ldots, c_{n,K_n}}^{\mathcal{C}_{n,K_n}}). \tag{6}
$$

The cycles-first mapping-walk *is obtained by concatenating the tree walks derived from M_n in this order. The* cycles-first search *of $[n]$ is the permutation $\sigma : [n] \to [n]$ where σ_j is the jth vertex of the digraph of M_n which is visited in the corresponding concatenation of tree searches.*

Definition 5 (basins-first ordering [1]). *If M_n has K_n cycles, first put the basins $\mathcal{B}_{n,j}$ in increasing order of their least elements, say $1 = b_{n,1} < b_{n,2} < \ldots < b_{n,K_n}$; let $c_{n,j} \in \mathcal{C}_{n,j}$ be the cyclic point at the root of the tree component containing $b_{n,j}$. Now list the trees around the cycles, just as in (6), but for the newly defined $c_{n,j}$ and $c_{n,j,i}$. Call the corresponding mapping-walk and search of $[n]$ the* basins-first mapping-walk *and* basins-first search.

Be aware that the meaning of $\mathcal{B}_{n,j}$ and $\mathcal{C}_{n,j}$ now depends on the ordering convention. Rather than introduce two separate notations for the two orderings, we use the same notation for both, and indicate nearby which ordering is meant. Whichever ordering, the definitions of $\mathcal{B}_{n,j}$ and $\mathcal{C}_{n,j}$ are always linked by $\mathcal{B}_{n,j} \supseteq \mathcal{C}_{n,j}$, and (4) holds.

Let us briefly observe some similarities between the two mapping-walks. For each given basin B of M_n with say b elements, the restriction of M_n to B is encoded in a segment of each walk which equals at 0 at some time, and returns again to 0 after $2b$ more steps. If the basin contains exactly c cyclic points, this walk segment of $2b$ steps will be a concatenation of c excursions away from 0. Exactly where this segment of $2b$ steps appears in the mapping-walk depends on the ordering convention, as does the ordering of excursions away from 0 within the segment of $2b$ steps. However, many features of the action of M_n on the basin B are encoded in the same way in the two different stretches of length $2b$ in the two walks, despite the permutation of excursions. One example is the number of elements in the basin whose height above the cycles is h, which is encoded in either walk as the number of upcrossings from h to $h+1$ in the stretch of walk of length $2b$ corresponding to that basin.

2.2 Symmetry properties of random mappings

We now apply the definitions above to a uniform random mapping M_n. Of course, the random partition $\{\mathcal{B}_{n,j}\}_{j=1,\dots,K_n}$ of $[n]$, and the random partition $\{\mathcal{C}_{n,j}\}_{j=1,\dots,K_n}$ of $\mathcal{C}_n$, are the same no matter which ordering convention is used. Each random partition is *exchangeable*, meaning its distribution is invariant under the action of a permutation of $[n]$. Let us spell out some further symmetry properties, each of which turns out to have some analog in the limiting Brownian scheme.

(a) The cycles-first ordering has the following very strong symmetry property: conditionally given $|\mathcal{C}_n| = m$ the tree components in cycles-first ordering form an exchangeable sequence of m random subsets of $[n]$; moreover this exchangeable sequence is independent of the sequence of cycle sizes $|\mathcal{C}_{n,j}|$ with $\sum_j |\mathcal{C}_{n,j}| = m$. Consequently, given $|\mathcal{C}_n| = m$, the cycles-first mapping-walk is a concatenation of m exchangeable excursions away from 0, and this mapping-walk is independent of $|\mathcal{C}_{n,j}|, j = 1, 2, \dots, K_n$.

(b) The basins-first ordering does not share the symmetry property above. But it has a different one: given that the basin $\mathcal{B}_{n,1}$ containing 1 has size $|\mathcal{B}_{n,1}| = b$, the action of M_n on $[n] - \mathcal{B}_{n,1}$ is that of a uniform random mapping of a set of $n - b$ elements. So given $|\mathcal{B}_{n,1}| = b$, the basins-first mapping-walk decomposes after $2b$ steps into two independent segments: the first $2b$ steps are distributed like the basins-first walk for a uniform mapping of $[b]$ conditioned to have a single basin, and the remaining $2(n - b)$ steps distributed like the basins-first walk associated with a uniform mapping of $[n - b]$.

(c) The sequence of basin sizes $(|\mathcal{B}_{n,j}|, 1 \leqslant j \leqslant K_n)$ does not have the same distribution for both orderings. For instance, if $|\mathcal{B}_{n,1}| = 1$ in the basins-first ordering then $|\mathcal{B}_{n,1}| = 1$ in the cycles-first ordering, but (for $n \geqslant 3$) not conversely. So the distribution of $|\mathcal{B}_{n,1}|$ must be different in the two orderings.

(d) For a given mapping M_n, the sequence of cycle sizes $(|\mathcal{C}_{n,j}|, 1 \leqslant j \leqslant K_n)$ may be different for the two different orderings. But for M_n with uniform distribution on $[n]^{[n]}$, the two sequences of cycle sizes have the same distribution: given $|\mathcal{C}_n| = m$, either sequence is distributed like the sizes of cycles of a uniform random permutation of $[m]$ in the (size-biased) order of least elements of the cycles. That is to say, given $|\mathcal{C}_n| = m$, the distribution of $|\mathcal{C}_{n,1}|$ is uniform on $[m]$; given $|\mathcal{C}_n| = m$ and $|\mathcal{C}_{n,1}|$ with $|\mathcal{C}_n| - |\mathcal{C}_{n,1}| = m_1$, the distribution of $|\mathcal{C}_{n,2}|$ is uniform on $[m_1]$, and so on. This is a well known property of uniform random permutations for the cycles-first ordering, and was shown for the basins-first ordering in [1, Lemma 22].

2.3 Brownian asymptotics for the mapping-walks

We now come to the main point of Section 2: the definitions of the interval partitions of Brownian bridge are motivated by the following theorem.

Theorem 6. *The scaled mapping-walk* $(M_u^{[n]}, 0 \leqslant u \leqslant 1)$, *with* $2n$ *steps of* $\pm 1/\sqrt{n}$ *per unit time, for either the cycles-first or the basins-first ordering of excursions corresponding to tree components, converges in distribution to* $2|B^{\mathrm{br}}|$ *jointly with*

$$\frac{|\mathcal{C}_n|}{\sqrt{n}} \xrightarrow{d} L_1^0 \tag{7}$$

where $(L_u^0, 0 \leqslant u \leqslant 1)$ *is the process of local time at 0 of* B^{br}, *normalized so that* $P(L_1^0 > \ell) = e^{-\ell^2/2}$. *Moreover,*
(i) *for the cycles-first ordering, with the cycles* $\mathcal{B}_{n,j}$ *in order of their least elements, these two limits in distribution hold jointly with*

$$\left(\frac{|\mathcal{B}_{n,j}|}{n}, \frac{|\mathcal{C}_{n,j}|}{\sqrt{n}} \right) \xrightarrow{d} \left(\lambda_{I_j}, L_{I_j}^0 \right) \tag{8}$$

as j *varies, where the limits are the lengths and increments of local time of* B^{br} *at 0 associated with the interval partition* $(I_j) := (I_j^T)$; *whereas*
(ii) [1] *for the basins-first ordering, with the basins* $\mathcal{B}_{n,j}$ *listed in order of their least elements, the same is true, provided the limiting interval partition is defined instead by* $(I_j) := (I_j^D)$.

The result for basins-first ordering is part of [1, Theorem 8]. The variant for cycles-first ordering can be established by a variation of the argument in [1], exploiting the exchangeability property of the cycles-first ordering (Section 2.2(a) instead of Section 2.2 (b)). See also [10] and [5] for alternate approaches to the basic result of [1].

We now explain how we first discovered some of the facts about Brownian bridge presented in Theorem 3 by consideration of Theorem 6 and the symmetry properties of Section 2.2. The arguments below are not part of the formal development in this paper. Indeed we show in Section 6 that the results of Theorem 3 hold much more generally, so these results do not really involve much of the rich combinatorial structure of mapping digraphs involved in Theorem 6.

(a) In the basins-first ordering, the first basin is by definition the basin containing element 1, and its walk-segment ends at the first time that the walk returns to 0 after the basins-first search has reached element 1. Suppose we could replace element 1 by a uniform random element, so the walk-segment corresponds asymptotically to the walk-segment ending at the first time of reaching 0 after a uniform random time on $[0, 2n]$. Rescaling, this corresponds to the time interval $[0, D_{V_1}]$ in Definition 1. Of course it is not obvious, and indeed is somewhat counter-intuitive, that replacing element 1 by a uniform random element will preserve length of walk-segment. But this is what eventually will emerge from our calculations.

Now consider the cycles-first ordering. The first basin is by definition the basin containing the smallest-numbered cyclic element $c_{n,1}$, and its walk-segment ends at the first time after reaching element $c_{n,1}$ that the walk returns to 0. Suppose as before (and again this is not obvious) one can replace element $c_{n,1}$ by a uniform random *cyclic* element, so the walk-segment corresponds asymptotically to the walk-segment ending at the first time of reaching 0 after visiting $U^*|\mathcal{C}_n|$ cyclic vertices, where U^* has uniform$[0,1]$ distribution. Rescaling, this corresponds to the time interval $[0, T_1]$ in Definition 2.

(b) The recursive property of the basins-first ordering in Section 2.2(b) plainly corresponds, under the asymptotics of Theorem 6, to the recursive decomposition of Brownian bridge at time D_{V_1} described later in Lemma 8.

(c) In Section 2.2(c) we observed that the distribution of $\mathcal{B}_{n,1}$ was different in the two orderings. This difference persists in the limit: Theorem 6 and the calculation below (26) imply

$$\lim_n n^{-1} E|\mathcal{B}_{n,1}| = \begin{cases} E(D_{V_1}) = 2/3 & \text{(for the basins-first ordering)} \\ E(T_1) = 1/2 & \text{(for the cycles-first ordering).} \end{cases}$$

(d) It is well known [41] that the asymptotic distribution as $n \to \infty$ of the fractions of elements in cycles of a random permutation of $[n]$, with the cycles in order of their least elements, (which amounts to a size-biased random order by exchangeability), is the *uniform stick-breaking sequence* $U_j \prod_{i=1}^{j-1}(1 - U_i)$. So the convergence in distribution (7) of $|\mathcal{C}_n|/\sqrt{n}$ to L_1^0, and the "uniform random permutation" feature of the cyclic decomposition (Section 2.2(d)), combine to show that with *either* ordering $|\mathcal{C}_{n,j}|/\sqrt{n} \xrightarrow{d} L_{I_j}^0$ with the same joint distribution:

$$\left(L_{I_j}^0, j \geq 1 \right) \overset{d}{=} \left(L_1^0 U_j \prod_{i=1}^{j-1} (1 - U_i), \; j \geq 1 \right) \tag{9}$$

for both $I_j = I_j^D$ and $I_j = I_j^T$. This is part (iv) of Theorem 3, which is generalized later by (27) and Theorem 25.

(e) Let $\mathcal{B}_{n,(j)}$ be the jth largest basin of M_n, with some arbitrary convention for breaking ties, and let $\mathcal{C}_{n,(j)}$ be the cycle contained in $\mathcal{B}_{n,(j)}$. It follows immediately from the convergence in distribution (8) that

$$\left(\frac{|\mathcal{B}_{n,(j)}|}{n}, \frac{|\mathcal{C}_{n,(j)}|}{\sqrt{n}} \right) \xrightarrow{d} \left(\lambda_{I_{(j)}}, L_{I_{(j)}}^0 \right) \tag{10}$$

jointly as j varies, where $I_{(j)}$ is the length-ranked interval partition derived from either (I_j^D) or (I_j^T). This is part (i) of Theorem 3. By exchangeability considerations, before passage to the limit the bivariate sequence in (8) as j varies is that in (10) biased by cycle-size in the cycles-first order and biased by basin-size in the basins-first ordering. Hence the conclusions of parts (ii) and (iii) of Theorem 3, by a straightforward passage to the limit.

(f) Due to Section 2.2(a), it makes no difference to anything if in the cycles-first ordering we replace the ordering within the jth cycle $c_{n,j,1}, c_{n,j,2}, \ldots, c_{n,j}$ by the possibly more natural $c_{n,j}, c_{n,j,1}, c_{n,j,2}, \ldots, c_{n,j,|C_{n,j}|-1}$. But in the basins-first ordering, this innocent looking change would spoil convergence to $2|B^{\mathrm{br}}|$. This is because in the basins-first ordering the tree with root $c_{n,1}$ is the tree containing 1, which is a size-biased choice from the exchangeable random partition of $[n]$ into tree components. As such, it tends to be a big tree. In fact, results from [1] imply that, if the mapping-walk is started by the excursion coding the tree rooted at $c_{n,1}$, the limit process will start with a zero free interval whose length is distributed as $D_U - G_U$ in Lemma 9 below for $\alpha = 1/2$. Such a process is obviously not $2|B^{\mathrm{br}}|$ or any other familiar Brownian process.

(g) The proof of Theorem 6 yields more information about the asymptotic sizes of tree components than can be deduced from the statement of that theorem. For instance, if $|\mathcal{T}_{n,(i)}|$ are the ranked sizes of the tree components of M_n, and $H_{n,i}$ are the corresponding maximal tree heights, as in (5), then $(|\mathcal{T}_{n,(i)}|/n, H_{n,i}/\sqrt{n})_{i=1,2,\ldots}$ converges in distribution to the sequence of ranked lengths and corresponding maximal heights of excursions of $2|B^{\mathrm{br}}|$, whose distribution was described in [35, Theorem 1 and Example 8]. If only the tree components of $\mathcal{B}_{n,j}$ were considered, the limit would be derived from excursions of B^{br} over the appropriate random interval I_j as in Theorem 6, with joint convergence as j varies.

3 The bridge decompositions

This section presents our main results for the D- and T-partitions. For ease of comparison, the results are presented together here, with outlines of the proofs. Some proofs and further details are deferred to Section 4 for the D-partition, and to Section 5 for the T-partition. Our primary interest is the analysis of the D- and T-partitions derived from a standard Brownian bridge, and the connections between these random partitions and the asymptotics of random mappings discussed in Section 2. But we find that our analysis applies just as well to the D- and T-partitions for a standardized bridge B^{br} derived from B a recurrent self-similar Markov process whose inverse local time process at 0 is a stable subordinator of index α for some $\alpha \in (0,1)$. Readers who don't care about this generalization can assume throughout this section that B is standard one-dimensional Brownian motion, and $\alpha = \beta = 1/2$.

3.1 General framework

Following Pitman–Yor [35, §2], we make the following basic assumptions:

- $B := (B_t, t \geqslant 0)$ is a real or vector-valued strong Markov process, started at $B_0 = 0$, with state space a cone contained in $\mathbb{R}^d$ for some $d = 1, 2, \ldots$, and càdlàg paths.
- B is β-self-similar for some real β. That is to say, if $B_*[0, t]$ now denotes the standardized process derived from B on $[0, t]$ as in (3), using λ_I^β instead of $\sqrt{\lambda_I}$ in the denominator, then $B_*[0, t] \overset{d}{=} B[0, 1]$ for all $t > 0$.
- The point 0 is a regular recurrent point for B, meaning that almost surely both 0 and ∞ are points of accumulation of the zero set of B.

As a well known consequence of these assumptions [14, 35], there exists a continuous local time process for B at 0, say $(L_t^0(B), t \geqslant 0)$, whose inverse process

$$\tau_\ell := \inf\{t : L_t^0(B) > \ell\} \qquad (\ell \geqslant 0)$$

is a stable subordinator of index α for some $\alpha \in (0, 1)$. That is

$$E \exp(-\xi \tau_\ell) = \exp(-\ell c \xi^\alpha) \qquad (\xi \geqslant 0) \tag{11}$$

for some $c > 0$, in which case

$$L_t^0(B) = \frac{\Gamma(1-\alpha)}{c} \lim_{\varepsilon \to 0} \varepsilon^\alpha N_{t,\varepsilon}(B) \tag{12}$$

uniformly for bounded t almost surely, where $N_{t,\varepsilon}(B)$ is the number of excursion intervals of B in $[0, t]$ whose length is greater than ε. Formula (12) can be then used with X instead of B to define $L_t^0(X)$ for various other processes X derived from B by conditioning or scaling, such as the standardized bridge B^{br} introduced in the next paragraph. As a consequence of (12) with X instead of B, there is following basic α-scaling rule for such local time processes: for $I = [G_I, D_I]$ a random subinterval of length $\lambda_I := D_I - G_I$ contained in the time domain of X, and $L_I^0(X) := L_{D_I}^0(X) - L_{G_I}^0(X)$,

$$L_I^0(X) = \lambda_I^\alpha L_1^0(X_*[I]). \tag{13}$$

Associated with the self-similar Markov process B are corresponding distributions of a *standard B-bridge* B^{br}, a *standard B-excursion* B^{ex}, and a *standard B-meander* B^{me}, defined by the following identities in distribution, valid for all $t > 0$:

$$B_*[0, G_t] \overset{d}{=} B^{\mathrm{br}}; \qquad B_*[G_t, D_t] \overset{d}{=} B^{\mathrm{ex}}; \qquad B_*[G_t, t] \overset{d}{=} B^{\mathrm{me}} \tag{14}$$

where $G_t := G_t(B)$, $D_t := D_t(B)$, and for any process X we use the notation

$$G_t(X) := \sup\{u < t : X_u = 0\} \tag{15}$$
$$D_t(X) := \inf\{u \geqslant t : X_u = 0\}. \tag{16}$$

See [9] for a review of properties of B^{br}, B^{ex} and B^{me} in the *Brownian case* when B is Brownian motion with state space $\mathbb{R}$, and $\beta = \alpha = 1/2$. See [25, §3] and [34] for some treatment of B^{br} and B^{ex} in the *Bessel case* when B with state space $\mathbb{R}_{\geqslant 0}$ is a recurrent Bessel process of dimension $\delta = 2 - 2\alpha \in (0,2)$, and $\beta = 1/2$. Other examples are provided by recurrent stable Lévy processes [8], symmetrized or skew Bessel processes [42], and Walsh processes [6, 7].

According to the Lévy–Itô theory of excursions of B, applied to the standard B-bridge as in [31, 35], if (I_j^{ex}) is the interval partition of $[0,1]$ defined by the excursion intervals of B^{br} in length-ranked order, then the processes $B_*^{\mathrm{br}}[I_j^{\mathrm{ex}}]$ are i.i.d. copies of B^{ex}, independent of (I_j^{ex}), which is an *exchangeable interval partition* in the sense of [16] recalled in Section 6.2. Moreover, the distribution of ranked lengths $(\lambda_{I_j^{\mathrm{ex}}})$ depends only on α, as described in [32, (16)] and [35, Example 8]. This general excursion decomposition of B^{br} implies that various results known for Bessel bridges hold also in the present general setting, and we take this for granted without further comment.

3.2 Main Results

All results of this section are presented with the notation and general framework of the previous section: B^{br} is the standard B-bridge derived from a self-similar recurrent Markov process B whose continuous local time process $(L_t^0(B), t \geqslant 0)$ is the inverse of a stable subordinator $(\tau_\ell, \ell \geqslant 0)$ of index $\alpha \in (0,1)$. The D- and T-partitions are defined in terms of B^{br} and its local time process at 0, according to Definitions 1 and 2.

Theorem 3, presented in the introduction in the Brownian case, is true in the more general framework of this section, as a consequence of the following theorem:

Theorem 7. *For a random interval $I \subseteq [0,1]$, let μ_I denote the random occupation measure induced by the path of $B^{\mathrm{br}}[I]$, so for each Borel subset A of the state space of B^{br}*

$$\mu_I(A) := \int_I 1(B_t^{\mathrm{br}} \in A)\,\mathrm{d}t.$$

(i) *The sequence of occupation measures (μ_{I_j}) has the same distribution for each of the two length-ranked partitions $(I_j) = (I_{(j)}^D)$ and $(I_j) = (I_{(j)}^T)$.*

(ii) *For $(I_j) = (I_j^D)$ the sequence of occupation measures (μ_{I_j}) is in λ-biased order, where λ_{I_j} is the total mass of the random measure μ_{I_j}.*

(iii) *For $(I_j) = (I_j^T)$ the sequence of occupation measures (μ_{I_j}) is in L^0-biased order, where $L_{I_j}^0 := L_{I_j}^0(B^{\mathrm{br}})$.*

(iv) *For (I_j) equal to any one of the four interval partitions (I_j^D), $(I_{(j)}^D)$, (I_j^T) or $(I_{(j)}^T)$, conditionally given $\lambda_{I_j} = \lambda_j$ and $L_{I_j}^0 = \ell_j$ for all $j = 1, 2, \ldots$, the random occupation measures μ_{I_j}, $j = 1, 2, \ldots$ are independent, with μ_{I_j}*

distributed like the random occupation measure of a process with the common conditional distribution of

$$(B[0,t] \mid B_t = 0, L_t^0 = \ell) \overset{d}{=} (B[0,\tau_\ell] \mid \tau_\ell = t) \tag{17}$$

for $t = \lambda_j$ and $\ell = \ell_j$.

Proof. Propositions 10, 11 and 14 provide more explicit descriptions of the law of $(\lambda_{I_j}, L_{I_j}^0, B_*^{\mathrm{br}}[I_j])_{j=1,2,\dots}$, for each of the four interval partitions (I_j). The above results for occupation measures are deduced from these propositions using Lemma 13. The fundamental *switching identity* (17) is well known [31, §5]. □

By general theory of local time processes for diffusions or continuous semi-martingales [15, 39, 38], in the Brownian and Bessel cases for each random subinterval I of $[0,1]$ the random occupation measure μ_I derived from B^{br} has an almost surely continuous density L_I^x relative to m at x, where m is a multiple of the speed measure of the one-dimensional diffusion B. To be precise about normalization of local times, in the Brownian case with state space $\mathbb{R}$, we take $m(\mathrm{d}x) = \mathrm{d}x$, so that (11) holds with $\alpha = 1/2$ and $c = \sqrt{2}$. In the Bessel(δ) case with state space $\mathbb{R}_{\geq 0}$, we take $m(\mathrm{d}x) = 2x^{\delta-1}\,\mathrm{d}x$, so that (11) holds with $\alpha = 1 - \delta/2$ and $c = 2^{1-\alpha}\Gamma(1-\alpha)/\Gamma(\alpha)$, by [31, (7.c)]. In either case, $L_{I_j}^0$ in (iii) and (iv) is recovered like λ_{I_j} as a measurable function of the random occupation measure μ_{I_j}. The distribution of the local time density of the conditional occupation measure in (iv) is described by a conditional form of the Ray–Knight theorem: see [20, 28] for details in the Brownian case.

Our analysis of the D-partition is the following expression of the decomposition of B^{br} at the times D_{V_j}, implicit in [1] in the Brownian case:

Lemma 8 ([1]). *For each j, the pre-D_{V_j} fragment of the bridge $B^{\mathrm{br}}[0, D_{V_j}]$ is independent of the standardized post-D_{V_j} fragment $B_*^{\mathrm{br}}[D_{V_j}, 1]$, which has the same distribution as B^{br}.*

This is easily verified, because the D_{V_j} are stopping times relative to a filtration with respect to which B^{br} has a strong Markov property.

To describe various distributions, let $(\Gamma_s, s \geq 0)$ denote a *standard gamma process*, that is the increasing Lévy process with marginal densities

$$P(\Gamma_s \in \mathrm{d}x)/\mathrm{d}x = \frac{1}{\Gamma(s)}\, x^{s-1}\, e^{-x} \qquad (x > 0), \tag{18}$$

so $\Gamma_t - \Gamma_s \overset{d}{=} \Gamma_{t-s}$ for $0 < s < t$. Recall that for $a, b > 0$ the beta(a, b) distribution is that of

$$\beta_{a,b} := \Gamma_a/\Gamma_{a+b}, \text{ which is independent of } \Gamma_{a+b}, \text{ with} \tag{19}$$

$$P(\beta_{a,b} \in du) = \frac{\Gamma(a+b)}{\Gamma(a)\Gamma(b)} u^{a-1}(1-u)^{b-1} du \qquad (0 < u < 1). \qquad (20)$$

It is well known [25, Lemma 3.7] that for $G_t = G_t(B)$,

the standard B-bridge $B_*[0, G_t]$ is independent of G_t with $G_t/t \overset{d}{=} \beta_{\alpha,1-\alpha}.$
$$\qquad (21)$$

Lemma 9 ([1, Prop. 2], [27, Prop. 15]). *Let U with uniform$[0,1]$ distribution be independent of B^{br}, and let $G_U := G_U(B^{\mathrm{br}})$, $D_U := D_U(B^{\mathrm{br}})$. Then*

$$(G_U, D_U - G_U, 1 - D_U) \overset{d}{=} (\Gamma_\alpha, \Gamma_1 - \Gamma_\alpha, \Gamma_{1+\alpha} - \Gamma_1)/\Gamma_{1+\alpha}.$$

Moreover, the random vector $(G_U, D_U - G_U, 1 - D_U)$ and the three standardized processes $B^{\mathrm{br}}_*[0, G_U]$, $B^{\mathrm{br}}_*[G_U, D_U]$ and $B^{\mathrm{br}}_*[D_U, 1]$ are independent, with

$$B^{\mathrm{br}}_*[0, G_U] \overset{d}{=} B^{\mathrm{br}}_*[D_U, 1] \overset{d}{=} B^{\mathrm{br}} \qquad and \qquad B^{\mathrm{br}}_*[G_U, D_U] \overset{d}{=} B^{\mathrm{ex}}. \qquad (22)$$

Proposition 10. *For the D-partition*
(i) *the sequence of lengths is such that*

$$\lambda_{I^D_j} = W_j \prod_{i=1}^{j-1}(1 - W_i) \qquad (23)$$

for a sequence of independent random variables W_j with $W_j \overset{d}{=} \beta_{1,\alpha}$.
(ii) *The corresponding sequence of local times at 0 can be expressed as*

$$L^0_{I^D_j} = \lambda^\alpha_{I^D_j} L^0_1(B^{\mathrm{br}}_*[I^D_j]) \qquad (24)$$

where the $L^0_1(B^{\mathrm{br}}_*[I^D_j])$ are independent random variables, independent also of the lengths $\lambda_{I^D_j}$, with

$$L^0_1(B^{\mathrm{br}}_*[I^D_j]) \overset{d}{=} L^0_1(B) \overset{d}{=} \tau_1^{-\alpha} \qquad (25)$$

for τ_1 with the stable distribution of index α defined by (11).
(iii) *The standardized path fragments $B^{\mathrm{br}}_*[I^D_j]$ are independent and identically distributed like $B^{\mathrm{br}}_*[0, D_U]$, and independent of the sequence of lengths $(\lambda_{I^D_j})$.*
(iv) *For the length-ranked D-intervals $I^D_{(j)}$ instead of I^D_j, the lengths $(\lambda_{I^D_{(j)}})$ have the Poisson–Dirichlet(α) distribution defined by ranking $(\lambda_{I^D_j})$ as in (i), while parts (ii) and (iii) hold without change.*

Proof. Parts (i)-(iii) are obtained by repeated application of Lemmas 8 and 9, using the α-scaling rule (13) for local times and (21), as in [25, Lemma 3.11], for part (ii). The second identity in (ii) is a well-known consequence of the inverse relation between $(L^0_t(B), t \geqslant 0)$ and $(\tau_\ell, \ell \geqslant 0)$, as discussed in [31]. Part (iv) follows immediately from (i)-(iii). $\qquad\square$

See [18, 32] and Lemma 15 for background on the Poisson–Dirichlet distribution appearing in (iv). Lévy [21] showed that in the Brownian case the common distribution of $L_1^0(B)$ and $\tau_1^{-1/2}$ appearing in (25) is simply the distribution of $|B_1|$, with B_1 standard Gaussian. But this does not generalize to the Bessel(δ) case for general $\delta = 2 - 2\alpha$. Then $B_1 \overset{d}{=} \sqrt{2\Gamma_{1-\alpha}}$, which is a simple transformation of the stable(α) distribution of τ_1 only for $\alpha = 1/2$.

The difficulty involved in Theorem 7 is that Definition 2 of the T_j involves the local time $L_1^0 := L_1^0(B^{\mathrm{br}})$, which depends on the path of B^{br} over the whole interval $[0, 1]$. While we can describe the finite-dimensional distributions of the bivariate sequence $(\lambda_{I_j^T}, L_{I_j^T}^0)_{j=1,2,\ldots}$ by conditioning on L_1^0 (see Proposition 20), this description is more complicated than our description of $(\lambda_{I_j^D}, L_{I_j^D}^0)_{j=1,2,\ldots}$ in Proposition 10.   In particular,

$$\lambda_{I_1^T} \overset{d}{\neq} \lambda_{I_1^D}. \tag{26}$$

Indeed, by (23) we have

$$E(\lambda_{I_1^D}) = E(W_1) = 1/(1+\alpha) > 1/2,$$

whereas (by symmetry of B^{br} with respect to time reversal in the Brownian or Bessel case) the distribution of T_1 is symmetric about $1/2$, so whatever $\alpha \in (0, 1)$

$$E(\lambda_{I_1^T}) = E(T_1) = 1/2.$$

Still, as explained combinatorially in the Brownian case around (9), the two partitions give rise to the same distribution for the sequence of local times:

$$\left(L_{I_j^D}^0 \right) \overset{d}{=} \left(L_{I_j^T}^0 \right) := \left(L_1^0 U_j \prod_{i=1}^{j-1} (1 - U_i) \right) \tag{27}$$

where the second equality by definition is read from (1). The first equality in distribution of sequences follows from Lemma 8 and the consequence of Lemma 9, noted in [1, (3)-(4)] in the Brownian case, that

$$L_{I_1^D}^0 / L_1^0 \text{ has uniform distribution on } (0,1), \text{ and is independent of } L_1^0. \tag{28}$$

As indicated in Section 6.2, this can also be checked in general using the exchangeability of the excursion interval partition.

According to Proposition 10, the standardized bridge fragments over intervals of the D-partition are i.i.d. copies of $B_*^{\mathrm{br}}[0, D_U]$, both for the intervals in their original order and for the intervals in length-ranked order. A subtle feature of the T-partition is that the standardized bridge fragments over its intervals are neither independent nor identically distributed in their original order, but these fragments become i.i.d. when put into length-ranked order. This and other parallels between the T- and D-partitions in length-ranked order are presented in the following Proposition:

Proposition 11. *For the T-partition in length-ranked order*
(i) *the sequence of lengths $(\lambda_{I^T_{(j)}})$ has the same Poisson–Dirichlet(α) distribution as $(\lambda_{I^D_{(j)}})$.*
(ii) *The corresponding sequence of local times at 0 can be expressed as*

$$L^0_{I^T_{(j)}} = \lambda^\alpha_{I^T_{(j)}} \, L^0_1\big(B^{\mathrm{br}}_*[I^T_{(j)}]\big) \tag{29}$$

where the $L^0_1(B^{\mathrm{br}}_*[I^T_{(j)}])$ are independent random variables, independent also of the lengths $L^0_{I^T_{(j)}}$, with

$$L^0_1\big(B^{\mathrm{br}}_*[I^T_{(j)}]\big) \overset{d}{=} L^0_1(B) \overset{d}{=} \tau_1^{-\alpha} \tag{30}$$

just as in (25).
(iii) *The standardized path fragments $B^{\mathrm{br}}_*[I^T_{(j)}]$ are independent and identically distributed like $B_*[0, \tau_1]$, and independent of the sequence of lengths $(\lambda_{I^T_{(j)}})$.*

The only difference between this description of the law of the sequence $(\lambda_{I_j}, L^0_{I_j}, B^{\mathrm{br}}_*[I_j])_{j=1,2,\ldots}$ for $I_j = I^T_{(j)}$, and the previous description in Proposition 10 for $I_j = I^D_{(j)}$, is that the common distribution of the standardized T-fragments is that of $B^{\mathrm{br}}_*[0, D_U]$, whereas the common distribution of the standardized D-fragments is that of $B_*[0, \tau_1]$. The standardized process $B_*[0, \tau_1]$ is known as the *pseudo-bridge* associated with the self-similar Markov process B. The following Lemma was established by Biane, Le Gall and Yor [11] in the Brownian case, and extended to the Bessel case in [31, Theorem 5.3].

Lemma 12 ([11, 31]). *The law of the pseudo-bridge $B_*[0, \tau_1]$ is mutually absolutely continuous with respect to the law of B^{br}, with density proportional to $1/L^0_1(B)$ relative to the law of B^{br}. That is, for all non-negative measurable path functionals F*

$$E\big[F([B_*[0, \tau_1]])\big] = \frac{1}{c\alpha\Gamma(\alpha)} \, E\left[\frac{F(B^{\mathrm{br}})}{L^0_1(B^{\mathrm{br}})}\right].$$

where c is determined by the normalization of local time via (11).

While the laws of $B^{\mathrm{br}}_*[0, D_U]$ and the pseudo-bridge $B_*[0, \tau_1]$ are mutually singular, their random occupation measures have the same distributions. In fact, the sample path of $B^{\mathrm{br}}_*[0, D_U]$ is simply a random rearrangement of the sample path of $B_*[0, \tau_1]$:

Lemma 13. *Let U be a uniform $(0, 1)$ variable independent of B^{br}, and independent of X distributed like $B_*[0, \tau_1]$. Then a process Y distributed like $B^{\mathrm{br}}_*[0, D_U]$ is created by the following rearrangement of the path of X, whereby the random occupation measures of X and Y are pathwise identical: let (G_U, D_U) be the excursion interval of X straddling time U, and let Y be* *derived from X by swapping the order of the path fragments $X[G_U, D_U]$ and $X[D_U, 1]$, say*

$$Y = X[0, G_U] : X[D_U, 1] : X[G_U, D_U] \tag{31}$$

with an obvious notation for concatenation of path fragments.

Proof. By construction, the path of Y ends with a B-excursion of length $1 - G_1(Y) = D_U - G_U$. The joint law of $Y[0, G_1(Y)]$ and $Y[G_1(Y), 1] := X[G_U, D_U]$ was described in [31, Theorem 1.3] and [25, Theorem 3.1 and (3.d)], and is identical to the joint law of $Z[0, G_1(Z)]$ and $Z[G_1(Z), 1]$ for $Z := B_*^{\mathrm{br}}[0, D_U(B^{\mathrm{br}})]$, which can be read from Lemma 9. To be explicit, the common distribution of $Y[0, G_1(Y)]$ and $Z[0, G_1(Z)]$ is that of $B[0, G_1(B)]$ described by (21), while both $Y_*[G_1(Y), 1] := X_*[G_U, D_U]$ and $Z_*[G_1(Y), 1] := B_*^{\mathrm{br}}[G_U(B^{\mathrm{br}}), D_U(B^{\mathrm{br}})]$ are standard B-excursions. Since the excursion is in each case independent of the preceding fragment, it follows that $Y \overset{d}{=} Z$.   □

Proposition 14. *Fix $\xi > 0$. Let G be a random variable independent of B^{br}, with $G \overset{d}{=} \Gamma_\alpha/\xi$. The distributions of the two bivariate sequences, defined by the lengths and bridge local time measures of intervals of the D-partition and the T-partition respectively, are determined as follows:*
(i) For $I_j = I_j^D$ the bivariate sequence

$$\left(G\lambda_{I_j}, G^\alpha L_{I_j}^{\mathrm{br}}\right)_{j=1,2,\dots} \tag{32}$$

is the sequence of points (X_j, Y_j), in X-biased random order, of a Poisson process on $\mathbb{R}_{>0}^2$ with intensity measure

$$\nu(\mathrm{d}t, \mathrm{d}\ell) := \alpha t^{-1} e^{-\xi t} \, \mathrm{d}t \, P(t^\alpha \tau_1^{-\alpha} \in \mathrm{d}\ell) = \ell^{-1} P(\tau_\ell \in \mathrm{d}t) e^{-\xi t} \tag{33}$$

for τ_1 as in (11), which makes

$$\Sigma_j X_j \overset{d}{=} \frac{\Gamma_\alpha}{\xi} \qquad \text{and} \qquad \Sigma_j Y_j \overset{d}{=} \frac{\Gamma_1}{c\xi^\alpha}. \tag{34}$$

(ii) If the points (X_j, Y_j) of a Poisson process with intensity ν on $\mathbb{R}_{>0}^2$ are listed in X-biased order then

$$\left(\lambda_{I_j^D}, L_{I_j^D}^0\right)_{j=1,2,\dots} \overset{d}{=} \left(\frac{X_j}{\Sigma_X}, \frac{Y_j}{\Sigma_X^\alpha}\right)_{j=1,2,\dots} \tag{35}$$

for $\Sigma_X := \sum_j X_j$ as in (34).
(iii) For $I_j = I_j^T$ the bivariate sequence in (32) is the sequence of points, say (X_j', Y_j'), in Y'-biased random order, of another Poisson process on $\mathbb{R}_{>0}^2$ with the same intensity measure ν. So if in (ii) the points (X_j, Y_j) are listed instead in Y-biased order, then (35) holds with the sequence of T-intervals instead of the sequence of D-intervals.

Proof. Part (i) is proved in Section 4. Part (ii) is just a restatement of part (i). Part (iii) is proved in Section 5.   □

Note that the normalization in (35) involves Σ_X and its αth power, both for the D-partition and for the T-partition. Obviously, this is easier to handle if the sampling is X-biased rather than Y-biased, which is one explanation of why various distributions associated with (I_j^D) are simpler than their counterparts for (I_j^T).

4 Analysis of the D-partition

As a preliminary for the proof of Proposition 14 (i), we recall the following well known lemma, which characterizes the distribution of a sequence (Q_j), known as the $GEM(\theta)$ distribution after Griffiths, Engen and McCloskey. The distribution of $(Q_{(j)})$ obtained by ranking (Q_j) is known as the *Poisson–Dirichlet distribution with parameter* θ. See [17], [18, §9.6], [32].

Lemma 15 (characterizations of $GEM(\theta)$ [23, 25, 26]). *Fix $\theta > 0$ and $\xi > 0$. Let G and $Q_j, j = 1, 2, \ldots$ be non-negative random variables. Then the following are equivalent:*
(i) the sequence (Q_j) admits the representation $Q_j = W_j \prod_{i=1}^{j-1}(1 - W_i)$ where the W_j are independent beta$(1, \theta)$ variables, and G is independent of (Q_j) with $G \stackrel{d}{=} \Gamma_\theta/\xi$;
(ii) $\sum_j Q_j = 1$ a.s. and (GQ_j) is the sequence of points of a Poisson point process on $\mathbb{R}_{>0}$ with intensity $\theta t^{-1} e^{-\xi t} \, dt$, listed in size-biased order.

The next well known result [18, §5.2], [37, Prop. 4.10.1], combined with the previous lemma, provides an efficient way to identify various Poisson processes.

Lemma 16 (Poisson marking). *Let $(S, \mathcal{S})$ and $(T, \mathcal{T})$ be two measurable spaces. Let (X_j) and (Y_j) be two sequences of random variables, with values in S and T respectively, such that the counting process $\sum_j 1(X_j \in \cdot)$ is Poisson with intensity measure μ on $\mathcal{S}$, and the Y_j are conditionally independent given (X_j), with*

$$P(Y_j \in \cdot \,|\, X_1, X_2, \ldots) = P'(X_j, \cdot)$$

for some Markov kernel P' from $(S, \mathcal{S})$ to $(T, \mathcal{T})$. Then the counting process $\sum_j 1((X_j, Y_j) \in \cdot)$ is a Poisson process on the product space $S \times T$ with intensity measure $\mu(dx)P'(x, dy)$ on the product σ-field.

Proof of Proposition 14 (i). Proposition 10 (i) and Lemma 15 (i) show that $(\lambda_{I_j^D}, j \geqslant 1)$ has $GEM(\alpha)$ distribution. By assumption, G is independent of this sequence with $G \stackrel{d}{=} \Gamma_\alpha/\xi$. Lemma 15 implies that $(G\lambda_{I_j^D})$ is the size-biased ordering of a Poisson point process of intensity $\alpha t^{-1} e^{-\xi t} \, dt$. Proposition 10 (ii) and Lemma 16 now identify the $(G\lambda_{I_j^D}, G^\alpha L_{I_j^D}^{\mathrm{br}})$ as the points of a Poisson process with intensity measure ν defined by the first expression in (33). To check the equality of the two expressions in (33), let

$$f_\ell(t) := P(\tau_\ell \in dt)/dt. \qquad (36)$$

Since $\tau_\ell \overset{d}{=} \ell^{1/\alpha}\tau_1$ by (11),

$$f_\ell(t) = \ell^{-1/\alpha} f_1(t/\ell^{1/\alpha}) \qquad (37)$$

whereas by another change of variables

$$P(t^\alpha \tau_1^{-\alpha} \in d\ell)/d\ell = \alpha^{-1} t \ell^{-1-1/\alpha} f_1(t/\ell^{1/\alpha}) = \alpha^{-1} t \ell^{-1} f_\ell(t) \qquad (38)$$

and the identity follows. By application of (11), the ℓ-marginal of ν is $\ell^{-1} e^{-c\xi^\alpha \ell} d\ell$. The distribution of $\sum_j Y_j$ is the infinitely divisible law with this Lévy measure, that is the exponential distribution with rate $c\xi^\alpha$. $\square$

Implicit in Lemma 12 and (38) is the following formula of [25, (3.u)] for the density of $L_1^0 := L_1^0(B^{br})$

$$P(L_1^0 \in d\ell) = c\alpha\Gamma(\alpha)\ell P(\tau_1^{-\alpha} \in d\ell) = c\Gamma(\alpha)f_\ell(1)\,d\ell \qquad (39)$$

for $f_\ell(x)$ as in (36) the stable(α) density of τ_ℓ determined by (11). That is to say, the distribution of $L_1^0(B^{br})$ is obtained by size-biasing the common distribution of $\tau_1^{-\alpha}$ and $L_1^0(B)$. In particular, the general formula (39) is consistent with Lévy's well known formulae in the Brownian case [21], with $\alpha = 1/2, c = \sqrt{2}$

$$f_\ell(x) = \frac{\ell}{\sqrt{2\pi}} x^{-3/2} e^{-\frac{1}{2}\ell^2/x} \qquad (40)$$

and

$$P(L_1^0 \in d\ell)/d\ell = \ell e^{-\frac{1}{2}\ell^2}. \qquad (41)$$

For general α, a series expression for $f_\ell(x)$ is known [36, 43, 44, 40]. If $\alpha = 1/n$ for some $n = 2, 3, \ldots$, integral expressions for $f_\ell(x)$ can be derived from a representation of $1/\tau_\ell$ as a product of $n-1$ independent gamma variables [43, Theorem 3.4.3]. To conclude this section, we record the following immediate consequence of Proposition 14 (i):

Corollary 1. *Let (σ_j) be a sequence of i.i.d. copies of τ_1 with the stable (α) distribution (11), and let (Q_j) with $GEM(\alpha)$ distribution of Lemma 15 be independent of (σ_j). Let $L_j := (Q_j/\sigma_j)^\alpha$ and $L := \sum_j L_j$. Then $L \overset{d}{=} L_1^0$ as in (39), and the sequence (L_j/L) has $GEM(1)$ distribution, independently of L.*

5 Analysis of the T-partition

We start by recalling the structure of a Markov process up to the last time it visits its initial state before an independent exponential time. This does not involve the self-similarity assumption.

Lemma 17 ([14]). *Let $(\tau_\ell, \ell \geqslant 0)$ be a drift-free subordinator which is the inverse of the continuous local time process $(L^0_t(B), t \geqslant 0)$ of a regular recurrent point 0, for a strong Markov process B started at 0. Let ε be an exponential variable with rate ξ, with ε independent of B, and let*

$$G := G_\varepsilon(B) \qquad and \qquad L := L^0_G(B) = L^0_\varepsilon(B). \tag{42}$$

(i) *The local time L has exponential distribution with rate $\psi(\xi)$, the Laplace exponent of the subordinator defined by $E(e^{-\xi\tau_\ell}) = e^{-\psi(\xi)\ell}$.*
(ii) *For $\ell > 0$, there is the equality in distribution of path fragments*

$$(B[0, G] \,|\, L = \ell) \overset{d}{=} (B[0, \tau_\ell] \,|\, \tau_\ell < \varepsilon). \tag{43}$$

(iii) *The joint distribution of (G, L) is*

$$P(G \in dt, L \in d\ell) = \psi(\xi) \, d\ell \, e^{-\xi t} P(\tau_\ell \in dt). \tag{44}$$

which is the distribution of the value at time 1 of a drift free bivariate subordinator with Lévy measure

$$\nu(dt, d\ell) = \ell^{-1} \, d\ell \, e^{-\xi t} P(\tau_\ell \in dt) \tag{45}$$

whose ℓ-marginal is the Lévy measure $\ell^{-1} e^{-\psi(\xi)\ell} \, d\ell$ of the exponential distribution of L.

Proof. These results are derived from Itô's theory of excursions of B, by letting $(N_t, t \geqslant 0)$ be a Poisson process with rate ξ, independent of B, and taking ε to be the time of the first point of N. To briefly recall the argument, say that a jump interval (τ_{y-}, τ_y) of the inverse local time process τ is *marked* if $N(\tau_{y-}, \tau_y) > 0$ and *unmarked* otherwise. Then, by basic theory of Poisson point processes, the sum of unmarked jumps

$$\tau^{\mathrm{u}}_\ell := \sum_{0 < y < \ell} (\tau_y - \tau_{y-}) 1(N(\tau_{y-}, \tau_y] = 0) \tag{46}$$

defines a subordinator with distribution

$$P(\tau^{\mathrm{u}}_\ell \in dt) = e^{\psi(\xi)\ell - \xi t} P(\tau_\ell \in dt). \tag{47}$$

The left end G of the first marked interval is $G = \tau_{L-} = \tau^{\mathrm{u}}_L$, and the subordinator τ^{u} summing unmarked jumps of τ is independent of L, the local time of the first marked jump. See also [14, 31, 35, 39]. □

To be more explicit, part (iii) of the Lemma states that

$$(G, L) \stackrel{d}{=} (\Sigma_j X_j, \Sigma_j Y_j) \tag{48}$$

for (X_j, Y_j) the points of a Poisson point process on $\mathbb{R}^2_{>0}$ with intensity measure ν defined by (45). In particular, for a self-similar B as in Section 3, this measure ν is identical to the measure ν featured in (33).

In the setting of Lemma 17, even with construction of the Poisson process of marks of rate ξ independent of B, more randomization is required to construct points (X_j, Y_j) such that (48) holds with equality almost surely rather than just in distribution. But this can be done by the following construction, which is the basis of our proofs of Proposition 11 and Proposition 14 (iii).

Lemma 18. *In the setting of Lemma 17, let $I := [G_I, D_I]$ be a random subinterval of $[0, 1)$, where the endpoint 1 is deliberately excluded, to avoid the jump of the inverse local time process (τ_ℓ) at time $\ell = L$ in the following construction. Suppose I is independent of B and ε, and define further random intervals*

$$IL := [G_I L, D_I L] \qquad and \qquad \tau(IL) := [\tau(G_I L), \tau(D_I L)] \tag{49}$$

where $\tau(\ell) := \tau_\ell$ for $\ell \geqslant 0$.
(i) For $y > 0$, there is the equality in distribution of path fragments

$$(B[\tau(IL)] \mid \lambda_{IL} = y) \stackrel{d}{=} (B[0, \tau_y] \mid \tau_y < \varepsilon) \tag{50}$$

where $\lambda_{IL} := D_I L - G_I L$ is the increment of local time of B over the time interval $\tau(IL)$.
(ii) If (I_j) is an interval partition of $[0, 1)$ which is independent of ε and B, and $(\tau_{I_j L})$ is the corresponding interval partition of $[0, G)$, then given the sequence of local time increments $(\lambda_{I_j L})$ the path fragments $B[\tau_{I_j L}]$ are conditionally independent with distributions described by (50) for I_j instead of I.
(iii) If $I_j := [\widehat{V}_{j-1}, \widehat{V}_j]$ with $\widehat{V}_j := 1 - \prod_{i=1}^{j}(1 - U_i)$ for independent uniform$(0, 1)$ variables U_i independent of B and ε, then the bivariate sequence of local time increments and path fragments

$$\left(\lambda_{I_j L}, B[\tau(I_j, L)]\right)_{j=1,2,\dots} \tag{51}$$

is the sequence of points of a Poisson point process on $\mathbb{R}_{>0} \times \Omega$, in local-time biased order, for a suitable space of path fragments Ω of arbitrary finite length, with intensity measure

$$\ell^{-1} P(\tau_\ell < \varepsilon, B[0, \tau_\ell] \in d\omega) \tag{52}$$

whose ℓ-marginal is the Lévy measure $\ell^{-1} e^{-\psi(\xi)\ell} d\ell$ of the exponential distribution of L with rate $\psi(\xi)$.

(iv) *Let* $X_j := \tau(G_{I_j}L) - \tau(G_{I_j}L)$ *be the length and* $Y_j := \lambda_{I_j L}$ *the local time increment associated with the random subinterval* $\tau(I_j L)$ *of* $[0, G)$. *Then the* (X_j, Y_j) *are the points of a Poisson point process on* $\mathbb{R}^2_{>0}$ *with intensity measure* ν *defined by* (45), *in* Y-*biased random order, and*

$$(G, L) = (\Sigma_j X_j, \Sigma_j Y_j) \qquad almost\ surely. \tag{53}$$

Proof. The first two assertions are straightforward consequences of the previous Lemma. Part (iii) follows from (ii), the Poisson representation of $GEM(1)$ in Lemma 15, and Poisson marking. Part (iv) follows from (iii) and Lemma 17 (iii). □

Proof of Proposition 14 (iii). We will exploit the following construction of the standard bridge B^{br} of the self-similar Markov process B by random scaling, as in [31] and [35, Lemma 4]. Let

$$B^{\mathrm{br}} := B_*[0, G_\varepsilon(B)] \qquad \text{for } \varepsilon \text{ independent of } B \text{ with} \qquad \varepsilon \overset{d}{=} \Gamma_1 \xi, \tag{54}$$

so ε is exponential with rate ξ. Then

$$G := G_\varepsilon(B) \overset{d}{=} \frac{\Gamma_\alpha}{\xi} \qquad \text{and} \qquad L := L^0_G(B) = G^\alpha L^0_1(B^{\mathrm{br}}) \overset{d}{=} \frac{\Gamma_1}{c\xi^\alpha} \tag{55}$$

by (21), (19), and α-scaling of local times, where the exponential distribution of L is read from Lemma 17. Suppose now that $I_j := [\widehat{V}_{j-1}, \widehat{V}_j]$, for $\widehat{V}_j$ as in Lemma 18. The T-sequence is now constructed as a function of these $\widehat{V}_j$ and $B^{\mathrm{br}} := B_*[0, G]$ as in (54) according to Definition 2, that is

$$T_j := \inf\{u : L^0_u/LB_1 > \widehat{V}_j\}. \tag{56}$$

By (56) and (55),

$$\lambda_{I_j L} = (\widehat{V}_j - \widehat{V}_{j-1})L = (\widehat{V}_j - \widehat{V}_{j-1})G^\alpha L^0_1(B^{\mathrm{br}}) = G^\alpha L^0_{I^T_j}(B^{\mathrm{br}})$$

$$\lambda_{\tau_{I_j}L} = \tau_{\widehat{V}_j L} - \tau_{\widehat{V}_{j-1}L} = G\lambda_{I^T_j}$$

$$B_*[\tau_{I_j L}] = B^{\mathrm{br}}_*[I^T_j].$$

Part (iii) of Proposition 14 can now be read from Lemma 18 (iv). □

Proof of Proposition 11. Parts (i) and (ii) follow immediately from the result of Proposition 14 (iii) proved above. Turning to consideration of the path fragments, we observe by switching identity (17) that

$$(B_*[0, \tau_\ell] \mid \tau_\ell = t) \overset{d}{=} (B^{\mathrm{br}} \mid L^0_1 = \ell t^{-\alpha}) \tag{57}$$

where $L^0_1 := L^0_1(B^{\mathrm{br}})$ as usual, and a regular conditional distribution for B^{br} given L^0_1 can be constructed as in [28, Lemma 12]. Hence from (52), if Ω_1

denotes a suitable space of paths of length 1, the trivariate sequence of local time increments, lengths of path fragments, and standardized path fragments

$$\left(\lambda_{I_j L}, \lambda_{\tau_{I_j L}}, B_*\left[\tau_{I_j L}\right]\right)_{j=1,2,\dots} = \left(G^\alpha L^0_{I^T_j}(B^{\mathrm{br}}), G\lambda_{I^T_j}, B^{\mathrm{br}}_*\left[I^T_j\right]\right)_{j=1,2,\dots} \quad (58)$$

is a Poisson process on $\mathbb{R}_{>0} \times \mathbb{R}_{>0} \times \Omega_1$ whose intensity measure is

$$\ell^{-1}\, \mathrm{d}\ell\, P(\tau_\ell \in \mathrm{d}t) e^{-\xi t} P(B^{\mathrm{br}} \in \mathrm{d}\omega_1 \mid L^0_1 = \ell t^{-\alpha}). \quad (59)$$

Using the first form of ν in (33) to integrate out ℓ in (59), we see that the lengths and standardized fragments

$$\left(\lambda_{\tau_{I_j L}}, B_*\left[\tau_{I_j L}\right]\right)_{j=1,2,\dots} = \left(G\lambda_{I^T_j}, B^{\mathrm{br}}_*\left[I^T_j\right]\right)_{j=1,2,\dots} \quad (60)$$

form a Poisson process on $\mathbb{R}_{>0} \times \Omega_1$ whose intensity measure is

$$\alpha t^{-1} e^{-\xi t}\, \mathrm{d}t\, Q(\mathrm{d}\omega_1) \quad (61)$$

where

$$Q(\mathrm{d}\omega_1) = \int_0^\infty P(B^{\mathrm{br}} \in \mathrm{d}\omega_1 \mid L^0_1 = y) P(\tau_1^{-\alpha} \in \mathrm{d}y) = P(B_*[0, \tau_1] \in \mathrm{d}\omega) \quad (62)$$

by the switching identity (57). The factorization in (61) shows that the $B^{\mathrm{br}}_*[I^T_j]$ are i.i.d. copies of $B_*[0, \tau_1]$ when listed in length-ranked order. That is part (iii) of Proposition 11. □

5.1 Further distributional results

We record in this section a number of further formulae related to the distribution of the lengths and local times defined by the T-partition.

Proposition 20. *For the T-partition the $(2n+1)$-variate joint density of the total bridge local time L^0_1, the lengths of the first n intervals, and the local times at 0 on these intervals, is given by the formula*

$$P(L^0_1 \in \mathrm{d}\ell, \lambda_{I^T_j} \in \mathrm{d}x_j, L^{\mathrm{br}}(I^T_j) \in \mathrm{d}y_j, 1 \leqslant j \leqslant n)$$

$$= c\Gamma(\alpha)\, \mathrm{d}\ell\, f_{\ell-y_1-\dots-y_n}(1 - x_1 - \dots - x_n) \prod_{j=1}^n \frac{\mathrm{d}x_j\, \mathrm{d}y_j\, f_{y_j}(x_j)}{\ell - y_1 - \dots - y_{j-1}}$$

for $f_y(x) := P(\tau_y \in \mathrm{d}x)/\mathrm{d}x$ the stable(α) density as in (36).

Proof. This follows from the switching identity (17) and the definition of the T-partition. □

While the distributions of the cut times T_k and interval lengths $(T_k - T_{k-1})$ in principle determined Proposition 20, formulae for these distributions are more easily obtained as follows. For $0 < u < 1$, let

$$\tau_u^{\mathrm{br}} := \inf\{t : L_t^0 / L_1^0 = u\}$$

where $(L_t^0, 0 \leqslant t \leqslant 1)$ is the local time process at 0 of B^{br}. Then by use of the switching identity (17) we can write down for $0 < x < 1$, $0 < \ell < \infty$,

$$P(\tau_u^{\mathrm{br}} \in \mathrm{d}x \mid L_1^0 = \ell)/\mathrm{d}x = \frac{f_{u\ell}(x) f_{\bar{u}\ell}(\bar{x})}{f_\ell(1)} \tag{63}$$

where $\bar{x} := 1 - x$. Integrating out with respect to the distribution (39) of L_1^0 gives the density

$$P(\tau_u^{\mathrm{br}} \in \mathrm{d}x)/\mathrm{d}x = c\Gamma(\alpha) \int_0^\infty f_{u\ell}(x) f_{\bar{u}\ell}(\bar{x}) \, \mathrm{d}\ell. \tag{64}$$

which can be simplified using Lévy's formula (40) in the Brownian case to give for $\alpha = 1/2$

$$P(\tau_u^{\mathrm{br}} \in \mathrm{d}x)/\mathrm{d}x = \frac{u\,\bar{u}}{2(\bar{x}u^2 + x\bar{u}^2)^{3/2}} \qquad (0 < u, x < 1). \tag{65}$$

In particular, for $u = 1/2$ we recover the the result of [9, Theorem 3.2] that $\tau_{1/2}^{\mathrm{br}}$ has uniform distribution on $[0,1]$ in the Brownian case.

According to Definition 2, $T_k := \tau_{\widehat{V}_k}^{\mathrm{br}}$, for $\widehat{V}_k$ independent of B^{br} with

$$1 - \widehat{V}_k \overset{d}{=} \widehat{V}_k - \widehat{V}_{k-1} \overset{d}{=} \Pi_k$$

for Π_k is a product of k independent uniform$(0,1)$ variables, with

$$\frac{P(\Pi_k \in \mathrm{d}u)}{\mathrm{d}u} = \frac{(-\log u)^{k-1}}{(k-1)!} \quad \text{and} \quad \sum_{k=1}^\infty \frac{P(\Pi_k \in \mathrm{d}u)}{\mathrm{d}u} = \frac{1}{u} \tag{66}$$

because $\log \Pi_k$ is the k'th point of a rate 1 Poisson process on $[0,\infty)$. Since the process $(\tau_u^{\mathrm{br}}, 0 \leqslant u \leqslant 1)$ has exchangeable increments, we find that $1 - T_k$ and the length of the kth T-interval have the common distribution

$$P(1 - T_k \in \mathrm{d}x) = P(\lambda_{I_k^T} \in \mathrm{d}x) = \int_0^1 P(\tau_u^{\mathrm{br}} \in \mathrm{d}x) P(\Pi_k \in \mathrm{d}u). \tag{67}$$

In particular, in the Brownian case $\alpha = 1/2$, (67) and (65) yield the curious formula

$$\frac{P(T_1 \in \mathrm{d}x)}{\mathrm{d}x} = \frac{h(x) + h(\bar{x})}{2} \quad \text{with} \quad h(x) := \frac{1}{\sqrt{x}} + \log\left(\frac{1}{\sqrt{x}} - 1\right). \tag{68}$$

Corollary 21. *The point process of lengths of T-intervals has mean density*

$$\sum_{k=1}^{\infty} P(\lambda_{I_k^T} \in dx) = \alpha x^{-1}(1-x)^{\alpha-1}\,dx = \int_0^1 P(\tau_u^{\mathrm{br}} \in dx)u^{-1}\,du \qquad (69)$$

for $x \in (0,1)$.

Proof. The first equality is read from part (i) of Proposition 11 and the well known formula for the mean density of points of a Poisson–Dirichlet(α) distributed sequence [32, (6)], which can be read from Lemma 15. The second equality is then read from (67) and (66). □

For general α, the second equality in (69) does not seem very obvious from (63) and (64). However, it can be checked for $\alpha = 1/2$ using (65), and it can also be verified by a very general argument, which we indicate in Section 6.3.

Path decompositions of B^{br} at the times T_k are more complicated than the corresponding decompositions for the times D_{V_j} expressed by Lemma 8. For the T-partition, the pieces are not pure B-bridges. Rather, when normalized they have density factors involving their local times at 0. Compare with similar constructions in [11, 13, 25, 33].

By the Poisson analysis of the previous section, conditionally given $(T_1, L_{T_1}^0, L_1^0)$ the pieces of B^{br} before and after time T_1 are independent B-bridges with prescribed lengths and local times at 0. The appearance of $h + k$ in formula (a) below shows that the right side does not factor into a function of (x, h) and a function of $(\bar{x}, k)$. So even in the Brownian case, $L_{T_1}^0$ and $L_1^0 - L_{T_1}^0$ are not conditionally independent given T_1, and hence the same can be said of the fragments of B^{br} before and after time T_1.

Proposition 22. *In the Brownian case with $\alpha = 1/2$, $c = \sqrt{2}$,*

$$\frac{P(T_1 \in dx, L_{T_1}^0 \in dh, L_1^0 - L_{T_1}^0 \in dk)}{dx\,dh\,dk} = \frac{h\,k}{\sqrt{2\pi}}\frac{(x\bar{x})^{-3/2}}{(h+k)}\exp\left(-\frac{h^2}{2x} - \frac{k^2}{\bar{x}}\right)$$

while for

$$X := \frac{L_{T_1}^0}{\sqrt{T_1}} = L_1^0(B_*^{\mathrm{br}}[0, T_1]) \qquad and \qquad Y := \frac{L_1^0 - L_{T_1}^0}{\sqrt{1-T_1}} = L_1^0(B_*^{\mathrm{br}}[T_1, 1]).$$

there is the joint density

$$\frac{P(X \in da, Y \in db)}{da\,db} = \frac{a\,b}{\sqrt{2\pi}}I(a,b)\exp\left(-\frac{a^2}{2} - \frac{b^2}{2}\right) \qquad (70)$$

where

$$I(a,b) := \int_0^1 \frac{(x\bar{x})^{-1/2}}{a\sqrt{x} + b\sqrt{\bar{x}}}\,dx = \frac{1}{r}\log\left(\frac{(r+a)(r+b)}{(r-a)(r-b)}\right)$$

for $r := \sqrt{a^2 + b^2}$.

Proof. The first formula is an instance of Proposition 20 which we now check. With notation as in (63),

$$P(T_1 \in \mathrm{d}x \mid L^0_{T_1} = h, L^0_1 - L^0_{T_1} = k)/\mathrm{d}x = f(x \mid u, \ell)$$

for $h = u\ell$ and $k = \bar{u}\ell$. We also know, by definition of T_1, that

$$P(L^0_{T_1} \in \mathrm{d}h, L^0_1 - L^0_{T_1} \in \mathrm{d}k) = \mathrm{d}h\,\mathrm{d}k\,e^{-\frac{1}{2}\ell^2}$$

where $\ell = h+k$ and an ℓ^{-1} has canceled the factor of ℓ in the density (41) of L^0_1. Combining these formulae gives the trivariate density of $(T_1, L^0_{T_1}, L^0_1 - L^0_{T_1})$, which rescales to give

$$\frac{P(T_1 \in \mathrm{d}x, X \in \mathrm{d}a, Y \in \mathrm{d}b)}{\mathrm{d}x\,\mathrm{d}a\,\mathrm{d}b} = \frac{a\,b}{\sqrt{2\pi}}\frac{(x\bar{x})^{-1/2}}{(a\sqrt{x}+b\sqrt{\bar{x}})}\exp\left(-\frac{a^2}{2}-\frac{b^2}{2}\right).$$

and (70) follows by integrating out x. $\qquad\qquad\qquad\qquad\qquad\qquad\qquad\square$

6 Complements

6.1 Mappings conditioned to have a single basin

In the Brownian case, a variation of the transformation from X to Y in Lemma 13, which further swaps the exchangeable pair of fragments $X[0, G_U]$ and $X[D_U, 1]$, is the continuous analog of the transformation, mentioned in Fact (2.2)(e) from the stretch of the cycles-first mapping walk for a given basin to the stretch of the basins-first walk for the same basin. As pointed out in the last section of [1], if the uniform mapping of $[n]$ is conditioned to have only one cycle, the scaled basins-first walk converges in distribution to the process $2|B^{\mathrm{br}}_*[0, D_U(B^{\mathrm{br}})]|$. The above argument yields:

Corollary 22. *For a uniform mapping of $[n]$ conditioned to have only one cycle, the scaled cycles-first walk converges in distribution to $2|B_*[0, \tau_1]|$ where $B_*[0, \tau_1]$ is the Brownian pseudo-bridge.*

The distributions of several basic functionals of pseudo-bridge $B_*[0, \tau_1]$ are known. In particular, the occupation density of the reflected process is governed by the same stochastic differential equation governing the occupation density process of a reflecting Brownian bridge or Brownian excursion [28]. According to Knight [19] (see also [33] and papers cited there), the law of the maximum of the reflected pseudo-bridge is identical to that of $1/(2\sqrt{H_1(R_3)})$ where $H_1(R_3)$ is the hitting time of 1 by the three-dimensional Bessel process, with transform $E(\exp(-\frac{1}{2}\theta^2 H_1(R_3))) = \theta/\sinh\theta$ for real θ. Thus we deduce:

Corollary 23. *For a uniform mapping of $[n]$ conditioned to have only one cycle, the asymptotic distribution of the maximum height of any tree above the cycle, normalized by $\sqrt{n}$, is the distribution of $1/\sqrt{H_1(R_3)}$.*

See also [12] for a survey of closely related distributions and their applications.

6.2 Exchangeable interval partitions

Suppose that (I_j^{ex}) is an exchangeable interval partition of $[0,1]$. That is (assuming for simplicity that the lengths $\lambda_{I_j^{\text{ex}}}$ are almost surely all distinct), for each $n = 2, 3, \ldots$ such that $\lambda_{I_{(n)}^{\text{ex}}} > 0$, where $(I_{(j)}^{\text{ex}})$ is the associated length-ranked interval partition, conditionally given $\lambda_{I_{(n)}^{\text{ex}}} > 0$ the ordering of the longest n sub-intervals $I_{(j)}^{\text{ex}}, 1 \leqslant j \leqslant n$ is equally likely to be any one of the $n!$ possible orders, independently of the lengths of these n intervals. Call (I_j^{ex}) *infinite* if $P(\lambda_{I_{(n)}^{\text{ex}}} > 0) = 1$ for all n. As shown by Kallenberg [16], for an infinite exchangeable interval partition (I_j^{ex}), for each $u \in [0,1]$ the fraction of the longest n intervals that lie to the left of u has an almost sure limit $\bar{L}_u^0$ as $n \to \infty$. The process $(\bar{L}_u^0, 0 \leqslant u \leqslant 1)$ is a continuous increasing process, the *normalized local time process* of (I_j^{ex}). It is easily shown that for B^{br} as in previous sections, and more generally for B^{br} the standard bridge of any nice recurrent Markov process, constructed as in [13], the interval partition (I_j^{ex}) defined by the excursions of B^{br} away from 0 is an infinite exchangeable interval partition of $[0,1]$, whose normalized local time process is $\bar{L}_u^0 = L_u^0/L_1^0, 0 \leqslant u \leqslant 1$ for any of the usual Markovian definitions of a bridge local time process $L_u^0 := L_u^0(B^{\text{br}})$. In particular, this remark applies to a self-similar recurrent process B as considered in previous sections.

Theorem 25. *The assertions of Theorem 3 remain valid for the D- and T-partitions defined by Definitions 1 and 2 for any infinite exchangeable interval partition (I_j^{ex}) instead of the excursion intervals of a standard Brownian bridge B^{br}, with the complement of $\bigcup_j I_j^{\text{ex}}$ in $[0,1]$ instead of the zero set of B^{br}, and the normalized local time process $(\bar{L}_u^0, 0 \leqslant u \leqslant 1)$ of (I_j^{ex}) instead of $(L_u^0/L_1^0, 0 \leqslant u \leqslant 1)$. Moreover, the sequence of normalized local times $(\bar{L}_{I_j}^0)$ has the same $GEM(1)$ distribution for $I_j = I_j^D$ as for $I_j = I_j^T$.*

Theorem 25 can be derived from a certain combinatorial analog, stated and proved as Lemma 26 below. Let us briefly outline the method of derivation, without details. Consider an infinite exchangeable interval partition (I_j). Take k independent uniform $(0,1)$ sample points, assign "weight" $1/k$ to each, and let $(I_j^{(k)})$ be the intervals containing at least one sample point. Each interval $I_j^{(k)}$ is thereby assigned weight $1/k \times$ (number of sample points in interval). For fixed k we can apply Lemma 26, interpreting "length" as "weight", and conditionally on the number of intervals in the partition. The conclusion of Lemma 26 is a variant of the desired Theorem 3 for (I_j), in which "position $x \in (0,1)$" of interval endpoint is replaced by "$1/k \times$ (number of sample points in $(0,x)$)", and in which "normalized local time at $u \in (0,1)$" is replaced by "relative number of sampled intervals in $(0,u)$". One can now argue that as $k \to \infty$ we have a.s. convergence of these variant quantities to the original quantities in Theorem 25.

Lemma 26. *Let* $(I_i^{\mathrm{ex}})_{1 \leqslant i \leqslant n}$ *be an exchangeable interval partition of* $[0, 1]$ *into* n *subintervals of strictly positive length. Define* D_{V_j} *as in Definition 1 for* $1 \leqslant j \leqslant J_n^D$, *where* J_n *is the first* j *such that* $D_{V_j} = 1$, *to create a D-partition* $(I_j^D)_{1 \leqslant j \leqslant J_n^D}$ *of* $[0, 1]$, *and define a T-partition* $(I_j^T)_{1 \leqslant j \leqslant J_n^T}$ *of* $[0, 1]$ *similarly using cut points* $T_j, 1 \leqslant j \leqslant J_n^T$ *determined as follows: given that the random set of endpoints of* $(I_j^{\mathrm{ex}})_{1 \leqslant j \leqslant n}$ *is* $\mathcal{U} := \{u_j\}_{0 \leqslant j \leqslant n}$ *with* $0 = u_0 < u_1 < \cdots < u_n = 1$, *let* T_1 *have uniform distribution on* $\mathcal{U} \cap (0, 1]$, *and given also* $T_1 = t_1 < 1$ *let* T_2 *have uniform distribution on* $\mathcal{U} \cap (t_1, 1]$, *and so on, until* $T_{J_n^T} = 1$. *For* I_j *an interval of either of the D- or T-partitions so defined, let* N_{I_j} *denote the number of intervals of* $(I_i^{\mathrm{ex}})_{1 \leqslant i \leqslant n}$ *which are contained in* I_j, *so* $1 \leqslant N_{I_j} \leqslant n$. *Then the assertions of Theorem 3 remain valid provided that* N_{I_j} *is substituted everywhere for* $L_{I_j}^0$.

Proof. We will check that part (i) of Theorem 3 holds in this setup, along with (72). The remaining claims are straightforward and left to the reader. By conditioning on the ranked lengths $\lambda_{I_{(j)}^{\mathrm{ex}}}$ of the intervals $(I_i^{\mathrm{ex}})_{1 \leqslant i \leqslant n}$, it suffices to consider the case when these ranked lengths are distinct constants. Let Π_n^D denote the random partition of $[n]$ defined by the random equivalence relation $i \sim j$ iff $I_{(i)}^{\mathrm{ex}}$ and $I_{(j)}^{\mathrm{ex}}$ are part of the same component interval of the D-partition, and define Π_n^T similarly in terms of the T-partition. Since each unordered collection of lengths and sub-interval counts is a function of the corresponding partition, it suffices to show that $\Pi_n^D \overset{d}{=} \Pi_n^T$. Due to the well known connection between the discrete stick-breaking scheme used to define the T-partition and the cycle structure of random permutations, which was recalled in Section 2.2 (d), we can write down the distribution of Π_n^T without calculation: for each unordered partition of $[n]$ into k non-empty subsets $\{A_1, \ldots, A_k\}$,

$$P\big(\Pi_n^T = \{A_1, \ldots, A_k\}\big) = \frac{1}{n!} \prod_{j=1}^k (|A_j| - 1)! \tag{71}$$

where $|A_i|$ is the number of elements of A_i. On the other hand, for the D-partition, for each ordered partition $(A_1, \ldots, A_k)$ and each choice of $a_j \in A_j, 1 \leqslant j \leqslant k$, with $\lambda(a)$ the length of $I_{(a)}^{\mathrm{ex}}$ and $\lambda(A) := \sum_{a \in A} \lambda(a)$, we can write down the probability

$$P\left(I_j^D = \bigcup_{a \in A_j} I_{(a)}^{\mathrm{ex}} \text{ and } I_{(a_j)}^{\mathrm{ex}} \text{ has right end } D_{V_j} \right) = \frac{1}{n!} \prod_{j=1}^k (|A_j| - 1)! \, \frac{\lambda(a_j)}{\sum_{i=j}^k \lambda(A_i)}$$

where the factors of $(|A_j| - 1)!$ come from the different possible orderings of all but the last $I_{(i)}^{\mathrm{ex}}$ to form I_j^D. If we now sum over all choices of $a_j \in A_j$, for each $1 \leqslant j \leqslant k$, we find that $\lambda(a_j)$ is simply replaced by $\lambda(A_j)$. If we then replace $(A_1, \ldots, A_k)$ by $(A_{\sigma(1)}, \ldots, A_{(\sigma(k)})$ and sum over all permutations σ of $[k]$, to consider all sequences of sets consistent with a given unordered partition

$\{A_1, \ldots, A_k\}$, we get precisely (71) for Π_n^D instead of Π_n^T, due to the identity

$$\sum_\sigma \prod_{j=1}^{k} \frac{\lambda(A_{\sigma(j)})}{\sum_{i=j}^{k} \lambda(A_{\sigma(i)})} = 1.$$

This is obvious, because the product is the probability of picking the sequence of sets $(A_{\sigma(j)}, 1 \leqslant j \leqslant k)$ in a process of $\lambda(A_i)$-biased sampling of blocks of the partition $\{A_1, \ldots, A_k\}$.   □

We note the consequence of the previous proof that the number of components J_n^D of the D-partition and the number of components J_n^T of the T-partition have the same distribution, which is the same for every exchangeable interval partition $(I_i^{\mathrm{ex}})_{1 \leqslant i \leqslant n}$ of $[0, 1]$ into n subintervals of strictly positive length:

$$J_n^D \overset{d}{=} J_n^T \overset{d}{=} K_n \overset{d}{=} \sum_{i=1}^{n} 1_{C_i} \tag{72}$$

where K_n is the number of cycles of a uniformly distributed random permutation of $[n]$, and the C_i are independent events with $P(C_i) = 1/i$. The second two of these equalities in distribution are well known and easily explained without calculation [26]. But the first is quite surprising, and we do not see how to explain it any more simply than by the previous proof.

6.3 Intensity measures

In this section we check Corollary 21 by showing it can be generalized and proved as follows:

Corollary 27. *In the setting of Theorem 25, the common intensity measure of the the point process of lengths of T-intervals and the point process of lengths of D-intervals is*

$$\sum_{k=1}^{\infty} P(\lambda_{I_k^T} \in dx) = \sum_{k=1}^{\infty} P(\lambda_{I_k^D} \in dx) = \frac{P(D_{V_1} \in dx)}{x} = \int_0^1 \frac{P(\bar\tau_u \in dx)}{u} \, du \tag{73}$$

where $\bar\tau_u := \inf\{t : \bar{L}_t^0 > u\}$ is the inverse of the normalized local time process of the exchangeable interval partition.

Proof. The equality of the first three measures displayed in (73) is read from the conclusion of Theorem 25, using the fact that the D-partition is in length-biased order. The equality of the first and fourth measures follows from the definition of the T_k, the exchangeable increments of $(\bar\tau_u, 0 \leqslant u \leqslant 1)$, and (66), just as in the proof of (69).   □

As a check on Theorem 25, let us verify the equality of the second and fourth measures in (73) in the following special case, which includes the setting of Corollary 21.

Let $(\tau_\ell, \ell \geq 0)$ be the inverse local time process of B at 0, for B as in Lemma 17 not necessarily self-similar. Note that we must explicitly assume $(\tau_\ell, \ell \geq 0)$ is drift free for the conclusion of part (iii) of that Lemma to be true. We assume that now. Assume that the Lévy measure of $(\tau_\ell, \ell \geq 0)$ has density $\rho(x)$. Let (I_j^{ex}) be the exchangeable partition of $[0, 1]$ generated by the excursion intervals of B conditional on $B_1 = 0$ and $L_1(B) = \ell$ for some fixed $\ell > 0$, or equivalently by the jumps of $(\tau_s, 0 \leq s \leq \ell)$ given $\tau_\ell = 1$. Then, formula (63) generalizes easily to show that the fourth measure in (73) has density at x

$$\int_0^1 u^{-1}\, du \, \frac{f_{u\ell}(x) f_{\bar{u}\ell}(\bar{x})}{f_\ell(1)} \tag{74}$$

for $f_\ell(x)$ as in (36). On the other hand, abbreviating $D := D_{V_1}$ and $G := G_{V_1}$ so $[G, D]$ is the interval I_j^{ex} which covers the independent uniform time V_1, we know from (75) that for $0 < w < 1$

$$P\big(1 - (D - G) \in dw\big) = \frac{\ell \rho(1 - w)(1 - w) f_\ell(w)\, dw}{f_\ell(1)}.$$

Also, it is easily seen that conditionally given $1 - (D - G) = w$, the normalized local time $\bar{L}_G^0$ is uniform on $(0, 1)$ and independent of the pair $(G, 1 - D)$, which is distributed like $(\tau_{u\ell}, \tau_{\bar{u}\ell})$ conditioned on $\tau_\ell = w$. Together with the previous formula for $w = y + 1 - x$, this gives the trivariate density

$$\frac{P\big(\bar{L}_G^0 \in du, G \in dy, D \in dx\big)}{du\, dy\, dx} = \frac{\ell \rho(x - y)(x - y) f_{u\ell}(y) f_{\bar{u}\ell}(\bar{x})}{f_\ell(1)}$$

$(0 < y < x < 1)$. Now (75) implies that

$$\int_0^y f_{u\ell}(y) \rho(x - y)(x - y)\, dy = \frac{x f_{u\ell}(x)}{u\ell}$$

so we deduce that

$$\frac{P\big(\bar{L}_G^0 \in du, D \in dx\big)}{du\, dx} = \frac{x f_{u\ell}(y) f_{\bar{u}\ell}(\bar{x})}{u f_\ell(1)}$$

and hence that the density displayed in (74) is indeed $x^{-1} P(D \in dx)/dx$.

6.4 Two orderings of a bivariate Poisson process

According to Proposition 14, for each $\alpha \in (0, 1)$ the Poisson point process with intensity measure $\nu(dx, dy) = \rho(x, y)\, dx\, dy$ displayed in (33) has the following paradoxical property:

(a) If the points (X_j, Y_j) are put in X-biased order, then the Y_j are in Y-biased order, whereas

(b) if the points (X_j, Y_j) are put in Y-biased order, then the X_j are not in X-biased random order; even the distribution of X_1 is wrong.

We first see this for $\alpha = 1/2$ by passage to the limit of elementary combinatorial properties of uniform random mappings. We then see it for general α from the bridge representations of Proposition 14. Other point processes of lengths and local times with these properties can be constructed from an exchangeable interval partition, as shown by Theorem 25 in the previous section and Lemma 15. This argument, shows that (a) holds for the bivariate Poisson process with intensity (45) featured in Lemma 17, for any drift free subordinator $(\tau_y, y \geq 0)$ with $E(e^{-\xi \tau_y}) = e^{-\psi(\xi)y}$. Then the Y_j normalized by their sum have $GEM(1)$ distribution, both for an X-biased and for a Y-biased ordering. We offer here a slightly different explanation of (a) in this case. That is, given some joint density $\rho(x, y)$, we indicate conditions on ρ which are necessary and sufficient for (a) to hold for the bivariate Poisson process with intensity ρ, and then check that these conditions are in fact satisfied in the case (45).

Let (X_j, Y_j) be the points of a Poisson process on $\mathbb{R}^2_{>0}$ with intensity $\rho(x, y) \, dx \, dy$, in X-biased order. Let $\Sigma_X := \sum_j X_j$ and $\Sigma_Y := \sum_j Y_j$. Let

$$f_X(x) := P(\Sigma_X \in dx)/dx; \qquad f_Y(y) := P(\Sigma_Y \in dy)/dy$$

$$\rho_X(x) := \int_0^\infty \rho(x, y) \, dy; \qquad \rho_Y(y) := \int_0^\infty \rho(x, y) \, dx.$$

By a basic Palm calculation, as in [25]

$$P(X_1 \in dx, \Sigma_X - X_1 \in dw) = \rho_X(x) \, dx \, f_X(w) \, dw \, \frac{x}{x + w} \qquad (75)$$

and similarly, with

$$f_{X,Y}(x, y) := P(\Sigma_X \in dx, \Sigma_Y \in dy)/(dx \, dy)$$

$$P(X_1 \in dx, Y_1 \in dy, \Sigma_X - X_1 \in dw, \Sigma_Y - Y_1 \in dv)$$
$$= \rho(x, y) \, dx \, dy \, f_{X,Y}(w, v) \, dw \, dv \, \frac{x}{x + w}. \qquad (76)$$

Now, a *necessary condition* for the Y_j to be in Y-biased order is that Y_1 should have the same joint distribution with Σ_Y as if Y_1 were a size-biased pick from the Y_i, that is like (75)

$$P(Y_1 \in dy, \Sigma_Y - Y_1 \in dv) = \rho_Y(y) \, dy \, f_Y(v) \, dv \, \frac{y}{y + v}. \qquad (77)$$

Thus a necessary condition on $\rho(x, y)$ for (a) to hold is that for all $y, v \geq 0$

$$\int_0^\infty dx \int_0^\infty dw \, \rho(x, y) f_{X,Y}(w, v) \, \frac{x}{x + w} = \rho_Y(y) f_Y(v) \, \frac{y}{y + v}. \qquad (78)$$

Moreover, by keeping track of the first k of the (X_j, Y_j) jointly with Σ_X and Σ_Y it is clear that we can write down a multivariate version of (78) whose truth for all k would be necessary and sufficient for (a).

In the special case (45), with $f_y(x) := P(\tau_y \in dx)/dx$, the subordination argument gives

$$\rho(x, y) = y^{-1} f_y(x) e^{-\xi x}.$$

Since the Y-marginal is exponential with rate $\psi(\xi)$,

$$f_Y(y) = \psi(\xi) y \rho_Y(y) = \psi(\xi) e^{-\psi(\xi)y}$$

and hence by generalization of (43), using $(\Sigma_X, \Sigma_Y) \overset{d}{=} (G, L)$,

$$f_{X,Y}(x, y) = \psi(\xi) f_y(x) e^{-\xi x} = \psi(\xi) y \rho(x, y).$$

If these expressions are substituted in (78), and we use the definition of $\psi(\xi)$ on the right side, we find that (78) reduces to the identity

$$E\left[\frac{\tau_y}{\tau_y + \tau_v} e^{-\xi(\tau_y + \tau_v)}\right] = \frac{y}{y + v} E\left[e^{-\xi(\tau_y + \tau_v)}\right].$$

But this is true by virtue of

$$E\left[\frac{\tau_y}{\tau_y + \tau_v} \,\middle|\, \tau_y + \tau_v\right] = \frac{y}{y + v}$$

which holds by exchangeability of increments of $(\tau_\ell, \ell \geqslant 0)$. Moreover, the multivariate form of (78) mentioned above is easily checked the same way.

Acknowledgement. We thank Grégory Miermont and an anonymous referee for careful reading and helpful comments.

References

1. D. Aldous and J. Pitman. Brownian bridge asymptotics for random mappings. *Random Structures and Algorithms*, 5: 487–512, 1994.
2. D. Aldous and J. Pitman. Invariance principles for non-uniform random mappings and trees. In V. Malyshev and A. M. Vershik, editors, *Asymptotic Combinatorics with Aplications in Mathematical Physics*, pages 113–147. Kluwer Academic Publishers, 2002.
3. D. Aldous and J. Pitman. The asymptotic distribution of the diameter of a random mapping. *C.R. Acad. Sci. Paris, Ser. I*, 334: 1021–1024, 2002.
4. D.J. Aldous. The continuum random tree III. *Ann. Probab.*, 21: 248–289, 1993.
5. D.J. Aldous, G. Miermont, and J. Pitman. Brownian bridge asymptotics for random p-mappings. Electron. J. Probab, 9: 37–56, 2004.
6. M. Barlow, J. Pitman, and M. Yor. On Walsh's Brownian motions. In *Séminaire de Probabilités XXIII*, pages 275–293. Springer, 1989. Lecture Notes in Math. 1372.

7. M. Barlow, J. Pitman, and M. Yor. Une extension multidimensionnelle de la loi de l'arc sinus. In *Séminaire de Probabilités XXIII*, pages 294–314. Springer, 1989. Lecture Notes in Math. 1372.

8. J. Bertoin. *Lévy Processes.* Cambridge University Press, 1996. Cambridge Tracts in Math. 126.

9. J. Bertoin and J. Pitman. Path transformations connecting Brownian bridge, excursion and meander. *Bull. Sci. Math. (2)*, 118: 147–166, 1994.

10. P. Biane. Some comments on the paper: "Brownian bridge asymptotics for random mappings" by D. J. Aldous and J. W. Pitman. *Random Structures and Algorithms*, 5: 513–516, 1994.

11. P. Biane, J.-F. Le Gall, and M. Yor. Un processus qui ressemble au pont brownien. In *Séminaire de Probabilités XXI*, pages 270–275. Springer, 1987. Lecture Notes in Math. 1247.

12. P. Biane, J. Pitman, and M. Yor. Probability laws related to the Jacobi theta and Riemann zeta functions, and Brownian excursions. *Bull. Amer. Math. Soc.*, 38: 435–465, 2001.

13. P. Fitzsimmons, J. Pitman, and M. Yor. Markovian bridges: construction, Palm interpretation, and splicing. In E. Çinlar, K.L. Chung, and M.J. Sharpe, editors, *Seminar on Stochastic Processes, 1992*, pages 101–134. Birkhäuser, Boston, 1993.

14. P. Greenwood and J. Pitman. Fluctuation identities for Lévy processes and splitting at the maximum. *Advances in Applied Probability*, 12: 893–902, 1980.

15. K. Itô and H. P. McKean. *Diffusion Processes and their Sample Paths.* Springer, 1965.

16. O. Kallenberg. The local time intensity of an exchangeable interval partition. In A. Gut and L. Holst, editors, *Probability and Statistics, Essays in Honour of Carl-Gustav Esseen*, pages 85–94. Uppsala University, 1983.

17. J. F. C. Kingman. Random discrete distributions. *J. Roy. Statist. Soc. B*, 37: 1–22, 1975.

18. J. F. C. Kingman. *Poisson Processes.* Clarendon Press, Oxford, 1993.

19. F. B. Knight. Inverse local times, positive sojourns, and maxima for Brownian motion. In *Colloque Paul Lévy sur les Processus Stochastiques*, pages 233–247. Société Mathématique de France, 1988. Astérisque 157–158.

20. C. Leuridan. Le théorème de Ray–Knight à temps fixe. In J. Azéma, M. Émery, M. Ledoux, and M. Yor, editors, *Séminaire de Probabilités XXXII*, pages 376–406. Springer, 1998. Lecture Notes in Math. 1686.

21. P. Lévy. Sur certains processus stochastiques homogènes. *Compositio Math.*, 7: 283–339, 1939.

22. J.-F. Marckert and A. Mokkadem. The depth first processes of Galton-Watson trees converge to the same Brownian excursion. *Ann. Probab.*, 31: 1655–1678, 2003.

23. J. W. McCloskey. A model for the distribution of individuals by species in an environment. Ph. D. thesis, Michigan State University, 1965.

24. C. A. O'Cinneide and A. V. Pokrovskii. Nonuniform random transformations. *Ann. Appl. Probab.*, 10(4): 1151–1181, 2000.

25. M. Perman, J. Pitman, and M. Yor. Size-biased sampling of Poisson point processes and excursions. *Probab. Th. Rel. Fields*, 92: 21–39, 1992.

26. J. Pitman. Some developments of the Blackwell-MacQueen urn scheme. In T.S. Ferguson et al., editor, *Statistics, Probability and Game Theory; Papers in*

honor of David Blackwell, volume 30 of *Lecture Notes-Monograph Series*, pages 245–267. Institute of Mathematical Statistics, Hayward, California, 1996.

27. J. Pitman. Partition structures derived from Brownian motion and stable subordinators. *Bernoulli*, 3: 79–96, 1997.

28. J. Pitman. The SDE solved by local times of a Brownian excursion or bridge derived from the height profile of a random tree or forest. *Ann. Probab.*, 27: 261–283, 1999.

29. J. Pitman. Random mappings, forests and subsets associated with Abel-Cayley-Hurwitz multinomial expansions. *Séminaire Lotharingien de Combinatoire*, Issue 46: 45 pp., 2001. http://www.mat.univie.ac.at/~slc/.

30. J. Pitman. Combinatorial stochastic processes. To appear in Springer lecture Notes in Mathematics, Vol. 1875, March 2006. Technical Report 621, Dept. Statistics, U.C. Berkeley, 2002. Lecture notes for St. Flour course, July 2002. Available at http://www.stat.berkeley.edu/tech-reports/.

31. J. Pitman and M. Yor. Arcsine laws and interval partitions derived from a stable subordinator. *Proc. London Math. Soc. (3)*, 65: 326–356, 1992.

32. J. Pitman and M. Yor. The two-parameter Poisson–Dirichlet distribution derived from a stable subordinator. *Ann. Probab.*, 25: 855–900, 1997.

33. J. Pitman and M. Yor. Random Brownian scaling identities and splicing of Bessel processes. *Ann. Probab.*, 26: 1683–1702, 1998.

34. J. Pitman and M. Yor. The law of the maximum of a Bessel bridge. *Electron. J. Probab.*, 4: Paper 15, 1–35, 1999.

35. J. Pitman and M. Yor. On the distribution of ranked heights of excursions of a Brownian bridge. *Ann. Probab.*, 29: 361–384, 2001.

36. H. Pollard. The representation of e^{-x^λ} as a Laplace integral. *Bull. Amer. Math. Soc.*, 52: 908–910, 1946.

37. S. Resnick. *Adventures in Stochastic Processes*. Birkhauser, 1992.

38. D. Revuz and M. Yor. *Continuous martingales and Brownian motion*. Springer, Berlin-Heidelberg, 1999. 3rd edition.

39. L.C.G. Rogers and D. Williams. *Diffusions, Markov Processes and Martingales, Vol. II: Itô Calculus*. Wiley, 1987.

40. V.V. Uchaikin and V. M. Zolotarev. *Chance and stability*. VSP, Utrecht, 1999. Stable distributions and their applications, With a foreword by V. Yu. Korolev and Zolotarev.

41. A.M. Vershik and A.A. Shmidt. Limit measures arising in the theory of groups, I. *Theor. Prob. Appl.*, 22: 79–85, 1977.

42. S. Watanabe. Generalized arc-sine laws for one-dimensional diffusion processes and random walks. In *Proceedings of Symposia in Pure Mathematics*, volume 57, pages 157–172. A. M. S., 1995.

43. V. M. Zolotarev. *One-dimensional stable distributions.*, volume 65 of *Translations of Mathematical Monographs*. Am. Math. Soc., 1986.

44. V. M. Zolotarev. On the representation of the densities of stable laws by special functions. *Theory Probab. Appl.*, 39: 354–362, 1994.

PÉNALISATIONS ET QUELQUES EXTENSIONS DU THÉORÈME DE PITMAN, RELATIVES AU MOUVEMENT BROWNIEN ET À SON MAXIMUM UNILATÈRE

Bernard Roynette[1], Pierre Vallois[1], Marc Yor[2]

[1] Institut E. Cartan, Université Henri Poincaré, BP 239, Bd des Aiguillettes,
F-54506 Vandœuvre Les Nancy Cedex, France
e-mail: Bernard.Roynette@iecn.u-nancy.fr
e-mail: Vallois@iecn.u-nancy.fr
[2] Université P. et M. Curie, Laboratoire de Probabilités et Modèles Aléatoires,
Case 188, 4 place Jussieu, F-75252 Paris Cedex, France

Le texte qui suit est composé de deux parties:

PARTIE 1: QUELQUES PÉNALISATIONS DU MOUVEMENT BROWN-
IEN FAISANT INTERVENIR, EN PARTICULIER, SON MAXIMUM UNI-
LATÈRE

PARTIE 2: SUR LE THÉORÈME DE PITMAN RELATIF À 2S-X ÉTENDU
À DES COUPLES MARKOVIENS $(X_., S_. \equiv \sup_{s \leqslant .} X_s)$

Ces deux parties ont été rédigées indépendamment l'une de l'autre, un lecteur
pouvant être intéressé soit par les exemples de pénalisations, soit par les extensions du Théorème de Pitman, qui sont présentés ici.
La rédaction que nous avons faite est, dans sa forme, du style de celles des
notes aux C. R. A. S., c'est-à-dire que nous présentons et discutons les résultats
principaux en donnant seulement les clés des démonstrations.

MOTS CLÉS: Pénalisations, Théorème de Pitman, Processus de Bessel de
dimension 3, Processus du maximum, minimum et temps local

1 QUELQUES PÉNALISATIONS DU MOUVEMENT BROWNIEN FAISANT INTERVENIR, EN PARTICULIER, SON MAXIMUM UNILATÈRE

Summary. On the canonical Wiener space $\Omega = \mathcal{C}([0, \infty[\to \mathbb{R})$, where $(X_t, t \geqslant 0)$
denotes the coordinate process, $\mathcal{F}_t = \sigma(X_s, s \leqslant t)$ its natural filtration and $(W_x, x \in \mathbb{R})$ the family of Wiener measures, we consider several kinds of adapted, $\mathbb{R}_+$-valued,
integrable functionals $(\Gamma_t, t \geqslant 0)$, for which we show that, for every fixed $s > 0$ and
$\Lambda_s \subset \mathcal{F}_s$, the quantity:
$$\frac{\mathrm{E}_x(1_{\Lambda_s} \Gamma_t)}{\mathrm{E}_x(\Gamma_t)}$$
admits a limit, as $t \to \infty$, and that this limit takes the form: $\mathrm{E}_x(1_{\Lambda_s} M_s^x)$ where
$(M_s^x, s \geqslant 0)$ is a $(\mathcal{F}_s, s \geqslant 0, W_x)$ martingale. This allows us to define, on $(\Omega, \mathcal{F}_\infty)$, a
probability Q_x via the formula:
$$Q_x(\Lambda_s) := \mathrm{E}_x(1_{\Lambda_s} M_s^x) \quad (\Lambda_s \in \mathcal{F}_s).$$
We then describe precisely the process $(X_t, t \geqslant 0)$ under the probability Q_x. In
general, this process is not Markovian on its own filtration, but several path decompositions and Markov properties are obtained.

Résumé. Soit $(\Omega = \mathcal{C}([0, \infty[\to \mathbb{R}), X_t, \mathcal{F}_t, t \geqslant 0, W_x, x \in \mathbb{R})$ l'espace de Wiener
canonique. Soit $\Gamma : \mathbb{R}_+ \times \Omega \to \mathbb{R}_+$ une fonctionnelle positive et adaptée. Pour
plusieurs classes de fonctionnelles Γ, nous montrons que, pour tout s fixé et $\Lambda_s \in \mathcal{F}_s$,
la quantité
$$\frac{\mathrm{E}_x(1_{\Lambda_s} \Gamma_t)}{\mathrm{E}_x(\Gamma_t)}$$

admet une limite lorsque $t \to \infty$ et que cette limite est de la forme $E_x(1_{\Lambda_s} M_s^x)$ où $(M_s^x, s \geqslant 0)$ est une $(\mathcal{F}_s, s \geqslant 0, W_x)$ martingale positive. Ceci permet de définir sur $(\Omega, \mathcal{F}_\infty)$ la probabilité Q_x par la formule:

$$Q_x(\Lambda_s) := E_x(1_{\Lambda_s} M_s^x) \quad (\Lambda_s \in \mathcal{F}_s).$$

Nous décrivons alors de manière précise le processus $(\Omega, X_t, \mathcal{F}_t, t \geqslant 0)$ sous la probabilité Q_x. En général, ce processus n'est pas markovien.

1.1 Introduction

Soit $\Omega = \mathcal{C}([0, \infty[\to \mathbb{R})$, $(X_t, t \geqslant 0)$ le processus des coordonnées de cet espace, $(\mathcal{F}_t, t \geqslant 0)$ la filtration naturelle associée et W_x $(x \in \mathbb{R})$ la famille des mesures de Wiener. Nous notons: $\mathcal{F}_\infty := \bigvee_{s \geqslant 0} \mathcal{F}_s$. Ces notations seront en vigueur dans toute cette partie. Soit $\Gamma : \mathbb{R}_+ \times \Omega \to \mathbb{R}_+$ une fonctionnelle adaptée et positive. Notre but est ici de montrer que, pour plusieurs classes de fonctionnelles Γ:

i) Pour tout s et tout $\Lambda_s \in \mathcal{F}_s$,

$$\frac{E_x(1_{\Lambda_s} \Gamma_t)}{E_x(\Gamma_t)} \tag{1}$$

admet une limite quand $t \to \infty$.

ii) Cette limite s'écrit: $E_x(1_{\Lambda_s} M_s^x)$, où $(M_s^x, s \geqslant 0)$ est une $(\mathcal{F}_s, W_x)$ martingale positive telle que $M_0^x = 1$.

Définissons alors, sur $(\Omega, \mathcal{F}_\infty)$ la probabilité Q_x $(x \in \mathbb{R})$ par:

$$Q_x(\Lambda_s) := E_x(1_{\Lambda_s} M_s^x) \quad (\Lambda_s \in \mathcal{F}_s).$$

Ainsi, la probabilité Q_x est une sorte de "mesure de Gibbs associée au mouvement brownien $(X_t, t \geqslant 0)$ et à la fonctionnelle Γ." Nous allons, dans ce qui suit, décrire de manière précise le processus $(X_t, \mathcal{F}_t, t \geqslant 0)$ sous Q_x. Pour des raisons de simplicité, nous nous limiterons le plus souvent à la description du processus $(X_t, t \geqslant 0)$ sous Q_0.

Dans un travail antérieur (cf [17] et [18]) nous avons mené à bien cette étude lorsque $\Gamma_t = \exp\left(-\frac{1}{2} \int_0^t u(X_s) ds\right)$, où $u : \mathbb{R} \to \mathbb{R}_+$ est une fonction positive (dans ce cas, le processus $(X_t, t \geqslant 0)$ sous Q_0 est markovien). Nous allons ici étudier cinq situations:

1. Γ_t est une fonction du maximum unilatère
2. Γ_t est une fonction de X_t et du maximum unilatère
3. Γ_t est une fonction de X_t^+, X_t^- et du temps local en 0
4. Γ_t est une fonction du maximum unilatère, du minimum unilatère et du temps local en 0

5. Γ_t est une fonction du nombre de descentes du processus $(X_t, t \geqslant 0)$ sur un intervalle $[a, b]$.

NOTATIONS: Si $(Z_t, t \geqslant 0)$ est un processus à valeurs réelles, nous notons:

$$S_t^Z = \sup_{u \leqslant t} Z_u \quad (S_t^Z \text{ est le maximum unilatère de Z}); \tag{2}$$

$$I_t^Z = -\inf_{u \leqslant t} Z_u \quad (-I_t^Z \text{ est le minimum unilatère de Z}); \tag{3}$$

L_t^Z : le temps local en 0 de Z au temps t (Z est supposée

être une semi-martingale); (4)

$$Z_t^* = \sup_{u \leqslant t} |Z_u| = S_t^Z \vee I_t^Z; \tag{5}$$

$$Z_t^+ = Z_t \vee 0, \quad Z_t^- = -(Z_t \wedge 0). \tag{6}$$

1.2 Pénalisation par une fonction du maximum unilatère

Soit $\varphi : \mathbb{R}_+ \to \mathbb{R}_+$ une fonction mesurable, bornée et positive. Nous supposons:

$$\int_0^\infty \varphi(x) dx = 1 \tag{7}$$

et nous notons

$$\phi(x) := \int_0^x \varphi(y) dy. \tag{8}$$

On suppose que, pour tout $x \geqslant 0$, $\phi(x) < 1$.

Théorème 1.1 *(cf [19])*

1. *Pour tout $s \geqslant 0$ et pour tout $\Lambda_s \in \mathcal{F}_s$, la limite quand $t \to \infty$ de*

$$\frac{E_0 \left(1_{\Lambda_s} \varphi(S_t^X) \right)}{E_0 \left(\varphi(S_t^X) \right)} \tag{9}$$

existe.
2. *Cette limite s'écrit:*

$$\lim_{t \to \infty} \frac{E_0 \left(1_{\Lambda_s} \varphi(S_t) \right)}{E_0 \left(\varphi(S_t) \right)} = E_0(1_{\Lambda_s} M_s^\varphi), \tag{10}$$

où

$$M_s^\varphi := (S_s^X - X_s)\varphi(S_s^X) + 1 - \phi(S_s^X) = 1 - \int_0^s \varphi(S_u^X) dX_u. \tag{11}$$

De plus, $(M_s^\varphi, s \geqslant 0)$ est une $(\mathcal{F}_s, s \geqslant 0, W_0)$ martingale. C'est une martingale d'Azéma-Yor (cf [1]).

3. *Soit la probabilité Q_0^φ définie sur $(\Omega, \mathcal{F}_\infty)$ par:*

$$Q_0^\varphi(\Lambda_s) := E_0(1_{\Lambda_s} M_s^\varphi) \qquad (\Lambda_s \in \mathcal{F}_s). \tag{12}$$

Sous Q_0^φ, il existe un mouvement brownien $(B_t, t \geqslant 0)$ tel que le processus $(X_t, t \geqslant 0)$ soit solution de:

$$X_t = B_t - \int_0^t \frac{\varphi(S_s^X) ds}{(S_s^X - X_s)\varphi(S_s^X) + 1 - \phi(S_s^X)}. \tag{13}$$

4. *Sous Q_0^φ,*

  a) *La variable aléatoire S_∞^X est finie p. s. et admet φ comme densité.*
  b) *Pour tout $x \geqslant 0$, conditionnellement à $S_\infty^X = x$, on a:*

$$T_x = \sup\{t \geqslant 0,\ X_t = x\} = \inf\{t \geqslant 0,\ X_t = x\} < \infty \quad p.\ s. \tag{14}$$

  et:
  i) *Le processus $(X_t, t \leqslant T_x)$ est un mouvement brownien arrêté quand il atteint le niveau x.*
  ii) *Le processus $(X_{T_x+t}, t \geqslant 0)$ satisfait à:*

$$x - X_{T_x+t} = \beta_t + \int_0^t \frac{ds}{(x - X_{T_x+s})} \tag{15}$$

  où $(\beta_t, t \geqslant 0)$ est un mouvement brownien. Ainsi, $(X_{T_x+t}, t \geqslant 0)$ est "un processus de Bessel de dimension 3 en dessous de x".
  iii) *Ces deux processus sont indépendants.*

Les points 1 et 2 du Théorème 1.1 reposent sur le:

Lemme 1.2 *Soit $s \geqslant 0$, $x \leqslant s$. Alors:*

$$E_0\left(\varphi\left(s \vee (x + S_u^X)\right)\right) \underset{u \to \infty}{\sim} \sqrt{\frac{2}{\pi u}} \left\{ (s - x)\varphi(s) + \int_s^\infty \varphi(y)dy \right\}. \tag{16}$$

Remarques 1.3   1. *Sous Q_0^φ, le processus $(X_t, t \geqslant 0)$ n'est pas markovien. Cela résulte de la présence du terme S_s^X dans (13). Par contre, le couple $(X_t, S_t^X, t \geqslant 0)$ est markovien, ce que nous exprimons en disant que le processus $(X_t, t \geqslant 0)$ est max-markovien.*
2. *Le point 2 du Théorème 1.1 s'étend pour un x quelconque sous la forme:*

$$\frac{E_x(1_{\Lambda_s}\varphi(S_t))}{E_x(\varphi(S_t))} \underset{t \to \infty}{\longrightarrow} E_x(1_{\Lambda_s} M_s^{x,\varphi}),$$

avec:

$$M_s^{x,\varphi} := \frac{1}{1 - \phi(x)}\left\{ (S_s^X - X_s)\varphi(S_s^X) + 1 - \phi(S_s^X) \right\}.$$

Il est possible de compléter et de préciser le Théorème 1.1 dans deux directions. C'est l'objet des deux Théorèmes suivants (cf [20]).

Théorème 1.4 *Définissons, pour tout $s \geqslant 0$, $\Lambda_s \in \mathcal{F}_s$ et $y > 0$, $t \geqslant s$*

$$Q_t^{(y)}(\Lambda_s) = P(\Lambda_s | S_t = y) \tag{17}$$

1. *Pour tout $s \geqslant 0$, $\Lambda_s \in \mathcal{F}_s$ et $y > 0$:*

$$\lim_{t \to \infty} Q_t^{(y)}(\Lambda_s) \quad existe$$

2. *Cette limite est égale à $Q^{(y)}(\Lambda_s)$ avec:*

$$Q^{(y)}(\Lambda_s) := e^{-\frac{y^2}{2s}} \sqrt{\frac{2}{\pi s}} \mathrm{E}\left(1_{\Lambda_s}(y - X_s)|S_s = y\right) + \mathrm{E}\left(1_{\Lambda_s} 1_{S_s < y}\right) \tag{18}$$

3. *La probabilité Q_0^φ définie par l'égalité (12) du Théorème 1.1 est un mélange des probabilités $(Q^{(y)}, y > 0)$. Plus précisement, pour tout $\Lambda \in \mathcal{F}_\infty$:*

$$Q_0^\varphi(\Lambda) = \int_0^\infty Q^{(y)}(\Lambda)\varphi(y)dy \tag{19}$$

et :

$$Q_0^\varphi(\Lambda | S_\infty^X = y) = Q^{(y)}(\Lambda). \tag{20}$$

Remarques 1.5 *Dans le même esprit que celui du Théorème 1.4, on peut montrer que, pour tout $s \geqslant 0$, $\Lambda_s \in \mathcal{F}_s$ et $y > a_+$:*

$$\lim_{t \to \infty} P(\Lambda_s | S_t = y, X_t = a) \quad existe.$$

La limite précédente définit sur $(\Omega, \mathcal{F}_\infty)$ une probabilité $Q^{y,a}$ telle que:

$$(2y - a)Q^{y,a} = (y - a)Q^{(y)} + \int_0^y Q^{(z)}dz \tag{21}$$

En particulier, $Q^{y,a}$ "dépend très peu de a". C'est une combinaison convexe des probabilités $(Q^{(z)}, z \in [0, y])$ (cf [20]).

Théorème 1.6 *Soit $\varphi : \mathbb{R}^+ \longrightarrow \mathbb{R}^+$ mesurable et bornée, telle que $\int_0^\infty \varphi(y)dy < \infty$ et soit $\phi(y) := \int_0^y \varphi(x)dx$, comme dans le Théorème 1.1. On suppose de plus qu'il existe un entier $n \geqslant 0$ tel que:*

$$\int_0^\infty x^{2(n+1)}\varphi(x)dx < \infty. \tag{22}$$

1. *Il existe, pour tout $i \leqslant n$, une famille de martingales $(\mathrm{M}_s^{\varphi,i}, s \geqslant 0)$ telles qu'on ait, quand $t \to \infty$, le développement suivant:*

$$\frac{\mathrm{E}\left(1_{\Lambda_s}\varphi(S_t)\right)}{\mathrm{E}\left(\varphi(S_t)\right)} = \sum_{i=0}^n \left(\frac{1}{2t}\right)^i \mathrm{E}(1_{\Lambda_s}\mathrm{M}_s^{\varphi,i}) + O\left(\frac{1}{t^{n+1}}\right). \tag{23}$$

2. *Les martingales* $(\mathrm{M}_s^{\varphi,i}, s \geqslant 0)$ *sont des fonctions de* s, S_s *et* X_s. *En d'autres termes, il existe, pour tout* $i \leqslant n$, *des fonctions (dépendant de φ)* $F_i : \mathbb{R}^+ \times \mathbb{R}^+ \times \mathbb{R} \to \mathbb{R}$ *telles que:*

$$\mathrm{M}_s^{\varphi,i} = F_i(s, \mathrm{S}_s, X_s) \tag{24}$$

avec:

$$\mathrm{M}_s^{\varphi,0} := \mathrm{M}_s^{\varphi} = (\mathrm{S}_s - X_s)\varphi(\mathrm{S}_s) + 1 - \phi(\mathrm{S}_s) \tag{25}$$

et

$$\mathrm{M}_s^{\varphi,1} = -\phi_1(\mathrm{S}_s, X_s) + (s + k)\mathrm{M}_s^{\varphi},$$

où:

$$\phi_1(\sigma, x) := \varphi(\sigma)\frac{(\sigma - x)^3}{3} + \int_\sigma^\infty \varphi(v)(v - x)^2 dv \quad (\sigma \geqslant x_+) \tag{26}$$

et

$$k := \phi_1(0,0) = \int_0^\infty \varphi(v)v^2 dv. \tag{27}$$

3. *En particulier, pour* $n = 0$, *on a:*

$$\frac{\mathrm{E}\left(1_{\Lambda_s}\varphi(\mathrm{S}_t)\right)}{\mathrm{E}\left(\varphi(\mathrm{S}_t)\right)} - \mathrm{E}\left(1_{\Lambda_s}\mathrm{M}_s^{\varphi}\right) \underset{t \to \infty}{\sim} \frac{\mathrm{E}(1_{\Lambda_s}\mathrm{M}_s^{\varphi,1})}{2t} \tag{28}$$

si bien que le Théorème 1.6 permet de préciser la vitesse de convergence de $\dfrac{\mathrm{E}(1_{\Lambda_s}\varphi(\mathrm{S}_t))}{\mathrm{E}(\varphi(\mathrm{S}_t))}$ *vers* $\mathrm{E}\left(1_{\Lambda_s}\mathrm{M}_s^{\varphi}\right)$ *quand* $t \to \infty$.

1.3 Pénalisation liée aux martingales de Kennedy

Soit $\psi : \mathbb{R}_+ \longrightarrow \mathbb{R}_+$ une fonction strictement positive et décroissante telle que:

$$\int_0^\infty \psi(x)dx < \infty. \tag{29}$$

Soit $\lambda > 0$ et $\varphi : [0, \infty[\longrightarrow \mathbb{R}$ définie par:

$$\varphi(x) := \psi(x) - \lambda e^{\lambda x}\int_x^\infty e^{-\lambda y}\psi(y)dy. \tag{30}$$

Il est facile de voir que:

$$\varphi(x) \geqslant 0 \quad \text{pour tout } x \geqslant 0 \quad \text{et} \quad \int_0^\infty \varphi(y)dy = \int_0^\infty e^{-\lambda y}\psi(y)dy. \tag{31}$$

On suppose, sans perte de généralité, que $\int_0^\infty e^{-\lambda y}\psi(y)dy = 1$. Soit, comme précédemment $\phi(x) = \int_0^x \varphi(s)ds$.

Théorème 1.7   *1. Pour tout $s \geqslant 0$ et pour tout $\Lambda_s \in \mathcal{F}_s$,*

$$\lim_{t \to \infty} \frac{\mathrm{E}_0\left(1_{\Lambda_s}\psi(\mathrm{S}_t^X)e^{\lambda(\mathrm{S}_t^X - X_t)}\right)}{\mathrm{E}_0\left(\psi(\mathrm{S}_t^X)e^{\lambda(\mathrm{S}_t^X - X_t)}\right)} \qquad \textit{existe.} \qquad (32)$$

2. Cette limite est égale à

$$\mathrm{E}_0(1_{\Lambda_s}\mathrm{M}_s^{\lambda,\varphi}) \qquad (33)$$

où

$$\mathrm{M}_s^{\lambda,\varphi} := e^{-\frac{\lambda^2 s}{2}}\left\{ \left(1 - \phi(\mathrm{S}_s^X)\right)\mathrm{ch}\left(\lambda(\mathrm{S}_s^X - X_s)\right) + \varphi(\mathrm{S}_s^X)\frac{\mathrm{sh}\left(\lambda(\mathrm{S}_s^X - X_s)\right)}{\lambda} \right\}.$$
$$(34)$$

De plus, $(\mathrm{M}_s^{\lambda,\varphi}, s \geqslant 0)$ est une $(\mathcal{F}_s, s \geqslant 0, \mathrm{W}_0)$ martingale. Cette martingale (à quelques modifications mineures près) est la martingale de Kennedy (cf [8], lorsqu'on choisit $\varphi(s) = e^{-\mu s}$).
On a de plus (cf [20]):

$$\mathrm{E}_0\left(1_{\Lambda_s}\mathrm{M}_s^{\lambda,\varphi}\right) = \mathrm{E}_0^{(-\lambda)}\left(1_{\Lambda_s}\varphi_\lambda(\mathrm{S}_\infty^X)\right)$$

où $P_0^{(-\lambda)}$ désigne la loi du mouvement brownien avec dérive $-\lambda$ et où :

$$\varphi_\lambda(x) = (1 - \phi(x))\,e^{\lambda x} + \frac{\varphi(x)}{\lambda}e^{\lambda x}$$

3. Soit la probabilité $Q_0^{\lambda,\varphi}$ définie sur $(\Omega, \mathcal{F}_\infty)$ par:

$$Q_0^{\lambda,\varphi}(\Lambda_s) := \mathrm{E}_0\left(1_{\Lambda_s}\mathrm{M}_s^{\lambda,\varphi}\right) \qquad (\Lambda_s \in \mathcal{F}_s). \qquad (35)$$

Alors, sous $Q_0^{\lambda,\varphi}$, il existe un mouvement brownien $(\mathrm{B}_t, t \geqslant 0)$ tel que $(X_t, t \geqslant 0)$ soit solution de:

$$X_t$$
$$= \mathrm{B}_t - \lambda \int_0^t \frac{\varphi(\mathrm{S}_s^X)\mathrm{ch}\left(\lambda(\mathrm{S}_s^X - X_s)\right) + \lambda\left(1 - \phi(\mathrm{S}_s^X)\right)\mathrm{sh}\left(\lambda(\mathrm{S}_s^X - X_s)\right)}{\lambda\left(1 - \phi(\mathrm{S}_s^X)\right)\mathrm{ch}\left(\lambda(\mathrm{S}_s^X - X_s)\right) + \varphi(\mathrm{S}_s^X)\mathrm{sh}\left(\lambda(\mathrm{S}_s^X - X_s)\right)}\,ds.$$
$$(36)$$

4. Sous $Q_0^{\lambda,\varphi}$,
   a) La variable aléatoire S_∞^X est finie p. s. et sa fonction de répartition $\mathrm{F}_{\mathrm{S}_\infty^X}(u) := Q_0^{\lambda,\varphi}(\mathrm{S}_\infty^X \leqslant u)$ est égale à:

$$\mathrm{F}_{\mathrm{S}_\infty^X}(u) = 1 - e^{-\lambda u}\left(1 - \phi(u)\right) \qquad (u \geqslant 0). \qquad (37)$$

   b) Conditionnellement à $\mathrm{S}_\infty^X = x \geqslant 0$, on a:

$$\mathrm{T}_x := \sup\{t \geqslant 0,\ X_t = x\} = \inf\{t \geqslant 0,\ X_t = x\} < \infty \quad \textit{p. s.,}$$
$$(38)$$

et:

i) *Le processus* $(X_t, t \leqslant T_x)$ *est un mouvement brownien avec dérive* λ *arrêté quand il atteint le niveau* x.

ii) *Le processus* $(X_{T_x+t}, t \geqslant 0)$ *est un "λ-processus de Bessel de dimension 3 en dessous de x". Plus précisément, il existe un mouvement brownien* $(\beta_t, t \geqslant 0)$ *issu de 0 tel que* $(X_{T_x+t}, t \geqslant 0)$ *soit solution de:*

$$x - X_{T_x+t} = \beta_t + \lambda \int_0^t \coth\left(\lambda(x - X_{T_x+u})\right) du. \qquad (39)$$

iii) *Ces deux processus sont indépendants, et leurs lois ne dépendent pas de* φ.

Les points 1 et 2 du Théorème 1.7 reposent sur le:

Lemme 1.8 *Soit* $\lambda > 0$, $s \geqslant 0$, *et* $s \geqslant x$. *Soit* $\psi : \mathbb{R}_+ \longrightarrow \mathbb{R}$ *telle que* $\int_0^\infty |\psi(s)| ds < \infty$. *Définissons:*

$$\rho_\lambda(s, x) = \psi(s) \operatorname{sh}\left(\lambda(s-x)\right) + \lambda e^{\lambda x} \int_s^\infty \psi(z) e^{-\lambda z} dz. \qquad (40)$$

Si $\rho_\lambda(s, x) \neq 0$, *alors:*

$$\mathrm{E}_0\left(\psi\left(s \vee (x + \mathrm{S}_t^X)\right) e^{\lambda(s \vee (x+\mathrm{S}_t^X) - x - X_t)}\right) \underset{t \to \infty}{\sim} 2\rho_\lambda(s, x) e^{\frac{\lambda^2 t}{2}}. \qquad (41)$$

Remarques 1.9 1. *Le processus* $(X_t, t \geqslant 0)$ *sous* $Q_0^{\lambda,\varphi}$ *n'est pas markovien, à cause de la présence du terme* S_s^X *dans (36). Par contre, le couple* $(X_t, \mathrm{S}_t^X, t \geqslant 0)$ *est markovien, et donc, avec notre terminologie,* $(X_t, t \geqslant 0)$ *est max-markovien.*

2. *Le Théorème 1.1 est le cas particulier du Théorème 1.7 lorsque* $\lambda = 0$.

3. *Les points 1 et 2 du Théorème 1.7 s'étendent pour x quelconque sous la forme:*

$$\frac{\mathrm{E}_x\left(1_{\Lambda_s} \psi(\mathrm{S}_t^X) e^{\lambda(\mathrm{S}_t^X - X_t)}\right)}{\mathrm{E}_x\left(\psi(\mathrm{S}_t^X) e^{\lambda(\mathrm{S}_t^X - X_t)}\right)} \underset{t \to \infty}{\longrightarrow} \mathrm{E}_x\left(1_{\Lambda_s} \mathrm{M}_s^{\lambda,\varphi,x}\right),$$

avec

$$\mathrm{M}_s^{\lambda,\varphi,x} := \frac{e^{-\frac{\lambda^2}{2} s}}{1 - \phi(x)} left\{ \left(1 - \phi(\mathrm{S}_s^X)\right) \operatorname{ch}\left(\lambda(\mathrm{S}_s^X - X_s)\right)$$
$$+ \varphi(\mathrm{S}_s^X) \frac{\operatorname{sh}\left(\lambda(\mathrm{S}_s^X - X_s)\right)}{\lambda} \right\}. \qquad (42)$$

4. Signalons les travaux de F. Knight ([10]) et T. Jeulin et M. Yor ([7]) qui étudient la transformée de Laplace de variables aléatoires de la forme

$$\int_0^{T_1} f(S_s^X, X_s)ds,$$

où $(X_s, s \geqslant 0)$ est un mouvement brownien. En fait, lors de leurs études, ces auteurs s'intéressent indirectement à des processus solutions de l'équation (36).

1.4 Pénalisation par une fonction de X_t^+, X_t^- et du temps local en 0

Soit h^+ et h^- deux fonctions de $\mathbb{R}^+$ dans $\mathbb{R}^+$, boréliennes et bornées telles que:

$$\frac{1}{2} \int_0^\infty \left(h^+(x) + h^-(x) \right) dx = 1. \tag{43}$$

Définissons:

$$H(l) := \frac{1}{2} \int_0^l \left(h^+(x) + h^-(x) \right) dx. \tag{44}$$

On suppose que, pour tout $l \geqslant 0$, $H(l) < 1$. Rappelons que, pour toute semi-martingale $(Z_t, t \geqslant 0)$, L_t^Z désigne son temps local en 0 au temps t.

Théorème 1.10   *1. Pour tout $s \geqslant 0$ et pour tout $\Lambda_s \in \mathcal{F}_s$,*

$$\lim_{t \to \infty} \frac{E_0 \left(1_{\Lambda_s} (h^+(L_t^X) 1_{X_t > 0} + h^-(L_t^X) 1_{X_t < 0}) \right)}{E_0 \left(h^+(L_t^X) 1_{X_t > 0} + h^-(L_t^X) 1_{X_t < 0} \right)} \quad existe. \tag{45}$$

2. Cette limite est égale à:

$$E_0 \left(1_{\Lambda_s} M_s^{h^+, h^-} \right) \tag{46}$$

où

$$M_s^{h^+, h^-} := \left(1 - H(L_s^X) \right) + X_s^+ h^+(L_s^X) + X_s^- h^-(L_s^X). \tag{47}$$

De plus, $(M_s^{h^+, h^-}, s \geqslant 0)$ est une $(\mathcal{F}_s, s \geqslant 0, W_0)$ martingale positive.

3. Soit la probabilité $Q_0^{h^+, h^-}$ définie sur $(\Omega, \mathcal{F}_\infty)$ par:

$$Q_0^{h^+, h^-}(\Lambda_s) := E_0 \left(1_{\Lambda_s} M_s^{h^+, h^-} \right) \quad (\Lambda_s \in \mathcal{F}_s). \tag{48}$$

Alors il existe sous $Q_0^{h^+, h^-}$ un mouvement brownien $(B_t, t \geqslant 0)$ tel que $(X_t, t \geqslant 0)$ soit solution de:

$$X_t = B_t + \int_0^t \frac{1_{X_s > 0} h^+(L_s^X) - 1_{X_s < 0} h^-(L_s^X)}{1 - H(L_s^X) + X_s^+ h^+(L_s^X) + X_s^- h^-(L_s^X)} ds. \tag{49}$$

4. *Sous $Q_0^{h^+,h^-}$,*

    a)) La variable aléatoire L_∞^X est finie p. s. et admet comme densité la fonction

$$\frac{1}{2}(h^+ + h^-). \tag{50}$$

    b)) Soit $g := \sup\{t \geqslant 0 \,;\, X_t = 0\}$. Alors, $g < \infty$ p. s. et conditionnellement à $L_\infty^X = x$:

        i.) $(X_t, t \leqslant g)$ est un mouvement brownien arrêté lorsque son temps local en 0 atteint le niveau x.

        ii.) Le processus $(X_{g+t}, t \geqslant 0)$ est soit un processus de Bessel issu de 0, soit l'opposé d'un processus de Bessel issu de 0. La première propriété a lieu avec la probabilité $\frac{1}{2}\int_0^\infty h^+(y)dy$, la seconde avec la probabilité $\frac{1}{2}\int_0^\infty h^-(y)dy$.

        iii.) Les deux processus figurant en i) et ii) sont indépendants.

Les points 1 et 2 du Théorème 1.10 reposent sur le:

Lemme 1.11 *Soit $h : \mathbb{R}^+ \longrightarrow \mathbb{R}^+$ une fonction mesurable et bornée telle que:*

$$\int_0^\infty h(x)dx < \infty. \tag{51}$$

Soit $a \geqslant 0$ et x réel. Alors:

    i) $\mathrm{E}_x\big[h(a + L_t^X)1_{X_t>0}\big] \underset{t\to\infty}{\sim} h(a)\sqrt{\frac{2}{\pi t}}\, x^+ + \frac{1}{\sqrt{2\pi t}}\int_a^\infty h(y)dy$ (52)

    ii) $\mathrm{E}_x\big[h(a + L_t^X)1_{X_t<0}\big] \underset{t\to\infty}{\sim} h(a)\sqrt{\frac{2}{\pi t}}\, x^- + \frac{1}{\sqrt{2\pi t}}\int_a^\infty h(y)dy.$ (53)

Remarques 1.12   *1. Le processus $(X_t, l \geqslant 0)$ sous $Q_0^{h^+,h^-}$ n'est pas markovien, à cause de la présence du terme L_t^X dans (49). Par contre, le couple $(X_t, L_t^X, t \geqslant 0)$ est markovien.*

  2. L'identité de Lévy, i. e. le fait que les deux processus, sous W_0, $(S_t^X, S_t^X - X_t, t \geqslant 0)$ et $(L_t^X, |X_t|, t \geqslant 0)$ ont même loi, permet de traduire le Théorème 1.10 en terme de maximum unilatère. En particulier, si l'on fait $h^+ = h^- = \varphi$ dans le Théorème 1.10, on obtient le Théorème 1.1.

  3. Un phénomène analogue à celui décrit au point 4, ii) du Théorème 1.10 apparait dans un article récent de C. Donati-Martin et Y. Hu (cf [3]).

1.5 Pénalisation par une fonction du maximum unilatère, du minimum unilatère et du temps local en 0

Soit ν une mesure de probabilité sur $\mathbb{R}^+ \times \mathbb{R}^+$ dont le support est inclus dans $[\alpha, +\infty[\times[\alpha, +\infty[$ pour un $\alpha > 0$. Définissons, pour $s, l, i \geqslant 0$:

$$A_\nu(s, i, l) = \int\int_{\mathbb{R}_+^2} e^{\frac{1}{2}\{\frac{1}{a}+\frac{1}{b}\}l} 1\!\!1_{\{a \geqslant s, b \geqslant i\}}\, \nu(da, db). \tag{54}$$

Théorème 1.13   *1. Pour tout $s \geqslant 0$ et pour tout $\Lambda_s \in \mathcal{F}_s$,*

$$\frac{\mathrm{E}_0\left(1_{\Lambda_s} \mathrm{A}_\nu(\mathrm{S}_t^X, \mathrm{I}_t^X, \mathrm{L}_t^X)\right)}{\mathrm{E}_0\left(\mathrm{A}_\nu(\mathrm{S}_t^X, \mathrm{I}_t^X, \mathrm{L}_t^X)\right)} \quad a \ une \ limite \ quand \ t \to \infty. \qquad (55)$$

2. Cette limite est égale à $\mathrm{E}_0\left(1_{\Lambda_s} \mathrm{M}_s^\nu\right)$, *où:*

$$\mathrm{M}_s^\nu := \int\!\!\int_{\mathbb{R}_+^2} (1 - \frac{X_s^+}{a})(1 - \frac{X_s^-}{b}) \exp\left\{\frac{1}{2}(\frac{1}{a} + \frac{1}{b})\mathrm{L}_s^X\right\} 1\!\!1_{\{\mathrm{S}_s^X \leqslant a, \mathrm{I}_s^X \leqslant b\}} \, \nu(da, db).$$

$$(56)$$

De plus, $(\mathrm{M}_s^\nu, s \geqslant 0)$ *est une* $(\mathcal{F}_s, s \geqslant 0, W_0)$ *martingale positive.*
3. Définissons la probabilité Q_0^ν *sur* $(\Omega, \mathcal{F}_\infty)$ *par:*

$$Q_0^\nu(\Lambda_s) := \mathrm{E}_0\left(1_{\Lambda_s} \mathrm{M}_s^\nu\right) \quad (\Lambda_s \in \mathcal{F}_s). \qquad (57)$$

Alors, sous Q_0^ν, *il existe un mouvement brownien* $(\mathrm{B}_t, t \geqslant 0)$ *issu de 0 tel que le processus* $(X_t, t \geqslant 0)$ *soit solution de:*

$$X_t = \mathrm{B}_t + \int_0^t \frac{m^\nu(s)}{\mathrm{M}_s^\nu} ds, \qquad (58)$$

avec

$$m^\nu(s)$$
$$:= \int\!\!\int_{\mathbb{R}_+^2} \left(-\frac{1}{a} 1_{X_s>0} + \frac{1}{b} 1_{X_s<0}\right) \exp\left\{\frac{1}{2}(\frac{1}{a} + \frac{1}{b})\mathrm{L}_s^X\right\} 1\!\!1_{\{\mathrm{S}_s^X \leqslant a, \mathrm{I}_s^X \leqslant b\}} \, \nu(da, db).$$

$$(59)$$

4. Sous Q_0^ν,
    a) Les variables aléatoires S_∞^X *et* I_∞^X *sont finies p. s. et le couple* $(\mathrm{S}_\infty^X, \mathrm{I}_\infty^X)$ *est de loi* ν.
    b) Soit $U_\infty := \inf\{t \geqslant 0, \ X_t = \mathrm{S}_\infty^X\}$; *alors* $U_\infty = \infty$ *p. s. De même,* $V_\infty := \inf\{t \geqslant 0, \ X_t = -\mathrm{I}_\infty^X\} = \infty$ *p. s.*
    c) Conditionnellement à $\mathrm{S}_\infty^X = a$ *et* $\mathrm{I}_\infty^X = b$ $(a, b > 0)$, *il existe un mouvement brownien* $(\beta_t, t \geqslant 0)$ *issu de 0 tel que* $(X_t, t \geqslant 0)$ *soit solution de:*

$$X_t = \beta_t - \int_0^t \frac{1_{X_s>0}}{a - X_s} ds + \int_0^t \frac{1_{X_s<0}}{b + X_s} ds. \qquad (60)$$

La démonstration des points 1 et 2 du Théorème 1.13 repose sur le:

Lemme 1.14 [1] *Soit* a *et* $b > 0$, *et soit* $c := \frac{1}{2}(\frac{1}{a} + \frac{1}{b})$. *Alors:*

1. a) si $T_a = \inf\{t \geqslant 0, X_t = a\}$ *et* $T_{-b} = \inf\{t \geqslant 0, X_t = -b\}$:

$$\lim_{t \to \infty} \mathrm{E}_0\left(e^{c\mathrm{L}_t^X} 1\!\!1_{\{t \leqslant T_a \wedge T_{-b}\}}\right) = \frac{3}{2} \qquad (61)$$

---

[1] Nous remercions vivement F. Petit qui nous a beaucoup aidés à prouver ce lemme.

b)

$$\sup_{t \geq 0} \mathrm{E}_0\left(e^{c\mathrm{L}_t^X}\, \mathbb{1}_{\{t \leq T_a \wedge T_{-b}\}}\right) < \infty. \tag{62}$$

2. *a)* Si $d > c = \frac{1}{2}(\frac{1}{a} + \frac{1}{b})$:

$$\lim_{t \to \infty} \mathrm{E}_0\left(e^{d\mathrm{L}_t^X}\, \mathbb{1}_{\{t < T_a \wedge T_{-b}\}}\right) = +\infty.$$

b) Si $d < c = \frac{1}{2}(\frac{1}{a} + \frac{1}{b})$:

$$\lim_{t \to \infty} \mathrm{E}_0\left(e^{d\mathrm{L}_t^X}\, \mathbb{1}_{\{t < T_a \wedge T_{-b}\}}\right) = 0.$$

Remarques 1.15 *1. Le Lemme 1.14 peut être généralisé de la manière suivante. Soit $(X_t, t \geq 0)$ l'araignée brownienne à N branches associée aux
nombres p_i $(i = 1, ..., N)$ $(p_i > 0, \sum_{i=1}^{N} p_i = 1)$ (cf [2]). Alors on a, pour
tout $a_1, ..., a_N$ $(a_i > 0$ pour $i = 1, ..., N)$:*

$$\mathrm{E}\left(e^{c\mathrm{L}_t} \prod_{i=1}^{N} \mathbb{1}_{S_t^i \leq a_i}\right) \xrightarrow[t \to \infty]{} \frac{3}{2}$$

*avec: $c = \sum_{i=1}^{N} \frac{p_i}{a_i}$. Le cas particulier où $N = 2$ correspond au skew mou-
vement brownien, et le lemme 1.14 au cas $N = 2$, $p_1 = p_2 = \frac{1}{2}$.*

*2. Une autre façon d'exprimer le point 4, c) du Théorème 1.13 consiste à
dire que, conditionnellement à $S_\infty^X = a$ et $I_\infty^X = b$ le processus $(X_t, t \geq 0)$
sous Q_0^ν est un mouvement brownien issu de 0 conditionné à rester dans
l'intervalle $]-b, a[$ (cf [9]).*

*3. Lorsque ν est la masse de Dirac au point (a, b) $(a, b > 0)$, la martingale
M_s^ν s'écrit:*

$$\mathrm{M}_s^\nu := \mathrm{M}_s^{a,b} = \left(1 - \frac{X_s^+}{a}\right)\left(1 - \frac{X_s^-}{b}\right) e^{c\mathrm{L}_s^X}$$

$$\times \mathbb{1}_{\{S_s^X \leq a, I_s^X \leq b\}} \quad \left(avec \ c = \frac{1}{2}\left(\frac{1}{a} + \frac{1}{b}\right)\right) \tag{63}$$

ou encore:

$$\mathrm{M}_s^{a,b} = \mathrm{N}_{s \wedge T_a \wedge T_{-b}}^{a,b}, \tag{64}$$

*avec $\mathrm{N}_s^{a,b} := (1 - \frac{X_s^+}{a})(1 - \frac{X_s^-}{b}) e^{c\mathrm{L}_s^X}$ puisque, si $s = T_a \wedge T_{-b}$, $\mathrm{N}_s^{a,b} = 0$.
Remarquons que, si l'on considère la martingale $\mathrm{M}_s^{h^+, h^-}$ définie par (47)
avec $h^+(l) = \frac{1}{a}e^{cl}$, $h^-(l) = \frac{1}{b}e^{cl}$ (et $c := \frac{1}{2}(\frac{1}{a} + \frac{1}{b})$), on a:*

$$\mathrm{H}(l) = e^{cl} - 1 \qquad (cf \ (44))$$

d'où:

$$\mathrm{M}_s^{h^+, h^-} = 2 - \left(1 - \frac{X_s^+}{a} - \frac{X_s^-}{b}\right) e^{c\mathrm{L}_s^X},$$

et puisque $X_s^+ \cdot X_s^- = 0$,

$$\mathrm{M}_s^{h^+,h^-} = 2 - \left(1 - \frac{X_s^+}{a}\right)\left(1 - \frac{X_s^-}{b}\right) e^{c\mathrm{L}_s^X}, \qquad (65)$$

soit encore: $\mathrm{M}_s^{h^+,h^-} = 2 - \mathrm{N}_s^{a,b}$. *Finalement:*

$$\mathrm{M}_s^{a,b} = 2 - \mathrm{M}_{s \wedge T_a \wedge T_{-b}}^{h^+,h^-}. \qquad (66)$$

Ainsi, la martingale $\mathrm{M}_s^{a,b}$ est, à une constante près, l'opposée de la martingale $\mathrm{M}_s^{h^+,h^-}$ arrêtée au temps $T_a \wedge T_{-b}$. La martingale $(\mathrm{M}_s^\nu, s \geqslant 0)$ définie par (56) est donc la martingale $2 - \mathrm{M}_s^{h^+,h^-}$ "randomisée". Observons enfin que, lorsque ν est la masse de Dirac en (a,b), le processus $(X_t, t \geqslant 0)$ sous Q_0^ν est un mouvement brownien conditionné à rester dans l'intervalle $]-b, a[$.

4. *La martingale $(\mathrm{M}_t^\nu, t \geqslant 0)$ définie par (56) peut encore s'écrire:*

$$\mathrm{M}_t^\nu = F(\mathrm{S}_t^X, \mathrm{I}_t^X, \mathrm{L}_t^X) - X_t^+ F^+(\mathrm{S}_t^X, \mathrm{I}_t^X, \mathrm{L}_t^X) - X_t^- F^-(\mathrm{S}_t^X, \mathrm{I}_t^X, \mathrm{L}_t^X), (67)$$

avec pour $s, i, l \geqslant 0$:

$$F(s,i,l) = \int\int_{\mathbb{R}_+^2} \mathbb{1}_{\{a \geqslant s, b \geqslant i\}} \exp\left\{\tfrac{1}{2}(\tfrac{1}{a} + \tfrac{1}{b})l\right\} \nu(da, db)$$

$$F^+(s,i,l) = \int\int_{\mathbb{R}_+^2} \mathbb{1}_{\{a \geqslant s, b \geqslant i\}} \tfrac{1}{a} \exp\left\{\tfrac{1}{2}(\tfrac{1}{a} + \tfrac{1}{b})l\right\} \nu(da, db)$$

$$F^-(s,i,l) = \int\int_{\mathbb{R}_+^2} \mathbb{1}_{\{a \geqslant s, b \geqslant i\}} \tfrac{1}{b} \exp\left\{\tfrac{1}{2}(\tfrac{1}{a} + \tfrac{1}{b})l\right\} \nu(da, db).$$

Si de plus ν possède une densité continue par rapport à la mesure de Lebesgue de $\mathbb{R}_+^2$, on a:

$$\frac{\partial F}{\partial l} = \frac{1}{2}(F^+ + F^-), \quad s\frac{\partial F^+}{\partial s} = \frac{\partial F}{\partial s}, \quad i\frac{\partial F^-}{\partial i} = \frac{\partial F}{\partial i}, \qquad (68)$$

et la propriété de martingale de $(\mathrm{M}_t^\nu, t \geqslant 0)$ découle, via la formule d'Itô, de (67).

5. *Supposons que ν ne charge que la diagonale de $\mathbb{R}_+^2$; alors M_s^ν s'écrit:*

$$\mathrm{M}_s^\nu = \int_{X_s^*}^\infty \left(1 - \frac{|X_s|}{a}\right) e^{\frac{1}{a}\mathrm{L}_s^X} \nu^*(da), \qquad (69)$$

ν^* étant alors une probabilité sur $\mathbb{R}^+$ à support dans $[\alpha, +\infty[$, pour un $\alpha > 0$. Comme au point 3 ci-dessus, on peut encore écrire

$$\mathrm{M}_s^\nu = F(X_s^*, \mathrm{L}_s^X) - |X_s| F^*(X_s^*, \mathrm{L}_s^X), \qquad (70)$$

avec pour $x, l \geqslant 0$:

$$F^*(x,l) = \int_x^\infty e^{\frac{l}{a}} \nu^*(da), \quad F(x,l) = \int_x^\infty \frac{1}{a} e^{\frac{l}{a}} \nu^*(da), \qquad (71)$$

et avec $x\frac{\partial F}{\partial x} = \frac{\partial F^*}{\partial x}$ si ν^* admet une densité continue.

1.6 Pénalisation liée à une fonction du nombre de descentes

Soit $a < b$. Pour simplifier les notations, nous supposerons que $0 < a < b$. Définissons:

$$\sigma_1 = \inf\{t \geqslant 0, X_t > b\} = T_b$$
$$\sigma_2 = \inf\{t \geqslant \sigma_1, X_t < a\}, \tag{72}$$

et, pour tout entier $n \geqslant 1$:

$$\sigma_{2n+1} = \inf\{t \geqslant \sigma_{2n}, X_t > b\}$$
$$\sigma_{2n+2} = \inf\{t \geqslant \sigma_{2n+1}, X_t < a\}. \tag{73}$$

Soit D_t le nombre de descentes, à l'instant t, du processus $(X_s, s \geqslant 0)$ sur l'intervalle $[a, b]$, i. e. (cf [21], p 57)

$$D_t = \sum_{n \geqslant 1} \mathbb{1}_{\{\sigma_{2n} \leqslant t\}} \qquad (t \geqslant 0). \tag{74}$$

Notons que

$$\sigma_{2D_t} = \sup\{n \leqslant 1, \sigma_{2n} \leqslant t\}. \tag{75}$$

Nous notons dans tout ce paragraphe $(\theta_t, t \geqslant 0)$ la famille des opérateurs de translation sur l'espace canonique $\Omega = \mathcal{C}([0, \infty[\to \mathbb{R})$:

$$X_s \circ \theta_t = X_{s+t} \qquad (s, t \geqslant 0). \tag{76}$$

Comme précédemment, $T_a = \inf\{t \geqslant 0, X_t = a\}$ et $T_b = \inf\{t \geqslant 0, X_t = b\}$. Soit $G : \mathbb{N} \longrightarrow \mathbb{R}^+$ positive et décroissante et telle que:

$$G(0) = 1, \quad G(\infty) = \lim_{n \to \infty} G(n) = 0. \tag{77}$$

Nous noterons $\Delta G : \mathbb{N} \longrightarrow \mathbb{R}^+$ la fonction définie par:

$$\Delta G(n) = G(n) - G(n+1) \qquad (n \geqslant 0). \tag{78}$$

Théorème 1.16   *1. Pour tout $s \geqslant 0$ et pour tout $\Lambda_s \in \mathcal{F}_s$:*

$$\frac{\mathrm{E}_0\left(1_{\Lambda_s} \Delta G(D_t)\right)}{\mathrm{E}_0\left(\Delta G(D_t)\right)} \qquad \textit{admet une limite quand } t \to \infty. \tag{79}$$

2. Cette limite est égale à

$$\frac{1}{\mathrm{M}_0^{\downarrow, G}} \mathrm{E}_0\left(1_{\Lambda_s} \mathrm{M}_s^{\downarrow, G}\right), \tag{80}$$

où:

$$\mathrm{M}_0^{\downarrow, G} = \frac{1}{2(b-a)}\left(G(0)(2b-a) - aG(1)\right) \tag{81}$$

et où:

$$\mathrm{M}_s^{\downarrow,G} = \sum_{n\geqslant 0}\left[\mathbb{1}_{[\sigma_{2n},\sigma_{2n+1}[}(s)\left\{\frac{G(n)}{2}(1+\frac{b-X_s}{b-a})+\frac{G(n+1)}{2}\left(\frac{X_s-a}{b-a}\right)\right\}\right.$$
$$\left.+\mathbb{1}_{[\sigma_{2n+1},\sigma_{2n+2}[}(s)\left\{\frac{G(n+1)}{2}(1+\frac{b-X_s}{b-a})+\frac{G(n)}{2}\left(\frac{X_s-a}{b-a}\right)\right\}\right].$$
$$(82)$$

De plus, $(\mathrm{M}_s^{\downarrow,G}, s\geqslant 0)$ *est une martingale positive et continue, qui s'écrit encore:*

$$\mathrm{M}_s^{\downarrow,G}$$
$$=\mathbb{1}_{\{\sigma_{2\mathrm{D}_s}+T_b\circ\theta_{\sigma_{2\mathrm{D}_s}}>s\}}\left(\frac{G(\mathrm{D}_s)}{2}\left(1+\frac{b-X_s}{b-a}\right)+\frac{G(1+\mathrm{D}_s)}{2}\left(\frac{X_s-a}{b-a}\right)\right)$$
$$+\mathbb{1}_{\{\sigma_{2\mathrm{D}_s}+T_b\circ\theta_{\sigma_{2\mathrm{D}_s}}\leqslant s\}}\left(\frac{G(1+\mathrm{D}_s)}{2}\left(1+\frac{b-X_s}{b-a}\right)+\frac{G(\mathrm{D}_s)}{2}\left(\frac{X_s-a}{b-a}\right)\right).$$
$$(83)$$

3. *Définissons la probabilité* Q_0^G *sur* $(\Omega,\mathcal{F}_\infty)$ *par:*

$$Q_0^G(\Lambda_s)=\frac{1}{\mathrm{M}_0^{\downarrow,G}}\mathrm{E}_0\left(1_{\Lambda_s}\mathrm{M}_s^{\downarrow,G}\right)\qquad(\Lambda_s\in\mathcal{F}_s).\qquad(84)$$

Alors, sous Q_0^G, *il existe un mouvement brownien* $(\mathrm{B}_t, t\geqslant 0)$ *issu de 0 tel que* $(X_t, t\geqslant 0)$ *soit solution de:*

$$X_t=\mathrm{B}_t+\int_0^t\left[\mathbb{1}_{\{\sigma_{2\mathrm{D}_s}+T_b\circ\theta_{2\mathrm{D}_s}>s\}}\frac{G(1+\mathrm{D}_s)-G(\mathrm{D}_s)}{G(\mathrm{D}_s)(2b-a-X_s)+G(1+\mathrm{D}_s)(X_s-a)}\right.$$
$$\left.-\mathbb{1}_{\{\sigma_{2\mathrm{D}_s}+T_b\circ\theta_{2\mathrm{D}_s}\leqslant s\}}\frac{G(1+\mathrm{D}_s)-G(\mathrm{D}_s)}{G(1+\mathrm{D}_s)(2b-a-X_s)+G(\mathrm{D}_s)(X_s-a)}\right]ds.$$
$$(85)$$

4. *Sous* Q_0^G,
   a) *La variable aléatoire* D_∞ *est finie p. s. et a pour loi* ΔG:

$$Q_0^G(\mathrm{D}_\infty=n)=\Delta G(n)\qquad(n\geqslant 0).\qquad(86)$$

   b) *Soit:*

$$T_\mathrm{D}:=\inf\{t\geqslant 0,\mathrm{D}_t=\mathrm{D}_\infty\}\quad et$$
$$\overline{T}_\mathrm{D}:=\inf\{t\geqslant T_\mathrm{D},X_t=b\}=T_\mathrm{D}+T_b\circ\theta_{T_\mathrm{D}}.\qquad(87)$$

   Alors T_D *est finie p. s.*
   c) *Conditionnellement à* $\mathrm{D}_\infty=n$:

i.) Le processus $(X_t, t \leqslant T_D)$ est un mouvement brownien arrêté lorsque son nombre de descentes sur $[a, b]$ atteint n.

ii.) Le processus $(X_{T_D+t}, t \geqslant 0)$ est un processus de Bessel de dimension 3 en dessous du niveau $(2b - a)$, issu de a. Plus précisément, il existe un mouvement brownien $(\beta_t, t \geqslant 0)$, issu de 0 tel que:

$$X_{T_D+t} = a + \beta_t - \int_0^t \frac{ds}{|2b - a - X_{T_D+s}|} \qquad (t \geqslant 0). \qquad (88)$$

On a alors, de deux choses l'une:

- *ou $\sup\limits_{t \geqslant T_D} X_t < b$ (et cet évènement arrive avec probabilité $\frac{1}{2}$,*

puisque, pour un processus de Bessel $(R_t, t \geqslant 0)$ de dimension 3 issu de a, la variable aléatoire $\inf\limits_{t \geqslant 0} R_t$ suit une loi uniforme sur l'intervalle $[0, a]$). On a alors $\overline{T}_D = +\infty$ p. s., et la description du processus $(X_{T_D+t}, t \geqslant 0)$ est donnée par (88).

- *ou $\sup\limits_{t \geqslant T_D} X_t \geqslant b$ (et cet évènement est aussi de probabilité $\frac{1}{2}$).*

On a alors $\overline{T}_D < \infty$ p. s. Le processus $(X_{\overline{T}_D+t}, t \geqslant 0)$ est un processus de Bessel de dimension 3, au dessus de a, issu de b. Plus précisément, il existe un mouvement brownien $(\overline{\beta}_t, t \geqslant 0)$ issu de 0, tel que:

$$X_{\overline{T}_D+t} = b + \overline{\beta}_t + \int_0^t \frac{ds}{X_{\overline{T}_D+s} - a}. \qquad (89)$$

iii.) Les deux processus considérés en i) et ii) sont indépendants.

La démonstration des points 1 et 2 du Théorème 1.16 repose sur le:

Lemme 1.17 *Soit* $H : \mathbb{N} \longrightarrow \mathbb{R}^+$ *telle que* $\sum\limits_{n \geqslant 0} H(n) < \infty$. *Alors, pour tout x réel:*

$$\lim_{t \to \infty} \sqrt{t}\, E_x\left(H(D_t)\right) = 2(b - a)\sqrt{\frac{2}{\pi}}\left\{ \sum_{n \geqslant 1} H(n) + H(0)\left(\frac{1}{2} + \frac{|x - b|}{2(b - a)}\right) \right\}.$$

$$(90)$$

2 QUELQUES GÉNÉRALISATIONS DU THÉORÈME DE PITMAN RELATIVES AU MOUVEMENT BROWNIEN ET À SON MAXIMUM UNILATÈRE

Summary To a real valued and continuous process $(X_t, t \geqslant 0)$, we associate its maximum process:

$$S_t^X = \sup_{s \leqslant t} X_s, \quad \text{and} \quad R_t^X = 2S_t^X - X_t, \quad t \geqslant 0.$$

We show that, for several processes $(X_t, t \geqslant 0)$ which are not Markovian, but such that the pair $(S_t^X, X_t; t \geqslant 0)$ is Markovian, the process $(R_t^X, t \geqslant 0)$ is a Markov process in its own filtration. This result generalizes a famous theorem of J. Pitman which establishes that, when $(X_t, t \geqslant 0)$ is a real-valued Brownian motion starting from 0, then $(R_t^X, t \geqslant 0)$ is a 3-dimensional Bessel process.

Résumé. Soit $(X_t, t \geqslant 0)$ un processus continu à valeurs réelles. Nous notons

$$S_t^X = \sup_{s \leqslant t} X_s, \qquad R_t^X = 2S_t^X - X_t.$$

Nous décrivons plusieurs processus $(X_t, t \geqslant 0)$ non markoviens tels que le processus $(R_t^X, t \geqslant 0)$ soit markovien dans sa propre filtration. Nous généralisons ainsi un célèbre Théorème de J. Pitman qui établit que, lorsque $(X_t, t \geqslant 0)$ est un mouvement brownien issu de 0, alors $(R_t^X, t \geqslant 0)$ est, dans sa propre filtration, un processus de Bessel de dimension 3 issu de 0.

2.1 Introduction

a) NOTATIONS. A un processus stochastique $\{(X_t, t \geqslant 0); (\mathcal{F}_t, t \geqslant 0, P)\}$, continu, à valeurs réelles, nous associons les processus

$$S_t^X = \sup_{s \leqslant t} X_s, \qquad R_t^X = 2S_t^X - X_t, \qquad J_t^X = \inf_{s \geqslant t} R_s^X, \qquad (1)$$

et nous notons $(\mathcal{R}_t^X, t \geqslant 0)$ la filtration naturelle du processus $(R_t^X, t \geqslant 0)$. Bien sûr, $\mathcal{R}_t^X \subset \mathcal{F}_t$.

b) Si $(X_t, \mathcal{F}_t, t \geqslant 0, P)$ est un mouvement brownien linéaire, issu de 0, un célèbre Théorème de J. Pitman (cf [14], voir aussi [6]) établit que $(R_t^X, t \geqslant 0)$, dans sa filtration naturelle $(\mathcal{R}_t^X, t \geqslant 0)$ est un processus de Bessel de dimension 3 issu de 0. De plus, $S_t^X = J_t^X$, $\mathcal{R}_t^X \subsetneq \mathcal{F}_t$ et la relation d'entrelacement suivante a lieu: si f est une fonction mesurable positive:

$$\mathrm{E}\left(f(S_t^X)|\mathcal{R}_t^X\right) = \mathrm{E}\left(f(S_t^X - X_t)|\mathcal{R}_t^X\right) = \Lambda f(R_t^X),$$

avec : $$\Lambda f(r) = \tfrac{1}{r} \int_0^r f(x)dx \qquad (r \geqslant 0).$$
$$(2)$$

c) Soit λ un réel. Si $(X_t^\lambda, \mathcal{F}_t, t \geqslant 0, P)$ est un mouvement brownien avec dérive λ, alors le processus $(R_t^{X^\lambda}, t \geqslant 0)$ est, dans sa propre filtration, un processus de Markov homogène (cf [15]). Sa loi $P^{(\lambda)}$ est celle de la solution $(R_t^\lambda, t \geqslant 0)$ de l'équation différentielle stochastique:

$$R_t = \beta_t + \int_0^t \lambda \coth(\lambda R_s)ds, \qquad (3)$$

où $(\beta_t, t \geqslant 0)$ est un mouvement brownien issu de 0. $P^{(0)}$ est ainsi la loi du processus de Bessel de dimension 3 issu de 0. La propriété d'entrelacement décrite en b) se généralise ici en:

$$E\left(f(S_t^{X^\lambda})|\mathcal{R}_t^{X^\lambda}\right) = \Lambda^{(\lambda)}f(R_t^{X^\lambda}),$$

avec:

$$\Lambda^{(\lambda)}f(r) = \frac{1}{\int_0^r e^{2\lambda x}dx}\int_0^r e^{2\lambda x}f(x)dx. \qquad (4)$$

d) Soit λ réel et $(Y_t^\lambda, t \geqslant 0)$ la solution de:

$$Y_t = \beta_t + \int_0^t \lambda\,\mathrm{th}(\lambda Y_s)ds, \qquad (5)$$

où $(\beta_{t,\geqslant 0})$ est un mouvement brownien issu de 0. Alors (cf [16]) le processus $(R_t^{Y^\lambda}, t \geqslant 0)$ est markovien dans sa propre filtration, et sa loi est $P^{(\lambda)}$, i. e. la même que celle de$(R_t^{X^\lambda}, t \geqslant 0)$ décrite au point précédent. En fait, cette propriété résulte très simplement du fait que les processus $(Y_t^\lambda, t \geqslant 0)$ et $(X_t + \epsilon\lambda t, t \geqslant 0)$ ont même loi (où, dans cette dernière expression, $(X_t, t \geqslant 0)$ est un mouvement brownien, et ϵ une variable aléatoire de Bernoulli symétrique, indépendante de $(X_t, t \geqslant 0)$).

e) L. C. G. Rogers (cf [16]) a établi une réciproque aux résultats précédents en montrant que, parmi toutes les "bonnes diffusions" markoviennes $(X_t, t \geqslant 0)$, les seules pour lesquelles $(R_t^X, t \geqslant 0)$ soit markovien dans sa propre filtration sont, essentiellement, celles décrites aux points b), c) et d) ci-dessus.

f) Notre but est ici de trouver des processus $(X_t, t \geqslant 0)$ tels que $(R_t^X, t \geqslant 0)$ soit, dans sa propre filtration, markovien et homogène. Nous dirons qu'un tel processus possède la propriété de Pitman. Bien sûr, d'après le résultat de Rogers, il va nous falloir chercher $(X_t, t \geqslant 0)$ non markovien. En fait, les processus X que nous allons trouver seront tels que le couple $(X_t, S_t^X, t \geqslant 0)$ soit markovien. Nous disons d'un tel processus X qu'il est max-markovien.

g) Expliquons comment nous allons construire certains processus $(X_t, t \geqslant 0)$ non markoviens mais possédant néanmoins la propriété de Pitman. Soit $(\Omega, X_t, \mathcal{F}_t, t \geqslant 0, W)$ le mouvement brownien canonique issu de 0, i. e. : $\Omega = \mathcal{C}([0, \infty[\to \mathbb{R})$, $(X_t, t \geqslant 0)$ sont les coordonnées de cet espace et W est la mesure de Wiener. Soit $(M_t, t \geqslant 0)$ une $(\mathcal{F}_t, W)$ martingale positive telle que $M_0 = 1$. Définissons une nouvelle probabilité Q sur $(\Omega, \mathcal{F}_\infty)$ par la formule:

$$Q(\Lambda_s) = E_W\left(1_{\Lambda_s}M_s\right) \qquad (\Lambda_s \in \mathcal{F}_s). \qquad (6)$$

Si nous choisissons convenablement $(M_t, t \geqslant 0)$, le processus canonique $(X_t, t \geqslant 0, \mathcal{F}_t, t \geqslant 0)$ sous Q est un bon candidat pour satisfaire la propriété de Pitman. En effet, pour toute fonctionnelle F, on a:

$$\begin{aligned}
E_Q\left(F(R_u^X, u \leqslant t)\right) &= E_W\left(F(R_u^X, u \leqslant t)M_t\right) \\
&= E_W\left(F(R_u^X, u \leqslant t)E(M_t|\mathcal{R}_t^X)\right), \qquad (7)
\end{aligned}$$

où $(R_t^X, t \geqslant 0)$ est, sous W, un processus de Bessel de dimension 3. Reste alors à calculer $\mathrm{E}(\mathrm{M}_t | \mathcal{R}_t^X)$. Bien sûr, $(\mathrm{E}(\mathrm{M}_t | \mathcal{R}_t^X), t \geqslant 0, \mathcal{R}_t^X, W)$ est une martingale. Nous allons distinguer trois situations:

i) CAS n°1. La martingale $(\mathrm{M}_t, t \geqslant 0)$ s'écrit $\mathrm{M}_t = f(\mathrm{S}_t^X, X_t)$, pour une fonction f régulière de $\mathbb{R}_+ \times \mathbb{R}$ dans $\mathbb{R}$. Dans ce cas, le processus $\mathrm{E}(\mathrm{M}_t | \mathcal{R}_t^X)$ est constant puisque, d'après (2), $\mathrm{E}(\mathrm{M}_t | \mathcal{R}_t^X)$ est une fonction de R_t^X, et les seules martingales positives de la forme $h(R_t^X)$ sont les constantes. D'après (7), le processus $(R_t^X, t \geqslant 0, \mathcal{R}_t^X, t \geqslant 0)$ est alors sous Q un processus de Bessel de dimension 3 (cf Théorème 2.1 ci-dessous).

ii) CAS n°2. La martingale $(\mathrm{M}_t, t \geqslant 0)$ est de la forme $\mathrm{M}_t = e^{-\frac{\lambda^2 t}{2}}$ $\times f(\mathrm{S}_t^X, X_t)$ pour un λ réel. Grâce à (2), nous savons calculer $\mathrm{E}(\mathrm{M}_t | \mathcal{R}_t^X)$ et l'on a:

$$\mathrm{E}(\mathrm{M}_t | \mathcal{R}_t^X) = e^{-\frac{\lambda^2}{2}t} \frac{\mathrm{sh}(\lambda R_t^X)}{\lambda R_t^X}.$$

La formule (7) définit alors le processus $(R_t^X, t \geqslant 0)$ sous Q comme une h-transformée de Doob du processus de Bessel $(R_t^X, t \geqslant 0)$ sous W, et la loi de $(R_t^X, t \geqslant 0)$ sous Q est $P^{(\lambda)}$ (cf Théorèmes 2.3 et 2.5 ci-dessous).

iii) CAS n°3. La martingale $(\mathrm{M}_t, t \geqslant 0)$ est de la forme $\mathrm{M}_t = f(\mathrm{S}_t^X, X_t)$ $\times \exp\left\{-\frac{1}{2} \int_0^t l(R_s^X) ds\right\}$. Nous savons dans ce cas encore calculer $\mathrm{E}(\mathrm{M}_t | \mathcal{R}_t^X)$. Le processus $(R_t^X, t \geqslant 0, \mathcal{R}_t^X, t \geqslant 0)$ sous Q est alors un processus de diffusion homogène dont nous savons calculer le générateur (cf Théorème 2.6 ci-dessous).

iv) CAS n°4. La martingale $(\mathrm{M}_t, t \geqslant 0)$ est une martingale quelconque. Bien que nous sachions, de façon abstraite, calculer $\mathrm{E}(\mathrm{M}_t | \mathcal{R}_t^X)$, nous ne sommes pas capables d'exploiter ce calcul pour décrire le processus $(R_t^X, t \geqslant 0, \mathcal{R}_t^X, t \geqslant 0)$ sous Q. Nous ne traiterons donc pas ce cas général dans ce travail. Un cas un peu moins général et néanmoins très intéressant est:

$$\mathrm{M}_t = f(\mathrm{S}_t^X, X_t) \exp\left\{-\frac{1}{2} \int_0^t g(\mathrm{S}_s^X, X_s) ds\right\}$$

L'étude complète de ce cas permettrait bien sûr d'unifier les études des cas n° 1, 2 et 3.

h) Signalons enfin une autre extension du Théorème de Pitman, dûe à H. Matsumoto et M. Yor (cf [11], [12], [13]). Soit $(X_t^{(\mu)} = X_t + \mu t, t \geqslant 0)$ un mouvement brownien avec dérive μ. Définissons:

$$Z_t^{(\mu)} = \left[\exp\left(-X_t^{(\mu)}\right)\right] \int_0^t e^{2X_s^{(\mu)}} ds \qquad (8)$$

H. Matsumoto et M. Yor prouvent que le processus $(Z_t^{(\mu)}, t \geqslant 0)$ est, dans sa propre filtration, markovien et homogène, de générateur L^μ:

$$L^{\mu}f(z) = \frac{1}{2}z^2 f''(z) + \left\{(\frac{1}{2} - \mu)z + \frac{K_{1+\mu}}{K_{\mu}}\left(\frac{1}{z}\right)\right\} f'(z) \qquad (9)$$

où K_{α} désigne la fonction de Bessel-Mc Donald d'indice α, et la relation d'entrelacement suivante a lieu:

$$P\left\{X_t^{(\mu)} \in dx \ / \ \mathcal{Z}_t^{(\mu)}, \ Z_t^{(\mu)} = z\right\} = \frac{\exp(\mu x)}{2K_{\mu}\left(\frac{1}{z}\right)} \exp\left(-\frac{\mathrm{ch}\,x}{z}\right) dx \qquad (10)$$

Ce résultat est, de plusieurs points de vue, une généralisation du Théorème de Pitman:
• Le processus $(Z_t^{(\mu)}, t \geqslant 0)$ est, comme $(R_t^{X^{\mu}}, t \geqslant 0)$, markovien dans sa propre filtration.
• Il y a perte d'information quand on passe de $(X_t^{(\mu)}, t \geqslant 0)$ à $(Z_t^{(\mu)}, t \geqslant 0)$ (i. e. $\mathcal{Z}_t^{(\mu)} \subsetneq \mathfrak{X}_t^{(\mu)}$) et cette perte d'information est mesurée par (10).
• Le Théorème de Pitman classique peut être retrouvé à partir du Théorème de Matsumoto-Yor, par passage à la limite après un changement d'échelle convenable portant sur le processus $(X_t^{(\mu)}, t \geqslant 0)$.

2.2 Le cas n°1. Généralisation du Théorème de Pitman liée aux martingales d'Azéma-Yor

Soit $\varphi : \mathbb{R}_+ \longrightarrow \mathbb{R}_+$ mesurable, bornée et telle que

$$\int_0^{\infty} \varphi(x)dx = 1. \qquad (11)$$

Soit

$$\phi(x) = \int_0^x \varphi(y)dy \qquad (x \geqslant 0). \qquad (12)$$

Soit $(B_t, t \geqslant 0, \mathcal{G}_t, t \geqslant 0, P)$ un mouvement brownien issu de 0 et soit $(Y_t, t \geqslant 0)$ la solution de:

$$Y_t = B_t - \int_0^t \frac{\varphi(S_s^Y)ds}{(S_s^Y - Y_s)\varphi(S_s^Y) + 1 - \phi(S_s^Y)}. \qquad (13)$$

Théorème 2.1 *1. Le processus $(R_t^Y := 2S_t^Y - Y_t, \mathcal{R}_t^Y, t \geqslant 0)$ est un processus de Bessel de dimension 3 issu de 0.*
 2. Pour toute f mesurable et positive:

$$E\left(f(S_t^Y)|\mathcal{R}_t^Y, R_t^Y = r\right) = \Lambda^{\varphi}f(r)$$

avec $\qquad \Lambda^{\varphi}f(r) = \frac{1}{r}\int_0^r f(z)\left\{(r-z)\varphi(z) + 1 - \phi(z)\right\}dz.$

$$(14)$$

Démonstration du Théorème 2.1: Soit $(\Omega, X_t, \mathcal{F}_t, t \geqslant 0, W)$ le mouvement brownien canonique issu de 0 et soit:

$$\mathrm{M}_t^\varphi := (\mathrm{S}_t^X - X_t)\varphi(\mathrm{S}_t^X) + 1 - \phi(\mathrm{S}_t^X).$$

$(\mathrm{M}_t^\varphi, t \geqslant 0)$ est la martingale d'Azéma-Yor (cf [1]). Elle s'écrit:

$$\mathrm{M}_t^\varphi = 1 - \int_0^t \varphi(\mathrm{S}_s^X) dX_s.$$

Définissons la probabilité Q^φ sur $(\Omega, \mathcal{F}_\infty)$ par:

$$Q^\varphi(\Lambda_s) = \mathrm{E}_W\left(1_{\Lambda_s}\mathrm{M}_s^\varphi\right) \qquad (\Lambda_s \in \mathcal{F}_s). \tag{15}$$

D'après le Théorème de Girsanov (cf [21], p $311 - 313$), on a:

$$\mathrm{E}\left(F(R_s^Y, s \leqslant t)\right) = \mathrm{E}_{Q^\varphi}\left(F(R_s^X, s \leqslant t)\right)$$
$$= \mathrm{E}_W\left(F(R_s^X, s \leqslant t) \cdot \mathrm{M}_t^\varphi\right) = \mathrm{E}_W\left(F(R_s^X, s \leqslant t)\mathrm{E}(\mathrm{M}_t^\varphi|\mathcal{R}_t^X))\right), \tag{16}$$

et $\mathrm{E}\left(\mathrm{M}_t|\mathcal{R}_t^X\right) = 1$ (en utilisant par exemple (2) ou, plus efficacement, le cas i) de la discussion suivant (7))). Enfin, la formule (14) découle de(2) et de :

$$E\left[f(S_t^Y)|\mathcal{R}_t^Y, R_t^Y = r\right] = E\left[f(S_t^X)M_t^\varphi|\mathcal{R}_t^X, R_t^X = r\right]. \qquad \square$$

Remarques 2.2 *1. La solution de (13) n'est pas markovienne puisque figure dans le terme de dérive de (13) le terme S_s^Y. Par contre, le couple $(Y_t, \mathrm{S}_t^Y, t \geqslant 0)$ est markovien. Ainsi, avec notre terminologie, le processus $(Y_t, t \geqslant 0)$ est max-markovien.*

2. Le processus $(Y_t, t \geqslant 0)$ a été obtenu dans la section 1.2 par un procédé de "pénalisation" et sa structure y est complétement décrite.

3. Il n'est pas difficile de voir que toutes les martingales M_t de la forme $\mathrm{M}_t = f(\mathrm{S}_t^X, X_t)$, avec f régulière, positive et $f(0,0) = 1$ sont égales à M_t^φ, pour φ bien choisie. Cela résulte du fait que si $f(\mathrm{S}_t^X, X_t)$ est une martingale, on a: $\frac{1}{2}f''_{x^2}(s, x) = 0$ et $f'_s(s, s) = 0$

4. Remarquons que, bien que φ figure dans l'équation (13) de manière importante, la loi du processus $(R_t^Y, t \geqslant 0)$ ne dépend pas de φ.

5. Définissons, pour $x \geqslant 0$:

$$Q^{(x)}(\Lambda_s) = e^{-\frac{x^2}{2s}}\sqrt{\frac{2}{\pi s}}\mathrm{E}_0\left(1_{\Lambda_s}(x - X_s)|\mathrm{S}_s = x\right) + \mathrm{E}_0\left(1_{\Lambda_s}1_{x>\mathrm{S}_s}\right).$$

Bien sûr, la loi $Q^{(x)}$ est celle décrite dans le Théorème 1.4, égalité (18). On a alors:

$$Q_0^\varphi(\Lambda_s) := \mathrm{E}_0(1_{\Lambda_s}\mathrm{M}_s^\varphi) = \int_0^\infty Q_0^{(x)}(\Lambda_s)\varphi(x)dx$$

et, sous Q_0^φ, S_∞^Y a comme loi $\varphi(y)dy$ et est indépendante du processus de Bessel de dimension 3 $\left(R_t^Y \equiv 2\mathrm{S}_t^Y - Y_t, t \geqslant 0\right)$ (cf Théorème 1.4).

2.3 Le cas n°2. Généralisation du Théorème de Pitman liée aux martingales de Kennedy

Soit λ réel, φ et ϕ satisfaisant à (11) et (12). Soit $(B_t, t \geqslant 0, \mathcal{G}_t, t \geqslant 0, P)$ un mouvement brownien issu de 0. Soit $(Y_t^\lambda, t \geqslant 0)$ la solution de:

$$Y_t = B_t - \lambda \int_0^t \frac{\lambda \left(1 - \phi(S_s^Y)\right) \operatorname{th}\left(\lambda(S_s^Y - Y_s)\right) + \varphi(S_s^Y)}{\lambda \left(1 - \phi(S_s^Y)\right) + \varphi(S_s^Y) \operatorname{th}\left(\lambda(S_s^Y - Y_s)\right)} ds. \qquad (17)$$

Théorème 2.3 *1. Le processus $\left(R_t^{Y^\lambda} := 2S_t^{Y^\lambda} - Y_t^\lambda, \mathcal{R}_t^{Y^\lambda}, t \geqslant 0\right)$ a pour loi $P^{(\lambda)}$.*

2. Pour toute f mesurable positive:

$$E\left(f(S_t^{Y^\lambda}) | \mathcal{R}_t^{Y^\lambda}, R_t^{Y^\lambda} = r\right) = \Lambda f(r)$$

avec

$$\Lambda f(r) = \frac{\lambda}{\operatorname{sh}(\lambda r)} \int_0^r f(z) \left\{ (1 - \phi(z)) \operatorname{ch}\left(\lambda(r - z)\right) + \varphi(z) \frac{\operatorname{sh}\left(\lambda(r - z)\right)}{\lambda} \right\} dz. \qquad (18)$$

Démonstration du Théorème 2.3: Soit $(\Omega, X_t, \mathcal{F}_t, t \geqslant 0, W)$ le mouvement brownien canonique issu de 0, et soit

$$M_t^{\lambda, \varphi} := e^{-\frac{\lambda^2}{2} t} \left\{ \left(1 - \phi(S_t^X)\right) \operatorname{ch}\left(\lambda(S_t^X - X_t)\right) + \varphi(S_t^X) \frac{\operatorname{sh}\left(\lambda(S_t^X - X_t)\right)}{\lambda} \right\}. \qquad (19)$$

$(M_t^{\lambda, \varphi}, t \geqslant 0)$ est une martingale (cette martingale est, à une petite modification près, la martingale de Kennedy (cf [8])) qui s'écrit:

$$M_t^{\lambda, \varphi} = 1$$
$$- \int_0^t \left[\lambda \left(1 - \phi(S_s^X)\right) \operatorname{sh}\left(\lambda(S_s^X - X_s)\right) + \varphi(S_s^X) \operatorname{ch}\left(\lambda(S_s^X - X_s)\right) \right] dX_s. \qquad (20)$$

Définissons, sur $(\Omega, \mathcal{F}_\infty)$ la probabilité $Q^{\lambda, \varphi}$ par:

$$Q^{\lambda, \varphi}(\Lambda_s) = E_W \left(1_{\Lambda_s} M_s^{\lambda, \varphi}\right) \qquad (\Lambda_s \in \mathcal{F}_s). \qquad (21)$$

On a alors, d'après le Théorème de Girsanov (cf [21], p 311-313)

$$E\left(F(R_s^{Y^\lambda}, s \leqslant t)\right) = E_{Q^{\lambda, \varphi}}\left(F(R_s^X, s \leqslant t)\right) = E_W\left(F(R_s^X, s \leqslant t) \cdot M_t^{\lambda, \varphi}\right)$$

$$= E_W\left(F(R_s^X, s \leqslant t) \, E\left(M_t^{\lambda, \varphi} | \mathcal{R}_t^X\right)\right). \qquad (22)$$

On calcule $E\left(M_t^{\lambda,\varphi}|\mathcal{R}_t^X\right)$ grâce à (2), et on trouve, en utilisant la forme explicite de $M_t^{\lambda,\varphi}$:

$$E\left(M_t^{\lambda,\varphi}|\mathcal{R}_t^X\right) = e^{-\frac{\lambda^2}{2}t} \frac{\text{sh}\left(\lambda R_t^X\right)}{\lambda R_t^X}. \tag{23}$$

Mais on peut aussi, pour prouver (23), utiliser le fait qu'il n'existe qu'une seule fonction φ croissante telle que $\varphi(0) = 1$ et $\left(\varphi(R_t)e^{-\lambda^2 t/2}, t \geqslant 0\right)$ soit une martingale.

(22) indique alors que $(R_s^{Y^\lambda}, s \geqslant 0)$ est la h-transformée de Doob du processus de Bessel de dimension 3 $(R_t^X, t \geqslant 0)$, avec $h(t,x) = e^{-\frac{\lambda^2}{2}t}\frac{\text{sh}(\lambda x)}{\lambda x}$. Ceci permet de calculer le générateur $L^{(\lambda)}$ du processus $(R_t^{Y^\lambda}, t \geqslant 0)$, qui vaut:

$$L^{(\lambda)}f(r) = \frac{1}{2}f''(r) + (\lambda\coth(\lambda r))f'(r). \tag{24}$$

$\square$

Remarques 2.4   *1. Le cas particulier du Théorème 2.3 où $\lambda = 0$ est le Théorème 2.1.*

2. Le processus $(Y_t^\lambda, t \geqslant 0)$ n'est pas markovien: il est max-markovien.

3. La loi du processus $(R_t^{Y^\lambda}, t \geqslant 0)$ ne dépend pas de φ.

4. Le processus $(Y_t^\lambda, t \geqslant 0)$ a été obtenu dans la section 1.2 par un procédé de "pénalisation" et la structure de ce processus y est complétement décrite.

5. Il n'est pas difficile de voir que toutes les martingales de la forme

$$M_t = e^{-\frac{\lambda^2}{2}t}f(S_t^X, X_t),$$

avec f régulière, positive et $f(0,0) = 1$ s'écrivent comme $M_t = M_t^{\lambda,\varphi}$, pour une fonction φ bien choisie. En effet, si $e^{-\frac{\lambda^2}{2}t}f(S_t^X, X_t)$ est une martingale, alors

$$\frac{1}{2}f''_{x^2}(s,x) - \frac{\lambda^2}{2}f(s,x) = 0 \quad et \quad f'_s(s,s) = 0.$$

Ce système différentiel s'intègre explicitement et a pour solution:

$$f(s,x) = (1 - \phi(s))\,\text{ch}\left(\lambda(s-x)\right) + \varphi(s)\frac{\text{sh}\left(\lambda(s-x)\right)}{\lambda}.$$

Un cas particulier:

Soit λ réel et $g : \mathbb{R}_+ \longrightarrow \mathbb{R}$ une fonction de classe C^1 telle que $g(0) = 0$. Soit $(Z_t, t \geqslant 0)$ la solution de:

$$Z_t = B_t + \lambda \int_0^t \text{th}\left(\lambda Z_s + g(S_s^Z)\right) ds. \tag{25}$$

Théorème 2.5 *La loi du processus $(R_t^Z := 2S_t^Z - Z_t, t \geqslant 0)$ est $P^{(\lambda)}$.*

Démonstration du Théorème 2.5: Le Théorème 2.5 est un cas particulier du Théorème 2.1: l'équation (25) est l'équation (17) pour un choix convenable de φ. On peut aussi prouver directement le Théorème 2.5. □

2.4 Le cas n°3. Une équation différentielle stochastique dont la solution possède la propriété de Pitman

Définissons l'espace $\mathcal{H}^0$ par:

$$\mathcal{H}^0 = \left\{ F :]0, \infty[\longrightarrow \mathbb{R},\ F \text{ de classe } C^2;\ \forall x > 0,\ \int_0^x e^{2F(y)}dy < \infty, \right.$$
$$\left. \int_0^\infty e^{2F(y)}dy = \infty \right\} \tag{26}$$

et soit T l'opérateur défini sur $\mathcal{H}^0$ par:

$$TF(x) = -F'(x) + \frac{e^{2F(x)}}{\int_0^x e^{2F(y)}dy} \qquad (x > 0). \tag{27}$$

Soit $(B_t, \mathcal{G}_t, t \geqslant 0, P)$ un mouvement brownien issu de 0, et soit $F \in \mathcal{H}^0$. Soit $(Y_t, t \geqslant 0)$ la solution de:

$$Y_t = B_t + \int_0^t F'(2S_s^Y - Y_s)ds. \tag{28}$$

Théorème 2.6 *1. Le processus $(R_t^Y := 2S_t^Y - Y_t, t \geqslant 0)$ est, dans sa propre filtration, markovien et homogène de générateur L^F:*

$$L^F f(r) = \frac{1}{2}f''(r) + TF(r)f'(r). \tag{29}$$

2. Conditionnellement à R_t^Y, S_t^Y est indépendante de $\mathcal{R}_t^Y$ et, pour toute f mesurable positive:

$$E\left(f(S_t^Y)|\mathcal{R}_t^Y\right) = \Lambda^F f(R_t^Y), \qquad avec: \quad \Lambda^F f(r) = \frac{\int_0^r f(x)e^{2F(x)}dx}{\int_0^r e^{2F(x)}dx}. \tag{30}$$

Démonstration du Théorème 2.6: Elle repose sur la remarque suivante. Soit $(M_t^F, t \geqslant 0)$ la densité de Girsanov de la loi de $(Y_t, t \geqslant 0)$ par rapport à la mesure de Wiener W. Cette densité M_t^F est de la forme:

$$M_t^F = g(S_t^X, X_t) \exp\left(-\frac{1}{2}\int_0^t l(R_s^X)ds\right),$$

et nous savons calculer $E\left(M_t^F|\mathcal{R}_t^X\right)$. □

Remarques 2.7 *1. Le processus $(R_t^Y, t \geqslant 0)$, dont le générateur infinité-simal est donné par (29) est à valeurs positives, transitoire, et éventuelle-ment explosif.*

2. Si F est régulière au voisinage de 0, $\mathrm{T}F(r) \underset{r \to \infty}{\sim} \frac{1}{r}$, si bien que le processus $(R_t^Y, t \geqslant 0)$ "se comporte au voisinage de 0 comme un processus de Bessel de dimension 3".

3. Soit $F_a(x) = \log x^a$ $(x > 0)$. La fonction F_a appartient à $\mathcal{H}^0$ si et seule-ment si $a > -\frac{1}{2}$. Dans ce cas, $\mathrm{T}F_a(r) = \frac{a+1}{r}$ et le processus $(R_t^X, t \geqslant 0)$ est un processus de Bessel de dimension $d = 2a + 3 > 2$. Le cas lim-ite $a = -\frac{1}{2}$, $d = 2$ correspond à un processus de Bessel qui n'est plus transitoire.

Le Théorème 2.6 admet une réciproque. Soit $C :]0, \infty[\longrightarrow \mathbb{R}$ de classe C^2 telle que:

$$\forall x > 0, \quad \int_x^\infty e^{-2C(y)} dy < \infty, \quad \int_0^\infty e^{-2C(y)} dy = \infty. \tag{31}$$

C étant fixée, définissons la fonction F par:

$$e^{F(x)} = \frac{e^{-C(x)}}{\int_x^\infty e^{-2C(y)} dy} \qquad (x > 0). \tag{32}$$

De la formule évidente:

$$\int_x^\infty e^{-2C(y)} dy = \frac{1}{\int_0^x e^{2F(y)} dy}, \tag{33}$$

nous déduisons que, si C satisfait à (31), alors $F \in \mathcal{H}^0$ (et réciproquement). Soit $(\mathrm{B}_t, \mathcal{G}_t, t \geqslant 0, P)$ un mouvement brownien issu de 0 et soit $(R_t, t \geqslant 0)$ la solution de:

$$R_t = \mathrm{B}_t + \int_0^t C'(R_s) ds. \tag{34}$$

Théorème 2.8 *(Une réciproque du Théorème 2.6)*
Supposons que C satisfait à (31). Alors:

1. Le processus $(R_t, t \geqslant 0)$ est à valeurs positives et $\lim_{t \to \infty} R_t = +\infty$ p. s.

2. Soit $\mathrm{J}_t := \inf_{s \geqslant t} R_t$ et $X_t := 2\mathrm{J}_t - R_t$. Il existe alors un mouvement brownien $(\beta_t, t \geqslant 0)$ issu de 0 tel que:

$$X_t = \beta_t + \int_0^t F'(2\mathrm{S}_s^X - X_s) ds. \tag{35}$$

Démonstration du Théorème 2.8: Le point 1 résulte du fait que la fonction $e : \mathbb{R} \longrightarrow \mathbb{R}$ définie par

$$e(x) = -\int_x^\infty e^{-2C(y)} dy,$$

est une fonction d'échelle pour le processus $(R_t, t \geqslant 0)$ et que $e(0_+) = -\infty$, $e(+\infty) = 0$, d'après (31). La démonstration du point 2 repose sur un résultat de Saisho-Tanemura (cf [22]) que l'on peut trouver aussi dans ([23], p 46). □

Deux généralisations du Théorème 2.6:

A) Examinons la conclusion du Théorème 2.6, et en particulier la forme du générateur infinitésimal L^F du processus $(R_t^Y, t \geqslant 0)$:

$$L^F f(r) = \frac{1}{2} f''(r) + \mathrm{T}F(r) f'(r). \tag{36}$$

Si maintenant F et G sont telles que $\mathrm{T}F = \mathrm{T}G$, les deux processus $(Y_t^F, t \geqslant 0)$ et $(Y_t^G, t \geqslant 0)$ associés par (28) respectivement à F et G seront tels que $(R_t^{Y^F}, t \geqslant 0)$ et $(R_t^{Y^G}, t \geqslant 0)$ ont la même loi. C'est ce que nous allons exploiter maintenant. Fixons quelques notations. Soit, pour $\alpha > 0$:

$$\mathcal{H}^\alpha = \left\{ F :]0, \infty[\longrightarrow \mathbb{R}, \text{ de classe } \mathrm{C}^2; \quad \int_0^\infty e^{2F(y)} dy = \frac{1}{\alpha} \right\}. \tag{37}$$

Rappelons que $\mathcal{H}^0$ a été définie en (26) et que cette définition est cohérente avec celle de $\mathcal{H}^\alpha$, $\alpha > 0$.

Définitions et Proposition 2.9 *Soit, pour $F \in \bigcup_{\beta \geqslant 0} \mathcal{H}^\beta$ et $x > 0$:*

$$A(F)(x) := \int_0^x e^{2F(y)} dy, \qquad Z(F)(x) := e^{-F(x)} A(F)(x) \tag{38}$$

et, pour tout $\alpha \geqslant 0$:

$$U_\alpha(F)(x) := F(x) - \log\left(1 + \alpha A(F)(x)\right). \tag{39}$$

Alors:

i)
$$\frac{1}{A(U_\alpha F)(x)} = \frac{1}{A F(x)} + \alpha \tag{40}$$

ii)   *pour tout $\alpha \geqslant 0$, $Z \circ U_\alpha = Z$, $\mathrm{T} \circ U_\alpha = \mathrm{T}$*
$$\tag{41}$$

iii) si $F \in \mathcal{H}^\beta$, alors $U_\alpha F \in \mathcal{H}^{\alpha+\beta}$; de plus $U_\alpha U_\beta = U_{\alpha+\beta}$.

Remarques 2.10 *Les quantités $A(F)(t)$, $Z(F)(t)$ et $U_\alpha F(t)$ ont été introduites par C. Donati-Martin, H. Matsumoto et M. Yor dans [4], lorsque $(F(t), t \geqslant 0)$ est une trajectoire brownienne (et non pas comme ici une fonction déterministe). L'analogie des formules de la Proposition 2.9, et utilisées dans le Théorème 2.11 avec celles de [4] est frappante, mais demeure mystérieuse.*

Démonstration de la Proposition 2.9: Cette Proposition est la Proposition 2.1 de [4]. La démonstration en est tout à fait élémentaire et repose sur l'identité immédiate:

$$\frac{d}{dx}\left(\frac{1}{A(F)}(x)\right) = -\frac{1}{Z^2(F)(x)}. \tag{42}$$

□

On déduit alors du Théorème 2.5 et de cette Proposition 2.9 (en particulier de la relation $T \circ U_\alpha = T$) le Théorème suivant:

Théorème 2.11 *Soit $\alpha \geqslant 0$ et $F \in \mathcal{H}^0$. Soit $(B_t, \mathcal{G}_t, t \geqslant 0, P)$ un mouvement brownien issu de 0 et soit $(Y_t^\alpha, t \geqslant 0)$ la solution de:*

$$Y_t = B_t + \int_0^t (U_\alpha F)' \left(2S_s^Y - Y_s\right) ds. \tag{43}$$

Alors le processus $(R_t^\alpha := 2S_t^{Y^\alpha} - Y_t^\alpha, t \geqslant 0)$ est markovien et homogène dans sa propre filtration, de générateur infinitésimal:

$$L^F f(r) = \frac{1}{2} f''(r) + TF(r) f'(r). \tag{44}$$

En particulier, la loi de $(R_t^\alpha, t \geqslant 0)$ ne dépend pas de α.

Remarques 2.12 *a) Le calcul explicite de $(U_\alpha F)'$, à partir de (39), conduit à:*

$$(U_\alpha F)'(x) = F'(x) - \frac{\alpha e^{2F(x)}}{1 + \alpha A(F)(x)} \tag{45}$$

b) Si F est constante, $F(x) = \lambda$, on a:

$$AF(x) = xe^{2\lambda}, \qquad U_\alpha F(x) = \lambda - \log\left(1 + \alpha x e^{2\lambda}\right)$$

et $(Y_t^\alpha, t \geqslant 0)$ est solution de:

$$Y_t = B_t - \int_0^t \frac{\alpha e^{2\lambda} ds}{1 + \alpha e^{2\lambda}\left(2S_s^Y - Y_s\right)}. \tag{46}$$

Puisque $T(F)(r) = \frac{1}{r}$, le processus $(R_t^\alpha, t \geqslant 0)$ est un processus de Bessel de dimension 3.

c) Si $F(x) = \lambda x$ ($\lambda > 0$), $AF(x) = \frac{1}{2\lambda}(e^{2\lambda x} - 1)$, $U_\alpha F(x) = \lambda x - \log\left(1 + \frac{\alpha}{2\lambda}(e^{2\lambda x} - 1)\right)$, $(Y_t^\alpha, t \geqslant 0)$ est solution de:

$$Y_t = B_t - \int_0^t \left(\lambda - \frac{2\lambda \alpha e^{2\lambda\left(2S_s^Y - Y_s\right)}}{2\lambda + \alpha\left(e^{2\lambda\left(2S_s^Y - Y_s\right)} - 1\right)}\right) ds \tag{47}$$

et puisque $TF(r) = -\lambda + 2\lambda \frac{e^{2\lambda r}}{e^{2\lambda r} - 1} = \lambda \coth(\lambda r)$, le processus $(R_t^{Y^\alpha}, t \geqslant 0)$ a comme loi: $P^{(\lambda)}$, pour tout $\alpha \geqslant 0$.

B) La démonstration du Théorème 2.6 repose sur le fait que la densité de Girsanov de la loi $(Y_t, t \geqslant 0)$, solution de (28), possède une densité M_t par rapport à la mesure de Wiener de la forme:

$$M_t = h(S_t^X, X_t) \exp\left(-\frac{1}{2}\int_0^t l(2S_u^X - X_u)du\right) \qquad (48)$$

où h est de classe C^2 en x, de classe C^1 en s et où l est positive. Une question naturelle est alors: quelles sont les martingales de la forme

$$h(S_t^X, X_t) \exp\left(-\frac{1}{2}\int_0^t l(2S_u^X - X_u)du\right),$$

avec h régulière? Nous allons répondre à cette question, et pour cela fixons quelques notations. Soient H et K deux solutions linéairement indépendantes de l'équation de Sturm-Liouville:

$$s''(x) = l(x)s(x) \qquad (49)$$

choisies telles que:

$i)$ H est décroissante sur $\mathbb{R}$, positive, et $H(0) = 1$

$ii)$ $\qquad K(x) = H(x) \int_0^x \frac{dy}{H^2(y)}$ $\qquad\qquad (50)$

$iii)$ $\qquad H(x)K'(x) - K(x)H'(x) = 1.$

Un tel choix est toujours possible (cf [5] ou [21], $3^{\text{ième}}$ éd., p 550).

Proposition 2.13 *Soit* $l : \mathbb{R}^+ \longrightarrow \mathbb{R}^+$ *et* $h : \mathbb{R}^+ \times \mathbb{R} \longrightarrow \mathbb{R}$, *de classe* C^1 *en s et de classe* C^2 *en x, telles que*

$$M_t := h(S_t^X, X_t) \exp\left(-\frac{1}{2}\int_0^t l(2S_s^X - X_s)ds\right) \qquad (51)$$

soit une martingale, avec $M_0 = h(0,0) = 1$. *Alors, il existe deux fonctions* $\lambda, \mu : \mathbb{R}^+ \longrightarrow \mathbb{R}$ *liées par la relation:*

$$\frac{(\lambda H^2)'}{H} + \frac{(\mu K^2)'}{K} = 0 \qquad (52)$$

telles que:

$$h(s,x) = \lambda(s)H(2s - x) + \mu(s)K(2s - x). \qquad (53)$$

$\left[\right.$ *Notons que la relation (52) permet, par exemple, d'exprimer* λ *en fonction de* μ:

$$\lambda(s) = \frac{-\mu(s)H(s)K(s) + 1 - \widetilde{K}(s)}{H^2(s)} \qquad (54)$$

où $\widetilde{K}(s)$ *est la primitive de K qui s'annule en 0.* $\left.\right]$

Démonstration de la Proposition 2.13: La formule d'Itô appliquée à (51) montre que $(M_t, t \geqslant 0)$ est une martingale, si et seulement si:

$$\begin{cases} h''_{x^2}(s,x) = l(2s-x)h(s,x) \\ h'_s(s,s) = 0. \end{cases} \tag{55}$$

On effectue alors le changement de fonction: $g(s,x) = h(s, 2s-x)$, et le système (55) s'intègre alors explicitement.   □

Remarques 2.14 *Dans le cadre du Théorème 2.6, la martingale* M_t *densité de la loi de* $(Y_t, t \geqslant 0)$ *par rapport à la mesure de Wiener correspond à:*

$$\mu \equiv 0, \quad \lambda(s) = \frac{1}{H^2(s)}, \quad H_1(x) = \exp\left(-F(x)\right), \quad h(s,x) = \frac{H_1(2s-x)}{H_1^2(s)},$$
$$l(x) = \left(F'^2 - F''\right)(x).$$

Nous pouvons alors généraliser le Théorème 2.6. Soit λ, μ, H, K satisfaisant à (50) et (52) et soit $(Y_t, t \geqslant 0)$ la solution de:

$$Y_t = B_t - \int_0^t \frac{\lambda(S_s^Y)H'(2S_s^Y - Y_s) + \mu(S_s^Y)K'(2S_s^Y - Y_s)}{\lambda(S_s^X)H(2S_s^Y - Y_s) + \mu(S_s^Y)K(2S_s^Y - Y_s)} ds. \tag{56}$$

Théorème 2.15 *Dans sa filtration naturelle, le processus* $(R_t^Y := 2S_t^Y - Y_t, t \geqslant 0)$ *est markovien homogène. Sa loi est celle de la solution de:*

$$R_t = Y_t + \int_0^t \frac{(\widetilde{\lambda}H + \widetilde{\mu}K)'}{(\widetilde{\lambda}H + \widetilde{\mu}K)}(R_s)ds \tag{57}$$

où $(\beta_t, t \geqslant 0)$ *est un mouvement brownien issu de 0 et où* $\widetilde{\lambda}$ *(resp.* $\widetilde{\mu}$*) désigne la primitive de* λ *(resp.* μ*) s'annulant en 0.*

Le Théorème 2.6 est le cas particulier du Théorème 2.15 avec $\mu(s) \equiv 0$, $\lambda(s) = \frac{1}{H^2(s)}$, $H(x) = \exp\left(-F(x)\right)$, $l(x) = \left(F'^2 - F''\right)(x)$.

Remarques 2.16 *La question suivante nous a été posée par J. Pitman: quels sont les processus* $(X_t, t \geqslant 0)$ *à trajectoires continues, tels que* $X_0 = 0$ *et tels que* $(R_t^X = 2S_t^X - X_t, t \geqslant 0)$ *soit un processus de Bessel de dimension 3? Une réponse (partielle) à cette question est donnée par la:*

Proposition 2.17   1. *Soit* $(R_t, t \geqslant 0)$ *un processus de Bessel de dimension 3 issu de 0 et U une v.a. positive, pouvant prendre éventuellement la valeur* $+\infty$. *Soit* $\tau = \sup \{t; \ R_t < U\}$. *Définissons:*

$$X_t = \begin{cases} 2J_t^R - R_t & si \ t \leqslant \tau \\ 2U - R_t & si \ t \geqslant \tau. \end{cases}$$

Alors $S_\infty^X = U$ *et* $(R_t^X = R_t, t \geqslant 0)$ *est un processus de Bessel de dimension 3 issu de 0 dans sa propre filtration.*

2. *Réciproquement, soit* $(X_t, t \geqslant 0)$ *un processus à trajectoires continues issu de* 0, *et supposons que* $(R_t^X, t \geqslant 0)$ *soit dans sa propre filtration un processus de Bessel de dimension* 3 *issu de* 0. *Alors, il existe:*
   - *un processus de Bessel de dimension* 3 *issu de* 0 $(R_t = R_t^X, t \geqslant 0)$
   - *une v.a.* $U \geqslant 0$ $(U = S_\infty^X)$
   - *un temps aléatoire* τ $(\tau = \inf \{t; \ X_t = S_\infty^X\} = \sup \{t; \ R_t^X < S_\infty^X\})$ *tels que*

$$X_t = \begin{cases} 2J_t^R - R_t & si \ t \leqslant \tau \\ 2U - R_t & si \ t \geqslant \tau. \end{cases}$$

Notons que:

1.) le cas $S_\infty^X = \infty$ p. s. est exactement le Théorème de Pitman ;
2.) le cas $U = S_\infty^X < \infty$ p. s. et S_∞^X indépendante de $(R_t^X, t \geqslant 0)$ est celui décrit au Théorème 2.1 (cf Remarque 2.2, 5)) avec $U = S_\infty^X$ de loi $\varphi(x)dx$.

Plus généralement, on peut poser la question suivante: décrire tous les processus $(X_t, t \geqslant 0)$ à trajectoires continues et issus de 0 tels que $(2S_t^X - X_t, t \geqslant 0)$ soit de loi donnée $\mathcal{L}$, markovienne ou non. Si l'on suppose que, sous $\mathcal{L}$, les trajectoires convergent p.s. vers $+\infty$ quand $t \to \infty$, on peut donner une réponse partielle à cette question. Plus précisément, si $(2S_t^X - X_t, t \geqslant 0)$ est de loi $\mathcal{L}$, il existe un processus $(Y_t, t \geqslant 0)$ de loi $\mathcal{L}$ et une v.a. $U \geqslant 0$ (finie ou infinie) définie sur le même espace de probabilité, tels que, avec: $\tau := \inf\{t \ / \ Y_t = U\}$, on ait, en loi:

$$X_t = \begin{cases} 2J_t^Y - Y_t & si \ t \leqslant \tau \\ 2U - Y_t & si \ t \geqslant \tau. \end{cases}$$

References

1. J. Azéma, M. Yor, *"Une solution simple au problème de Skorokhod"* LNM 721, Sém. Proba. XIII Springer, (1979), p 90-115 et 625–633
2. M. Barlow, J. Pitman, M. Yor, *"On Walsh's Brownian motions"*, LNM 1372, Sém. Proba. XXIII Springer, (1989), p 275-293 et 294–314
3. C. Donati-Martin, Y. Hu, *"Penalization of the Wiener measure and principal values"*, LNM 1801, Sém. Proba XXXVI Springer, (2003), p 251–269
4. C. Donati-Martin, H. Matsumoto, M. Yor, *"Some absolute continuity relationships for anticipative transformations of geometric Brownian motions"*, Publ. Res. Ins. Math. Sci. Kyoto **37**, (2001), p 295–326
5. M. Jeanblanc, J. Pitman, M. Yor, *"The Feynman Kac formula and decomposition of brownian paths"*, Matématica Aplic. e Comp. **16**, Birkhäuser (1997), n°1, p 27–52
6. T. Jeulin, *"Un théorème de J. W. Pitman"*, LNM 721, Sém. Proba. XIII Springer, (1979) p 521–532
7. T. Jeulin, M. Yor, *"Sur les distributions de certaines fonctionnelles du mouvement brownien"*, LNM 850, Sém. Proba. XV Springer, (1981), p 210–226
8. D. Kennedy, *"Some martingales related to cumulative sum tests and single server queues"*, Stoch. Proc. and their Appl., (1976), **4** p 261–269

9. F. B. Knight, *"Brownian local times and taboo processes"*, Trans. Amer. Math. Soc. 143, (1969), p 173–185

10. F. B. Knight, *"On the sojourn times of killed Brownian motion"*, LNM 649, Sém. Proba. XII Springer, (1978) p 428–445

11. H. Matsumoto, M. Yor, *"A version of Pitman's 2M-X Theorem for geometric Brownian motions"*, CRAS Paris, Série I, 328 (1999), p 1067–1074

12. H. Matsumoto, M. Yor, *"An analogue of Pitman's 2M-X Theorem for exponential Wiener functionals. Part I: A time inversion approach"*, Nagoya Math. J. Vol. **159**, (2000), p 125–166

13. H. Matsumoto, M. Yor, *"An analogue of Pitman's 2M-X Theorem for exponential Wiener functionals. Part II: The role of the generalized inverse Gaussian laws"*, Nagoya Math. J. Vol. **162**, (2001), p 65–86

14. J. Pitman, *"One dimensional Brownian motion and the three dimensional Bessel process"*, Adv. Appl. Prob., (1975), **7** p 511–526

15. J. Pitman, L.C.G. Rogers, *"Markov functions"*, Annals of Proba., (1981), Vol. 9, n°4, p 573–582

16. L.C.G. Rogers, *"Characterizing all diffusions with 2M-X property"*, Annals of Proba., (1981), Vol. 9, n°4, p 561–572

17. B. Roynette, P. Vallois, M. Yor, *"Limiting laws associated with Brownian motion perturbed by normalized exponential weights"*, CRAS Série I **337**, (2003), p 667–673

18. B. Roynette, P. Vallois, M. Yor, *"Limiting laws associated with Brownian motion perturbed by normalized exponential weights, I"*, à paraître dans Studia Math. Hungarica (2005)

19. B. Roynette, P. Vallois, M. Yor, *"Limiting laws associated with Brownian motion perturbed by its maximum, minimum and local times, II"*, soumis pour publication à Studia Math. Hungarica

20. B. Roynette, P. Vallois, M. Yor, *Limiting laws for long Brownian bridges perturbed by their one-sided maximum, III*, soumis pour publication à Periodica Math. Hungarica (2005)

21. D. Revuz, M. Yor, *"Continuous Martingales and Brownian motion"*, Grundlehren der Math. Wis. Springer Verlag Berlin, (1999), $3^{\text{ième}}$ édition

22. Y. Saisho, H. Tanemura, *"Pitman type theorem for one dimensional diffusion processes"*, Tokyo J. Math, (1990), n°2, p 425–440

23. M. Yor, *"Some aspects of Brownian motion"*, Part II, Some recent martingale problems, Lect. in Math. ETH Zürich, Birkhäuser Verlag Basel, (1997)

Some Remarkable Properties of the Dunkl Martingales

Léonard Gallardo[1] and Marc Yor[2]

[1] Université de Tours, Laboratoire de Mathématiques, et Physique Théorique-UMR 6083, Parc de Grandmont, 37200 Tours, France
e-mail: gallardo@univ-tours.fr
[2] Université Pierre et Marie Curie, Laboratoire de Probabilités et Modèles, Aléatoires, 4, Place Jussieu 75252, 75252 Paris Cedex 05, France
e-mail: deaproba@proba.jussieu.fr

Abstract In this paper, we study a class, depending on a parameter $k \geqslant 0$, of real valued Feller processes $X^{(k)} = (X_t^{(k)})_{t>0}$, the so called Dunkl processes which are martingales satisfying the Brownian scaling property. These processes are the only martingales whose absolute value is a Bessel process. Moreover, the absolute continuity and intertwining relations valid for some pairs of Bessel processes may be generalized to Dunkl processes. The main result of the paper is a mixed chaotic representation property for the L^2 space of the martingale $X^{(k)}$ in terms of its continuous part (which is a Brownian motion) and its purely discontinuous part, a martingale γ with bracket $< \gamma >_t = 2kt$.

2000 Mathematics subject classification: 60G17, 60G44, 60J25, 60H05.

Keywords and phrases: Dunkl operator, Dunkl processes, Bessel processes, intertwining semigroups, Wiener chaos decomposition.

1 Introduction

Various motivations, originating either from physical problems, or purely mathematical ones, may lead to the study of the class of real valued Feller processes $\{(X_t)_{t\geqslant0}; (P_x)_{x\in\mathbb{R}}\}$ which, moreover, satisfy the properties:

(i) under P_x, $(X_t, t \geqslant 0)$ is a martingale;
(ii) the process $(X_t)_{t\geqslant0}$ satisfies the Brownian scaling property, i.e.:

$$\forall c > 0, \{(X_{ct}, t \geqslant 0); P_x\} \overset{(d)}{=} \{(\sqrt{c}X_t, t \geqslant 0), P_{x/\sqrt{c}}\}. \tag{1}$$

The Feller processes which also satisfy (ii) have been studied in depth by Lamperti [17], who calls them semi-stable Markov processes, and whose work has now found many applications. One should also note, however, that in [17], the processes being considered take their values in $\mathbb{R}_+$.

Among the real valued Feller processes, which also satisfy the property (i), one should cite, obviously, Brownian motion, and also the celebrated Azéma martingale (see, e.g., [3]) and some of its generalizations which we shall call the Azéma-Emery martingales ([12], [13], [14]), some of which enjoy the chaotic representation property.

In the present paper, we are interested in another class of Feller processes which satisfy (i) and (ii), and which we shall call the Dunkl processes. These processes are parametrized by $k \geqslant 0$, and have infinitesimal generator $\frac{1}{2}L_k$, where for $f \in C^2(\mathbb{R})$,

$$L_k f(x) = f''(x) + k\left(\frac{2}{x}f'(x) - \frac{f(x) - f(-x)}{x^2}\right). \tag{2}$$

Initiated by the analytical work of Dunkl [10], a study of the corresponding Feller process has been made by Rösler [23] and Rösler-Voit [24] (see also [15]). These processes may be considered as a generalization of the classical Brownian motion (obtained with $k = 0$) as L_k is the square of the first order operator:

$$T_k : f(\in C^1(\mathbb{R})) \to T_k f(x) = f'(x) + k\left(\frac{f(x) - f(-x)}{x}\right), \tag{3}$$

which is the derivation operator, perturbed by a term involving only differences : $k\left(\frac{f(x) - f(-x)}{x}\right)$.

A general theory, which was initiated by Dunkl ([9], [10]) considers the perturbations of the usual partial derivatives in $\mathbb{R}^d$ by terms with differences, associated to a root system. In this paper, we shall only discuss the case $d = 1$, for which already a number of remarkable properties of the Dunkl processes are exhibited.

In section 2, we recall the main analytical results already known about the Dunkl processes (from [23], [24], mainly), and we put some emphasis about the martingale properties of the Dunkl processes which we then call Dunkl martingales. In particular, we show the following:

Theorem 1. *The Dunkl process $X^{(k)}$, started at x, is the unique martingale whose absolute value is a Bessel process , started at $r = |x|$, with dimension $\delta = 1 + 2k$.*

At this point, it is quite natural to recall Gilat's theorem ([16]; see also [4], [5] and [7], p.153 and p.356):

Every positive submartingale is the absolute value of a martingale,

a statement which should be understood, as usual, in terms of distributions of processes. Thus, according to Theorem 1, the Dunkl martingales solve Gilat's "problem" when the submartingale is a Bessel process.

The proof of Theorem 1 necessitates to write Itô's formula for regular functions of $(X_t^{(k)})$ in a convenient manner; this is developed further in section 2, and will be useful in the sequel of the paper.

We also note, following a remark by M. Emery, that if we write $X_t^{(k)} = x + \beta_t + \sqrt{2k}\,\tilde{\gamma}_t$, where (β_t) and $(\sqrt{2k}\,\tilde{\gamma}_t)$ denote respectively the continuous and purely discontinuous martingale parts of $X^{(k)}$, then we obtain :

Theorem 2. *The pair* $(\beta_t\ ,\ \tilde{\gamma}_t)_{t \geqslant 0}$ *is a two-dimensional martingale which satisfies the Meyer structure equation :*

$$d[\beta, \beta]_t = dt \ ; \ \ d[\beta, \tilde{\gamma}]_t = 0$$

$$d[\tilde{\gamma}, \tilde{\gamma}]_t = dt - \sqrt{\tfrac{2}{k}}(x + \beta_{t-} + \sqrt{2k}\tilde{\gamma}_{t-})d\tilde{\gamma}_t.$$

Note that the last equation may be verified by identifying the jumps of both sides :

$$(\Delta\tilde{\gamma}_t)^2 = \tfrac{1}{2k}(\Delta X_t^{(k)})^2 = \tfrac{2}{k}(X_{t-}^{(k)})^2 \mathbf{1}_{(\Delta X_t^{(k)} \neq 0)}$$

$$= -\sqrt{\tfrac{2}{k}}(X_{t-}^{(k)})\Delta\tilde{\gamma}_t$$

$$= -\tfrac{1}{k}(X_{t-}^{(k)})\Delta X_t^{(k)} = \tfrac{2}{k}(X_{t-}^{(k)})^2 \mathbf{1}_{(\Delta X_t^{(k)} \neq 0)}.$$

In section 3, inspired by the constructive proofs of Gilat's Theorem (see Protter-Sharpe [21], in particular), we give a skew-product and a Lamperti representation of the Dunkl processes. In fact, we generalize this representation to a family of real valued Feller processes containing the Dunkl processes as particular cases.

In section 4, we show that the absolute continuity and intertwining relations which have been obtained previously ([6], [28], [30]) for some pairs of Bessel processes may be "lifted" to the level of their Dunkl counterparts.

In section 5, we prove one of the main results of this paper, which is the mixed chaotic representation property of any of the Dunkl martingales, with parameter $k > 0$; we explain this property, and our terminology: the Dunkl martingale $X^{(k)}$ starting at x say, may be decomposed as: $X_t^{(k)} = x + \beta_t + \gamma_t$, where (β_t) is a standard Brownian motion, and (γ_t) a purely discontinuous martingale , with $< \gamma >_t= 2kt$ (i.e.: $\gamma_t = \sqrt{2k}\,\tilde{\gamma}_t$, with our previous notation). We denote by Z^ϵ, either β (: for $\epsilon = 0$), or γ (for $\epsilon = 1$). With this notation, we define the mixed n^{th} order chaos $K_{\epsilon_1,...,\epsilon_n}$, which consists of all the random variables in $L^2(\mathcal{F}_\infty)$ of the form:

$$\int_0^\infty dZ_{u_1}^{\epsilon_1} \int_0^{(u_1)-} dZ_{u_2}^{\epsilon_2} \ldots \int_0^{(u_{n-1})-} dZ_{u_n}^{\epsilon_n} f_n(u_1, ..., u_n), \tag{4}$$

for f_n such that: $\int_{\Delta_n} du_1...du_n f_n^2(u_1, ..., u_n) < \infty$, where

$$\Delta_n = \{(u_1, \ldots, u_n); 0 < u_n < u_{n-1} \ldots < u_1\}.$$

(We note that such mixed integrals are also discussed in [7], Chap. XXI, p. 261). We now say that $X^{(k)}$ enjoys the mixed chaos decomposition property if:

$$L^2(\mathcal{F}_\infty) = \oplus_{n=0}^\infty \left(\oplus_{\{\epsilon_1,\dots,\epsilon_n\}\in\{0,1\}^n} K_{\epsilon_1,\dots,\epsilon_n} \right). \quad (5)$$

In other words, for any $Y \in L^2(\mathcal{F}_\infty)$, one can write

$$Y = E[Y] + \sum_{n=1}^\infty \left(\sum_{\epsilon\in\{0,1\}^n} K_\epsilon(f_\epsilon) \right),$$

where $K_\epsilon(f_\epsilon)$ is of the form (4) and the series converges in L^2.

Finally, we remark that, although in [1], [2] and [25] one finds situations where the solutions of several dimensional structure equations enjoy the mixed chaos decomposition property, these situations always concern processes which are a.s. uniformly bounded on compact sets of time. Thus the above result is new within the class of vector martingales satisfying a structure equation.

2 Generalities about Dunkl processes

2.1 An explicit formula for the Dunkl semigroup

M. Rösler [23] has obtained an explicit form of the semigroup $P_t^{(k)}(x,dy) = p_t^{(k)}(x,y)dy$ of the Dunkl (Markov) process, with parameter $k \geq 0$, precisely:

$$p_t^{(k)}(x,y) = \frac{\exp(-(x^2+y^2)/2t))}{c_k t^{k+1/2}} D_k(\frac{x}{\sqrt{t}}, \frac{y}{\sqrt{t}})|y|^{2k}, \quad (6)$$

where $c_k = 2^{k+1/2}\Gamma(k+1/2)$ and $D_k(x,y)$ denotes the Dunkl kernel, which is given by the formula:

$$D_k(x,y) = \frac{1}{B(1/2,k)} \int_{-1}^1 e^{uxy}(1-u)^{k-1}(1+u)^k du, \quad (7)$$

with $B(1/2,k) = \Gamma(1/2)\Gamma(k)/\Gamma(k+1/2)$.

The Dunkl kernel may also be written as:

$$D_k(x,y) = j_{k-1/2}(ixy) + \frac{xy}{2k+1}j_{k+1/2}(ixy), \quad (8)$$

where j_α is the normalized Bessel function of the first kind, with index $\alpha > -1$, which is defined for $u \in \mathbb{C}$ as:

$$j_\alpha(u) = \Gamma(\alpha+1)\sum_{n=0}^\infty (-1)^n \frac{(u/2)^{2n}}{n!\Gamma(n+\alpha+1)}. \quad (9)$$

Note that $j_\alpha(u) = 2^\alpha\Gamma(\alpha+1)\frac{J_\alpha(u)}{u^\alpha}$, where J_α is the ordinary Bessel function of the first kind, with index $\alpha > -1$. Denoting as usual by $I_\alpha(u) = i^{-\alpha}J_\alpha(iu)$

the modified Bessel function of the first kind, with index $\alpha > -1$, we remark that the Dunkl kernel may be written as:

$$D_k\left(\frac{x}{\sqrt{t}}, \frac{y}{\sqrt{t}}\right) = \frac{2^\nu \Gamma(\nu + 1)}{z^\nu}(I_\nu(z) + I_{\nu+1}(z)), \tag{10}$$

where $z = \frac{xy}{t}$, and $\nu = k - 1/2$ $(x, y \in \mathbb{R}, t > 0)$. This expression of the Dunkl kernel will be useful later when comparing the Dunkl semigroup $(P_t^{(k)})$ with a corresponding Bessel semigroup.

2.2 The Dunkl processes as martingales

Before going any further in our detailed study of the Dunkl processes, we point out the following simple but fundamental property

Proposition 1 *For any starting point $x \in \mathbb{R}$, the Dunkl process $X^{(k)}$ is a martingale under P_x, and its absolute value $|X^{(k)}|$ is a Bessel process with "dimension" $\delta = 1 + 2k$ (or index $\nu = k - 1/2$).*

Proof: The fact that $|X^{(k)}|$ is a Bessel process follows immediately from the expression of the infinitesimal generator of $X^{(k)}$ when it is restricted to even functions. Moreover, taking $f_1(x) \equiv x$, we obtain $L_k(f_1) = 0$, which implies that $X^{(k)}$ is a local martingale. But, since $|X^{(k)}|$ is a Bessel process, then, for every $t \geqslant 0$, $E_x\left[\sup_{s \leqslant t} |X_s^{(k)}|\right] < \infty$, hence $X^{(k)}$ is a martingale. □

Definition 1 *When the emphasis in our study of $X^{(k)}$ is about its martingale property, we speak of the Dunkl martingale, with parameter k or index $\nu = k - 1/2$, or dimension $\delta = 1 + 2k$.*

2.3 An intertwining property

For $k > 0$, we introduce the Dunkl intertwining operator V_k defined for $f \in C(\mathbb{R})$ by

$$V_k f(x) = \frac{1}{B(1/2, k)} \int_{-1}^{1} f(xu)(1 - u)^{k-1}(1 + u)^k du. \tag{11}$$

We note that the Dunkl kernel which is given in (7) may be written as:

$$D_k(x, y) = V_k(e_y)(x), \tag{12}$$

where $e_y(u) = \exp(yu)$.

We now describe some important intertwining properties of V_k.

Dunkl [9] has shown that V_k is an algebraic isomorphism from $\mathbb{C}[\mathbb{R}]$ onto itself which intertwines the operator T_k defined in (3) and the derivation operator $\frac{d}{dx}$, i.e.:

$$T_k V_k = V_k \frac{d}{dx}. \tag{13}$$

In [26] Trimèche has extended V_k to a topological automorphism of $C^\infty(\mathbb{R})$ which also continues to satisfy (13). As a consequence, we obtain:

$$L_k V_k = V_k \frac{d^2}{dx^2}, \tag{14}$$

which, in terms of semigroups, yields the following result:

Proposition 2 *(i) The semigroup $(P_t^{(k)})$ of the Dunkl operator of parameter $k > 0$, and the Brownian semigroup $(P_t^{(0)})$ are intertwined via the operator V_k, i.e.:*

$$P_t^{(k)} V_k = V_k P_t^{(0)}, \tag{15}$$

(ii) the semigroup $(\tilde{P}_t^{(k)})$ of the Bessel process of dimension $\delta = 1 + 2k$, and the semigroup $(\tilde{P}_t^{(0)})$ of the reflecting Brownian motion are intertwined via the operator $\tilde{V}_k$, i.e.:

$$\tilde{P}_t^{(k)} \tilde{V}_k = \tilde{V}_k \tilde{P}_t^{(0)}, \tag{16}$$

where $\tilde{V}_k \colon f \to \tilde{V}_k f(x) = \int_0^1 f(xu)(1 - u^2)^{k-1} du$.

Proof: Part (i) of the Proposition follows from (14) and is a known result, but it is worth emphasizing the relationship between (15) and (16). Integrate an even function φ on both sides of (15) ; then $V_k \varphi(x) = \tilde{V}_k \varphi(x)$ with the notation in (16), and $P_t^{(0)} \varphi \equiv \tilde{P}_t^{(0)} \varphi$ also, as well as $P_t^{(0)} \psi \equiv \tilde{P}_t^{(0)} \psi$, for ψ an even function. Finally, we have obtained (16), which is a particular case of the intertwinings between the Bessel processes, as will be discussed more completely in subsection 4.2.   □

2.4 A converse to Proposition 1

The aim of this subsection is to prove Theorem 1. For this purpose, and also for those of section 5, it will be quite convenient to dispose of an adequate form of Itô's formula, which is provided by the following result (in which we denote $X := X^{(k)}$ for the Dunkl process with parameter k):

Proposition 3 *Let X be a Dunkl martingale with parameter k starting at x. It decomposes as: $X_t = x + \beta_t + \gamma_t$, where (β_t) is a standard Brownian motion, and (γ_t) is a purely discontinuous martingale , with $< \gamma >_t = 2kt$. Then for every regular function $f : \mathbb{R} \times \mathbb{R}_+ \to \mathbb{R}$, there is the identity:*

$$f(X_t, t) = f(x, 0) + \int_0^t f_x'(X_s, s) d\beta_s$$

$$+ \int_0^t \frac{f(X_{s-}, s) - f(-X_{s-}, s)}{2X_{s-}} d\gamma_s \tag{17}$$

$$+ \int_0^t ds \left(f_s' + \frac{1}{2} L_k f \right)(X_s, s).$$

Remark: We would also like to refer to the general Itô type formula developed by Emery ([12], p. 71, Proposition 2), of which (17) is a particular case.

Proof of Proposition 3: (i) By Proposition 1, the process $Y_t = X_t^2$ is the square of a Bessel process with dimension δ, hence it satisfies both identities:

$$Y_t = x^2 + 2 \int_0^t X_{s-}(dX_s^c + dX_s^d) + [X, X]_t$$

$$= x^2 + 2 \int_0^t X_{s-}dX_s^c + \delta t, \tag{18}$$

where X^c and X^d denote respectively the continuous and purely discontinuous parts of X.

On the other hand, it is well known that (Y_t) may be written in the form:

$$Y_t = x^2 + 2 \int_0^t \sqrt{Y_s}d\tilde{\beta}_s + \delta t, \tag{19}$$

where $(\tilde{\beta}_s)$ denotes a Brownian motion. Comparing (18) and (19), we obtain: $\int_0^t X_{s-}dX_s^c = \int_0^t \sqrt{Y_s}d\tilde{\beta}_s$, so that $\int_0^t Y_s d < X^c >_s = \int_0^t Y_s ds$, for all t i.e.:

$$Y_s d < X^c >_s = Y_s ds. \tag{20}$$

However, it is a general fact that: $1_{(X_s=0)}d < X^c >_s = 0$, and, on the other hand the set $\{s : Y_s = 0\}$ has 0-Lebesgue measure. Thus it follows from (20) that $< X^c >_t = t$, hence as a consequence of Lévy's Theorem, $(X_t^c, t \geqslant 0)$ is a Brownian motion. It then follows that $< X^d >_t = (\delta - 1)t$, hence following P. A. Meyer's terminology, X^d is also a "normal" martingale. Part (i) is then proved.

(ii) To prove formula (17), we only need to identify the continuous and purely discontinuous parts of the local martingale:

$$M_t^f = f(X_t, t) - f(x, 0) - \int_0^t ds \left(f_s' + \frac{1}{2}L_k f \right)(X_s, s).$$

From the ordinary Itô's formula, we deduce that its continuous martingale part is that of $\int_0^t f_x'(X_{s-}, s)dX_s$ which yields the desired result. Then to show that its purely discontinuous martingale part is given as the stochastic integral w.r.t. (γ_s) as indicated in (17), it suffices to identify the jumps of both (M_t^f) and the stochastic integral, which is immediate. □

We now give **a proof of Theorem 1**:
Let X be a martingale whose absolute value is a Bessel process with dimension $\delta = 1 + 2k$ $(k \geqslant 0)$. It suffices to show that X is a Markov process

with generator $\frac{1}{2}L_k$, given by (2). Let $X = X^c + X^d$ be its martingale decomposition, with respective continuous and purely discontinuous parts $X^c = \beta$ and $X^d = \gamma$, as shown in Proposition 3. Since $|X|$ is continuous, the jump $\Delta X_s = X_s - X_{s-} \equiv \Delta X_s^d$ is equal to $-2X_{s-}$ when $\Delta X_s \neq 0$, and for every regular function f, the general Itô's formula writes:

$$f(X_t) = f(X_0) + \int_0^t f'(X_{s-})dX_s + \frac{1}{2}\int_0^t f''(X_s)ds$$
$$+ \sum_{s \leqslant t} \left(f(X_s) - f(X_{s-}) - f'(X_{s-})\Delta X_s \right).$$

(21)

The sum $\sum_{s \leqslant t}$ may be written as:

$$\sum_{s \leqslant t} \frac{f(-X_{s-}) - f(X_{s-}) + 2f'(X_{s-})X_{s-}}{4X_{s-}^2}(\Delta X_s^d)^2,$$

which may be compensated (since $< X^d >_t = (\delta - 1)t$) by

$$\int_0^t \frac{f(-X_{s-}) - f(X_{s-}) + 2f'(X_{s-})X_{s-}}{4X_{s-}^2}(\delta - 1)ds$$

(of course, in this integral, we could replace X_{s-} by X_s). Since $\int_0^t f'(X_{s-})dX_s$ is a local martingale, it follows immediately from (21) that:

$$f(X_t) - f(X_0) - \frac{1}{2}\int_0^t f''(X_s)ds - (\delta - 1)\int_0^t ds \frac{f(-X_s) - f(X_s) + 2f'(X_s)X_s}{4X_s^2}$$

$$\equiv f(X_t) - f(X_0) - \int_0^t \frac{1}{2}L_k f(X_s)ds$$

is a local martingale which proves that X is a Markov process with infinitesimal generator $\frac{1}{2}L_k$.   □

Proof of Theorem 2 : It follows immediately from (18) that : $2\int_0^t X_{s-}dX_s^d + [X,X]_t = \delta t$, from which, with the help of Theorem 1, Theorem 2 follows.   □

3 A skew-product and a Lamperti representation of the Dunkl process

3.1 The representation theorem

The following theorem gives a representation of a Dunkl process in terms of its absolute value - a Bessel process - and an independent Poisson process.

Theorem 3. *A Dunkl process $(X_t, t \geqslant 0)$ with index $\nu = k - \frac{1}{2} \geqslant 0$, starting from $x \neq 0$, may be represented as:*

$$X_t \overset{(a)}{=} Y_{A_t} \overset{(b)}{=} |X_t| \, (-1)^{\mathcal{N}_{A_t}^{(\lambda)}}, \; t \geqslant 0, \tag{22}$$

where $A_t = \int_0^t \frac{ds}{X_s^2}$, and $Y_u = (\exp(\beta_u + \nu u)) \, (-1)^{\mathcal{N}_u^{(\lambda)}}$ with $(\mathcal{N}_u^{(\lambda)}, u \geqslant 0)$ a Poisson process of parameter $\lambda = \frac{k}{2}$ independent from the Brownian motion β.

We shall call the representation (22) (a) a Lamperti type representation, since Lamperti's Theorem for semi-stable Markov processes taking values in $\mathbb{R}_+$, is of the same kind with, instead of Y_u here, the process $\exp(\xi_u)$, where ξ denotes a real valued Lévy process. Here we can also write trivially

$$Y_u = \exp\left(\beta_u + \nu u + i\pi \mathcal{N}_u^{(\lambda)}\right), \tag{23}$$

which looks even more like Lamperti's representation!

In fact these Lamperti representations $(22)(a)$ and $(22)(b)$ are valid for a more general class of semi stable Markov processes (which are no longer martingales) depending on two parameters $\nu \geqslant \frac{1}{2}$ and $\lambda > 0$, with infinitesimal generator

$$\frac{1}{2} L^{(\nu, \lambda)} : f \longmapsto \frac{1}{2} f''(x) + \frac{\nu + \frac{1}{2}}{x} f'(x) + \lambda \left(\frac{f(-x) - f(x)}{x^2} \right) \tag{24}$$

More precisely, we have the following result which reduces to Theorem 2 in the special case $\lambda = \frac{k}{2}$.

Theorem 4 (Theorem 3 generalized). *A Markov process $(X_t, t \geqslant 0)$, with infinitesimal generator given by (24), starting from $x \neq 0$, may be represented as:*

$$X_t = Y_{A_t} = |X_t| \, (-1)^{\mathcal{N}_{A_t}^{(\lambda)}}, \; t \geqslant 0, \tag{25}$$

where $A_t = \int_0^t \frac{ds}{X_s^2}$, and $Y_u = \exp(\beta_u + \nu u + i\pi \mathcal{N}_u^{(\lambda)})$ with $(\mathcal{N}_u^{(\lambda)}, u \geqslant 0)$ a Poisson process with parameter $\lambda > 0$, independent from the Brownian motion β.

Proof of Theorem 4: Let $(Y_u, u \geqslant 0)$ be the process defined by $X_t = Y_{A_t}$ or equivalently $Y_u = X_{\tau_u}$, where $\tau_u = \inf\{t : A_t > u\}$ (note that once Y has been defined this way, we also find $\tau_u = \int_0^u Y_s^2 ds$). It is well-known since Doeblin [8], Volkonski [27], Dynkin [11] that such time-changes respect the Markovian character of the transformed process, and that furthermore (Y_u) admits as infinitesimal generator (acting on $f \in C_b^2(\mathbb{R})$):

$$\frac{1}{2} \tilde{L}^{(\nu, \lambda)}(f)(x) = \frac{x^2}{2} L^{(\nu, \lambda)}(f)(x)$$

$$= \frac{x^2}{2} f''(x) + (\nu + \frac{1}{2})x f'(x) + \lambda(f(-x) - f(x)). \tag{26}$$

Hence the theorem will be proven once we show that, given a Brownian motion $\beta \equiv (\beta_u)_{u \geqslant 0}$, and an independent Poisson process $\mathcal{N} \equiv (\mathcal{N}_u^{(\lambda)}, u \geqslant 0)$, with parameter $\lambda > 0$, the process $Y_u = \exp(\beta_u + \nu u + i\pi\mathcal{N}_u^{(\lambda)})$ is Markovian and admits as infinitesimal generator the operator $\tilde{L}^{(\nu,\lambda)}$. But a simple application of Itô's formula to:

$$f(Y_u) \equiv f(\exp(\beta_u + \nu u + i\pi\mathcal{N}_u^{(\lambda)})),$$

proves that

$$f(Y_u) - f(Y_0) - \int_0^u \frac{1}{2}\tilde{L}^{(\nu,\lambda)}f(Y_s)ds$$

is a local martingale and the result follows.   □

It will be interesting, in the sequel of this paper, to keep generalizing results obtained first for the Dunkl processes to the two parameters family of semi-stable Markov processes with infinitesimal generators given by (24).

3.2 An application of the skew-product representation to some asymptotics of the Dunkl processes

It is well known (see, e.g., Revuz-Yor [22] chap XI) that, if $(R_u, u \geq 0)$ denotes a Bessel process with dimension $\delta \geq 2$, with $R_0 = r > 0$, then

$$\text{if } \delta = 2, \quad \frac{4}{(\log t)^2}\int_0^t \frac{ds}{R_s^2} \xrightarrow{t\to\infty} T \quad \text{in law}, \tag{27}$$

where T denotes a stable$\left(\frac{1}{2}\right)$ random variable whose Laplace transform equals $e^{-\sqrt{2\lambda}}$ $(\lambda \geq 0)$, whereas,

$$\text{if } \delta > 2, \quad \frac{2}{\log t}\int_0^t \frac{ds}{R_s^2} \xrightarrow{t\to\infty} \frac{1}{\delta - 2} \quad \text{a.s.} \tag{28}$$

With the help of the two previous results for Bessel processes, we may obtain some asymptotics for $(N_t, t \geq 0)$ the process of changes of sign for any Dunkl process with parameter $k \geq \frac{1}{2}$. Indeed, from the representation (b) in Theorem 3, we have for $\lambda = \frac{k}{2}$,

$$N_t = \mathcal{N}_{A_t}^{(\lambda)},$$

and as a consequence of the law of large numbers for $(\mathcal{N}_u^{(\lambda)}, u \geq 0)$, we have $\frac{1}{u}\mathcal{N}_u^{(\lambda)} \xrightarrow{u\to\infty} \lambda$ a.s., hence

$$\frac{N_t}{A_t} \xrightarrow{t\to\infty} \lambda \quad \text{a.s.}$$

Consequently, for $\delta = 1 + 2k \equiv 2$, we have

$$\frac{4}{(\log t)^2}N_t = \frac{4}{(\log t)^2}A_t\frac{N_t}{A_t} \xrightarrow{t\to\infty} \lambda T \quad \text{in law},$$

where $\lambda = \frac{k}{2} = \frac{1}{4}$; hence, we have precisely:

$$\frac{16}{(\log t)^2} N_t \stackrel{t \to \infty}{\longrightarrow} T \quad \text{in law.} \tag{29}$$

For $\delta = 1 + 2k > 2$, the same arguments yield:

$$\frac{2}{\log t} N_t = \frac{2}{\log t} A_t \frac{N_t}{A_t} \stackrel{t \to \infty}{\longrightarrow} \left(\frac{1}{\delta - 2}\right) \frac{k}{2} \equiv \frac{k}{2(2k - 1)} \quad \text{a.s., i.e.}$$

$$\frac{4}{\log t} N_t \stackrel{t \to \infty}{\longrightarrow} \frac{k}{2k - 1} \quad \text{a.s.} \tag{30}$$

In this case $(k > \frac{1}{2})$, it should also be possible to complete this result with a second order result of the kind:

$$\frac{1}{\sqrt{\log t}} \left(\frac{4}{\log t} N_t - \frac{k}{2k - 1}\right) \stackrel{t \to \infty}{\longrightarrow} G \quad \text{in distribution,}$$

with G a Gaussian random variable since there is also such a result for the case mentioned in (28).

3.3 An application of the skew-product representation to the computation of the Dunkl semigroup from the knowledge of the Bessel semigroup

We shall obtain an even more general result than that mentioned in the title of this subsection, as we shall compute the semigroup $(P_t^{(\nu,\lambda)})_{t>0}$ of the "generalized (ν, λ) Dunkl process" with infinitesimal generator $\frac{1}{2} L^{(\nu,\lambda)}$ given by (24) from the knowledge of $\tilde{P}_t^\nu$, the Bessel semigroup with index ν. For simplicity, we take $x > 0$, and we compute for $f \in \mathcal{C}_c(\mathbb{R})$,

$$P_t^{(\nu,\lambda)} f(x) = \mathbb{E}_x \left[f(X_t^{(\nu,\lambda)})\right]$$

$$= \mathbb{E}_x \left[f(|X_t^{(\nu,\lambda)}|) 1_{\left(\mathcal{N}_{A_t}^{(\lambda)} \text{ is even }\right)}\right]$$

$$+ \mathbb{E}_x \left[f(-|X_t^{(\nu,\lambda)}|) 1_{\left(\mathcal{N}_{A_t}^{(\lambda)} \text{ is odd }\right)}\right].$$

Since $\mathbb{P}\left(\mathcal{N}_u^{(\lambda)} \text{ is even }\right) = \frac{1}{2}(1 + \exp(-2\lambda u))$, we get:

$$P_t^{(\nu,\lambda)} f(x) = \mathbb{E}_x \left[f(X_t^{(\nu,\lambda)})\right]$$

$$= \mathbb{E}_x \left[f(|X_t^{(\nu,\lambda)}|) \tfrac{1}{2}(1 + \exp(-2\lambda A_t))\right] \tag{31}$$

$$+ \mathbb{E}_x \left[f(-|X_t^{(\nu,\lambda)}|) \tfrac{1}{2}(1 - \exp(-2\lambda A_t))\right]$$

Now denoting by $\mathbb{E}_x^\mu$ the expectation relative to the Bessel process of index μ starting from $x > 0$ and using the formula

$$\mathbb{E}_x^\mu \left(\exp(-\frac{\theta^2}{2} A_t) \Big| |X_t| = y \right) = \left(\frac{I_{\sqrt{\theta^2 + \mu^2}}}{I_\mu} \right) \left(\frac{xy}{t} \right) \tag{32}$$

given by Yor in [29] (Theorem 4.7. p.80), we deduce from (31) that the semi-group $\{P_t^{(\nu,\lambda)}\}$ may be written in the form $P_t^{(\nu,\lambda)}(x, dy) = p_t^{(\nu,\lambda)}(x, y)dy$, with

$$p_t^{(\nu,\lambda)}(x, y) = \tilde{p}_t^\nu(x, |y|) \left(1_{(y \in \mathbb{R}_+)} \frac{1}{2}(1 + \left(\frac{I_{\sqrt{\nu^2 + 4\lambda}}}{I_\nu} \right) \left(\frac{xy}{t} \right)) \right. \tag{33}$$
$$\left. + 1_{(y \in \mathbb{R}_-)} \frac{1}{2}(1 - \left(\frac{I_{\sqrt{\nu^2 + 4\lambda}}}{I_\nu} \right) \left(\frac{-xy}{t} \right)) \right)$$

where $\tilde{p}_t^\nu(x, y)$ denotes the transition density function of the Bessel process of index ν which is given in explicit form (see [22], [29]) by

$$\tilde{p}_t^\nu(x, y) = \frac{1}{t} \left(\frac{y}{x} \right)^\nu y \exp(-\frac{x^2 + y^2}{2t}) I_\nu \left(\frac{xy}{t} \right), \quad x, y > 0. \tag{34}$$

Note that in the particular case where $2\lambda = \nu + \frac{1}{2}$, and taking into account (10) and (34), we recover the expression (6) of the transition density of the Dunkl process of index $\nu = k - \frac{1}{2}$. We may in fact extend this expression (6) to $p_t^{(\nu,\lambda)}$ at least for certain values of (ν, λ), as follows:

For the expressions (33) and (34), using the notation: $z = \frac{xy}{t}$ and if $\sqrt{\nu^2 + 4\lambda} - \nu = n$ is an odd integer (i.e.: $4\lambda = n(2\nu + n)$ with n odd), we may write:

$$p_t^{(\nu,\lambda)}(x, y) = \frac{1}{t^{\nu+1}} |y|^{2\nu+1} \exp(-\frac{x^2 + y^2}{2t}) D^{(\nu,\lambda)}(z),$$

where

$$D^{(\nu,\lambda)}(z) = \frac{1}{2z^\nu}(I_\nu + I_{\nu+n})(z), \tag{35}$$

which can be easily deduced from (33) if we write $\frac{I_{\nu+n}(\xi)}{\xi^\nu} = \xi^n \frac{I_{\nu+n}(\xi)}{\xi^{\nu+n}}$ and use the fact that for all $\mu > -1$, the function $\xi \to \frac{I_\mu(\xi)}{\xi^\mu}$, is even. The case $n = 1$ corresponds to the classical Dunkl process.

In fact for any $n \in \mathbb{N}$, there is the formula (see [19], p 80)

$$\frac{I_{\nu+n}(z)}{z^\nu} = \frac{2^{-\nu} \pi^{-1/2}(-1)^n(n!)\Gamma(2\nu)}{\Gamma(\nu + 1/2)\Gamma(2\nu + n)} \int_0^\pi \exp(-z\cos t)C_n^\nu(\cos t)(\sin t)^{2\nu} dt,$$

where C_n^ν denotes the Gegenbauer polynomial of index ν, and degree n. As a consequence, we obtain, after some elementary manipulation:

$$\frac{(I_\nu + I_{\nu+n})(z)}{z^\nu} = \frac{2^{-\nu} \pi^{-1/2} \Gamma(2\nu)}{\Gamma(\nu + 1/2)} \int_{-1}^1 e^{-uz} f_{\nu,n}(u)(1 - u^2)^{\nu-1/2} du,$$

where for $u \in [-1, +1]$,

$$f_{\nu,n}(u) = \frac{1}{\Gamma(2\nu)} + (-1)^n (n!) (\Gamma(2\nu + n))^{-1} C_n^\nu(u). \tag{36}$$

Moreover it may be shown that $f_{\nu,n}(u) \geqslant 0$, from the following expression of the Gegenbauer polynomials (see [19]) :

$$C_n^\nu(\cos t) = \sum_{m=0}^{n} \frac{\Gamma(\nu + m)\Gamma(\nu + n - m)}{m!(n-m)!(\Gamma(\nu))^2} \cos((n - 2m)t).$$

In particular $C_0^\nu(\cos t) \equiv 1$, and

$$|C_n^\nu(\cos t)| \leqslant C_n^\nu(1) = \frac{(2\nu)(2\nu + 1)\ldots(2\nu + n - 1)}{n!},$$

from which we deduce

$$f_{\nu,n}(u) \geqslant \frac{1}{\Gamma(2\nu)} - (n!)(\Gamma(2\nu + n))^{-1} C_n^\nu(1) = 0.$$

Thus we have obtained the following expression for the generalized Dunkl kernel (35) (for $x, y \in \mathbb{R}$ and $z = \frac{xy}{t}$):

$$D^{(\nu,\lambda)}(z) = \frac{2^{-\nu} \pi^{-1/2} \Gamma(2\nu)}{\Gamma(\nu + 1/2)} \int_{-1}^{1} e^{-uz} f_{\nu,n}(u)(1 - u^2)^{\nu - 1/2} du,$$

where $f_{\nu,n}(u)$ is given explicitly in (36).

4 Absolute continuity and intertwining relations between two Dunkl processes

4.1 The case of two Bessel processes

We first recall some absolute continuity relations between the laws of two Bessel processes. The law of a Bessel process with index ν starting at $r > 0$ is denoted $\tilde{P}_r^\nu$, on $\mathcal{C}(\mathbb{R}_+, \mathbb{R}_+)$, R_t denotes the process of coordinates and $\mathcal{R}_t = \sigma(R_s, s \leqslant t)$ is the natural filtration. The following formulae are found in, e.g. Yor [29] and Revuz-Yor ([22], chap.XI) :

$$\tilde{P}_r^\nu|_{\mathcal{R}_t} = \left(\frac{R_t}{r}\right)^\nu \exp(-\frac{\nu^2}{2} \int_0^t \frac{ds}{R_s^2}) . \tilde{P}_r^0|_{\mathcal{R}_t}, \tag{37}$$

for $\nu \geqslant 0$ and $r > 0$, whereas for $\nu \in]-1, 0[$, there is the h-transform relation:

$$\tilde{P}_r^\nu|_{\mathcal{R}_t \cap \{t < T_0\}} = \left(\frac{R_t}{r}\right)^{2\nu} . \tilde{P}_r^{-\nu}|_{\mathcal{R}_t}, \tag{38}$$

where $T_0 = \inf\{t : R_t = 0\}$.

The following notion will be frequently used in the sequel:

Definition 2 : *If m is a bounded real random variable, the operator M : $f \longrightarrow Mf$, defined on $C(\mathbb{R})$ by*

$$Mf(x) = \mathbb{E}(f(xm)) \quad (x \in \mathbb{R}),$$

will be called the multiplication operator associated with m.

We now present intertwining relations between the semigroups of two Bessel processes, or rather between the semigroups of their squares (see e.g. Yor [28], Carmona-Petit-Yor [6]):

$$Q_t^{\delta + \delta'} \Lambda_{\delta,\delta'} = \Lambda_{\delta,\delta'} Q_t^{\delta}, \tag{39}$$

where $\{Q_t^{\alpha}\}$ denotes the semigroup of the square of the Bessel process with dimension α and $\Lambda_{\delta,\delta'}$ is the multiplication operator

$$\Lambda_{\delta,\delta'} f(x) = \mathbb{E}[f(x\beta_{\frac{\delta}{2},\frac{\delta'}{2}})], \tag{40}$$

where $\beta_{a,b}$ denotes a beta-random variable with parameters a and b. Consequently, if $\{\tilde{P}_t^{\delta}\}$ denotes the semigroup of the Bessel process with dimension δ, then we immediately deduce from (39) that

$$\tilde{P}_t^{\delta + \delta'} \tilde{\Lambda}_{\delta,\delta'} = \tilde{\Lambda}_{\delta,\delta'} \tilde{P}_t^{\delta}, \tag{41}$$

where

$$\tilde{\Lambda}_{\delta,\delta'} = \mathbb{E}[f(x\sqrt{\beta_{\frac{\delta}{2},\frac{\delta'}{2}}})]. \tag{42}$$

Comparing this formula with that in (16) (Proposition 2), we see that $\tilde{V}_k$ in (16) is the multiplication operator associated with the random variable $\sqrt{\beta_{\frac{1}{2},k}}$, which of course agrees with the integral formula for $\tilde{V}_k$ given in Proposition 2. We now study the counterpart for Dunkl processes of both the absolute continuity and the intertwining relations between two Bessel processes.

We will use the notation of section 2. More precisely, $\{P_t^{(k)}, t > 0\}$ is the semigroup of the Dunkl process $X^{(k)}$ of parameter $k > 0$, whereas for $x \in \mathbb{R}$, $P_x^{(k)}$ is the law of $X^{(k)}$ (starting at x) on $\mathcal{D}(\mathbb{R}_+, \mathbb{R})$ and $\{\mathcal{F}_t, t > 0\}$ is the natural filtration of the process.

4.2 Absolute continuity relations

Proposition 4 *For $\frac{1}{2} \leqslant k \leqslant k'$ and $x > 0$, there is the absolute continuity relationship:*

$$P_x^{(k')}|_{\mathcal{F}_t} = \left(\frac{|X_t|}{|x|}\right)^{k'-k} \left(\frac{k'}{k}\right)^{N_t} \exp(-\frac{(k')^2 - k^2}{2} \int_0^t \frac{ds}{X_s^2}).P_x^{(k)}|_{\mathcal{F}_t}, \tag{43}$$

where N_t denotes the number of sign changes of X over the time interval $[0,t]$.

Proof: Just as for the absolute continuity relationship (38), (43) may be reduced to the absolute continuity relationship between Brownian motions with two different drifts (the Cameron-Martin absolute continuity relationship), it is most natural here to deduce from (43), once the time change $\tau_u = \inf\{t : A_t > u\}$ has been performed, so that (43) is equivalent to:

$$
P_x^{(k')}|_{\mathcal{F}_{\tau_u}} = \left(\frac{|X_{\tau_u}|}{|x|}\right)^{k'-k} \left(\frac{k'}{k}\right)^{\mathcal{N}_u} \exp(-\frac{(k')^2 - k^2}{2}u).P_x^{(k)}|_{\mathcal{F}_{\tau_u}}. \tag{44}
$$

Now, the absolute continuity relationship (44) is holding between, - on the left, a pair $(\beta_u^{(\mu)}, \mathcal{N}_u^{(\tau)})$ and - on the right, a pair $(\beta_u^{(\nu)}, \mathcal{N}_u^{(\lambda)})$), both of which are made up of a Brownian motion with drift $\beta^{(\cdot)}$ and a Poissson process $\mathcal{N}^{(\cdot)}$, with respective parameters $(\mu = k' - \frac{1}{2}, \tau = k'/2)$ and $(\nu = k - 1/2, \lambda = k/2)$.

Let us denote $\mathbb{P}^{(\mu,\tau)} = W^{(\mu)} \otimes \Pi^{(\tau)}$, where $W^{(\mu)}$ is the Wiener measure (of Brownian motion with drift μ) and $\Pi^{(\tau)}$ the law of the Poisson process with parameter τ. Likewise, we have $\mathbb{P}^{(\nu,\lambda)} = W^{(\nu)} \otimes \Pi^{(\lambda)}$. It is well known that:

$$
W^{(\mu)}|_{\mathcal{G}_u} = \exp((\mu - \nu)\theta_u - (\frac{\mu^2 - \nu^2}{2})u).W^{(\nu)}|_{\mathcal{G}_u}, \tag{45}
$$

where (θ_u) denotes the coordinate process on $\mathcal{C}(\mathbb{R}_+, \mathbb{R})$, and $\mathcal{G}_u = \sigma(\theta_s, s \leqslant u)$ the natural filtration. On the other hand, we have also

$$
\Pi^{(\tau)}|_{\mathcal{H}_u} = \left(\frac{\tau}{\lambda}\right)^{\mathcal{N}_u} \exp(-(\tau - \lambda)u).\Pi^{(\lambda)}|_{\mathcal{H}_u}, \tag{46}
$$

where $\mathcal{H}_u = \sigma(\mathcal{N}_v, v \leqslant u)$ and $(\mathcal{N}_v)$ is the coordinate process.

It then remains to take the tensor product of the RN densities found in (45) and (46), and we obtain (44). □

4.3 Intertwining relations

Proposition 5 *For* $k' > k$, *the Dunkl semigroups* $\{P_t^{(k)}\}$ *and* $\{P_t^{(k')}\}$ *are intertwined as follows:*

$$
P_t^{(k')}V_{k,k'} = V_{k,k'}P_t^{(k)}, \tag{47}
$$

where $V_{k,k'}$ *is the multiplication operator associated with the random variable*

$$
v_{k,k'} = \beta_{(k,k')}\exp(i\pi\mathcal{N}_{\log 1/\beta_{(k,k')}}), \tag{48}
$$

where $(\mathcal{N}_u, u \geqslant 0)$ *is a standard Poisson process independent of the random variable*

$$
\beta_{(k,k')} \equiv \sqrt{\beta_{\frac{1}{2}+k,k'-k}}. \tag{49}
$$

Finally, formula (48) is equivalent to the following integral formula for $V_{k,k'}$:

$$V_{k,k'}f(x) = \frac{1}{B(\frac{1}{2}+k,k'-k)} \int_{-1}^{1} duf(xu)u^{2k}(1-u)^{k'-k-1}(1+u)^{k'-k}. \quad (50)$$

Proof. Assume that $V_{k,k'}$ exists and satisfies (47). Operating V_k on the right of both terms in (47), we obtain

$$P_t^{(k')}V_{k,k'}V_k = V_{k,k'}P_t^{(k)}V_k = V_{k,k'}V_kP_t^{(0)}. \quad (51)$$

Since $\{P_t^{(k')}\}$ and $\{P_t^{(0)}\}$ are intertwined with $V_{k'}$, it suffices that $V_{k'} = V_{k,k'}V_k$ holds for (47) to be valid.

We now remark that V_k is the multiplication operator associated with the random variable

$$v_k = \beta_{(k)}\exp(i\pi\mathcal{N}_{log(1/\beta_{(k)})}), \quad (52)$$

where $\beta_{(k)} \equiv \sqrt{\beta_{\frac{1}{2},k}}$.

The beta-gamma algebra and the independence of increments of $(\mathcal{N}_u, u \geqslant 0)$ now ensure that

$$v_{k'} = v_{k,k'}v_k, \quad \text{in distribution.}$$

Finally, it remains to show formula (50), which follows easily from the expression (48) of $v_{k,k'}$. □

5 A Dunkl martingale enjoys the mixed chaotic decomposition property

As announced in the Introduction, the aim of this section is to show the following result:

Theorem 5. *Let $X^{(k)}$ be a Dunkl martingale with parameter $k > 0$, natural filtration $\{\mathcal{F}_t\}$ and terminal σ-field $\mathcal{F}_\infty$. If $X_t^{(k)} = x + \beta_t + \gamma_t$ is the decomposition given by Proposition 3, then for every $Y \in L^2(\mathcal{F}_\infty)$, one can write*

$$Y = E[Y] + \sum_{n=1}^{\infty}\left(\sum_{\epsilon \in \{0,1\}^n} K_\epsilon(f_\epsilon)\right),$$

where for every $\epsilon = (\epsilon_1, \ldots, \epsilon_n)$,

$$K_\epsilon(f_\epsilon) = \int_0^\infty dZ_{u_1}^{\epsilon_1} \int_0^{(u_1)-} dZ_{u_2}^{\epsilon_2} \ldots \int_0^{(u_{n-1})-} dZ_{u_n}^{\epsilon_n} f_\epsilon(u_1, \ldots, u_n),$$

with Z^{ϵ_i} denoting, either β (: for $\epsilon_i = 0$), or γ (for $\epsilon_i = 1$) and f_ϵ is such that $\int_{\Delta_n} du_1 \ldots du_n f_\epsilon^2(u_1, \ldots, u_n) < \infty$, and

$$\Delta_n = \{(u_1, \ldots, u_n); 0 < u_n < u_{n-1} \ldots < u_1\}.$$

Reduction and principle of the proof: It suffices to obtain the representation for a family of functionals Y, which are total in $L^2(\mathcal{F}_\infty)$.
The family of polynomials

$$\Pi := X_{t_1}^{p_1} \ldots X_{t_N}^{p_N} \quad , \tag{53}$$

for $t_1 < \ldots < t_N$ and $p_1, \ldots, p_N$ integers, satisfies this totality property. Indeed, the closed vector space in $L^2(\mathcal{F}_\infty)$ generated by these polynomials Π, contains the set

$$\mathcal{E} = \{\exp(\sum_{j=1}^N \lambda_j X_{t_j}); N \in \mathbb{N}, 0 < t_1 < \ldots < t_N, \lambda_j \in \mathbb{R}\},$$

since, as $(|X_t| ; t \geqslant 0)$ is a Bessel process, these functionals are integrable, as $|X_{t_j}|$ admits all exponential moments. Now clearly, $\mathcal{E}$ is total in $L^2(\mathcal{F}_\infty)$.

In order to obtain the representation result for every Π, we shall make use of a recurrence argument, both with respect to the exponents (p_i) and to the time arguments (t_i).

Even with the preceding ideas in mind, we shall use two slightly different methods.

First proof of Theorem 5 : We simply use the form of the Itô formula for $f(X_t)$, as given in (17) above, to first develop $X_t^{p_N}$ between the times t_{N-1} and t_N. This gives an expression of the monomial $X_t^{p_N}$ involving stochastic integrals of the monomials $X_{s-}^{p_N-1}$ relative to $d\beta_s$ or $d\gamma_s$ and we proceed by induction on decreasing degrees.

Then, concerning the representation property of Π, some new terms appear; indeed, denoting by Π' the product : $X_{t_1}^{p_1} \ldots X_{t_{N-1}'}^{p_N-1}$, we get:

$$\Pi' X_{t_{N-1}}^{p_N}$$

$$\Pi' \int_{t_{N-1}}^{t_N} p_N X_s^{p_N-1} d\beta_s$$

$$\Pi' \int_{t_{N-1}}^{t_N} (1 - (-1)^{p_N}) \frac{X_{s-}^{p_N-1}}{2} d\gamma_s \tag{54}$$

$$\Pi' \int_{t_{N-1}}^{t_N} \frac{1}{2}(L_k(x^{p_N}))(X_s) ds.$$

For each of the last 3 terms, the last exponent is strictly less than p_N, and this clearly allows the recurrence argument to be developed, while the first one now concerns only $(N-1)$ times.

Second proof of Theorem 5 : : This proof is even more closely related to the proof given in the Zürich ETH volume ([30]) for the Azéma-Emery martingales chaotic decomposition property. It hinges on the following

Lemma 1. : *For every $p \in \mathbb{N}^*$, there exists a space-time harmonic polynomial $P_p(x,t)$, of degree p in the variable x such that for $s < t$:*

$$E[X_t^p | \mathcal{F}_s] = P_p(X_s, s - t)$$

Precisely, we have

$$P_p(x,t) = \frac{1}{\mathbb{E}(v_k^p)} V_k H_p(.,t)(x),$$

where $H_p(x,t)$ is the classical space-time Hermite polynomial of degree p (see [30], p.82), V_k is the intertwining operator (11) acting on the space variable and v_k is the random variable (52).

Moreover, there is the representation property :

$$P_p(X_s, s - t) = P_p(0, -t) + \int_0^s P_p^{(c)}(X_{u-}, u - t)d\beta_u \tag{55}$$

$$+ \int_0^s P_p^{(d)}(X_{u-}, u - t)d\gamma_u,$$

where we use the notation:

$$P_p^{(c)}(x, h) = \frac{\partial}{\partial x} P_p(x, h) \tag{56}$$

$$P_p^{(d)}(x, h) = \frac{P_p(x,h) - P_p(-x,h)}{2x}.$$

Proof of Lemma 1: Denoting simply y^p the function $y \longrightarrow y^p$, we remark that:

$$V_k(y^p) = \mathbb{E}((xv_k)^p) = \mathbb{E}(v_k^p)x^p. \tag{57}$$

Then the Markov property yields:

$$E[X_t^p | \mathcal{F}_s] = P_{t-s}^{(k)}(y^p)(X_s), \quad \text{a.s.}$$

But from (57) and (15), it follows that:

$$P_u^{(k)}(y^p) = \frac{1}{\mathbb{E}(v_k^p)} P_u^{(k)} V_k(y^p) = \frac{1}{\mathbb{E}(v_k^p)} V_k P_u^{(0)}(y^p),$$

which implies the first claim of the lemma, since by ([30], p.83):

$$P_u^{(0)}(y^p) = H_p(., u).$$

The identities (55) and (56) are consequences of the Itô formula as presented in Proposition 3, formula (17). □

Now to obtain the representation property for Π , we use the martingale representation of $E[X_{t_N}^{p_N}|\mathcal{F}_t]$, for t in the interval $[t_{N-1}, t_N]$.

This yields :

$$
X_{t_N}^{p_N} = P_{p_N}(X_{t_{N-1}}, t_{N-1} - t_N) + \int_{t_{N-1}}^{t_N} P_{p_N}^{(c)}(X_{s-}, s - t_N)d\beta_s
$$

$$
+ \int_{t_{N-1}}^{t_N} P_{p_N}^{(d)}(X_{s-}, s - t_N)d\gamma_s. \tag{58}
$$

Clearly, using the same notation Π' as in the first proof, we have now decomposed $\Pi' X_{t_N}^{p_N}$ into three terms, and the recurrence arguments apply. □

Remark: Note that one advantage of the second method over the first one is that formula (58) contains only 3 terms, instead of 4 in (54) for the first method... This may be significant if one is interested in a "quick" method of computation of the chaotic decomposition of a given functional.

Acknowledgment: We are grateful to M. Emery for pointing out the parenthood between our present study and [1, 2, 25].

References

[1] S. Attal, M. Emery: Equations de structure pour des martingales vectorielles. Sém.Prob. XXVIII, LNM 1583, p. 256–278, Springer (1994).

[2] S. Attal, M. Emery: Martingales d'Azéma bidimensionnelles. In: Hommage à P.A. Meyer et J. Neveu, Astérisque 236, p. 9–21 (1996).

[3] J. Azéma, M. Yor: Etude d'une martingale remarquable. Sém.Prob. XXIII, LNM 1372, p. 88–130, Springer (1989).

[4] M.T. Barlow: Construction of a martingale with given absolute value. Ann.Prob., vol. 9, (2), p. 314–320 (1981).

[5] M.T. Barlow, M. Yor: Sur la construction d'une martingale continue de valeur absolue donnée. Sem. Prob. XIV, LNM 784, p. 62–75, Springer (1980).

[6] Ph. Carmona, F. Petit, M. Yor: Beta-gamma random variables and intertwining relations between certain Markov processes. Revista Ibero Americana, 14(2), p. 311–367 (1998).

[7] C. Dellacherie, B. Maisonneuve, P.A. Meyer: Probabilités et potentiel, Chapitres XVII à XXIV. Hermann, Paris (1992).

[8] W. Doeblin: Sur l'équation de Kolmogoroff. Comptes Rendus Acad. Sc. Paris. Special issue, vol 331, p. 1059–1100, December 2000.

[9] C. Dunkl: Differential-difference operators associated to reflection groups. Trans. Amer. Math-Soc 311 (1), 167–183, (1989).

[10] C. Dunkl: Hankel Transforms Associated to Finite Reflection Groups. Contemp. Math. vol. 138, p. 123–138, (1992).

[11] E.B. Dynkin: Markov processes. Springer (1965).

[12] M. Emery: On the Azéma martingales. Sém. Prob. XXIII, LNM 1372, p. 66–87, Springer (1989).

[13] M. Emery: Sur les martingales d'Azéma (suite), Sém. Prob. XXIV, LNM 1426, p. 442–447, Springer (1990).

[14] M. Emery: Quelques cas de représentation chaotique, Sém. Prob. XXV, LNM 1485, p. 10–23, Springer (1991).

[15] L. Godefroy: Frontière de Martin sur les hypergroupes et principe d'invariance relatif au processus de Dunkl. Thèse. Université de Tours. December 2003.

[16] D. Gilat: Every non-negative submartingale is the absolute value of a martingale. Ann. Prob. 5, p. 475–481 (1977).

[17] J. Lamperti: Semi-stable Markov processes I, Zeit. für Wahr. 22, p. 205–225 (1972).

[18] B. Maisonneuve: Martingales de valeur absolue donnée, d'après Protter-Sharpe, Sem. Prob. XIII, LNM 721, p. 642–645, Springer (1979).

[19] W. Magnus, F. Oberhettinger, R.P. Soni: Formulas and Theorems for the Special Functions of Mathematical Physics. Springer-Verlag, (1966).

[20] P.A. Meyer: Construction de solutions d'équations de structure, Sem. Prob. XXIII, LNM 1372, p. 142–145, Springer (1989).

[21] Ph. Protter, M.J. Sharpe: Martingales with given absolute value, Ann Prob. 7, p. 1056–1058 (1979).

[22] D. Revuz, M. Yor: Continuous Martingales and Brownian Motion. Third ed. Springer (1999).

[23] M. Rösler: Generalized Hermite polynomials and the heat equation for Dunkl operators. Comm. Math. Phys. 192 (3), p. 519–542, (1998).

[24] M. Rösler, M. Voit: Markov processes related with Dunkl operators. Adv. in App Math. , 21 (4), p. 575–643, (1998).

[25] G. Taviot: Martingales et équations de structure : étude géométrique. Ph. D. Thesis presented in Strasbourg, March 1999.

[26] K. Trimèche: The Dunkl intertwining operator on spaces of functions and distributions and integral representation of its dual. Int. Transforms and Sp. Functions, 12 (4), p.349–374, (2002).

[27] V.A. Volkonski: Random time changes in strong Markov processes. Teor. Vero. Prim. vol.3, p.310–326, (1958).

[28] M. Yor: Une extension markovienne de l'algèbre des lois beta-gamma. C.R. Acad. Sci. Paris, Ser. I, 308 (8), p. 257–260, (1989).

[29] M. Yor: Loi de l'indice du lacet brownien, et distribution de Hartman-Watson. Zeit. für Wahr. 53, p.71–95 (1980).

[30] M. Yor: Some Aspects of Brownian Motion. Part II : Some recent martingale problems. ETH Zürich Lect. in Maths. Birkhäuser (1997).

Enroulements Browniens et Subordination dans les Groupes de Lie

Nathanaël Enriquez[1], Jacques Franchi[2], and Yves Le Jan[3]

[1] Laboratoire de Probabilités de Paris 6, 4, place Jussieu, 75252 Paris cedex 05, France
e-mail: enriquez@ccr.jussieu.fr
[2] I.R.M.A., Université Louis Pasteur, 7, rue René Descartes, 67084 Strasbourg, France
e-mail: franchi@math.u-strasbg.fr
[3] Université Paris Sud, Mathématiques, Bâtiment 425, 91405 Orsay, France
e-mail: yves.lejan@math.u-psud.fr

Résumé. L'objet de ce travail est de faire apparaître le lien entre les enroulements browniens et l'opération de subordination, et de montrer que ce lien peut être étendu à des groupes de Lie non abéliens.

1 Introduction

Le résultat de Spitzer [22], sur l'enroulement du mouvement brownien plan autour de l'origine, a suscité de multiples travaux (cf. par exemple [15, 19, 5]). Notre propos est d'une part, par l'emploi d'une échelle de temps adéquate permettant d'obtenir des résultats non asymptotiques, de faire apparaître clairement le lien entre ce type de théorème et l'opération de subordination, d'autre part de montrer qu'il est susceptible d'être étendu à des groupes de Lie non abéliens. Dans les deux premières sections, sont respectivement étudiées l'opération de subordination pour un mouvement brownien dans un groupe de Lie, et son application au processus d'enroulement. Des exemples sont présentés dans la troisième section, notamment celui de l'enroulement dans des pointes hyperboliques complexes, qui conduit à un processus de Lévy sur le groupe d'Heisenberg.

2 Subordination d'un mouvement brownien sur un groupe de Lie

Soit (X_t) un mouvement brownien gauche sur un groupe de Lie G, défini par l'équation

$$X_0 = \mathrm{Id}, \qquad (\circ\, \mathrm{d}X_t)\, X_t^{-1} = \mathrm{d}W_t,$$

où (W_t) est un mouvement brownien sur l'algèbre de Lie $\mathcal{G}$; de sorte que les accroissements à gauche $X_{t_{j+1}} X_{t_j}^{-1}$, $0 \leqslant t_0 < \cdots < t_n$, sont indépendants et homogènes.

Soit (ν_t) le semi-groupe de convolution sur G donnant la loi de (X_t).

Considérons un subordinateur (Λ_t) sans dérive de mesure de Lévy π_Λ, indépendant de X. Le processus $(s, \Delta \Lambda_s)$ est un processus de Poisson ponctuel sur $\mathbb{R}_+ \times \mathbb{R}_+$ d'intensité $\mathrm{d}s \otimes \pi_\Lambda(\mathrm{d}\ell)$.

Posons $\mathcal{F}_t := \sigma(\Lambda_s,\, s \leqslant t) \vee \sigma(W_s,\, s \leqslant \Lambda_t)$, et introduisons le processus $(Y_t := X_{\Lambda_t})$, qui est $(\mathcal{F}_t)$-adapté.

Introduisons une métrique sur $\mathcal{G}$ et notons d la distance sur G qui lui est associée. Soit $\rho > 0$ tel que $\exp|_{B(0,\rho)}$ soit injective. L'image de $B(0,\rho)$ par l'exponentielle est une boule de G centrée en l'identité et de rayon ρ.

Nous pouvons alors exprimer le générateur de (Y_t) comme suit, ce qui précise dans notre contexte la formule de Lévy–Khintchine sur le groupe G établie dans [8] (Theorem 5.1) : la mesure de Lévy du processus (Y_t) est égale à $\int_0^\infty \nu_s(\mathrm{d}y)\, \pi_\Lambda(\mathrm{d}s)$.

Proposition 1. *Soit f une fonction de G dans $\mathbb{R}$, de classe $\mathcal{C}^2$ à support compact. Alors*

$$
f(Y_.) - \int_0^{\cdot} \left(\int_{G \times \mathbb{R}_+} \left(f(Y_u\, y) - f(Y_u) - \langle \mathrm{d}f(Y_u), Y_u \exp^{-1}(y) \rangle 1_{\{d(\mathrm{Id},y)<\rho\}} \right) \right.
$$

$$
\left. \nu_s(\mathrm{d}y)\, \pi_\Lambda(\mathrm{d}s) \right) \mathrm{d}u
$$

est une $(\mathcal{F}_t)$-martingale.

Démonstration. Remarquons d'abord que $f(Y_t) - f(\mathrm{Id}) = \int_0^{\Lambda_t} \mathrm{d}(f \circ X)_u$, et découpons cette intégrale suivant les sauts de Λ de taille supérieure à $\varepsilon > 0$. Pour cela introduisons :

— $T_n^\varepsilon :=$ le $n^{\text{ième}}$ temps de saut de Λ de taille supérieure à ε.

— $N_t^\varepsilon := \mathrm{Sup}\{n \in \mathbb{N} \mid T_n^\varepsilon \leqslant t\}$.

(N_t^ε) est un processus de Poisson sur $\mathbb{R}_+$ d'intensité $\pi_\varepsilon(1)$, où $\pi_\varepsilon(\mathrm{d}x) := 1_{\{x \geqslant \varepsilon\}} \pi_\Lambda(\mathrm{d}x)$.

Nous avons $f(Y_t) - f(\mathrm{Id}) = A_t^\varepsilon + B_t^\varepsilon$, avec

$$
A_t^\varepsilon := \int_{[0, \Lambda_t] \setminus \bigcup_{\{n < N_t^\varepsilon\}} [\Lambda_{T_n^\varepsilon -}, \Lambda_{T_n^\varepsilon}]} \mathrm{d}(f \circ X)_u
$$

et

$$
B_t^\varepsilon := \sum_{n < N_t^\varepsilon} \left(f(Y_{T_n^\varepsilon}) - f(Y_{T_n^\varepsilon -}) \right).
$$

Le processus (A_t^ε) tend vers 0 dans L^1. En effet la semi-martingale $f \circ X$ se décompose en une martingale M^f et un processus à variation bornée A^f. f ayant ses dérivées bornées, $\mathrm{d}\langle M^f, M^f \rangle_t / \mathrm{d}t$ et $\mathrm{d}A_t^f / \mathrm{d}t$ sont bornées par des constantes, et donc

$$\mathbb{E}[|A_t^\varepsilon|] \leqslant \mathcal{O}(1) \times \mathbb{E}\left[\int_{\mathbb{R}_+} 1_{[0,\Lambda_t]\setminus\bigcup_{\{n<N_t^\varepsilon\}}[\Lambda_{T_n^\varepsilon-},\Lambda_{T_n^\varepsilon}]}(x)\,\mathrm{d}x\right],$$

quantité qui tend vers 0 lorsque ε tend vers 0, par convergence dominée.

Ensuite $B_t^\varepsilon = C_t^\varepsilon + D_t^\varepsilon$, avec

$$C_t^\varepsilon := \sum_{n<N_t^\varepsilon} \left(f(Y_{T_n^\varepsilon}) - f(Y_{T_n^\varepsilon}-) - \mathbb{E}[f(Y_{T_n^\varepsilon}) - f(Y_{T_n^\varepsilon}-) \,|\, \mathcal{F}_{T_n^\varepsilon}-]\right)$$

et

$$D_t^\varepsilon := \sum_{n<N_t^\varepsilon} \mathbb{E}[f(Y_{T_n^\varepsilon}) - f(Y_{T_n^\varepsilon}-) \,|\, \mathcal{F}_{T_n^\varepsilon}-].$$

Il apparaît que (C_t^ε) est une martingale. En effet C_t^ε s'écrit sous la forme

$$C_t^\varepsilon = \sum_{n\in\mathbb{N}} Z_n 1_{\{n<N_t^\varepsilon\}} = \sum_{n\in\mathbb{N}} Z_n 1_{\{T_n^\varepsilon<t\}},$$

où

$$Z_n := f(Y_{T_n^\varepsilon}) - f(Y_{T_n^\varepsilon}-) - \mathbb{E}[f(Y_{T_n^\varepsilon}) - f(Y_{T_n^\varepsilon}-) \,|\, \mathcal{F}_{T_n^\varepsilon}-]$$

vérifie $\mathbb{E}[Z_n \,|\, \mathcal{F}_{T_n^\varepsilon}-] = 0$. Le processus (C_t^ε) est donc une somme dénombrable de martingales, qui converge dans L^1 (on vérifie en effet que le reste de la somme des moments d'ordre 1 des variables ajoutées se majore par une constante fois le reste de la série des $\mathbb{P}\{N_t^\varepsilon > n\}$, qui est convergente puisque N_t^ε est une variable de Poisson).

Quant à D_t^ε, il s'écrit :

$$D_t^\varepsilon = \sum_{n<N_t^\varepsilon} \int_{\mathbb{R}_+}\int_G \left(f(Y_{(T_n^\varepsilon-)}y) - f(Y_{T_n^\varepsilon}-)\right) \nu_s(\mathrm{d}y)\,\frac{\pi_\varepsilon(\mathrm{d}s)}{\pi_\varepsilon(1)}.$$

Si on pose $H_t^\varepsilon := \int_{G\times\mathbb{R}_+} (f(Y_{t_-}y) - f(Y_{t_-})) \nu_s(\mathrm{d}y)\,\pi_\varepsilon(\mathrm{d}s)/\pi_\varepsilon(1)$, alors $D_t^\varepsilon = \int_0^t H_s^\varepsilon\,\mathrm{d}N_s^\varepsilon$ se décompose par la formule de compensation relative aux processus de Poisson composés en la somme d'une $(\mathcal{F}_t)$-martingale et de

$$E_t^\varepsilon := \int_0^t\int_{G\times\mathbb{R}_+} \left(f(Y_{u_-}y) - f(Y_{u_-})\right) \nu_s(\mathrm{d}y)\,\pi_\varepsilon(\mathrm{d}s)\,\mathrm{d}u$$

$$= \int_0^t\int_{G\times\mathbb{R}_+} \left(f(Y_u\,y) - f(Y_u)\right) \nu_s(\mathrm{d}y)\,\pi_\varepsilon(\mathrm{d}s)\,\mathrm{d}u.$$

Pour faire tendre ε vers 0, il convient d'ajouter un contre-terme.

Notons que $\int_G \langle\mathrm{d}f(Y_{u_-}), Y_{u_-}\exp^{-1}(y)\rangle 1_{\{d(\mathrm{Id},y)<\rho\}}\,\nu_s(\mathrm{d}y)$ est nul, car ν_s est invariant par $y\mapsto y^{-1}$ et $\exp^{-1}(y^{-1}) = -\exp^{-1}(y)$. Donc

$$E_t^\varepsilon = \int_0^t\int_{G\times\mathbb{R}_+} \left(f(Y_u\,y) - f(Y_u) - \langle\mathrm{d}f(Y_u), Y_u\exp^{-1}(y)\rangle\, 1_{\{d(\mathrm{Id},y)<\rho\}}\right)$$

$$\nu_s(\mathrm{d}y)\,\pi_\varepsilon(\mathrm{d}s)\,\mathrm{d}u.$$

Or

$$\left| f(Y_u\, y) - f(Y_u) - \langle \mathrm{d}f(Y_u), Y_u \exp^{-1}(y)\rangle \right| \leqslant \mathrm{Cste} \times d(\mathrm{Id}, y)^2$$

et

$$\left| f(Y_u\, y) - f(Y_u) - \langle \mathrm{d}f(Y_u), Y_u \exp^{-1}(y)\rangle\, 1_{\{d(\mathrm{Id},y)<\rho\}} \right|$$
$$\leqslant \mathrm{Cste} \times \min\{d(\mathrm{Id}, y)^2, 1\},$$

et

$$\int_G \min\{d(\mathrm{Id}, y)^2, 1\}\, \nu_s(\mathrm{d}y) = \mathbb{E}[\min\{d(X_s, \mathrm{Id})^2, 1\}] \leqslant \mathrm{Cste} \times \min\{s, 1\}\,;$$

donc

$$\int_{G\times\mathbb{R}_+} \big(f(Y_u\, y) - f(Y_u) - \langle \mathrm{d}f(Y_u), Y_u \exp^{-1}(y)\rangle\, 1_{\{d(\mathrm{Id},y)<\rho\}} \big)\, \nu_s(\mathrm{d}y)$$

est dans $L^1\big(\mathbb{P}(\mathrm{d}\omega) \otimes \pi_\Lambda(\mathrm{d}s) \otimes 1_{[0,t]}(u)\,\mathrm{d}u\big)$. Le théorème de convergence dominée donne alors la convergence dans L^1 de E_t^ε vers

$$\int_0^t \int_{G\times\mathbb{R}_+} \big(f(Y_u\, y) - f(Y_u) - \langle \mathrm{d}f(Y_u), Y_u \exp^{-1}(y)\rangle\, 1_{\{d(\mathrm{Id},y)<\rho\}} \big)$$
$$\nu_s(\mathrm{d}y)\, \pi_\Lambda(\mathrm{d}s)\, \mathrm{d}u.$$

Donc les martingales $(f(Y_t) - f(\mathrm{Id}) - A_t^\varepsilon - E_t^\varepsilon)$ convergent dans L^1.   □

3 Processus d'enroulement dans un groupe

Fixons un groupe de Lie G d'algèbre de Lie $\mathcal{G}$, un sous-groupe fermé Γ de G, une variété riemannienne $\mathcal{M}$, et un ouvert U de $\mathcal{M}$ tel que $\overline{U}$ soit différent de $\mathcal{M}$ et difféomorphe à $(\Gamma\backslash G) \times \mathbb{R}_+$. Notons $(x(z), y(z)) \in (\Gamma\backslash G) \times \mathbb{R}_+$ les coordonnées de $z \in \overline{U}$.

Considérons une diffusion récurrente (Z_t) sur $\mathcal{M}$, de générateur $\mathcal{L}$ (au sens des problèmes de martingales). Lorsque (Z_t) est récurrente positive, nous noterons μ sa mesure invariante normalisée. Notons $(\mathcal{F}_t)$ la filtration canonique de (Z_t).

Faisons l'hypothèse que dans U le générateur $\mathcal{L}$ se décompose en produit semi-direct : il existe une fonction mesurable g de $\mathbb{R}_+$ dans $\mathbb{R}_+^\star$, un générateur $\mathcal{L}_y$ sur $\mathbb{R}_+$, et un générateur $\mathcal{L}_G$ sur G tels que $\mathcal{L} = \mathcal{L}_y + g \circ y \times \mathcal{L}_G$ dans U. Précisément, $\mathcal{L}_y$ s'écrit $\mathcal{L}_y h(y) = a(y)h''(y) + b(y)h'(y)$, pour a, b boréliennes bornées sur $\mathbb{R}_+$, $\mathcal{L}_G$ s'écrit $\mathcal{L}_G = \sum_{j=1}^r Y_j^2$ pour certains champs de vecteurs lisses et invariants à gauche $Y_1, \ldots, Y_r$ sur G, et enfin pour toutes fonctions de classe C^2, f sur $\mathbb{R}_+$ et F sur $\Gamma\backslash G$, nous avons dans U :

$$\mathcal{L}\big((F \circ x) \times (f \circ y)\big) = (F \circ x) \times \big((\mathcal{L}_y f) \circ y\big) + \big((gf) \circ y\big) \times \big((\mathcal{L}_G F) \circ x\big).$$

L'enroulement dans G associé au passage de la diffusion (Z_t) dans U est naturellement défini comme la solution (θ_t) de l'équation différentielle stochastique :

$$d\theta_t = \theta_t \circ dV_t, \qquad \text{où} \qquad V_t := \int_0^t 1_U(Z_s)\, x(Z_s)^{-1} \circ dx(Z_s).$$

Posons $a_t := \int_0^t 1_U(Z_s)\, g(y(Z_s))\, ds$, et soit $\tau_s := \inf\{t > 0 \mid a_t > s\}$ son inverse à droite.

Le processus $y_{\tau_t} := y(Z_{\tau_t})$ est alors une diffusion sur $\mathbb{R}_+$. Notons L_t son temps local en 0, d'inverse à droite Λ_s. Ce dernier processus est un subordinateur dont l'exposant caractéristique sera noté $\psi : \psi(\alpha) := -\frac{1}{s} \log(\mathbb{E}[e^{-\alpha \Lambda_s}])$. Le processus $L_t^0 = L_{a_t}$ est un temps local de ∂U pour la diffusion Z_t ; son inverse à droite est évidemment $\Lambda_s^0 := \tau_{\Lambda_s}$. Notons $X_t := \theta_{\tau_t}$ l'enroulement dans l'échelle de temps τ_t, et $\Phi_s := \theta_{\Lambda_s^0} = X_{\Lambda_s}$ l'enroulement dans l'échelle de temps Λ_s^0.

Proposition 2. *La semi-martingale (X_t) est un $(\mathcal{F}_{\tau_t})$-mouvement brownien gauche indépendant de (Λ_t), à valeurs dans G, de générateur $\mathcal{L}_G$.*

Démonstration. Les processus (V_{τ_t}) et (y_{τ_t}) sont des diffusions indépendantes : (y_{τ_t}) est une diffusion sur $\mathbb{R}_+$ de générateur $(1/g(y))\mathcal{L}_y$ réfléchie en 0, et (V_{τ_t}) est un mouvement brownien sur $\mathcal{G}$ de générateur $\sum_{j=1}^r (D_{y_i})^2$ où $y_i := Y_i(\text{Id})$. Or, du fait que (V_t) est constant là où (a_t) l'est, il est clair que $dX_t = X_t \circ dV_{\tau_t}$. Le processus (X_t) est donc indépendant de (y_{τ_t}), et donc de (Λ_t). $\qquad\square$

Proposition 3. *Notons $\hat{P}$ la mesure d'Itô des excursions dans U du processus (Z_t) (ou dans $\mathbb{R}_+$ du processus $(y_t := y(Z_t))$), et notons ζ le temps de retour en 0 de (y_t) et la durée de vie d'une excursion. Fixons la normalisation de la mesure $\hat{P}$ (et donc du temps local) de sorte que $\hat{E}(\zeta) = 1$. Nous avons alors pour tout $\alpha \in \mathbb{R}_+$:*

$$\psi(\alpha) = -C^{-1} \left.\frac{d}{d\eta}\right|_{\eta=0} \mathbb{E}_\eta\left[\exp\left(-\alpha \int_0^\zeta g(y_s)\, ds\right)\right]$$

$$\text{avec} \quad C := -\lim_{\alpha \searrow 0} \alpha^{-1}\, \psi_0(\alpha)$$

en notant

$$\psi_0(\alpha) = \left.\frac{d}{d\eta}\right|_{\eta=0} \mathbb{E}_\eta(e^{-\alpha \zeta}).$$

Nota Bene. Le coefficient $\psi_0(\alpha)$ est l'exposant caractéristique du subordinateur inverse du temps local en zéro de la diffusion (y_{σ_t}), où $\sigma_t := \inf\{s \geq 0 \mid \int_0^s 1_U(Z_v)\, dv \geq t\}$.

Démonstration. La formule exponentielle pour le processus ponctuel de Poisson des excursions entraîne aussitôt (voir par exemple ([2], proposition 9.1)) :

$$\psi(\alpha) = \hat{E}\left[1 - \exp\left(-\alpha \int_0^\zeta g(y_s)\,\mathrm{d}s\right)\right].$$

Notant τ^η le temps d'atteinte de η, nous avons (voir par exemple ([20], VI.48.1))

$$\hat{E}\left[1 - \mathrm{e}^{-\alpha\zeta}\right] = \lim_{\eta\searrow 0} \hat{E}\left[1_{\{\tau^\eta < \infty\}}\,\mathbb{E}_\eta\left[1 - \mathrm{e}^{-\alpha\zeta}\right]\right]$$

$$= \lim_{\eta\searrow 0} \hat{P}\{\tau^\eta < \infty\} \times \left(1 - \mathbb{E}_\eta\left[\mathrm{e}^{-\alpha\zeta}\right]\right)$$

$$= \lim_{\eta\searrow 0}(C\,\eta)^{-1} \times \left(1 - \mathbb{E}_\eta\left[\mathrm{e}^{-\alpha\zeta}\right]\right) = -C^{-1}\left.\frac{\mathrm{d}}{\mathrm{d}\eta}\right|_{\eta=0}\mathbb{E}_\eta\left[\mathrm{e}^{-\alpha\zeta}\right],$$

pour une certaine constante C dépendant de la normalisation de $\hat{P}$, et de même

$$\hat{E}\left[1 - \exp\left(-\alpha \int_0^\zeta g(y_s)\,\mathrm{d}s\right)\right] = -C^{-1}\left.\frac{\mathrm{d}}{\mathrm{d}\eta}\right|_{\eta=0}\mathbb{E}_\eta\left[\exp\left(-\alpha\int_0^\zeta g(y_s)\,\mathrm{d}s\right)\right].$$

Enfin, le théorème de convergence monotone donne

$$1 = \hat{E}[\zeta] = \lim_{\alpha\searrow 0}\hat{E}\left[\frac{1 - \mathrm{e}^{-\alpha\zeta}}{\alpha}\right] = -\lim_{\alpha\searrow 0}(C\alpha)^{-1}\left.\frac{\mathrm{d}}{\mathrm{d}\eta}\right|_{\eta=0}\mathbb{E}_\eta\left[\mathrm{e}^{-\alpha\zeta}\right]. \qquad \square$$

Remarque 1. La relation $\mathbb{E}\left[\int_0^t 1_{\{Z_s \in U\}}\,\mathrm{d}s\right] = \mathbb{E}[L_t^0] \times \hat{E}[\zeta]$ précise la normalisation du temps local résultant de celle fixée pour $\hat{P}$ dans la proposition 3 ci-dessus, (L_t^0) étant l'inverse de (Λ_s^0) à droite, c'est à dire le temps local correspondant à la mesure d'excursions $\hat{P}$: nous avons $\mathbb{E}[L_t^0] = \mathbb{E}\left[\int_0^t 1_U(Z_s)\,\mathrm{d}s\right]$. En particulier, nous avons dans le cas ergodique : $\lim_{t\to\infty} L_t^0/t = \mu(U)$ presque sûrement.

Notons π_Λ la mesure de Lévy du subordinateur (Λ_t), de sorte que, selon la formule de Lévy–Khintchine pour les subordinateurs (voir [1], III.1), dans le cas de sauts purs, nous avons pour tout $r \in \mathbb{R}_+$

$$\psi(r) = \int_{\mathbb{R}_+}(1 - \mathrm{e}^{-ru})\,\pi_\Lambda(\mathrm{d}u), \qquad \text{et} \qquad \int_{\mathbb{R}_+}\min\{1, u\}\,\pi_\Lambda(\mathrm{d}u) < \infty.$$

La théorie de Krein fournit, dans le cas des diffusions réelles (voir [12], et aussi [13], ([2], corollaire 9.7)) une unique mesure m_Λ sur $\mathbb{R}_+$ telle que

$$\pi_\Lambda(\mathrm{d}s) = \left(\int_0^\infty \mathrm{e}^{-us}\,m_\Lambda(\mathrm{d}u)\right)\mathrm{d}s \qquad \text{et} \qquad \int_0^\infty \frac{m_\Lambda(\mathrm{d}u)}{u(1 + u)} < \infty.$$

Le résultat qui suit est maintenant une conséquence des propositions 1 et 2 (rappelons que nous avons posé $\Phi_t := \theta_{\Lambda_t^0} = X_{\Lambda_t}$) :

Théorème 1. *Le processus* (Φ_t) *est un processus de Lévy, dont la mesure de Lévy est :*

$$\int_{\mathbb{R}_+} \nu_s(\mathrm{d}g)\, \pi_\Lambda(\mathrm{d}s) = \int_0^\infty R_u(\mathrm{d}g)\, m_\Lambda(\mathrm{d}u),$$

où (ν_t) *et* (R_u) *désignent respectivement le semi-groupe et la résolvante du générateur* $\mathcal{L}_G$.

4 Exemples

4.1 Le plan

La variété $\mathcal{M}$ est ici le plan privé de 0, U est la boule ouverte (pointée) de centre 0 et de rayon R, de sorte que $G = \mathbb{R}$, $\Gamma = \mathbb{Z}$, $\Gamma \backslash G \equiv \mathbb{S}^1$.

Prenons pour (Z_t) le mouvement brownien de $\mathbb{R}^2$, i.e. pour $\mathcal{L}$ le demi-laplacien $\frac{1}{2}\Delta$ de $\mathbb{R}^2$. En coordonnées polaires

$$\Delta = \mathcal{L}_r + r^{-2}\, \frac{\partial^2}{\partial \theta^2}, \qquad \text{où} \qquad \mathcal{L}_r = \frac{\partial^2}{\partial r^2} + \frac{1}{r}\, \frac{\partial}{\partial r}.$$

Changeant $\mathbb{R}_+$ en $]0, R]$, nous sommes dans le cadre envisagé dans cet article, avec $g(r) = r^{-2}$.

D'après la section 3, (Φ_t) est le processus de Lévy à valeurs dans $\mathbb{R}$ qui a pour exposant caractéristique $\lambda \mapsto \psi(\lambda^2/2)$.

Proposition 4. *Nous avons pour* α, $R \in \mathbb{R}_+$ *(*I_ν *désignant la fonction de Bessel usuelle) :*

$$\psi^0(\alpha) = R^{-1}\sqrt{2\alpha}\, \frac{I_1\big(R\sqrt{2\alpha}\,\big)}{I_0\big(R\sqrt{2\alpha}\,\big)} \qquad et \qquad \psi(\alpha) = R^{-2}\sqrt{2\alpha}\,.$$

Remarque. Les rapports $I_{\nu+1}/I_\nu$ interviennent dans la dérive de processus de Bessel conditionnés et dans le calcul de la loi de certains temps d'occupation de processus de Bessel ; voir ([18] ; § 4, p. 310 et § 9, p. 338).

Démonstration. Nous nous ramenons au cadre de la section précédente par le changement de variable $y = \log(R/r)$, et nous appliquons la proposition 3 avec $\zeta = \inf\{t > 0 \,|\, r_t = R\}$:

$$\psi^0(\alpha) = -C^{-1}\, \frac{\mathrm{d}}{\mathrm{d}\eta}\bigg|_{\eta=0} \mathbb{E}_{Re^{-\eta}}\big[e^{-\alpha\zeta}\big],$$

et

$$\psi(\alpha) = -C^{-1}\, \frac{\mathrm{d}}{\mathrm{d}\eta}\bigg|_{\eta=0} \mathbb{E}_{Re^{-\eta}}\bigg[\exp\bigg(-\alpha \int_0^\zeta \frac{\mathrm{d}s}{r_s^2}\bigg)\bigg].$$

Pour $\alpha > 0$, cherchons f lisse sur $]0, +\infty[$, bornée près de 0, telle que

$$M_t := f(r_t) \times \exp\left(-\alpha \int_0^t r_s^{-2}\, \mathrm{d}s\right)$$

soit une martingale. Or nous avons $\mathrm{d}r_s = \mathrm{d}b_s + \frac{1}{2r_s}\,\mathrm{d}s$, pour un brownien réel (b_s), et donc la formule d'Itô montre aussitôt que (M_t) est une martingale locale ssi

$$f''(r) + \frac{1}{r}\,f'(r) - \frac{2\,\alpha}{r^2}\,f(r) = 0.$$

On trouve ainsi $f(r) = r^{\sqrt{2\alpha}}$ (de sorte que les martingales M_t qu'on retrouve ici sont celles qui interviennent dans les relations d'absolue continuité entre les différents processus de Bessel, voir ([21], chapitre XI)), et donc en appliquant le théorème d'arrêt :

$$\mathbb{E}_{R\,\mathrm{e}^{-\eta}}\left[\exp\left(-\alpha \int_0^{\inf\{t>0\,|\,r_t=R\}} \frac{\mathrm{d}s}{r_s^2}\right)\right] = \frac{f(R\,\mathrm{e}^{-\eta})}{f(R)} = \mathrm{e}^{-\eta\sqrt{2\alpha}}.$$

De même, pour le calcul de $\psi^0(\alpha)$ nous voulons h lisse sur $]0,\infty[$ et bornée près de 0 telle que $h(r_t)\,\mathrm{e}^{-\alpha t}$ soit une martingale, et donc telle que $h''(r) + h'(r)/r - 2\,\alpha\,h(r) = 0$. Les solutions bornées près de 0 sont proportionnelles à $h(r) = I_0(r\sqrt{2\alpha})$, I_0 désignant la fonction de Bessel usuelle, de sorte que le théorème d'arrêt nous donne (voir aussi [18]) :

$$\mathbb{E}_{R\,\mathrm{e}^{-\eta}}\left[\exp\left(-\alpha \inf\{t>0\,|\,r_t=R\}\right)\right] = \frac{I_0\left(R\,\mathrm{e}^{-\eta}\sqrt{2\alpha}\right)}{I_0\left(R\sqrt{2\alpha}\right)},$$

et donc

$$-\frac{\mathrm{d}}{\mathrm{d}\eta}\bigg|_{\eta=0} \mathbb{E}_{R\,\mathrm{e}^{-\eta}}\left[\mathrm{e}^{-\alpha\zeta}\right] = R\sqrt{2\alpha}\,\frac{I_0'\left(R\sqrt{2\alpha}\right)}{I_0\left(R\sqrt{2\alpha}\right)}$$

$$= R\sqrt{2\alpha}\,\frac{I_1\left(R\sqrt{2\alpha}\right)}{I_0\left(R\sqrt{2\alpha}\right)} = R^2\alpha + \mathcal{O}(\alpha^2).$$

Finalement nous avons $C = R^2$, d'où le résultat.   □

On en déduit (notons qu'un manuscrit de D. Williams, non publié mais repris dans [17], contenait déjà la remarque que $\theta_{\inf\{s;|Z_s|>e^t\}}$ a une loi de Cauchy) :

Corollaire 1. *Le processus (Φ_t) est un processus de Cauchy (de paramètre R^{-2} avec notre normalisation, voir la proposition 3).*

4.2 La sphère

La variété $\mathcal{M}$ est ici la sphère $\mathbb{S}^2$ (de rayon 1) privée d'un point o, U est la boule ouverte (pointée) de centre o et de rayon $\varepsilon \in\,]0,\pi[$, de sorte que $G = \mathbb{R}$, $\Gamma = \mathbb{Z}$, $\Gamma\backslash G \equiv \mathbb{S}^1$.

Prenons pour (Z_t) le mouvement brownien de $\mathbb{S}^2$, i.e. pour $\mathcal{L}$ le demi-laplacien $\frac{1}{2}\Delta$ de $\mathbb{S}^2$.

Introduisons les coordonnées polaires de pôle o sur $\mathbb{S}^2 \subset \mathbb{R}^3$: $(\cos\varphi \times \sigma, \sin\varphi)$, où σ décrit $\mathbb{S}^1$ et φ décrit $[0,\pi]$ et représente la distance rie-mannienne à o. Dans ces coordonnées $\Delta = \mathcal{L}_\varphi + (\sin\varphi)^{-2}\,\partial^2/\partial\sigma^2$, où $\mathcal{L}_\varphi$ est l'opérateur de Legendre $\mathcal{L}_\varphi = \partial^2/\partial\varphi^2 + \cotg\varphi\,\partial/\partial\varphi$.

Changeant $\mathbb{R}_+$ en $]0,\varepsilon]$, nous sommes dans le cadre envisagé dans cet article, avec $g(\varphi) = (\sin\varphi)^{-2}$.

D'après la section 3, (Φ_t) est le processus de Lévy à valeurs dans $\mathbb{R}$ qui a pour exposant caractéristique $\lambda \mapsto \psi(\lambda^2/2)$.

Proposition 5. *Nous avons pour tous* $\varepsilon \in\,]0,\pi[$ *et* $\alpha \in \mathbb{R}_+$:

$$\psi(\alpha) = \frac{\sqrt{2\alpha}}{4\sin^2(\varepsilon/2)}$$

de sorte que (Φ_t) *est un processus de Cauchy (de paramètre* $(2\sin(\varepsilon/2))^{-2}$ *dans la normalisation de la proposition 3).*

Démonstration. Nous nous ramenons au cadre de la section précédente par le changement de variable $y = \log(\varepsilon/\varphi)$, et nous appliquons la proposition 3 avec $\zeta = \inf\{t > 0 \mid \varphi_t = \varepsilon\}$ (où (φ_t) est le processus de Legendre, à valeurs dans $[0,\pi]$) :

$$\psi^0(\alpha) = -\,C^{-1}\,\frac{\mathrm{d}}{\mathrm{d}\eta}\bigg|_{\eta=0}\,\mathbb{E}_{\varepsilon e^{-\eta}}\big[e^{-\alpha\zeta}\big],$$

et

$$\psi(\alpha) = -\,\frac{\mathrm{d}}{\mathrm{d}\eta}\bigg|_{\eta=0}\,\mathbb{E}_{\varepsilon\,e^{-\eta}}\bigg[\exp\bigg(-\alpha\int_0^\zeta \frac{\mathrm{d}s}{\sin^2\varphi_s}\bigg)\bigg].$$

Pour $\alpha > 0$, cherchons f lisse sur $]0,\pi[$, bornée près de 0, telle que

$$M_t := f(\varphi_t) \times \exp\bigg(-\alpha\int_0^t \sin^{-2}\varphi_s\,\mathrm{d}s\bigg)$$

soit une martingale. Or nous avons $\mathrm{d}\varphi_s = \mathrm{d}b_s + \frac{1}{2}\cotg\varphi_s\,\mathrm{d}s$, pour un brownien réel (b_s), et donc la formule d'Itô montre aussitôt que M_t est une martingale locale ssi

$$f''(\varphi) + \cotg\varphi \times f'(\varphi) - 2\,\alpha\,(\sin\varphi)^{-2} \times f(\varphi) = 0.$$

Le changement de variable $f(\varphi) = H(\mathrm{tg}(\varphi/2))$ donne $t^2 H''(t) + t H'(t) - 2\alpha H(t) = 0$, et donc les solutions bornées près de 0 sont proportionnelles à $f(\varphi) = (\mathrm{tg}(\varphi/2))^{\sqrt{2\alpha}}$.

Le théorème d'arrêt nous donne ensuite :

$$\mathbb{E}_{\varepsilon\,e^{-\eta}}\bigg[\exp\bigg(-\alpha\int_0^{\inf\{t>0\,\mid\,\varphi_t=\varepsilon\}} \frac{\mathrm{d}s}{\sin^2\varphi_s}\bigg)\bigg] = \frac{f(\varepsilon\,e^{-\eta})}{f(\varepsilon)},$$

et donc

$$\psi(\alpha) = \frac{\varepsilon\sqrt{2\alpha}}{C\sin\varepsilon}.$$

De même, pour le calcul de $\psi^0(\alpha)$ nous voulons h lisse sur $]0,\pi[$ et bornée près de 0 telle que $h(\varphi_t)\,\mathrm{e}^{-\alpha t}$ soit une martingale, et donc telle que

$$h''(\varphi) + \cot g\,\varphi \times h'(\varphi) - 2\,\alpha\,h(\varphi) = 0.$$

Classiquement, h doit être proportionnelle à $G(\cos\varphi)$, avec $G = P_\nu^0$ fonction de Legendre solution de $(1 - u^2)G''(u) - 2uG'(u) - 2\alpha G(u) = 0$ et ν tel que $\nu(\nu + 1) + 2\alpha = 0$. Nous avons donc (avec [16] page 171)

$$\psi^0(\alpha) = -C^{-1}\left.\frac{\mathrm{d}}{\mathrm{d}\eta}\right|_{\eta=0}\frac{h(\varepsilon\,\mathrm{e}^{-\eta})}{h(\varepsilon)} = \frac{\varepsilon\,h'(\varepsilon)}{C\,h(\varepsilon)} = -\frac{\varepsilon\,(\sin\varepsilon)\,(P_\nu^0)'(\cos\varepsilon)}{C\,P_\nu^0(\cos\varepsilon)}$$

$$= \nu\,\varepsilon\,\frac{(\cos\varepsilon)\,P_\nu^0(\cos\varepsilon) - P_{\nu-1}^0(\cos\varepsilon)}{C\,(\sin\varepsilon)\,P_\nu^0(\cos\varepsilon)}.$$

Puisque

$$1 = \hat{E}[\zeta] = \lim_{\alpha\searrow 0}\alpha^{-1}\psi^0(\alpha) = -2\,\varepsilon\,\frac{(\cos\varepsilon)\,P_0^0(\cos\varepsilon) - P_{-1}^0(\cos\varepsilon)}{C\,(\sin\varepsilon)\,P_0^0(\cos\varepsilon)} = \frac{2\varepsilon}{C}\,\mathrm{tg}\Big(\frac{\varepsilon}{2}\Big),$$

nous obtenons $C = 2\varepsilon\,\mathrm{tg}(\varepsilon/2)$, d'où le résultat. Nous avons en outre

$$\psi^0(\alpha) = \nu\cot g\,\varepsilon \times \frac{(\cos\varepsilon)\,P_\nu^0(\cos\varepsilon) - P_{\nu-1}^0(\cos\varepsilon)}{2\,(\sin\varepsilon)\,P_\nu^0(\cos\varepsilon)}. \qquad \Box$$

Lemme 1 ([5], lemme 8). *Si (Φ_t) est un processus de Cauchy de paramètre p et si (A_t) est un processus tel que A_t/t converge en probabilité vers une constante c, alors les lois marginales de dimension finie de $(\Phi_{A_{st}}/t,\ s \geqslant 0)$ convergent vers celles d'un processus de Cauchy de paramètre pc.*

On en déduit le théorème de Spitzer relatif à la sphère (voir par exemple [5], ou aussi [14]).

Corollaire 2. *La loi de θ_t/t converge lorsque $t \to \infty$ vers la loi de Cauchy de paramètre $1/4$ (indépendamment de ε). Il en est de même pour les lois marginales de dimension finie de $(\theta_{st}/t,\ s \geqslant 0)$, et conjointement pour les enroulements autour d'un nombre fini de points de la sphère, qui sont asymptotiquement indépendants.*

Démonstration. Prenant $\alpha = \lambda^2/2t^2$, nous avons pour tout $\lambda \in \mathbb{R}$:

$$\mathbb{E}\left[\exp\Big(\sqrt{-1}\,\frac{\lambda}{t}\,\Phi_t\Big)\right] = \exp\Big(-\frac{|\lambda|}{4\sin^2(\varepsilon/2)}\Big).$$

Pour revenir à l'échelle de temps usuelle, utilisons la remarque 1 : nous avons par ergodicité convergence presque sûre de L_t^0/t vers $\mu(U) = \sin^2(\varepsilon/2)$, de sorte que selon le lemme 1 ci-dessus le comportement asymptotique de l'enroulement θ_t est celui de $\Phi_{t\,\sin^2(\varepsilon/2)}$. Donc pour tout $s \in \mathbb{R}_+$

$$\lim_{t \to \infty} \mathbb{E}\left[\exp\left(\sqrt{-1}\,\frac{\lambda}{t}\,\theta_{st}\right)\right] = \exp(-|\lambda|\,s/4).$$

Pour l'indépendance asymptotique des enroulements θ_t^j autour d'un nombre fini de points $o_1, \ldots, o_k$ de la sphère, il suffit de noter que les processus de Cauchy correspondants (Φ_t^j) sont indépendants conditionnellement aux temps locaux $(L_t^{0,j})$, conditionnement qui disparaît à la limite. $\qquad\square$

4.3 Pointes hyperboliques réelles

La variété $\mathcal{M}$ est ici une variété hyperbolique réelle de dimension $d + 1$ et de volume fini, (Z_t) est son mouvement brownien, μ est la mesure de volume normalisée, G est le groupe $\mathbb{R}^d$, que nous identifions avec son algèbre de Lie, le sous-groupe parabolique Γ est un réseau de $\mathbb{R}^d$, et U est un voisinage horocyclique d'une pointe $\mathcal{P}$ de $\mathcal{M}$. Choisissons pour modèle le demi-espace de Poincaré, de façon que la pointe $\mathcal{P}$ soit en ∞, et notons h la hauteur (> 0 suffisamment grande) à laquelle se trouve la section ∂U dans le modèle choisi. L'opérateur $\mathcal{L}$ est le demi-laplacien hyperbolique, et donc $\mathcal{L}_G$ est le demi-laplacien euclidien de $\mathbb{R}^d$, $g(y) = y^2$, et $\mathcal{L}_y = y^2/2\,\partial^2/\partial y^2 + (1 - d)y/2\,\partial/\partial y$, de sorte que lors de chaque excursion dans $[h, \infty[$, $y_t = h\,e^{w_t - d\,t/2}$, où (w_t) désigne un mouvement brownien réel issu de 0. Le processus (X_t) est un brownien de $\mathbb{R}^d$ issu de 0, indépendant de (y_t), et $\theta_t = \int_0^t y_s\,dW_s$, lors de chaque excursion.

Nota Bene. Par rapport aux notations générales introduites dans la section 3, nous avons changé $\mathbb{R}_+$ en $[h, \infty[$, afin de respecter les notations usuelles du demi-espace de Poincaré.

Exposants caractéristiques, mesures de Lévy et de Krein

Nous pouvons encore dans le cas présent calculer les exposants caractéristiques ψ^0 et ψ, et en déduire la mesure de Krein (introduite juste avant le théorème 1 dans la section 3 ci-dessus ; voir aussi de nouveau [12], ([2], 9.2.3)).

Proposition 6. *Nous avons pour tout* $\alpha \in \mathbb{R}_+$:

$$\psi^0(\alpha) = \left(\sqrt{2\alpha + \frac{d^2}{4}} - \frac{d}{2}\right)\frac{d}{2} \qquad et \qquad \psi(\alpha) = \frac{d\,h\sqrt{2\alpha}}{2} \times \frac{K_{d/2-1}\big(h\sqrt{2\alpha}\big)}{K_{d/2}\big(h\sqrt{2\alpha}\big)}.$$

La mesure de Krein m_Λ *du subordinateur* (Λ_t) *est donnée par*

$$m_\Lambda(ds) = \frac{8d}{\pi^2}\left(\big(J_{d/2}^2 + Y_{d/2}^2\big)\big(\tfrac{h}{2}\sqrt{s}\big)\right)^{-1}ds,$$

où $J_{d/2}$ *et* $Y_{d/2}$ *désignent les fonctions de Bessel usuelles.*

Remarque. Les rapports $K_{\nu-1}/K_\nu$ interviennent dans les lois de processus de Bessel arrêtés à un dernier temps de passage ; le rapport $\psi(\alpha/(2h^2))/(\alpha\,d)$ est en effet la transformée de Laplace d'un certain temps d'occupation d'un tel processus ; voir [18, § 9, p. 339].

Démonstration. Changeons y en $y-h$, et appliquons la proposition 3, ζ désignant le temps d'atteinte de h par le processus (y_t) (issu de $h+\eta$).

$$\psi^0(\alpha) = -C^{-1}\frac{\mathrm{d}}{\mathrm{d}\eta}\Big|_{\eta=0}\mathbb{E}_{h+\eta}\big(\mathrm{e}^{-\alpha\zeta}\big),$$

et

$$\psi(\alpha) = -C^{-1}\frac{\mathrm{d}}{\mathrm{d}\eta}\Big|_{\eta=0}\mathbb{E}_{h+\eta}\left[\exp\left(-\alpha\int_0^\zeta (y_s)^2\,\mathrm{d}s\right)\right].$$

Puisque $y_s = (h+\eta)\,\mathrm{e}^{w_s-\frac{d}{2}s}$ lors de chaque excursion partant du niveau $(h+\eta)$, la formule de Cameron–Martin donne :

$$\mathbb{E}_{(h+\eta)}\big[\mathrm{e}^{-\alpha\zeta}\big] = \mathbb{E}\big[\exp\big(-\alpha\inf\{s>0\,|\,w_s-\tfrac{d}{2}s < \log h - \log(h+\eta)\}\big)\big]$$

$$= \left(1+\frac{\eta}{h}\right)^{d/2}\mathbb{E}\left[\exp\big(-(\alpha+d^2/8)\inf\{s>0\,|\,w_s<-\log(1+\eta/h)\}\big)\right]$$

$$= \left(1+\frac{\eta}{h}\right)^{d/2-\sqrt{2\alpha+d^2/4}},$$

d'où $\psi^0(\alpha) = \big(\sqrt{2\alpha+d^2/4}-d/2\big)/(hC)$, et par suite $C=2/hd$.

D'autre part cherchons une fonction régulière f bornée près de $+\infty$ et telle que

$$M_s := \exp\left(-\alpha\int_0^s (y_u)^2\,\mathrm{d}u\right)\times f(y_s)$$

définisse une martingale. Nous avons

$$M_s = f(y_0) + \text{martingale}$$

$$+ \frac{1}{2}\int_0^s \mathrm{e}^{-\alpha\int_0^r (y_u)^2\,\mathrm{d}u}\left(f''(y_r) + \frac{1-d}{y_r}f'(y_r) - 2\alpha\,f(y_r)\right)\times (y_r)^2\,\mathrm{d}r.$$

Nous devons donc avoir $f''(y) + (1-d)\,f'(y)/y - 2\alpha\,f(y) = 0$. Posant $H(y) := y^{-d/2}f(y)$, nous obtenons $H''(y) + H'(y)/y - 2\alpha\,H(y) = 0$. Par conséquent H est une fonction de Bessel modifiée, bornée près de $+\infty$. Donc $f(y) = c\,y^{d/2}K_{d/2}(y\sqrt{2\alpha})$. Finalement nous obtenons

$$\mathbb{E}_{(h+\eta)}\left[\exp\left(-\alpha\int_0^\zeta (y_s)^2\,\mathrm{d}s\right)\right] = \frac{f(h+\eta)}{f(h)} = (1+\eta/h)^{d/2}$$

$$\times \frac{K_{d/2}\big((h+\eta)\sqrt{2\alpha}\big)}{K_{d/2}\big(h\sqrt{2\alpha}\big)}.$$

La valeur de ψ en découle, étant donné que $\frac{d}{2}K_{d/2}(y)+yK'_{d/2}(y)+yK_{d/2-1}(y) \equiv 0$:

$$\psi(\alpha) = C^{-1}\sqrt{2\alpha} \times \frac{K_{d/2-1}\big(h\sqrt{2\alpha}\big)}{K_{d/2}\big(h\sqrt{2\alpha}\big)}.$$

Selon Ismail [11], nous avons

$$|\lambda|\,\frac{K_{d/2-1}(|\lambda|)}{K_{d/2}(|\lambda|)} = \frac{16}{\pi^2}\int_0^\infty \frac{|\lambda|^2\,\mathrm{d}t}{t(|\lambda|^2+8t)\big(J_{d/2}^2+Y_{d/2}^2\big)\big(\sqrt{t}\big)}$$

$$= \frac{16}{\pi^2}\int_0^\infty\int_0^\infty \frac{\big(1-\mathrm{e}^{-|\lambda|^2 s/8}\big)\,\mathrm{e}^{-st}\,\mathrm{d}s\,\mathrm{d}t}{\big(J_{d/2}^2+Y_{d/2}^2\big)\big(\sqrt{t}\big)}$$

$$= \frac{16}{\pi^2}\int_0^\infty\int_0^\infty \frac{\big(1-\mathrm{e}^{-|\lambda|^2 s/2}\big)\mathrm{e}^{-st}\,\mathrm{d}s\,\mathrm{d}t}{\big(J_{d/2}^2+Y_{d/2}^2\big)\big(\tfrac{1}{2}\sqrt{t}\big)}.$$

Ainsi

$$\psi(\alpha) = \frac{8d}{\pi^2}\int_0^\infty\int_0^\infty \frac{\big(1-\mathrm{e}^{-\alpha h^2 s}\big)\mathrm{e}^{-st}\,\mathrm{d}s\,\mathrm{d}t}{\big(J_{d/2}^2+Y_{d/2}^2\big)\big(\tfrac{1}{2}\sqrt{t}\big)},$$

et donc la mesure de Lévy π_Λ est donnée par

$$\pi_\Lambda(\mathrm{d}s) = \left(\frac{8d}{\pi^2}\int_0^\infty \frac{\mathrm{e}^{-st}\,\mathrm{d}t}{\big(J_{d/2}^2+Y_{d/2}^2\big)\big(\tfrac{h}{2}\sqrt{t}\big)}\right)\mathrm{d}s,$$

d'où le résultat. $\square$

Remarque. Un calcul un peu plus sophistiqué donne la loi conjointe de ζ et de $\int_0^\zeta (y_s)^2\,\mathrm{d}s$:

$$\mathbb{E}_{(h+\eta)}\left[\exp\left(-\rho\zeta - \alpha\int_0^\zeta (y_s)^2\,\mathrm{d}s\right)\right] = (1+\eta/h)^{d/2}$$

$$\times \frac{K_{\sqrt{2\rho+d^2/4}}\big((h+\eta)\sqrt{2\alpha}\big)}{K_{\sqrt{2\rho+d^2/4}}\big(h\sqrt{2\alpha}\big)}.$$

Du fait que $\mathcal{L}_G$ est le laplacien euclidien de $\mathbb{R}^d$, nous déduisons par simple composition la première partie de la proposition suivante.

Proposition 7. (i) *Le processus des enroulements* (Φ_t) *du brownien* (Z_t) *dans un voisinage horocyclique* U *d'une pointe de la variété hyperbolique réelle* $\mathcal{M}$ *(sectionnée à la hauteur* h*) est le processus de Lévy à valeurs dans* $\mathbb{R}^d$ *ayant pour exposant caractéristique* $\mathbb{R}^d \ni \lambda \mapsto \frac{d}{2}h|\lambda| \times \frac{K_{d/2-1}(h|\lambda|)}{K_{d/2}(h|\lambda|)}$.

(ii) *La mesure de Lévy* ν *du processus de Lévy* (Φ_t) *est donnée par :*

$$\nu(\mathrm{d}x) = \frac{16d}{\pi^2}(2\pi)^{-d/2}\left(\int_0^\infty \frac{K_{\frac{d}{2}-1}(|x|t)\,t^{d/2}}{\big(J_{d/2}^2+Y_{d/2}^2\big)\big(\tfrac{ht}{2\sqrt{2}}\big)}\,\mathrm{d}t\right)|x|^{1-\frac{d}{2}}\,\mathrm{d}x,$$

où $J_{d/2}$ *et* $Y_{d/2}$ *désignent les fonctions de Bessel usuelles.*

Démonstration de (ii). D'après le théorème 1, la mesure de Lévy du processus des enroulements (Φ_t) est donnée par

$$\nu(\mathrm{d}x) = \frac{8d}{\pi^2}\left(\int_0^\infty\int_0^\infty \frac{\mathrm{e}^{-\frac{|x|^2}{2s}}\,\mathrm{e}^{-st}\,(2\pi s)^{-d/2}}{\left(J_{d/2}^2 + Y_{d/2}^2\right)\left(\frac{h}{2}\sqrt{t}\right)}\,\mathrm{d}s\,\mathrm{d}t\right)\mathrm{d}x$$

$$= \frac{8d}{\pi^2}\left(\int_0^\infty\left(\left(\frac{|x|}{\sqrt{2t}}\right)^{1-d/2}\int_0^\infty \frac{\mathrm{e}^{-\frac{|x|\sqrt{2t}}{2}\left(\frac{1}{s}+s\right)}}{(2\pi s)^{d/2}}\,\mathrm{d}s\right)\frac{\mathrm{d}t}{\left(J_{d/2}^2 + Y_{d/2}^2\right)\left(\frac{h}{2}\sqrt{t}\right)}\right)\mathrm{d}x$$

$$= \frac{16d}{\pi^2}\left(\int_0^\infty\left((2\pi)^{-d/2}\left(\frac{|x|}{\sqrt{2t}}\right)^{1-d/2}K_{d/2-1}\left(|x|\sqrt{2t}\right)\right)\frac{\mathrm{d}t}{\left(J_{d/2}^2 + Y_{d/2}^2\right)\left(\frac{h}{2}\sqrt{t}\right)}\right)\mathrm{d}x$$

$$= \frac{16d}{\pi^2}\,(2\pi)^{-d/2}\left(\int_0^\infty \frac{K_{\frac{d}{2}-1}(|x|t)\,t^{d/2}}{\left(J_{d/2}^2 + Y_{d/2}^2\right)\left(\frac{h\,t}{2\sqrt{2}}\right)}\,\mathrm{d}t\right)|x|^{1-\frac{d}{2}}\,\mathrm{d}x. \qquad \square$$

Remarque 2. Lorsque $d = 1$,

$$\frac{1}{\left(J_{1/2}^2 + Y_{1/2}^2\right)(z)} = \frac{\pi\,z}{2}\,,\ \ K_{d/2-1}(x)$$

$$= \mathrm{e}^{-x}\sqrt{\frac{\pi}{2x}}\,,\ \text{et donc }\nu(\mathrm{d}x) = \frac{h\sqrt{2}}{\pi\,x^2}\,\mathrm{d}x:$$

nous retrouvons la mesure de Lévy d'un processus de Cauchy.

Résultat asymptotique

Particularisons suivant la dimension, et utilisons la remarque 1.

Notons que dans les coordonnées $(x_1,\ldots,x_d,y)$ du demi-espace de Poincaré la mesure de volume est proportionnelle à $y^{-d-1}\,\mathrm{d}y\,\mathrm{d}x_1\ldots\mathrm{d}x_d$, et que U est isométrique à $(\Gamma\backslash\mathbb{R}^d)\times[h,\infty[$, h représentant la hauteur à laquelle on a sectionné la pointe $\mathcal{P}$ pour obtenir le voisinage horocyclique U. De sorte que

$$\lim_{t\to\infty}L_t^0/t = \mu(U) = \kappa\int_h^\infty y^{-d-1}\,\mathrm{d}y = \kappa\,h^{-d}/d,$$

κ étant une constante dépendant du volume global de la variété $\mathcal{M}$.

Le comportement asymptotique de l'enroulement (θ_t) produit par (Z_t) dans U est donc (en loi) celui de $(\Phi_{\kappa\,h^{-d}\,t/d})$.

Si $d = 1$, étant donné que $K_{d/2-1}/K_{d/2}\equiv 1$, nous voyons que le processus des enroulements (Φ_t) est un processus de Cauchy de paramètre $h/2$. Nous retrouvons ainsi via le lemme 1 que $\lim_{t\to\infty}\mathbb{E}[\mathrm{e}^{\sqrt{-1}\lambda t^{-1}\theta_{ts}}] = \exp(-|\lambda|\kappa\,s/2)$, et donc que $t^{-1}\theta_{ts}$ converge au sens des marginales de dimension finie vers un processus de Cauchy de paramètre $\kappa/2$. Nous retrouvons ainsi le résultat donné dans [4] (et appliqué ensuite au flot géodésique).

Ceci n'est plus vrai en dimension supérieure.

Si $d = 2$:

$$\frac{K_{d/2-1}}{K_{d/2}}(y) = y \log\left(\frac{1}{y}\right) + \mathcal{O}(y), \quad \text{d'où} \lim_{t\to\infty} \mathbb{E}\left[e^{\sqrt{-1}\,\lambda \cdot \frac{\Phi_{st}}{\sqrt{t \log t}}}\right] = e^{-|\lambda|^2 h^2 d\, s/4}.$$

Nous retrouvons le résultat [6] de convergence (au sens des marginales de dimension finie) de l'enroulement normalisé $(t \log t)^{-1/2}\theta_{ts}$ vers un mouvement brownien plan isotrope de variance $\kappa/4$, indépendant de la hauteur h.

Si $d \geqslant 3$:

$$\frac{K_{d/2-1}}{K_{d/2}}(y) \sim \frac{y}{(d-2)}, \qquad \text{et donc} \qquad \lim_{t\to\infty} \mathbb{E}\left[e^{\sqrt{-1}\,\lambda \cdot \frac{\Phi_{st}}{\sqrt{t}}}\right] = e^{-\frac{|\lambda|^2 h^2 d\, s}{2(d-2)}}.$$

De sorte que l'enroulement normalisé $t^{-1/2}\theta_{ts}$ converge (au sens des marginales de dimension finie) vers un mouvement brownien isotrope (de $\mathbb{R}^d$) de variance $\kappa\, h^{2-d}/(d-2)$; ce qui correspond à des enroulements de variance asymptotique proportionnelle à h^{2-d}, logiquement dépendants de la hauteur h, étant donné que nous sommes ici dans le cas d'un théorème central limite.

Par ailleurs nous avons $K_a/K_b \sim 1$ en $+\infty$ pour tout (a,b), et donc pour tout $d \in \mathbb{N}^*$:

$$\lim_{h\nearrow\infty} \mathbb{E}\left[e^{\sqrt{-1}\,\lambda \cdot (h^{-1}\Phi_t)}\right] = e^{-|\lambda|\, d\, t/2}.$$

Ainsi, la loi asymptotique lorsque $h \to \infty$ du processus d'enroulements normalisés $(h^{-1}\Phi_t)$ (rappelons que Φ_t dépend de la hauteur h) est toujours celle du processus de Cauchy isotrope de paramètre $d/2$.

4.4 Pointes hyperboliques complexes

Nous supposons ici que $\mathbb{H} = \mathbb{H}_{\mathbb{C}}^d$ est l'espace hyperbolique complexe de dimension (complexe) d, avec $d \geqslant 2$. Nos références sont surtout [9], [10], [7]. Voir aussi [3].

Les deux modèles les plus classiques sont celui de Siegel et le projectif :

$$\mathbb{H}_{\mathbb{C}}^d = \left\{(w,z) \in \mathbb{C}^{d-1} \times \mathbb{C} \mid 2\,\mathrm{Re}(z) > |w|^2\right\}$$
$$= \left\{P(w,w_1,w_0) \in P\mathbb{C}^d \mid Q(w,w_1,w_0) < 0\right\},$$

où $Q(w,w_1,w_0) := |w|^2 - w_1\overline{w}_0 - w_0\overline{w}_1$, pour $(w,w_1,w_0) \in \mathbb{C}^{d-1} \times \mathbb{C} \times \mathbb{C}$.

Le bord de $\mathbb{H}_{\mathbb{C}}^d$ est

$$\partial\mathbb{H}_{\mathbb{C}}^d = \left\{(w,z) \in \mathbb{C}^{d-1} \times \mathbb{C} \mid 2\,\mathrm{Re}(z) = |w|^2\right\} \cup \{\infty\}.$$

Les coordonnées horosphériques de $(w,z) \in \overline{\mathbb{H}_{\mathbb{C}}^d}\backslash\{\infty\}$ sont $(\zeta, v, y) \in \mathbb{C}^{d-1} \times \mathbb{R} \times \mathbb{R}_+$ définies par : $\zeta = w$, $v = -2\,\mathrm{Im}(z)$, $y = 2\,\mathrm{Re}(z) - |w|^2$. (Donc $z = (y + |\zeta|^2 - \sqrt{-1}v)/2$).

La norme de Cygan est définie par : $\|(\zeta, v, y)\| := \sqrt{|y + |\zeta|^2 - \sqrt{-1}v|}$.

Le groupe de Heisenberg $\mathcal{H}_d$ est $\mathbb{C}^{d-1} \times \mathbb{R}$ muni de la loi :

$$(\zeta, v) \cdot (\zeta', v') := \big(\zeta + \zeta', \, v + v' + 2\,\mathrm{Im}(\zeta\overline{\zeta'})\big).$$

On l'identifie à $\partial\mathbb{H}^d_{\mathbb{C}} \backslash \{\infty\}$ via $(\zeta, v) \equiv (\zeta, v, 0)$. Il conserve ∞ et agit sur $\overline{\mathbb{H}^d_{\mathbb{C}}} \backslash \{\infty\}$ par translation à gauche sur chaque horocycle basé en ∞ :

$$(\zeta, v) \cdot (\zeta', v', y) := \big(\zeta + \zeta', \, v + v' + 2\,\mathrm{Im}(\zeta\overline{\zeta'}), y\big).$$

Repassant aux coordonnées (w, z) de Siegel, cela donne :

$$(\zeta, v) \equiv \big(\zeta, (|\zeta|^2 - \sqrt{-1}v)/2\big)$$

et

$$(\zeta, v) \cdot (w, z) := \big(w + \zeta, \, z + w\overline{\zeta} + (|\zeta|^2 - \sqrt{-1}v)/2\big).$$

Le groupe unitaire $U(d-1)$ conserve ∞ et agit sur $\overline{\mathbb{H}^d_{\mathbb{C}}} \backslash \{\infty\}$ par rotation de la coordonnée ζ. Le produit semi-direct $\mathcal{H}(d)$ des deux groupes $\mathcal{H}_d$ et $U(d-1)$ constitue le groupe des isométries de $\mathbb{H}^d_{\mathbb{C}}$ qui conservent l'orientation, ∞, et chaque horosphère $\{y = y_0\}$.

Fixons le système de coordonnées usuel sur $\mathbb{C}^{d-1}$:

$$\zeta = (\zeta_1, \ldots, \zeta_{d-1}), \qquad \text{avec } \zeta_j = x_j + \sqrt{-1}\,y_j \text{ pour } 1 \leqslant j < d.$$

La mesure de volume de $\mathbb{H}$ est proportionnelle à

$$y^{-d-1}\,\mathrm{d}y\,\mathrm{d}v\,\mathrm{d}x_1 \ldots \mathrm{d}x_{d-1}\,\mathrm{d}y_1 \ldots \mathrm{d}y_{d-1}.$$

L'algèbre de Lie de $\mathcal{H}_d$ est engendrée par $\partial/\partial x_1$, $\partial/\partial y_1$, $\ldots$, $\partial/\partial x_{d-1}$, $\partial/\partial y_{d-1}, \partial/\partial v$, avec pour seuls crochets non triviaux $[\partial/\partial x_j, \partial/\partial y_j] = 2\partial/\partial v$. Pour tout $s > 0$ notons H_s l'homothétie complexe définie par :

$$H_s(\zeta, v, y) = \left(\frac{\zeta}{\sqrt{s}}, \frac{v}{s}, \frac{y}{s}\right)$$

C'est une isométrie, qui vérifie $H_s((\zeta, v) \cdot (\zeta', v')) = H_s(\zeta, v) \cdot H_s(\zeta', v')$.

Donnons-nous $2d$ mouvements browniens réels standard indépendants :

$$x_s^1, \ldots, x_s^{d-1}, \quad y_s^1, \ldots, y_s^{d-1}, \quad w_s, \quad \beta_s.$$

Le mouvement brownien sur le groupe d'Heisenberg est la diffusion dégénérée de générateur $\frac{1}{2}\Delta$, où

$$\Delta := y^2 \frac{\partial^2}{\partial y^2} + (1-d)y\,\frac{\partial}{\partial y} + b\,y\,\Delta^K,$$

$$\Delta^K := \sum_{j=1}^{d-1} \left(\left(\frac{\partial}{\partial x_j} + 2y_j\frac{\partial}{\partial v}\right)^2 + \left(\frac{\partial}{\partial y_j} - 2x_j\frac{\partial}{\partial v}\right)^2\right)$$

$$= \sum_{j=1}^{d-1} \left(\frac{\partial^2}{\partial x_j^2} + \frac{\partial^2}{\partial y_j^2} + 4y_j\frac{\partial^2}{\partial x_j\partial v} - 4x_j\frac{\partial^2}{\partial y_j\partial v} + 4\big(x_j^2 + y_j^2\big)\frac{\partial^2}{\partial v^2}\right)$$

étant le laplacien de Kohn sur $\mathcal{H}_d$, et $b > 0$ étant un paramètre arbitraire.

Nous retrouvons les notations de la section 3, en prenant

$$\mathcal{L} = \frac{1}{2}\,\Delta, \qquad \mathcal{L}_y = \frac{y^2}{2}\,\frac{\partial^2}{\partial y^2} + \left(\frac{1-d}{2}\right) y\,\frac{\partial}{\partial y}, \qquad g(y) = y, \qquad \mathcal{L}_G = \frac{b}{2}\,\Delta^K.$$

Lors de chaque excursion dans $[h, \infty[$ nous avons $y_s = h\exp\left(w_s - \frac{d}{2}\,s\right)$.
Soit

$$B_s = (\zeta_s\,,\, 2\mathcal{A}_s)$$

$$= \left(x_s^1 + \sqrt{-1}\,y_s^1, \ldots, x_s^{d-1} + \sqrt{-1}\,y_s^{d-1},\, 2\sum_{j=1}^{d-1}\int_0^s \left(y_t^j\,\mathrm{d}x_t^j - x_t^j\,\mathrm{d}y_t^j\right)\right)$$

le mouvement brownien du groupe $G = \mathcal{H}_d$, où $\mathcal{A}_s = \sum_{j=1}^{d-1}\mathcal{A}_s^j$ et

$$\mathcal{A}_s^j := \int_0^s \left(y_t^j\,\mathrm{d}x_t^j - x_t^j\,\mathrm{d}y_t^j\right).$$

Rappelons la formule d'aire de Paul Lévy : pour tous ϱ, x, $y \in \mathbb{R}$ et $s > 0$

$$\mathbb{E}\left[\exp\left(\sqrt{-1}\,\varrho\,\mathcal{A}_s^j\right) \mid x_s^j = x, y_s^j = y\right] = \frac{\varrho s}{\mathrm{sh}(\varrho s)} \times \exp\left(\frac{x^2 + y^2}{2s}\left[1 - \frac{\varrho s}{\mathrm{th}(\varrho s)}\right]\right).$$

La formule d'Itô montre directement que

$$Z_s \equiv \left(H_{b^{-1}}\left(B\left(\int_0^s y_t\,\mathrm{d}t\right)\right),\, y_s\right)$$

est le mouvement brownien (d'Heisenberg) de $\mathbb{H}$, en coordonnées horocycliques.

Excursions près d'une pointe de $\mathbb{H}_{\mathbb{C}}$

La variété $\mathcal{M}$ est ici une variété hyperbolique complexe de dimension $2d$ et de volume fini, (Z_t) est son mouvement brownien, μ est la mesure de volume normalisée, G est le groupe de Heisenberg $\mathcal{H}_d$, d'algèbre de Lie $\mathcal{G} \equiv \mathbb{R}^{2d-1}$, et U est un voisinage horocyclique d'une pointe $\mathcal{P}$ de $\mathcal{M}$, que nous plaçons en $\{y = \infty\}$. Le voisinage U est isométrique à $(\Gamma\backslash\mathcal{H}_d) \times [h, \infty[$, h représentant la hauteur à laquelle on a sectionné la pointe pour obtenir le voisinage horocyclique U. Comme dans le cas hyperbolique réel, nous avons changé $\mathbb{R}_+$ en $[h, \infty[$, pour respecter les notations usuelles du demi-espace de Siegel. Nous avons encore $\mu(U) = \kappa'\,h^{-d}/d$, pour la même raison que dans le cas réel (voir la section 4.3).

Nous pouvons de nouveau calculer les exposants caractéristiques ψ^0 et ψ.

Lemme 2. *La loi de $\Theta^1 = \displaystyle\int_0^{\inf\{s\,|\,y_s=h\}} y_t\,dt$ est donnée par : pour tous η, $\alpha \geqslant 0$, nous avons (K_d étant l'usuelle fonction de Bessel modifiée) :*

$$\mathbb{E}_{(h+\eta)}\big[\exp(-\alpha\,\Theta^1)\big] = (1+\eta/h)^{d/2} \times \frac{K_d\big(2\sqrt{2\alpha\,(h+\eta)}\,\big)}{K_d\big(2\sqrt{2\alpha\,h}\,\big)}.$$

Démonstration. Pour α, y réels positifs, soient $\zeta := \inf\{s>0\,|\,y_s=h\}$ et

$$f(y) := \mathbb{E}_y\bigg[\exp\Big(-\alpha\int_0^\zeta y_t\,dt\Big)\bigg].$$

La fonction f est une fonction positive décroissante sur $[h,+\infty[$, égale à 1 en h, et qui est harmonique par rapport au semi-groupe de la diffusion (y_s) avec potentiel $-\alpha\,\mathrm{Id}$; elle doit donc vérifier :

$$f''(y) + \tfrac{1-d}{y}\,f'(y) - \tfrac{2\alpha}{y}\,f(y) = 0.$$

Effectuons le changement : $f(y) := y^{d/2}\,F\big(2\sqrt{2\alpha\,y}\,\big)$. Cela donne :

$$t^2\,F''(t) + t\,F'(t) - (t^2+d^2)\,F(t) = 0.$$

C'est une équation de Bessel, dont une solution positive décroissante sur $\mathbb{R}_+^*$ est K_d. Le résultat s'ensuit aussitôt. □

Proposition 8. *Nous avons pour tous $h>0$, $\alpha\geqslant 0$:*

$$\psi^0(\alpha) = \bigg(\sqrt{2\alpha+\tfrac{d^2}{4}} - \tfrac{d}{2}\bigg)\tfrac{d}{2} \qquad et \qquad \psi(\alpha) = \frac{d\,\sqrt{2\alpha\,h}}{2} \times \frac{K_{d-1}\big(2\sqrt{2\alpha\,h}\,\big)}{K_d\big(2\sqrt{2\alpha\,h}\,\big)}.$$

Remarque. Comme dans la proposition 6, on voit intervenir la transformée de Laplace de temps de passage de processus de Bessel; voir [18, § 9, p. 339].

Démonstration. De même que dans la preuve de la proposition 6, changeons y en $y-h$ et appliquons la proposition 3. Cela donne d'abord

$$\psi^0(\alpha) = \big(\sqrt{2\alpha+d^2/4} - d/2\big)/(hC), \qquad et \qquad C = 2/hd,$$

comme dans la proposition 6, puisque le processus (y_s) a la même expression exactement dans les cas réel et complexe ; ensuite, utilisant le lemme 2 ci-dessus, on obtient :

$$\psi(\alpha) = -C^{-1}\,\frac{d}{d\eta}\bigg|_{\eta=0}\bigg((1+\eta/h)^{d/2} \times \frac{K_d\big(2\sqrt{2\alpha\,(h+\eta)}\,\big)}{K_d\big(2\sqrt{2\alpha\,h}\,\big)}\bigg)$$

$$= \frac{\sqrt{2\alpha\,h}}{hC} \times \frac{K_{d-1}\big(2\sqrt{2\alpha\,h}\,\big)}{K_d\big(2\sqrt{2\alpha\,h}\,\big)}. \qquad\qquad □$$

Corollaire 3. *La mesure de Lévy π_Λ du subordinateur (Λ_t) (autrement dit la loi de Θ^1 sous la mesure d'excursion $\hat{P}$) a pour densité sur $\mathbb{R}_+^*$:*

$$s \longmapsto \frac{4d}{\pi^2} \int_0^\infty \frac{\mathrm{e}^{-st}\,\mathrm{d}t}{(J_d^2 + Y_d^2)(\sqrt{ht})},$$

où J_d et Y_d désignent les fonctions de Bessel usuelles. Donc la mesure de Krein m_Λ du subordinateur (Λ_t) est donnée par

$$m_\Lambda(\mathrm{d}s) = \frac{4d}{\pi^2}\left(\left(J_d^2 + Y_d^2\right)(\sqrt{hs})\right)^{-1}\mathrm{d}s.$$

Démonstration. Nous avons en effet d'après la proposition 8 et d'après Ismail [11], pour tout $\varrho > 0$:

$$\psi(\alpha) = \frac{4d}{\pi^2}\int_0^\infty \frac{\varrho\,\mathrm{d}t}{t(\varrho+t)(J_d^2+Y_d^2)(\sqrt{ht})} = \frac{4d}{\pi^2}\int_0^\infty\int_0^\infty \frac{(1-\mathrm{e}^{-\varrho s})\mathrm{e}^{-st}\,\mathrm{d}s\,\mathrm{d}t}{(J_d^2+Y_d^2)(\sqrt{ht})}$$

$$= \frac{4d}{\pi^2}\int_0^\infty\int_0^\infty \frac{(1-\mathrm{e}^{-\varrho s})\mathrm{e}^{-st}\,\mathrm{d}s\,\mathrm{d}t}{(J_d^2+Y_d^2)(\sqrt{ht})}. \qquad \square$$

Résultat asymptotique

Lemme 3. *Nous avons pour tout a réel :*

$$\lim_{t\to\infty} t\,\psi\left(\frac{\alpha}{t}\right) = \frac{\alpha hd}{d-1},$$

ce qui signifie que Λ_t/t converge en probabilité lorsque $t \to \infty$ vers $hd/(d-1)$.

Démonstration. Nous déduisons aussitôt de la proposition 8, pour tout $d \geqslant 2$:

$$t\,\psi\left(\frac{\alpha}{t}\right) = \frac{d}{2}\sqrt{2\alpha ht} \times \frac{K_{d-1}\left(2\sqrt{2\alpha h/t}\right)}{K_d\left(2\sqrt{2\alpha h/t}\right)}$$

$$\sim \frac{d}{2}\sqrt{2\alpha ht} \times \frac{\left(\sqrt{2\alpha h/t}\right)^{1-d} \times (d-2)!}{\left(\sqrt{2\alpha h/t}\right)^{-d} \times (d-1)!} = \frac{\alpha hd}{d-1}. \qquad \square$$

Corollaire 4. *La variable $H_t(\Phi_t)$ des enroulements normalisés dans $\mathcal{H}_d$ converge en loi lorsque $t \to \infty$ vers*

$$\left(\sqrt{b}\,\zeta_{\frac{hd}{d-1}}\,,\, 2b\,\mathcal{A}_{\frac{hd}{d-1}}\right) = H_{b-1}\left(B_{\frac{hd}{d-1}}\right) \equiv H_{\frac{d-1}{bhd}}(B_1).$$

Donc le processus normalisé $(H_t(\theta_{st}))$ des enroulements dans $\mathcal{H}_d$ produits par les excursions du brownien (Z_t) dans la pointe U (sectionnée à la hauteur h) converge en loi (au sens des distributions marginales de dimension finie) lorsque $t \to \infty$ vers le brownien de Heisenberg $\left(H_{\frac{(d-1)h^{d-1}}{b\kappa}}(B_s)\right)$.

Démonstration. La proposition 2 assure que $\Phi_t = H_{b^{-1}}(B_{\Lambda_t})$, avec les notations de la section 4.4, le brownien B et le subordinateur Λ étant indépendants. Par conséquent $H_t(\Phi_t) = H_{t/b}(B_{\Lambda_t})$ a même loi que $H_{t/(b\Lambda_t)}(B_1)$, qui selon le lemme 3 converge vers $H_{\frac{d-1}{bhd}}(B_1)$. Enfin nous savons que $H_t(\theta_{st})$ se comporte asymptotiquement comme $H_t(\Phi_{\kappa'h^{-d}st/d})$, et donc converge en loi vers $H_{\frac{(d-1)h^{d-1}}{b\kappa's}}(B_1)$, i.e. vers $H_{\frac{(d-1)h^{d-1}}{b\kappa'}}(B_s)$. □

Ce résultat signifie qu'il n'y a pas pour $\mathbb{H}_{\mathbb{C}}^d$ de résultat d'enroulements singuliers (analogue au théorème classique de Spitzer), quelque soit $d \geqslant 2$, mais seulement un théorème central limite, expliquant la dépendance en h du processus limite.

Appendice : une extension au cas du Brownien hyperbolique complexe non dégénéré

Étendons ce qui précède au cas où notre mouvement Brownien sur le groupe d'Heisenberg est non dégénéré, par ajout d'une excitation indépendante sur la composante centrale v.

Fixons la métrique de type Bergmann : $\mathrm{d}\ell^2 := y^{-2}\,\mathrm{d}y^2 + \mathrm{d}\ell_y^2$, où

$$\mathrm{d}\ell_y^2 := (b\,y)^{-1}\sum_{j=1}^{d-1}\bigl(\mathrm{d}x_j^2 + \mathrm{d}y_j^2\bigr) + (k\,y)^{-2}\biggl(\mathrm{d}v + 2\sum_{j=1}^{d-1}\bigl(x_j\mathrm{d}y_j - y_j\mathrm{d}x_j\bigr)\biggr)^2,$$

avec $b > 0$ et $k > 0$.

La mesure de volume riemannien associée à $\mathrm{d}\ell^2$ est proportionnelle à

$$y^{-d-1}\,\mathrm{d}y\,\mathrm{d}v\,\mathrm{d}x_1\ldots\mathrm{d}x_{d-1}\,\mathrm{d}y_1\ldots\mathrm{d}y_{d-1}\,.$$

L'opérateur de Laplace–Beltrami associé est

$$\Delta := y^2\frac{\partial^2}{\partial y^2} + (1-d)\,y\,\frac{\partial}{\partial y} + \Delta_{\mathcal{H}}, \qquad \text{avec} \qquad \Delta_{\mathcal{H}} := (k\,y)^2\,\frac{\partial^2}{\partial v^2} + b\,y\,\Delta^K,$$

Δ^K étant toujours le laplacien de Kohn sur $\mathcal{H}_d$.

La formule d'Itô montre directement que

$$Z_s \equiv \left(H_{b^{-1}}\left(B\left(\int_0^s y_t\,\mathrm{d}t\right)\right) + k\left(0, \beta\left(\int_0^s y_t^2\,\mathrm{d}t\right)\right), y_s\right)$$

est le mouvement brownien de $\mathbb{H}$, en coordonnées horocycliques et relativement à la métrique $\mathrm{d}\ell^2$.

Ce mouvement brownien ne se met plus sous la forme d'un produit semi-direct comme dans la section 3, mais cependant nous pouvons le considérer de façon assez analogue.

Le lemme 2 et la proposition 8 doivent être modifiés comme suit.

Lemme 4. *La loi de* $\left(\Theta^1 = \displaystyle\int_0^{\inf\{s\,|\,y_s = h\}} y_t\,\mathrm{d}t\,,\ \Theta^2 = \displaystyle\int_0^{\inf\{s\,|\,y_s = h\}} y_t^2\,\mathrm{d}t\right)$ *est donnée par : pour tous* $\eta,\ \varrho \geq 0,\ \sigma > 0,$ *notant* $a := \frac{d+1}{2} + \frac{\varrho}{\sigma},$

$$\mathbb{E}_{(h+\eta)}\big[\exp\big(-\varrho\,\Theta^1 - \sigma^2\,\Theta^2/2\big)\big]$$
$$= (1 + \eta/h)^d\,\mathrm{e}^{-\sigma\,\eta} \times \frac{\int_0^\infty \mathrm{e}^{-2\sigma(h+\eta)s}\,s^{a-1}\,(1+s)^{d-a}\,\mathrm{d}s}{\int_0^\infty \mathrm{e}^{-2\sigma h s}\,s^{a-1}\,(1+s)^{d-a}\,\mathrm{d}s}.$$

Démonstration. Pour $\varrho,\ \sigma,\ y$ réels positifs, soient $\zeta := \inf\{s > 0 \,|\, y_s = h\}$ et

$$f(y) := \mathbb{E}_y\left[\exp\left(-\varrho \int_0^\zeta y_t\,\mathrm{d}t - \frac{\sigma^2}{2}\int_0^\zeta y_t^2\,\mathrm{d}t\right)\right].$$

La fonction f est une fonction positive décroissante sur $[1, +\infty[$, égale à 1 en h, et qui est harmonique par rapport au semi-groupe de la diffusion y_s avec potentiel $-\varrho\,\mathrm{Id} - \sigma^2\mathrm{Id}^2/2$; elle doit donc vérifier :

$$f''(y) + \left(\frac{1-d}{y}\right)f'(y) - \left(\frac{2\varrho}{y} + \sigma^2\right)f(y) = 0.$$

Effectuons le changement : $f(y) := \mathrm{e}^{-\sigma y}\,F(2\sigma\,y)$. Cela donne :

$$t\,F''(t) + (1 - d - t)\,F'(t) + \left(\frac{d-1}{2} - \frac{\varrho}{\sigma}\right)F(t) = 0.$$

C'est une équation différentielle confluente hypergéométrique (de Kummer), dont une solution positive sur $\mathbb{R}_+^*$ est

$$F(t) = t^d \int_0^\infty \mathrm{e}^{-ts}\,s^{a-1}\,(1+s)^{d-a}\,\mathrm{d}s, \qquad \text{avec} \qquad a := \frac{d+1}{2} + \frac{\varrho}{\sigma}.$$

Ceci se vérifie en observant que

$$\left(t\,\frac{\partial^2}{\partial t^2} + (1 - d - t)\,\frac{\partial}{\partial t} + (d - a)\right)\left(t^d\mathrm{e}^{-ts}s^{a-1}(1+s)^{d-a}\right)$$
$$+ \frac{\partial}{\partial s}\left(t^d\mathrm{e}^{-ts}s^a(1+s)^{d-a+1}\right) = 0.$$

De plus nous avons

$$F(t) = t^{d-a}\int_0^\infty \mathrm{e}^{-s}\,s^{a-1}\,(1+s/t)^{d-a}\,\mathrm{d}s$$
$$= \Gamma(a)\,t^{d-a} + t^{d-a}\int_0^\infty \mathrm{e}^{-s}\,s^{a-1}\left((1+s/t)^{d-a} - 1\right)\mathrm{d}s$$
$$= \Gamma(a)\,t^{d-a} + t^{d-a}\int_0^\infty \mathrm{e}^{-s}\,s^{a-1}\,\mathcal{O}\left((s/t)(1+s/t)^{(d-a-1)^+}\right)\mathrm{d}s$$
$$= \Gamma(a)\,t^{d-a} \times \left(1 + \mathcal{O}(1/t)\right).$$

Une solution de l'équation de Kummer non colinéaire à F est

$$t^d \times \sum_{j \in \mathbb{N}} \frac{\Gamma(a+j)}{j!(d+j)!} \, t^j = e^t \times t^{a-1} \times \big(1 + \mathcal{O}(1/t)\big),$$

ce qui fait que seule (à une constante près) F correspond à une fonction f bornée en $+\infty$. Donc nous avons $f(y) = c\,e^{-\sigma y} F(2\sigma y)$, et

$$\mathbb{E}_{h+\eta}\big[\exp\big(-\varrho\,\Theta^1 - \sigma^2\Theta^2/2\big)\big] = f(h+\eta)/f(h). \qquad \square$$

Corollaire 5. *Nous avons pour tous* $h > 0$, $\varrho \geqslant 0$, $\sigma > 0$, *notant* $a = (d+1)/2 + \varrho/\sigma$:

$$\hat{E}\Big[\exp\Big(-\varrho\,\Theta^1 - \frac{\sigma^2}{2}\,\Theta^2\Big) - 1\Big]$$
$$= \frac{hd}{2} \times \Big(\frac{d}{h} - \sigma - 2\sigma \times \frac{\int_0^\infty e^{-2\sigma hs}\, s^a\,(1+s)^{d-a}\,\mathrm{d}s}{\int_0^\infty e^{-2\sigma hs}\, s^{a-1}\,(1+s)^{d-a}\,\mathrm{d}s}\Big).$$

En particulier

$$\hat{E}\Big[1 - e^{-\frac{\sigma^2}{2}\Theta^2}\Big] = \frac{h\sigma d}{2} \times \frac{K_{d/2-1}(h\sigma)}{K_{d/2}(h\sigma)}.$$

Démonstration. De même que dans la preuve de la proposition 8, nous avons

$$\frac{2}{hd} \times \hat{E}\Big[\exp\Big(-\varrho\,\Theta^1 - \frac{\sigma^2}{2}\,\Theta^2\Big) - 1\Big] = \frac{\partial}{\partial\eta}\Big|_{\eta=0} \mathbb{E}_{(h+\eta)}\Big[\exp\Big(-\varrho\,\Theta^1 - \frac{\sigma^2}{2}\,\Theta^2\Big)\Big]$$

$$= \frac{\partial}{\partial\eta}\Big|_{\eta=0}\Big[(1+\eta/h)^d\, e^{-\sigma\eta} \times \frac{\int_0^\infty e^{-2\sigma(h+\eta)s}\, s^{a-1}\,(1+s)^{d-a}\,\mathrm{d}s}{\int_0^\infty e^{-2\sigma hs}\, s^{a-1}\,(1+s)^{d-a}\,\mathrm{d}s}\Big]$$

$$= \frac{d}{h} - \sigma - 2\sigma \times \frac{\int_0^\infty e^{-2\sigma hs}\, s^a\,(1+s)^{d-a}\,\mathrm{d}s}{\int_0^\infty e^{-2\sigma hs}\, s^{a-1}\,(1+s)^{d-a}\,\mathrm{d}s}$$

$$= \frac{d}{h} - \sigma + 2\sigma \times \frac{\phi'(a, d+1\,;\, 2\sigma h)}{\phi(a, d+1\,;\, 2\sigma h)},$$

où $\phi(a, d+1\,;\, x) := \Gamma(a)^{-1}\int_0^\infty e^{-xs}s^{a-1}(1+s)^{d-a}\,\mathrm{d}s = \Gamma(a)^{-1}x^{-d}F(x)$ (F étant celle de la preuve du lemme 4) est la fonction hypergéométrique confluente (solution de $x\phi''(x) + (d+1-x)\phi'(x) - a\phi(x) = 0$) de Tricomi. Lorsque $\varrho = 0$, nous retombons exactement sur le cas de la proposition 6. $\quad\square$

Les définitions de la section 3 et la proposition 2 admettent un analogue :

$$\Phi_t := H_{b-1}\Big[B\big(\int_0^{\Lambda_t^0} y_t\,\mathrm{d}t\big)\Big] + k\,\big(0\,,\, \beta\big(\int_0^{\Lambda_t^0} y_t^2\,\mathrm{d}t\big)\big)$$

$$= H_{b-1}\big(B_{\Lambda_t^1}\big) + k\,\big(0\,,\, \beta_{\Lambda_t^2}\big) = \Big(\sqrt{b}\,\zeta_{\Lambda_t^1}\,,\, 2b\mathcal{A}_{\Lambda_t^1} + k\beta_{\Lambda_t^2}\Big),$$

où $(\Lambda_t^1, \Lambda_t^2) := \Big(\int_0^{\Lambda_t^0} y_t\,\mathrm{d}t\,,\, \int_0^{\Lambda_t^0} y_t^2\,\mathrm{d}t\Big)$. Rappelons que (ζ_t, β_t) est le brownien standard de $\mathbb{C}^{d-1} \times \mathbb{R}$, que $\mathcal{A}_t$ est l'aire de Lévy associée à ζ_t, et que

(ζ_t, β_t) est indépendant de $(\Lambda_t^1, \Lambda_t^2)$. En outre la version bidimensionnelle de la formule exponentielle utilisée précédemment (voir la proposition 3 et sa preuve) est

$$\Psi(\varrho, \alpha) = -\frac{1}{t} \log\big(\mathbb{E}\big[\exp(-\varrho\,\Lambda_t^1 - \alpha\,\Lambda_t^2)\big]\big)$$
$$= \hat{E}\big[1 - \exp\big(-\varrho\,\Theta^1 - \alpha\,\Theta^2\big)\big].$$

La mesure de Lévy associée est π_Λ telle que

$$\Psi(\varrho, \alpha) = \int_{(\mathbb{R}_+)^2} (1 - e^{-\varrho u - \alpha v})\, \pi_\Lambda(\mathrm{d}u, \mathrm{d}v).$$

Le lemme 3 est modifié comme suit.

Lemme 5. *Nous avons pour tous ϱ, σ réels :*

$$\lim_{t \to \infty} t\,\Psi\left(\frac{\varrho}{t}, \frac{\sigma^2}{2t^2}\right) = \frac{\varrho h d}{d-1},$$

ce qui signifie que $(t^{-1}\Lambda_t^1,\ t^{-2}\Lambda_t^2)$ converge en probabilité lorsque $t \to \infty$ vers $(hd/(d-1), 0)$.

Démonstration. Nous déduisons du corollaire 5, pour $\sigma \neq 0$:

$$\frac{2}{hd} \times t\,\Psi\Big(\frac{\varrho}{t}, \frac{\sigma^2}{2t^2}\Big) = \sigma - t\,\frac{d}{h} + 2t\,\sigma \times \frac{\int_0^\infty e^{-2\sigma h s}\, s^a\, (t^{-1} + s)^{d-a}\, \mathrm{d}s}{\int_0^\infty e^{-2\sigma h s}\, s^{a-1}\, (t^{-1} + s)^{d-a}\, \mathrm{d}s}$$

$$= \sigma - t\,\frac{d}{h} + 2t\,\sigma \times \frac{(2\sigma h)^{-d-1}\,d! + (2\sigma h)^{-d}\,(d-1)!\,(d-a)/t + \mathcal{O}(t^{-2})}{(2\sigma h)^{-d}\,(d-1)! + (2\sigma h)^{1-d}\,(d-2)!\,(d-a)/t + \mathcal{O}(t^{-2}\log t)}$$

$$= \sigma - t\,\frac{d}{h} + \frac{t}{h} \times \frac{d + 2\sigma h\,(d-a)/t + \mathcal{O}(t^{-2})}{1 + 2\sigma h\,(d-1)^{-1}(d-a)/t + \mathcal{O}(t^{-2}\log t)}$$

$$= \sigma - 2\sigma\,(d-a)/(d-1) + \mathcal{O}(t^{-1}\log t) = 2\varrho/(d-1) + \mathcal{O}(t^{-1}\log t). \qquad \square$$

Le corollaire 4 et la fin de la section 4.4 restent enfin valables tels quels.

Références

1. J. Bertoin, Lévy Processes. Cambridge Univ. Press, 1996.
2. J. Bertoin, Subordinators: examples and applications. École d'été de Saint-Flour XXVII, 1997. Lecture Notes in Math. n° 1717, Springer 1998.
3. N. Enriquez, J. Franchi, Masse des pointes, temps de retour et enroulements en courbure négative. Bull. Soc. Math. France, volume 130, n° 3, 349–386, 2002.
4. N. Enriquez, Y. Le Jan, Statistic of the winding of geodesics on a Riemann surface with finite area and constant negative curvature. Rev. Mat. Iberoamericana, Vol. 13, 2, 377–401, 1997.

5. J. Franchi, Théorème des résidus asymptotiques pour le mouvement brownien sur une surface riemannienne compacte. Annales de l'Institut Henri Poincaré, vol. 27, n° 4, 445–462, 1991.

6. J. Franchi, Asymptotic singular homology of a complete hyperbolic 3-manifold of finite volume. Proc. London Math. Soc. (3) n° 79, 451–480, 1999.

7. W.M. Goldman, Complex hyperbolic geometry. Oxford Univ. Press, 1999.

8. G.A. Hunt, Semi-groups of measures on Lie groups. Trans. A.M.S. 81, 264–293, 1956.

9. S. Hersonsky, F. Paulin, On the volumes of complex hyperbolic manifolds. Duke Math. J. vol. 84, n° 3, 719–737, 1996.

10. S. Hersonsky, F. Paulin, On the rigidity of discrete isometry groups of negatively curved spaces. Comment. Math. Helv. 72, 349–388, 1997.

11. M.E.H. Ismail, Bessel functions and the infinite divisibility of the Student t-distribution. Annals of Proba. 5, n° 4, 582–585, 1977.

12. F.B. Knight, Characterization of the Lévy measure of inverse local times of gap diffusions. In : Seminar on Stochastic Processes, 53-78, Birkhäuser 1981.

13. S. Kotani, S. Watanabe, Krein's spectral theory of strings and general diffusion processes. In: Functional Analysis in Markov Processes, 235–259, Proceeding Katata and Kyoto 1981 (éditeur Fukushima M.), Lecture Notes in Math. n° 923, Springer 1981.

14. J.-F. Le Gall, M. Yor, Étude asymptotique de certains mouvements browniens complexes avec drift. PTRF 71, 183–229, 1986.

15. T. Lyons, H.P. McKean, Windings of the plane Brownian motion. Adv. Math. 51, 212–225, 1984.

16. W. Magnus, F. Oberhettinger, R.P. Soni, Formulas and theorems for the special functions of mathematical physics. Springer, Berlin 1966.

17. P. Messulam, M. Yor, On D. Williams' "pinching method" and some applications. J. London Math. Soc. 26, 348–364, 1982.

18. J. Pitman, M. Yor, Bessel processes and infinitely divisible laws. In "Stochastic Integrals", Durham 1980, 285–370, ed. D. Williams, LNM 851, Springer 1981.

19. J. Pitman, M. Yor, Further asymptotic laws of planar Brownian motion. Annals of Proba. 17, n° 3, 965–1011, 1989.

20. L.C.G. Rogers, D. Williams, Diffusions, Markov Processes, and Martingales. Volume 2 : Itô Calculus. John Wiley & Sons, 1986.

21. D. Revuz, M. Yor, Continuous martingales and Brownian motion. Springer, Berlin, 1999.

22. F. Spitzer, Some theorems concerning 2-dimensional Brownian motion. Trans. A.M.S. 87, 187–197, 1958.

Stochastic Covariant Calculus with Jumps and Stochastic Calculus with Covariant Jumps

Laurence Maillard-Teyssier

Laboratoire LAMA, Université de Versailles Saint Quentin en Yvelines
e-mail: maillard@math.uvsq.fr

Summary. We propose a stochastic covariant calculus for càdlàg semimartingales in the tangent bundle TM over a manifold M. In ordinary differential geometry, a connection on M is needed to define the covariant derivative of a C^1 curve in TM; by applying the transfer principle, Norris has defined a stochastic covariant integration along a continuous semimartingale in TM. We extend this to the case when the semimartingale jumps, using Norris' work and Cohen's results on stochastic calculus with jumps on manifolds. Depending on the order in which the function giving the jumps and the connection are composed, one obtains a "stochastic covariant calculus with jumps" or a "stochastic calculus with covariant jumps", which are in general not equivalent. Under suitable conditions, Norris' results for the continuous case are recovered. This case can be described by a covariant continuous calculus of order two, which involves the notion of a connection of order two.

1 Introduction

Our aim is to define a stochastic covariant calculus for processes with jumps in the tangent bundle TM over a differentiable manifold M. Our motivations come from both differential geometry and probability. On the one hand, covariant differential equations are frequently used, in physical sciences for example, to intrinsically describe in a moving frame the evolution of some process. Moreover, some problems, for instance in gauge theories, require studying covariant differential equations in a stochastic framework (see [Elw82]). On the other hand, discretization theorems for stochastic differential equations (s.d.e.) can be written for manifold-valued processes with jumps. Therefore, one main application of the formalism we develop will be to discretize covariant s.d.e. Our work uses, on the one hand, Cohen's results in [Coh96-1] and [Coh96-2], concerning stochastic calculus with jumps on a manifold, and, on the other hand, Norris' formalism proposed in [Nor92], defining a covariant stochastic calculus for continuous processes. We extend these works in two directions: the first one, called *stochastic covariant calculus with jumps*, extends Norris' covariant calculus to the non continuous case, whereas the second one, called *stochastic calculus with covariant jumps*, extends Cohen's calculus with jumps to the covariant case. We will see that, in general, these two approches are not equivalent. A comparison will be done.

Geometric assumptions and notations

Throughout the paper, M and U are C^∞ manifolds without boundary. They are assumed to have a countable atlas, so Whitney's imbedding theorem allows to imbed these paracompact manifolds in $\mathbb{R}^m$. All formulas given in coordinates are expressed in such an imbbeding (but never depend on it). The Einstein summation convention is used. We work on the manifold called M; however, every time we recall general results, intended to be applied later to M or to TM, the manifold is called U. The projection of TM onto M is denoted by π. For every V in TM, the notation $V = (x, v)$ means that x is the projection πV of V on M and v is the part of V in the fibre $T_x M$ above x.

For z in U, $T_z U$ denotes the tangent space to U at z, and $TU = \cup_{z \in U} T_z U$ the tangent bundle. Recall that α is a first order form on U if, for every z in U, α_z belongs to $T_z^* U$ (for instance, the differential df of a smooth function f on U is a first order form on U). For every smooth $\phi : U \to W$, for every z in U, we denote by $d\phi(z) : T_z U \to T_{\phi(z)} W$ and call the differential of ϕ at z, the linear tangent map defined, for all A in $T_z M$ and every smooth f, by $d\phi(z)(A)(f) = A(f \circ \phi)$.

For all z in U, $\tau_z U$ denotes the tangent space of order two to U at z. It is the set of second order tangent vectors at z on U, that is, differential operators of order at most two, with no constant term. $\tau U = \cup_{z \in U} \tau_z U$ is called the tangent bundle of order two. Θ is a second order form on U if, for every z in U, Θ_z belongs to $\tau_z^* U$ (for instance, the 2-jet $d^2 f$ of a smooth function f on U is a second order form on U). For every smooth $\phi : U \to W$, for every z in U, we denote by $d^2 \phi(z) : \tau_z U \to \tau_{\phi(z)} W$ and call the 2-jet of ϕ at z, the linear tangent map of order two defined, for all $\mathbb{A}$ in $\tau_z M$ and every smooth f, by $d^2 \phi(z)(\mathbb{A})(f) = \mathbb{A}(f \circ \phi)$ (see [Em89] for more details). We will often use the following result (given in paragraph (6.19) of [Em89]) for the composition of 2-jets:

Let $f : F \to \mathbb{R}$ and $\phi : E \to F$ be smooth maps. Then the 2-jet of $f \circ \phi : E \to \mathbb{R}$ is given by $d^2(f \circ \phi) = (d^2 \phi)^*(d^2 f)$. In an imbedding, one has

$$d^2(f \circ \phi)(z) = \frac{\partial f}{\partial z^i}(\phi(z))d^2\phi^i(z) + \frac{\partial^2 f}{\partial z^i \partial z^j}(\phi(z))d\phi^i.d\phi^j(z). \qquad (1)$$

For all z in U, $\overset{\triangle}{\tau^*}_z U$ denotes the set of real functions θ_z on U, twice differentiable at z and such that $\theta_z(z) = 0$. $\overset{\triangle}{\tau^*} U$ is defined by $\overset{\triangle}{\tau^*} U = \cup_{z \in U} \overset{\triangle}{\tau^*}_z U$ (see [Coh96-1] for more details).

A connection on a manifold U permits to define exponential maps and geodesics. U is said to be simply convex if any two points x and y can be joined by a unique geodesic on U. We will suppose in that case that this geodesic can be written with the exponential map on M as $(c_t)_{t\in[0,1]} = (\exp_x(t \exp_x^{-1} y))_{t\in[0,1]}$, with $c_0 = x$, $c_1 = y$, and $\dot{c}_0 = \exp_x^{-1}(y)$.

Note on the coordinates used in formulas: we use coordinates on M, TM and τM, arising from a given system of coordinates on M. For every point x on M, $(x^i)_{i=1...m}$ denotes a system of coordinates around x in M. The standard basis of $T_x M$ is $(\frac{\partial}{\partial x^i})_{i=1...m}$ and the corresponding dual basis of $T_x^* M$ is $(dx^i)_{i=1...m}$. It follows that

- every u in $T_x M$ can be written $u = dx^i(u)\frac{\partial}{\partial x^i}$,
- every L in $\tau_x M$ can be written $L = d^2 x^i(L)\frac{\partial}{\partial x^i} + dx^i.dx^j(L)\frac{\partial^2}{\partial x^i \partial x^j}$.

For every $V = (x, v)$ in TM, the systems of coordinates $(x^i)_{i=1...m}$ around x in M and $(v^i)_{i=1...m} = (dx^i(v))_{i=1...m}$ around v in $T_x M$, yield a system $(V^i)_{i=1...m} = ((x^i, v^i))_{i=1...m}$ of coordinates around V in TM. The standard basis of $T_V TM$ is $(\frac{\partial}{\partial V^i})_{i=1...m} = ((\frac{\partial}{\partial x^i}, \frac{\partial}{\partial v^i}))_{i=1...m}$ and the corresponding dual basis of $T_V^* TM$ is $(dV^i)_{i=1...m} = (dx^i, dv^i)_{i=1...m}$. It follows that

- every $\mathbb{U}$ in $T_V TM$ can be written $\mathbb{U} = dx^i(\mathbb{U})\frac{\partial}{\partial x^i} + dv^i(\mathbb{U})\frac{\partial}{\partial v^i}$,
- every $\mathbb{L}$ in $\tau_V TM$ can be written

$$\mathbb{L} = d^2 x^i(\mathbb{L})\frac{\partial}{\partial x^i} + d^2 v^i(\mathbb{L})\frac{\partial}{\partial v^i} + dx^i.dx^j(\mathbb{L})\frac{\partial^2}{\partial x^i \partial x^j}$$
$$+ 2\, dx^i.dv^j(\mathbb{L})\frac{\partial^2}{\partial x^i \partial v^j} + dv^i.dv^j(\mathbb{L})\frac{\partial^2}{\partial v^i \partial v^j}.$$

Probabilistic assumptions and notations

$(\Omega, \mathcal{F}, P)$ is a standard probability space, endowed with a filtration $(\mathcal{F}_t)_{t\geqslant 0}$, satisfying the usual hypotheses. Any process z on U is denoted by (z_t), where the time t is always assumed to be positive. Recall that a process is said to be "càdlàg" if it is right continuous and left limited. (z_t^c) denotes the continuous martingale part of the càdlàg semimartingale (z_t). We say that a U-valued process (z_t) is a continuous (resp. càdlàg) semimartingale on U if, for any C^2 function f on U, $(f(z_t))$ is a real continuous (resp. càdlàg) semimartingale. The notions of locally bounded and predictable processes on U will be used. We refer the reader to [Em89] and [Coh96-1] for precise definitions.

2 Reminders of geometry and probability

These reminders of geometry and probability will help replacing our results in their context and explain why there are two natural ways to define a stochastic calculus, simultaneously "covariant" and "with jumps". First, we summarize results about stochastic calculus of order two on a manifold: a formalism was proposed in [Mey82] and [Sch82] by Meyer and Schwartz for continuous semimartingales, then extended by Cohen in [Coh96-1] to the case of semimartingales with jumps. Next, we introduce the definition of a linear connection on M, yielding the notion of a covariant derivative. Then we recall the formalism given by Norris in [Nor92] for a continuous stochastic covariant calculus on manifolds. For the proofs of these reminders, the reader is referred to [Ma03].

2.1 Stochastic continuous calculus of order two ($d\!I$)

To integrate on a manifold U, first order forms α are usually considered. The transfer principle permits to define Stratonovich integrals on manifolds, providing an analogy between the integration of α against the tangent vector $\dot{z}_t \in T_{z_t} U$ to a C^1 curve (z_t) on U and the integration against the Stratonovich infinitesimal variation ∂z_t of a semimartingale (z_t). The Itô calculus on U is then usually deduced from a Stratonovich to Itô conversion formula, for which we need a torsion-free connection on U. However, the stochastic infinitesimal variations ∂z_t (Stratonovich) and dz_t (Itô) are not intrinsic. The Itô formula rather incites us to consider a second order stochastic infinitesimal variation, similar to a second order tangent vector on U,

$$d\!I z_t = dz_t^i \frac{\partial}{\partial z^i} + \frac{1}{2} d < z^i, z^j >_t \frac{\partial^2}{\partial z^i \partial z^j}.$$

Therefore, the integrands are second order forms Θ on U.

This second order formalism was developed by Meyer in [Mey82] and Schwartz in [Sch82]. It provides compact and geometrically intrinsic formulas. The stochastic integral of order two of Θ along (z_t) is defined in a unique way as follows (see [Em89]).

Definition 1. *Let (z_t) be a continuous semimartingale in U and (Θ_t) a τ^*U-valued predictable locally bounded process above (z_t). There exists a unique linear mapping $\Theta \to \int \Theta d\!I z$, such that $\int \Theta d\!I z$ is a real semimartingale, called the stochastic integral of Θ along (z_t), satisfying, for every smooth f and for every locally bounded predictable processes K and Θ,*

$$(i) \int d^2 f(z) \, d\!I z = f(z) - f(z_0), \quad (ii) \int K\Theta d\!I z = \int K \, d\left(\int \Theta d\!I z\right).$$

Note that (i) corresponds to the Itô formula. In an imbedding, one has

$$\int_0^t d^2 f(z_s) d\!I z_s = \int_0^t \frac{\partial f}{\partial z^i}(z_s) dz_s^i + \frac{1}{2} \int_0^t \frac{\partial^2 f}{\partial z_s^i \partial z_s^j}(z_s) d < z^i, z^j >_s .$$

In [Em89], Emery presents the Stratonovich and the Itô calculus as particular cases of the stochastic calculus of order 2. For this, a map has to be introduced (d_s for Stratonovich, G for Itô), transforming a first order form into a second order form.

Stratonovich continuous calculus (∂)

Definition 2. *The map d_s is the unique linear map from the set of first order forms on U to the set of second order ones that verifies, for every function f*

and every first order form α on U,

$$d_s(df) = d^2 f \quad and \quad d_s(f\alpha) = df.\alpha + f d_s\alpha$$

In an imbedding, one has $d_s\alpha = \alpha_i d^2 z^i + \frac{\partial \alpha_i}{\partial z^j}(z)dz^i.dz^j$.

Definition 3 (Meyer, Schwartz). *For every continuous semimartingale (z_t) on U and every first order form α on U, the Stratonovich integral of α along (z_t) is defined by*

$$\int \alpha_{z_t} \, \partial z_t = \int (d_s\alpha)_{z_t} \, d\!\!I z_t.$$

In an imbedding, one has $\int \alpha_{z_t} \, \partial z_t = \int (\alpha_{z_t})_i dz_t^i + \frac{1}{2} \int \frac{\partial \alpha_i}{\partial z^j}(z_t)d < z^i, z^j >_t$.

Itô continuous calculus (d)

The Itô integral along a continuous semimartingale can be defined from a Stratonovich to Itô conversion formula. The manifold U is required to be endowed with a connection, which is assumed to be torsion-free. This connection induces the following map G. Note that, by abuse of language, we will also call G a connection.

Definition 4. *G is the linear mapping from the set of first order forms on U to the set of second order ones defined, for every first order form α on U, by*

$$G(\alpha) = d_s\alpha - H(\nabla\alpha), \tag{2}$$

where d_s is given by definition 2, and H is the unique linear mapping from bilinear forms to second order ones such that its restriction to first order forms is null and thut, for all forms α and β, we have $H(\alpha \otimes \beta) = \alpha.\beta$.

In an imbedding, one has

$$G(\alpha)_z = (\alpha_z)_i d^2 z^i + (\alpha_z)_k \Gamma^k_{ij}(z)dz^i.dz^j,$$

with (Γ^k_{ij}) the Christoffel symbols of the connection G.

Definition 5 (Meyer). *Let G be a connection on U. For every continuous semimartingale (z_t) on U and every first order form α on U, the Itô integral of α along (z_t) is defined by*

$$\int \alpha_{z_t} \, dz_t = \int (G\alpha)_{z_t} \, d\!\!I z_t.$$

Note that formula (2) is the Stratonovich to Itô conversion formula. It yields $\int G(\alpha) \, d\!\!I z = \int d_s\alpha \, d\!\!I z - \int H(\nabla\alpha) \, d\!\!I z$, that is $\int \alpha_{z_t} dz_t = \int \alpha_{z_t} \partial z_t - \frac{1}{2} \int \nabla\alpha(\partial z_t, \partial z_t)$.

It can be shown (see [Ma03]) that definition 5 is equivalent to the one proposed by Norris in [Nor92], using horizontal lifts.

2.2 Stochastic calculus of order two with jumps $(d\overset{\triangle}{I})$

In [Coh96-1], Cohen extends the formalism of second order calculus to the case when the semimartingale may jump. The integrands θ are now in $\tau^* \overset{\triangle}{U}$, that is, for all z in $\check{U}$, θ_z is a real function on U, twice differentiable at z and such that $\theta_z(z) = 0$.

The stochastic integral of order two of θ along a càdlàg semimartingale (z_t) is defined in a unique and intrinsic way as follows.

Definition 6 (Cohen). *Let (z_t) be a càdlàg semimartingale on U and (θ_t) a $\tau^* \overset{\triangle}{U}$-valued predictable locally bounded process above (z_{t-}). There exists a unique linear mapping $\theta \to \int \theta \, d\overset{\triangle}{I} \, z$ such that $\int \theta \, d\overset{\triangle}{I} \, z$ is a real semimartingale, called the stochastic integral of θ along (z_t), which is null at 0 and satisfies, for every f in $C^2(U)$ and every real locally bounded predictable process K,*

(i) $\forall t > 0, \ \theta_t(z) = f(z) - f(z_{t-}) \ \Rightarrow \ \int_0^t \theta_s \, d\overset{\triangle}{I} \, z_s = f(z_t) - f(z_0),$

(ii) $\int K_s \theta_s \, d\overset{\triangle}{I} \, z_s = \int K_s d(\int \theta_u \, d\overset{\triangle}{I} \, z_u)_s,$

(iii) $d^2\theta_t = 0 \ \Rightarrow \ \int_0^t \theta_s \, d\overset{\triangle}{I} \, z_s = \sum_{0 \leqslant s \leqslant t} \theta_s(z_s).$

In an imbedding, one has

$$\int_0^t \theta_s \, d\overset{\triangle}{I} \, z_s = \int_0^t \frac{\partial \theta_s}{\partial z^i}(z_{s-}) dz_s^i + \frac{1}{2} \int_0^t \frac{\partial^2 \theta_s}{\partial z_s^i \partial z_s^j}(z_{s-}) d < z^{i^c}, z^{j^c} >_s$$

$$+ \sum_{0 < s \leqslant t} (\theta_s(z_s) - \frac{\partial \theta_s}{\partial z^i}(z_{s-}) \Delta z_s^i). \qquad (3)$$

Note that we recover the change of variable formula for $\theta_t = f(z_t) - f(z_{t-})$, with $f \in C^2(V)$.

By points (i) and (ii) of the definition, the continuous case defined in the previous paragraph is recovered when the semimartingale is continuous, since one has

$$\int_0^t d^2\theta_s d\!I \, z_s = \int_0^t \frac{\partial \theta_s}{\partial z^i}(z_s) dz_s^i + \frac{1}{2} \int_0^t \frac{\partial^2 \theta_s}{\partial z_s^i \partial z_s^j}(z_s) d < z^i, z^j >_s . \qquad (4)$$

Point (iii) describes the case of a finite number of jumps. In particular, the jump of the integral $\int \theta \, d\overset{\triangle}{I} \, z$ when the semimartingale (z_t) jumps from z_{t-} to z_t is given by $\theta_t(z_{t-})$.

The map θ contains all the information: it describes the jumps of the integral and provides the drift and diffusion parts from its first and second derivatives. As a consequence, we will be able to define a stochastic covariant integral of order two if we describe one covariant jump θ of the stochastic integral along (z_t).

Cohen defines a Stratonovich and an Itô integral along a càdlàg semimartingale as second order integrals. To describe the jumps on the manifold, the notions of interpolation rule (for Stratonovich) and connection rule (for Itô) are required.

Stratonovich calculus with jumps ($\overset{\triangle}{\partial}$)

Definition 7 (Cohen). *A mapping $I : [0, 1] \times U \times U \to U$ is an interpolation rule if, for all z, y in U, $I(., x, y)$ is smooth, for all s in $[0, 1]$, $I(s, ., .)$ is measurable, C^3 on a neighbourhood of the diagonal of $U \times U$, and if I satisfies*

(i) $I(s, z, z) = z$,   (ii) $I(0, z, y) = z$, $I(1, z, y) = y$,   (iii) $\frac{\partial I(s, z, y)}{\partial y}\big|_{y=z} = \lambda(s) Id_{T_z U}$ *for a smooth λ.*

For instance, if U is a vector space, the linear interpolation $I(s, z, y) = z + s(y - z)$ is an interpolation rule on U. The fundamental example we will consider is the following.

Example 1 *(Geodesic interpolation rule) If U is simply convex, the geodesic joining two points, $c : [0, 1] \times U \times U \to U$, defined by $\forall s \in [0, 1], \forall z, y \in U$, $c(s, z, y) = \exp_z(s. \exp_z^{-1}(y))$, is an interpolation rule, called the geodesic interpolation rule on U.*
Indeed, (i) and (ii) are obvious. For (iii), we write $dc(s, z, .)(z) = d \exp_z(s. \exp_z^{-1}(z))(s.d \exp_z^{-1}(z))$. But $\exp_z^{-1}(z) = 0$, so we have $dc(s, z, .)(z) = d \exp_z(0)(s.d \exp_z^{-1}(z))$. Moreover $d \exp_z(0) = Id_{T_0 T_z U}$ and $d \exp_z^{-1}(z)) = Id_{T_z U}$, so (iii) is proved.

When (z_t) jumps, its path is replaced by an interpolation curve between the points z_{t-} and z_t on M. Therefore, the jump of the Stratonovich integral is given by the integration of the form α along this interpolation curve.

Definition 8 (Cohen). *Let I be an interpolation rule on U. For every càdlàg semimartingale (z_t) on U and every first order form α on U, the Stratonovich integral of α along (z_t) is defined by*

$$\int \alpha_{z_{t-}} \overset{\triangle}{\partial} z_t = \int \overset{\triangle}{I}(\alpha)_{z_{t-}} \overset{\triangle}{dI} z_t,$$

where the jump of the integral, when (z_t) jumps from z_{t-} to z_t, is

$$\overset{\triangle}{I}(\alpha)_{z_{t-}}(z_t) = \int_0^1 \alpha_{I(s, z_{t-}, z_t)} \dot{I}(s, z_{t-}, z_t) \, ds. \tag{5}$$

Note that, for all fixed z in U, the differential and the 2-jet at point $u = z$ of the map

$$\overset{\triangle}{I}(\alpha)_z : u \in U \to \overset{\triangle}{I}(\alpha)_z(u) = \int_0^1 \alpha_{I(s, z, u)} \dot{I}(s, z, u) \, ds,$$

satisfy

$$d \overset{\triangle}{I} (\alpha)_z(z) = \alpha_z, \quad d^2 \overset{\triangle}{I} (\alpha)_z(z) = (d_s\alpha)_z \tag{6}$$

(where d_s is the map of definition 2). This yields the following proposition, explaining how we recover the known continuous case.

Proposition 1 (Cohen) *If (z_t) is continuous, $\int \alpha \overset{\triangle}{\partial} z$ is independent of the interpolation rule and it coincides with the continuous Stratonovich integral*

$$\int \alpha \overset{\triangle}{\partial} z = \int \alpha \, \partial z, \quad that \ is \quad \int \alpha \overset{\triangle}{\partial} z = \int (d_s\alpha)_z \, d\!\!\!/ z.$$

Itô calculus with jumps $(\overset{\triangle}{\partial})$

Definition 9 (Picard). *A mapping $\gamma : U \times U \to TU$, is a connection rule if it is measurable, C^2 on a neighbourhood of the diagonal of $U \times U$, and if it satisfies, for all z, y in U,*

(i) $\gamma(z,y) \in T_zU$, $\quad (ii)$ $\gamma(z,z) = 0$, $\quad (iii)$ $\frac{\partial \gamma(z,y)}{\partial y}|_{y=z} = Id_{T_zU}$.

For instance, if U is a vector space, the linear map $\gamma(z,y) = y - z$ is a connection rule on U. The fundamental example we will consider is the following.

Example 2 *(Geodesic connection rule) If U is simply convex, the initial tangent vector to the geodesic joining the points z and y on U, $\gamma_0 : U \times U \to TU$, defined by $\gamma_0(z,y) = \exp_z^{-1}(y)$, is a connection rule, called the geodesic connection rule on U.*

Indeed, (i) is obvious, (ii) comes from $\exp_z^{-1}(z) = 0$, and (iii) from $d \exp_z^{-1}(z) = Id_{T_zU}$.

The vector $\gamma(z_{t-}, z_t)$ represents the jump of (z_t) in $T_{z_{t-}}U$. Therefore, the jump of the Itô integral is given by the application of the form α to $\gamma(z_{t-}, z_t)$.

Definition 10 (Cohen). *Let γ be a connection rule on U. For every càdlàg semimartingale (z_t) on U and every first order form α on U, the Itô integral of α along (z_t) is defined by*

$$\int \alpha_{z_{t-}} \overset{\triangle}{d} z_t = \int \overset{\triangle}{\gamma} (\alpha)_{z_{t-}} \, d\!\!\!/ z_t,$$

where the jump of the integral, when (z_t) jumps from z_{t-} to z_t, is

$$\overset{\triangle}{\gamma} (\alpha)_{z_{t-}} (z_t) = \alpha_{z_{t-}} \gamma(z_{t-}, z_t). \tag{7}$$

Note that, for all fixed z in U, the differential and the 2-jet at point $u = z$, of the map

$$\overset{\triangle}{\gamma} (\alpha)_z : u \in U \to \overset{\triangle}{\gamma} (\alpha)_z(u) = \alpha_z \circ \gamma(z,u),$$

satisfy

$$d \overset{\triangle}{\gamma} (\alpha)_z(z) = \alpha_z, \quad d^2 \overset{\triangle}{\gamma} (\alpha)_z(z) = G(\alpha)_z \tag{8}$$

(where G is the map of definition 4). This yields the following proposition, explaining how we recover the known continuous case.

Proposition 2 (Cohen) *If (z_t) is continuous, $\int \alpha \overset{\triangle}{d} z$ only depends on the 2-jet of the connection rule γ, giving a connection G on U, defined for every first order form α on U, for z in U, by $G(\alpha)_z = \alpha_z d^2\gamma(z,.)(z)$, and it coincides with the continuous Itô integral*

$$\int \alpha \overset{\triangle}{d} z = \int \alpha dz, \quad \text{that is} \quad \int \alpha \overset{\triangle}{d} z = \int G(\alpha)_z \, d\hspace{-0.3em}I z.$$

2.3 Covariant calculus

Connection of order one on M

There are many ways to define a connection on a manifold. We specify here our approach. Starting from the general definition of a connection in a principal fibre bundle, following [KN63] and [Spi79], we work in the particular case of a linear connection in the principal fibre bundle $L(M)$ of linear frames on M, with the linear group $Gl_m(\mathbb{R})$. More precisely, we consider a connection on the associated vector bundle TM with standard fibre $\mathbb{R}^m$.

Note that our presentation of connection may slightly differ from the usual ones. More details and proofs needed to get the equivalence with classical definitions can be found in [Ma03].

Note also that, in all the paper, TM may be replaced by a principal bundle E over M.

For all $V = (x,v)$ in TM, the fibre through V is T_xM. The tangent space T_VTM to TM at V contains a canonical subset $T_v(T_xM)$ (rigorously $di(V)(T_V(x,T_xM))$ with the inclusion $i : (x,T_xM) \hookrightarrow TM$). This space is called the vertical space at V, and denoted by $\mathcal{V}_VTM$. A connection on M is a way of choosing for each V a supplementary space $\mathcal{H}_VTM$ to this vertical space, called the horizontal space at V.

Definition 11. *A connection on M is a smooth map $p : V \in TM \to p_V$ such that, for every $V = (x,v)$ in TM*

1. p_V projects T_VTM onto the vertical space $\mathcal{V}_VTM = T_v(T_xM)$.
2. $\forall \lambda \in \mathbb{R}, \quad p_{\bar{\lambda}V} \circ d\bar{\lambda}(V) = \lambda p_V$ where $\bar{\lambda}(V) = (x, \lambda.v)$.

Every vector $U = dx^i(U)\frac{\partial}{\partial x^i} + dv^i(U)\frac{\partial}{\partial v^i}$ on T_VTM can be uniquely written as $U = U^{\mathcal{V}} + U^{\mathcal{H}}$ with a vertical component $U^{\mathcal{V}}$ and a horizontal one $U^{\mathcal{H}}$, expressed as follows

$$U^{\mathcal{V}} = p_V(U) = \left[dv^i(U) + \Gamma^i_{jk}(x)v^k dx^j(U)\right] \frac{\partial}{\partial v^i}, \tag{9}$$

$$U^{\mathcal{H}} = U - p_V(U) = dx^i(U)\frac{\partial}{\partial x^i} - dx^k(U)\Gamma_{kj}^i(x)v^k\frac{\partial}{\partial v^i}. \tag{10}$$

The connection p is said to be torsion-free if the functions (Γ_{ik}^j) are symmetric $(\Gamma_{ik}^j = \Gamma_{ki}^j)$. We shall use connections for two different purposes. Connections used for covariant calculus are not supposed to be torsion-free. Other connections are required for Itô calculus on a manifold. Emery explained in [Em89] that only the symmetric part of the connection is involved for the Itô calculus. Thus, without loss of generality, these connections are assumed to be torsion-free.

Covariant derivative of a C^1 curve

Let p be a connection on M and $(Y_t) = ((x_t, y_t))$ a curve of class C^1 in TM. Its covariant derivative is the usual derivative seen in a frame moving parallelly along its projection curve (x_t) on M. It can also be seen as a projection of the usual derivative onto the vertical space as follows.

Definition 12. *The covariant derivative of the curve* $(Y_t) = ((x_t, y_t))$ *is defined by*

$$\frac{\partial^{\mathcal{V}} Y_t}{\partial t} = p_{Y_t}(\dot{Y}_t). \tag{11}$$

In an imbedding, one has $\frac{\partial^{\mathcal{V}} Y_t}{\partial t} = [\,\dot{y}_t^i + \Gamma_{jk}^i(x_t)\,y_t^k\,\dot{x}_t^j\,]\,\frac{\partial}{\partial y^i},\quad \frac{\partial}{\partial y^i} \in T_{y_t}(T_{x_t}M)$

Notice our (non usual) notation for the covariant derivative: the exponent $\mathcal{V}$ stands for "vertical", since the covariant derivative is valued in the vertical space $T_{y_t}(T_{x_t}M)$.

2.4 Stochastic covariant calculus

Stratonovich continuous covariant calculus ($\partial^{\mathcal{V}}$)

Let p be a connection on M. According to (11), we can consider the covariant case as the image of the classical one by the projection p. With regard to the integration of a form α, it corresponds to the pull-back of α by p.

Definition 13. *The pull-back of the first order form* α *by the map* p *is the first order form* $p^*(\alpha)$ *on* TM*, defined by:* $\forall V \in TM, \forall U \in T_V TM, \quad p^*(\alpha)_V(U) = \alpha_V(p_V(U))$

Note that, since the projection p_V is $T_v(T_xM)$-valued, only the part of $p^*(\alpha)$ acting on this vertical space is involved. The vector $p_V(U) \in T_v(T_xM)$ is seen as an element of $T_V TM$.

We use the transfer principle to define a Stratonovich covariant integral as follows.

Definition 14. *Let p be a connection on M. For every continuous semi-martingale $(Y_t) = ((x_t, y_t))$ on TM and every first order form α on TM, we define the Stratonovich covariant integral of α along (Y_t) by*

$$\int \alpha_{Y_t} \partial^{\mathcal{V}} Y_t = \int p^*(\alpha)_{Y_t} \ \partial Y_t, \quad \textit{that is} \quad \int \alpha_{Y_t} \partial^{\mathcal{V}} Y_t = \int d_s(p^*(\alpha))_{Y_t} \ dY_t.$$

In an imbedding, one has

$$\int \alpha_{Y_t} \partial^{\mathcal{V}} Y_t = \int (\alpha_{Y_t})_i [\partial y_t^i \ + \ \Gamma_{jk}^i(x_t) \ y_t^k \ \partial x_t^j \ ],$$

where $(\alpha_{Y_t})_i = \alpha_{Y_t}(\frac{\partial}{\partial y^i})$, $\frac{\partial}{\partial y^i} \in T_{y_t}(T_{x_t} M)$.

Itô continuous covariant calculus $(d^{\mathcal{V}})$

A connection p is given on M to describe the covariant calculus. This covariant calculus concerns semimartingales in TM. Therefore, a connection $\mathbb{G}$ on TM is required for Itô calculus, with *a priori* no link with the connection p on M. We define an Itô covariant integral as follows.

Definition 15. *Let p be a connection on M and $\mathbb{G}$ a connection on TM. For every continuous semimartingale $(Y_t) = ((x_t, y_t))$ on TM and every first order form α on TM, we define the Itô covariant integral of α along (Y_t) by*

$$\int \alpha_{Y_t} d^{\mathcal{V}} Y_t = \int p^*(\alpha)_{Y_t} \ dY_t, \quad \textit{that is,} \quad \int \alpha_{Y_t} d^{\mathcal{V}} Y_t = \int \mathbb{G}(p^*(\alpha))_{Y_t} \ dY_t.$$

Link with Norris' formalism

Stratonovich and Itô covariant integrals along a continuous semimartingale $(Y_t) = ((x_t, y_t))$ in TM were already defined by Norris in [Nor92]. Here is how our definitions 14 and 15 relate to Norris' formalism. Norris applies the transfer principle to the following formula, equivalent to (11),

$$\frac{\partial^{\mathcal{V}} Y_t}{\partial t} = u_t \frac{\partial}{\partial t}(u_t^{-1} y_t), \tag{12}$$

where u_t is the horizontal lift starting from a given u_0 of the part (x_t) of (Y_t) on M (note that the formula does not depend on the choice of u_0).

For a continuous semimartingale (x_t), the horizontal lift, starting from a given $u_o \in GL(\mathbb{R}^m, T_{x_0} M)$, can be defined with the transfer principle as the unique curve $((x_t, u_t))$ in $GL(\mathbb{R}^m, T_{x_t} M)$ (for all t, $u_t \in GL(\mathbb{R}^m, T_{x_t} M)$), satisfying the stochastic differential equation $p_{(x_t, u_t)}(\partial x_t, \partial u_t) = 0, u_0 = u_o$. Therefore, Norris proposes the following definitions.

Definition 16 (Norris). *Let p be a connection on M. For every continuous semimartingale $(Y_t) = ((x_t, y_t))$ on TM and every first order form α on M,*

- the Stratonovich covariant integral of α along (Y_t) is defined by Norris as

$$\int \alpha_{x_t} u_t \partial(u_t^{-1} y_t)$$

- the Itô covariant integral of α along (Y_t) is defined by Norris as

$$\int \alpha_{x_t} u_t d(u_t^{-1} y_t)$$

Note on the identification of the vertical space:

In definition 16, it is a form on M that is integrated, whereas a form on TM is considered in definitions 14 and 15. Indeed, Norris applies the transfer principle to formula (12), where the covariant derivative $\frac{\partial^V Y_t}{\partial t}$ is seen as an element of the tangent space $T_{x_t} M$, although it belongs to $T_{y_t}(T_{x_t} M)$, according to definition 12. Actually, an identification is done between the vertical space $T_{y_t}(T_{x_t} M)$ and the tangent space $T_{x_t} M$ (since $T_{y_t}(T_{x_t} M)$ is the tangent space at y_t to the vector space $T_{x_t} M$, it can be identified with the vector space itself). The difference of nature of both Stratonovich covariant integrals is best understood in the particular following case. Assume that (Y_t) is a continuous semimartingale in $\mathbb{R}^{2m}$ and p the flat connection on $\mathbb{R}^m$.

For every form α on M, set $(\alpha_x)_i = \alpha(\frac{\partial}{\partial x^i})$ (with $\frac{\partial}{\partial x^i} \in T_x M$). Then the Stratonovich covariant integral (given by definition 16) yields, in that particular case:

$$\int \alpha_{x_t} \partial^V Y_t = \int (\alpha_{x_t})_i \partial y_t^i = \int (\alpha_{x_t})_i dy_t^i + \frac{1}{2} \int \frac{\partial \alpha_i}{\partial x^j}(x_t) d < x^i, y^j >_t . \quad (13)$$

For every form α on TM, set $(\alpha_Y)_i = \alpha_Y(\frac{\partial}{\partial y^i})$ (with $\frac{\partial}{\partial y^i} \in T_y(T_x M)$). Then the Stratonovich covariant integral (given by definition 14) yields, in that particular case:

$$\int \alpha_{Y_t} \partial^V Y_t = \int (\alpha_{Y_t})_i \partial y_t^i = \int (\alpha_{Y_t})_i dy_t^i + \frac{1}{2} \int \frac{\partial \alpha_i}{\partial y^j}(Y_t) d < y^i, y^j >_t . \quad (14)$$

The main difference between (13) and (14) comes from the quadratic variation terms. The integral (14) is the one expected for a covariant integral when the manifold is $\mathbb{R}^m$ with the flat connection, since it corresponds to the Stratonovich integration of the semimartingale (y_t), part of (Y_t) in the fibre. Moreover, there is no canonical analoguous identification of the vertical space for the jumps and at order two (it would require a connection). For these reasons, we shall not make the identification in the paper. As a consequence, we will integrate forms on TM.

Norris' integrals can be written with a form on TM. Therefore, we have an equivalence between the Stratonovich covariant integrals of definitions 14 and 16, since they both come from the transfer principle, applied to the equivalent formulas for the covariant derivative.

With regard to the Itô covariant integrals, our definition includes Norris' one. Indeed, recall that definition 15 requires two connections, p describing the covariant calculus and $\mathbb{G}$ allowing to define Itô calculus on TM. Norris' definition only uses the connection p, to define the horizontal lift (u_t): the semimartingale is brought by (u_t) in $\mathbb{R}^m$, where the Itô calculus is applied. Thus, the second connection corresponds to the flat connection on $\mathbb{R}^m$. Actually, the connection $\mathbb{G}$ needed in definition 15 only acts on the vertical space which is a vector space, so that $\mathbb{G}$ may naturally be a flat connection on this space. Under this assumption, we recover Norris' integral (a proof is given in [Ma03]).

Part I: Stochastic covariant calculus with jumps

We now study the case where the semimartingale is càdlàg in TM. According to the foregoing, the covariant calculus is obtained from the classical one by considering the form $p^*\alpha$ instead of α.

2.5 A Stratonovich covariant calculus with jumps ($\overset{\triangle}{\partial^{\mathcal{V}}}$)

On the one hand, the covariant calculus requires a connection p on M. On the other hand, the Stratonovich calculus with jumps in TM requires an interpolation rule on TM. We give meaning to a Stratonovich covariant integral with jumps, by applying results of paragraph 2.2 to the form $p^*\alpha$, in the manifold $U = TM$.

Definition 17. *Let p be a connection on M and $\mathbb{I}$ an interpolation rule on TM. For every càdlàg semimartingale $(Y_t) = ((x_t, y_t))$ on TM and every first order form α on TM, the Stratonovich covariant integral of α along (Y_t) is defined by*

$$\int \alpha_{Y_{t-}} \overset{\triangle}{\partial^{\mathcal{V}}} Y_t = \int \overset{\triangle}{\mathbb{I}} (p^*\alpha)_{Y_{t-}} \overset{\triangle}{d\mathbb{I}} Y_t,$$

where the jump of the integral, when (Y_t) jumps from Y_{t-} to Y_t, is

$$\overset{\triangle}{\mathbb{I}} (p^*\alpha)_{Y_{t-}}(Y_t) = \int_0^1 (p^*\alpha)_{\mathbb{I}(s, Y_{t-}, Y_t)} \dot{\mathbb{I}}(s, Y_{t-}, Y_t)\, ds. \qquad (15)$$

Note that, by (11), formula (15) is also

$$\overset{\triangle}{\mathbb{I}} (p^*\alpha)_{Y_{t-}}(Y_t) = \int_0^1 \alpha_{\mathbb{I}(s, Y_{t-}, Y_t)} \frac{\partial^{\mathcal{V}}\mathbb{I}}{\partial s}(s, Y_{t-}, Y_t)\, ds.$$

Example 3 *Let $\mathbb{I}$ be the geodesic interpolation rule on TM (see example 1 with $U = TM$). It is given by $\mathbb{I}(s, (x, v), (y, w)) = (\exp_x(s\exp_x^{-1} y), I(s, v, w))$, where $I(s, v, w) = \tau_{0s}^{//}[s(\tau_{10}^{//}(w) - v) + v]$, with $(\tau_{0s}^{//})$ the parallel transport* *along the geodesic* $(\pi\mathbb{I}(s,(x,v),(y,w)) = (\exp_x(s\exp_x^{-1}(y))))$. *We can compute the covariant derivative, as* $\frac{\partial^{\mathcal{V}}\mathbb{I}}{\partial s}(s,Y_{t-},Y_t) = \tau_{0s}^{//}\frac{\partial}{\partial s}(\tau_{s0}^{//}I(s,y_{t-},y_t))$, *that is* $\frac{\partial^{\mathcal{V}}\mathbb{I}}{\partial s}(s,Y_{t-},Y_t) = \tau_{0s}^{//}\frac{\partial}{\partial s}(s(\tau_{x_t,x_{t-}}^{//}(y_t) - y_{t-}) + y_{t-}) = \tau_{0s}^{//}(\tau_{x_t,x_{t-}}^{//}(y_t) - y_{t-})$. *Then the jump* (15) *is expressed by*

$$\overset{\triangle}{\mathbb{I}}(p^*\alpha)_{Y_{t-}}(Y_t) = \int_0^1 \alpha_{\mathbb{I}(s,Y_{t-},Y_t)}(0,\tau_{x_{t-},\pi\mathbb{I}(s,Y_{t-},Y_t)}^{//}(\tau_{x_t,x_{t-}}^{//}y_t - y_{t-}))ds \quad (16)$$

where $\tau_{x_{t-},\pi\mathbb{I}(s,Y_{t-},Y_t)}^{//}(\tau_{x_t,x_{t-}}^{//}y_t - y_{t-})$ *is seen as an element of* $T_{\mathbb{I}(s,y_{t-},y_t)}$ $\times (T_{\pi\mathbb{I}(s,Y_{t-},Y_t)}M)$, *since it is the covariant derivative of* $\mathbb{I}(s,Y_{t-},Y_t)$.

Proposition 3 *If* (Y_t) *is a continuous semimartingale on* TM, *the covariant Stratonovich integral is independent of the choice of the interpolation rule* $\mathbb{I}$ *and it coincides with the continuous Stratonovich covariant integral*

$$\int \alpha_Y \overset{\triangle}{\partial^{\mathcal{V}}} Y = \int p^*(\alpha)_Y \partial Y.$$

Proof : Apply proposition 1 to the form $p^*(\alpha)$. □

2.6 An Itô covariant calculus with jumps $(\overset{\triangle}{d^{\mathcal{V}}})$

On the one hand, the covariant calculus requires a connection p on M. On the other hand, the Itô calculus with jumps requires a connection rule on TM (yielding a connection on TM, which has *a priori* no link with p on M). We give meaning to a covariant Itô integral with jumps by applying results of paragraph 2.2 to the form $p^*\alpha$, in the manifold $U = TM$.

Definition 18. *Let* p *be a connection on* M *and* γ *a connection rule on* TM. *For every càdlàg semimartingale* $(Y_t) = ((x_t,y_t))$ *on* TM *and every first order form* α *on* TM, *the covariant Itô integral of* α *along* (Y_t) *is defined by*

$$\int \alpha_{Y_{t-}} \overset{\triangle}{d^{\mathcal{V}}} Y_t = \int \overset{\triangle}{\gamma}(p^*\alpha)_{Y_{t-}} \overset{\triangle}{d\mathbb{I}} Y_t,$$

where the jump of the integral, when (Y_t) *jumps from* Y_{t-} *to* Y_t, *is*

$$\overset{\triangle}{\gamma}(p^*\alpha)_{Y_{t-}}(Y_t) = (p^*\alpha)_{Y_{t-}} \circ \gamma(Y_{t-},Y_t) = \alpha_{y_{t-}} \circ p_{Y_{t-}}(\gamma(Y_{t-},Y_t)). \quad (17)$$

Example 4 *Let* γ *be the geodesic connection rule on* TM *(see example 2). It is given by* $\gamma(V,W) = \dot{\mathbb{I}}(0,V,W)$, *where* $\mathbb{I}(s,V,W)$ *is the geodesic interpolation rule on* TM. *Then we have* $p_{Y_{t-}}(\gamma(Y_{t-},Y_t)) = p_{\mathbb{I}(0,Y_{t-},Y_t)}(\dot{\mathbb{I}}(0,Y_{t-},Y_t))$ $= \frac{\partial^{\mathcal{V}}\mathbb{I}}{\partial s}(s,Y_{t-},Y_t))|_{s=0}$. *The covariant derivative has been computed in example 3 as* $\frac{\partial^{\mathcal{V}}\mathbb{I}}{\partial s}(s,Y_{t-},Y_t) = \tau_{0s}^{//}(\tau_{x_t,x_{t-}}^{//}(y_t) - y_{t-})$. *At time* $s = 0$, *we get* $p_{Y_{t-}}(\gamma(Y_{t-},Y_t)) = \tau_{x_t,x_{t-}}^{//}(y_t) - y_{t-}$. *Then the jump* (17) *is expressed by*

$$\overset{\triangle}{\gamma}(p^*\alpha)_{Y_{t-}}(Y_t) = \alpha_{Y_{t-}}\left(0, \tau^{//}_{x_t, x_{t-}}(y_t) - y_{t-}\right) \tag{18}$$

where $\tau^{//}_{x_t, x_{t-}} y_t - y_{t-}$ is seen as an element of $T_{y_{t-}}(T_{x_{t-}}M)$, since it is projected by $p_{Y_{t-}}$.

Proposition 4 *If (Y_t) is a continuous semimartingale on TM, the covariant Itô integral depends only on the 2-jet of the connection rule γ, giving a connection $\mathbb{G}$ on TM, defined for every first order form α on TM, by*

$$\forall Y \in TM, \ \mathbb{G}(\alpha)_Y = \alpha_Y d^2 \overset{\triangle}{\gamma}(Y,.)(Y);$$

this integral coincides with the continuous Itô covariant integral

$$\int \alpha_{Y_{t-}} \overset{\triangle}{d^\nabla} Y_t = \int \alpha_{Y_{t-}} d^\nabla Y_t.$$

Proof : Apply proposition 2 to the form $p^*(\alpha)$.□

Part II: Stochastic calculus with covariant jumps

In part I, we have described separately the Stratonovich covariant integral and the Itô covariant integral. In the case with jumps described by Cohen as in the continuous case described by Emery, Stratonovich and Itô integrals are particular cases of an integral of order two. We would like to include in the same way the covariant integrals in the framework of a covariant stochastic calculus of order two. In this second part, we propose such a general formalism in the case with jumps. In the continuous case, this formalism will give a continuous stochastic covariant calculus of order two, which has never been defined as far as we know. It has to include, as particular cases, the covariant continuous integrals given by Norris. We propose here one definition. Our formalism is compatible with the Schwartz principle, according to which any stochastic calculus on a manifold is a calculus of order two, that is, it involves differential operators of order two. For the covariant case, we need to specify a notion of connection of order two on M, as a way of decomposing at each V in TM the tangent bundle $\tau_V TM$ of order two in a second order vertical space and an horizontal one.

2.7 Connection of order two

For all $V = (x, v)$ in TM, the fibre containing V is T_xM. The tangent space of order two, $\tau_V TM$, to TM at V, contains a canonical subset $\tau_v(T_xM)$ (rigorously, $d^2i(V)(\tau_V(x, T_xM))$ with the inclusion $i : (x, T_xM) \hookrightarrow TM$). We call this space the vertical space of order two at V, and denote it by $\tilde{\mathcal{V}}_V TM$. We call connection of order two on M, a way of choosing for each V a supplementary space $\tilde{\mathcal{H}}_V TM$ of this vertical space.

Definition 19. *A connection of order two on M is a smooth map $\bar{p} : V \in TM \to \bar{p}_V$ such that for all $V = (x, v)$ in TM,*

1. *$\bar{p}_V$ projects $\tau_V TM$ onto the vertical space of order two $\tilde{\mathcal{V}}_V TM = \tau_v(T_x M)$,*
2. *$\forall \lambda \in \mathbb{R}, \quad \bar{p}_{\bar{\lambda}V} \circ d^2 \bar{\lambda}(V) = d^2 \bar{\lambda}(V) \circ \bar{p}_V$ where $\bar{\lambda}(x, v) = (x, \lambda.v)$,*
3. *the restriction of $\bar{p}$ to first order vectors is a connection p of order one on M.*

2.8 Continuous covariant calculus of order two (dI^ν)

Let $\bar{p}$ be a connection of order two on M. By analogy with the first order, we define the covariant calculus of order two as the image of the (non covariant) calculus of order two by the projection $\bar{p}$. With regard to the integration of a second order form Θ, it corresponds to the pull-back of Θ by $\bar{p}$.

Definition 20. *If Θ is a second order form on TM, the pull-back of Θ by the map $\bar{p}$ is the second order form $\bar{p}^*(\Theta)$ on TM defined by $\forall V \in TM, \forall \mathbb{L} \in \tau M$: $\bar{p}^*(\Theta)_V(\mathbb{L}) = \Theta_V \circ \bar{p}_V(\mathbb{L})$.*

Definition 21. *Let $\bar{p}$ be a connection of order two on M. For every continuous semimartingale (Y_t) in TM and every second order form Θ on TM, the integral*

$$\int \Theta dI^\nu Y = \int \bar{p}^*(\Theta) dIY$$

is called the stochastic continuous covariant integral of Θ along (Y_t).

We show in [Ma03] that the covariant Stratonovich and Itô continuous integrals of definitions 14 and 15 can be written with a connection of order two.

3 Transport on M

We propose a formalism, called stochastic calculus with covariant jumps, which corresponds to the stochastic covariant calculus of order two of definition 21 when the integrator is a continuous semimartingale. We need to give an expression for a covariant jump. This requires a notion of transport.

3.1 Definition and properties of a transport

As is well known, the parallel transport along a curve joining x to y is a map bringing a vector from the tangent space $T_x M$ to M at x to the tangent space $T_y M$ to M at y. It has many interesting properties; in particular it is linear and invertible. Picard has defined a more general linear transport in [Pic91]. We still extend the notion of transport as a way of mapping at each x in M the tangent bundle TM onto the tangent space $T_x M$. Moreover, a transport is allowed to depend on a tangent vector v based on x. In order to recover the stochastic covariant calculus of order two in the continuous case, the 2-jet of a transport yields a connection of order two on M. Thus, the definition of transport requires properties on its 2-jet. When the transport itself verifies similar properties, we get a stronger notion, which is the one given by Picard. Such a transport can be seen as an analogue to a linear connection for the jumps.

Definition 22. *Let τ be a map yielding, for every $V = (x,v)$ in TM, a map $\tau_V : TM \to T_xM$ such that $(V, W) \to \tau_V(W)$ is a C^2 map on a neighbourhood of the diagonal of $TM \times TM$. Let p and $\tilde{p}$ denote the differential and the 2-jet of τ_V at V, that is*

$$p : V \in TM \to p_V = d\tau_V(V) : T_V TM \to T_v(T_xM)$$

$$\tilde{p} : V \in TM \to \tilde{p}_V = d^2\tau_V(V) : \tau_V TM \to \tau_v(T_xM)$$

The map τ is called a transport on M if $\tilde{p}$ is a connection of order two on M. It implies in particular that p is a connection of order one on M

Many transports satisfy this definition. Note that, for all $V = (x,v)$ in TM, τ_V may depend on a vector $v \in T_xM$ based at x. If it does not, we shall call it a basic transport.

The following notion of linear and idempotent transport corresponds to the one proposed by Picard in [Pic91] when the transport is basic.

Proposition 5 *Let τ be a map on M yielding, for all $V = (x,v)$ in TM, a map $\tau_V : TM \to T_xM$, satisfying*

- *$(V, W) \to \tau_V(W)$ is a C^2 map on a neighbourhood of the diagonal of $TM \times TM$,*
- *(i) $\forall w \in T_xM$, $\tau_V(x, w) = w$,*
  (ii) $\forall \lambda \in \mathbb{R}$, $\forall W = (y, w) \in TM$, $\tau_{(x,\lambda v)}(y, \lambda w) = \lambda \tau_{(x,v)}(y, w)$,

then τ is a transport on M, called idempotent (property (i)) and linear (property (ii)).

Proof : Property (i) of idempotence implies that the map $w \to d^2\tau_V(x, .)(w)$ yields, for $w = v$, a projection onto $\tau_v(T_xM)$. Moreover, property (ii) of linearity and formula (1) imply that, for every λ in $\mathbb{R}$, $d^2\tau_{\bar{\lambda}V}(\bar{\lambda}V) \circ d^2\bar{\lambda}(V) = d^2\bar{\lambda}(V) \circ d^2\tau_V(V)$. Hence, $\tilde{p}_V = d^2\tau_V(V)$ is a connection of order two on M and so τ is a transport on M.$\square$

Such a transport yields for all V in TM a projection map from TM onto T_xM, which may represent a vertical jump space at V. Then the transport yields a way of projecting onto this vertical space, so is an analogue for the jumps case to a linear connection on M, a connection "of order 0" on M.

We now turn to invertible transports. Invertibility will be instrumental in uniqueness of the solution to a s.d.e. with covariant jumps, defined with a transport.

Definition 23. *A transport τ is said to be invertible if, for all V in TM, for all y in M, the map $\tau_V(y,.) : T_yM \to T_xM$ is invertible. τ is said to be strongly invertible if it is basic and satisfies moreover, for every x and y in M, $\tau_x(y,.)^{-1} = \tau_y(x,.)$.*

Actually, local invertibility of the transport is often sufficient. It can be shown (see [Ma03]) that a linear transport is always locally invertible (there exists a neighbourhood N_x of x such that, for all y in N_x, $\tau_V(y,.) : T_yM \to T_xM$ is invertible), so in several cases we won't need to require the invertibility of a transport.

The property of linearity and strongly invertibility allows in particular to write in coordinates the connection of order two $\tilde{p}$ with an expression only depending on the Christoffel symbols of the connection p and their partial derivatives.

Proposition 6 *Let τ be a transport on M and (Γ^i_{jk}) be the Christoffel symbols of the associated connection p. Then one has*

$$(a)\ \frac{\partial \tau^i_V}{\partial x^j}(V) = \Gamma^i_{jk}(x)\, v^k, \quad (b)\ \frac{\partial \tau^i_V}{\partial v^j}(V) = \delta^i_j, \quad (c)\ \frac{\partial^2 \tau^i_V}{\partial v^j \partial v^l}(V) = 0.$$

Moreover, if τ is a linear and strongly invertible transport, one gets

$$(d)\ \frac{\partial \tau^i_x}{\partial y^j}\left(y, \frac{\partial}{\partial y^k}\right)|_{y=x} = \Gamma^i_{jk}(x), \quad (d')\ \frac{\partial \tau^i_x}{\partial x^j}\left(y, \frac{\partial}{\partial y^k}\right)|_{y=x} = -\Gamma^i_{jk}(x)$$

and also

$$(e)\ \frac{\partial^2 \tau^i_x}{\partial x^j \partial v^k}(x, v) = \Gamma^i_{jk}(x), \quad (f)\ \frac{\partial^2 \tau^i_x}{\partial x^k \partial x^l}\left(x, \frac{\partial}{\partial x^m}\right)$$

$$= \frac{1}{2}\left[\frac{\partial \Gamma^i_{lm}}{\partial x^k}(x) + \frac{\partial \Gamma^i_{km}}{\partial x^l}(x) + \Gamma^i_{kj}(x)\Gamma^j_{lm}(x) + \Gamma^i_{lj}(x)\Gamma^j_{km}(x)\right].$$

Proof : Let $V = (x, v)$ and $W = (y, w)$ be in TM. Recall that τ and p are related by $d\tau_V(V) = p_V$. This yields $\frac{\partial \tau^i_V}{\partial y^j}(y, w)|_{W=V} = \frac{\partial \tau^i_V}{\partial x^j}(V) = dv^i(p_V(\frac{\partial}{\partial x^j}))$ and $\frac{\partial \tau^i_V}{\partial w^j}(y, w)|_{W=V} = \frac{\partial \tau^i_V}{\partial v^j}(V) = dv^i(p_V(\frac{\partial}{\partial v^j}))$. Then, formula (9) for the connection p implies (a) and (b).

By (1), we have

$$d^2\tau_V(V)\left(\frac{\partial^2}{\partial v^j \partial v^l}\right) = \frac{\partial^2 \tau^i_V}{\partial v^j \partial v^l}(V)\frac{\partial}{\partial v^i} + \frac{\partial \tau^i_V}{\partial v^j}(V)\frac{\partial \tau^k_V}{\partial v^l}(V)\frac{\partial^2}{\partial v^k \partial v^i}.$$

Using (b), it gives

$$d^2\tau_V(V)\left(\frac{\partial^2}{\partial v^j \partial v^l}\right) = \frac{\partial^2 \tau^i_V}{\partial v^j \partial v^l}(V)\frac{\partial}{\partial v^i} + \frac{\partial^2}{\partial v^j \partial v^l}.$$

But $d^2\tau_V(V) = \tilde{p}_V$ is a projection onto $\tau_v(T_xM)$, so it satisfies $\tilde{p}_V(\frac{\partial^2}{\partial v^j \partial v^l})$ $= \frac{\partial^2}{\partial v^j \partial v^l}$. As a consequence, we get (c).

Now, assume that τ is linear and strongly invertible. This implies in particular that τ is basic. Since the map $w \to \tau_x(y, w)$ is linear, we have $\tau_x(y, w) = \tau_x(y, \frac{\partial}{\partial y^k})w^k$. It implies $\frac{\partial \tau_x^i}{\partial w^k}(y, w) = \tau_x^i(y, \frac{\partial}{\partial y^k})w^k$, and we get (d) for $y = x$. Since τ is strongly invertible, it satisfies $\tau_x^i(x, \tau_x(z, u)) = u^i$. Using the linearity of τ, it yields $\tau_z^i(x, \frac{\partial}{\partial x^j})\tau_x^j(z, \frac{\partial}{\partial z^r}) = \delta_r^i$. Deriving this relation with respect to x^l gives

$$\frac{\partial}{\partial x^l}\left(\tau_z^i\left(x, \frac{\partial}{\partial x^j}\right)\right)\tau_x^j\left(z, \frac{\partial}{\partial z^r}\right) + \tau_z^i\left(x, \frac{\partial}{\partial x^j}\right)\frac{\partial}{\partial x^l}\left(\tau_x^j\left(z, \frac{\partial}{\partial z^r}\right)\right) = 0. \quad (*)$$

Using (d) and the property $\tau_z(z, u) = u$ which yields $\tau_z^i(z, \frac{\partial}{\partial z^r}) = \delta_r^i$, the formula gives (d') for $z = x$. By the foregoing, for all $V = (x, v)$ and $W = (y, w)$ in TM, $\frac{\partial}{\partial y^j}(\frac{\partial \tau_x^i}{\partial w^k})(W) = \frac{\partial}{\partial y^j}(\tau_x^i(y, \frac{\partial}{\partial y^k}))$, which is $\Gamma_{jk}^i(x)$ for $W = V$, by (d). This proves (e).

To prove (f), let us start by computing the partial derivative $\frac{\partial \Gamma_{lm}^j}{\partial x^k}(x)$. Note that, since τ is strongly invertible, it is in particular basic. By (d), $\Gamma_{lm}^i(x)$ can be seen as a map $f_{lm}^i(x, x)$ where $f_{lm}^i(z, y) = \frac{\partial \tau_z^i}{\partial y^l}(y, \frac{\partial}{\partial y^m})$. Hence we write

$$\frac{\partial \Gamma_{lm}^i}{\partial x^k}(x) = \frac{\partial f_{lm}^i(z, x)}{\partial z^k}\Big|_{z=x} + \frac{\partial f_{lm}^i(x, y)}{\partial y^k}\Big|_{y=x}$$

that is,

$$\frac{\partial \Gamma_{lm}^i}{\partial x^k}(x) = \frac{\partial^2}{\partial z^k \partial x^l}\left(\tau_z^i\left(x, \frac{\partial}{\partial x^m}\right)\right)\Big|_{z=x} + \frac{\partial^2 \tau_x^i}{\partial y^k \partial y^l}\left(y, \frac{\partial}{\partial y^m}\right)\Big|_{y=x}.$$

Then we get

$$\frac{\partial^2 \tau_x^i}{\partial y^k \partial y^l}\left(y, \frac{\partial}{\partial y^m}\right)\Big|_{y=x} = \frac{\partial \Gamma_{lm}^i}{\partial x^k}(x) - \frac{\partial^2}{\partial z^k \partial x^l}\left(\tau_z^i\left(x, \frac{\partial}{\partial x^m}\right)\right)\Big|_{z=x}.$$

We need to compute the partial derivative of $\tau_z^i(x, .)$ with respect to z^k and x^l. Differentiating $(*)$ with respect to z^k gives

$$\frac{\partial^2}{\partial z^k \partial x^l}\left(\tau_z^i\left(x, \frac{\partial}{\partial x^j}\right)\right)\tau_x^j\left(z, \frac{\partial}{\partial z^r}\right) + \frac{\partial}{\partial x^l}\left(\tau_z^i(x, \frac{\partial}{\partial x^j})\right)\frac{\partial}{\partial z^k}\left(\tau_x^j\left(z, \frac{\partial}{\partial z^r}\right)\right)$$

$$+ \frac{\partial \tau_z^i}{\partial z^k}\left(x, \frac{\partial}{\partial x^j}\right)\frac{\partial \tau_x^j}{\partial x^l}\left(z, \frac{\partial}{\partial z^r}\right) + \tau_z^i\left(x, \frac{\partial}{\partial x^j}\right)\frac{\partial^2}{\partial z^k \partial x^l}\left(\tau_x^j\left(z, \frac{\partial}{\partial z^r}\right)\right) = 0$$

Using (d), (d') and the property $\tau_z^i(z, \frac{\partial}{\partial z^r}) = \delta_r^i$, we get, for $z = x$,

$$\frac{\partial^2}{\partial z^k \partial x^l}\left(\tau_z^i\left(x, \frac{\partial}{\partial x^r}\right)\right)\Big|_{z=x} + \Gamma_{lj}^i(z)\Gamma_{kr}^j(z)$$

$$+ (-\Gamma_{kj}^i(z))\left(-\Gamma_{lr}^j(z)\right) + \frac{\partial^2}{\partial z^k \partial x^l}\left(\tau_x^i\left(z, \frac{\partial}{\partial z^r}\right)\right)\Big|_{z=x} = 0.$$

Note that the map $(z, x) \rightarrow \tau_z(x, .)$ is C^2 for z and x nearby, so the Schwarz lemma implies $\frac{\partial^2}{\partial z^k \partial x^l}(\tau_z^j(x, \frac{\partial}{\partial x^i}))|_{x=z} = \frac{\partial^2}{\partial x^l \partial z^k}(\tau_z^j(x, \frac{\partial}{\partial x^i}))|_{x=z}$. But $\frac{\partial^2}{\partial x^l \partial z^k}(\tau_z^j(x, \frac{\partial}{\partial x^i}))|_{x=z}$ is equivalently written $\frac{\partial^2}{\partial z^l \partial x^k}(\tau_x^j(z, \frac{\partial}{\partial z^i}))|_{x=z}$. Hence we get the sum

$$\frac{\partial^2}{\partial z^k \partial x^l} \tau_z^j(x, \frac{\partial}{\partial z^r})|_{x=z} + \frac{\partial^2}{\partial z^l \partial x^k} \tau_x^j(z, \frac{\partial}{\partial z^r})|_{x=z} = -[\Gamma_{li}^j(z)\Gamma_{kr}^i(z) + \Gamma_{ki}^j(z)\Gamma_{lr}^i(z)].$$

Now use the symmetry $\frac{\partial^2 \tau_V^j}{\partial z^k \partial z^l}(V) = \frac{\partial^2 \tau_V^j}{\partial z^l \partial z^k}(V)$ and write

$$\frac{\partial^2 \tau_V^j}{\partial z^k \partial z^l}(V) = \frac{1}{2}\left[\frac{\partial^2 \tau_V^j}{\partial z^k \partial z^l}(V) + \frac{\partial^2 \tau_V^j}{\partial z^l \partial z^k}(V)\right].$$

Then we get

$$\frac{\partial^2 \tau_V^j}{\partial z^k \partial z^l}(V) = \frac{1}{2}\left[\frac{\partial \Gamma_{lm}^j}{\partial z^k}(z) + \frac{\partial \Gamma_{km}^j}{\partial z^l}(z)\right.$$

$$\left. - \frac{\partial^2}{\partial z^k \partial x^l}(\tau_z^j(x, \frac{\partial}{\partial x^m}))|_{x=z} - \frac{\partial^2}{\partial z^l \partial x^k}(\tau_z^j(x, \frac{\partial}{\partial x^m}))|_{x=z}\right],$$

and the result is proved. $\square$

As a consequence, p_V and $\tilde{p}_V$ can be written as follows in an imbedding.

Proposition 7 *For every transport τ, p is expressed in coordinates as follows. For every $\mathbb{W}$ in $T_V TM$,*

$$p_V(\mathbb{W}) = [\ dv^i(\mathbb{W}) + \Gamma_{jk}^i(x)v^k\ dx^j\ (\mathbb{W})]\frac{\partial}{\partial v^i}. \tag{19}$$

For every linear and strongly invertible transport τ, $\tilde{p}$ is expressed in coordinates as follows.

For every $\mathbb{L}$ in $\tau_V TM$,

$$\tilde{p}_V(\mathbb{L}) = [d^2v^i(\mathbb{L}) + \Gamma_{ml}^i(x)v^l d^2x^m(\mathbb{L})$$

$$+ (\frac{\partial \Gamma_{lm}^i}{\partial x^k}(x) + \Gamma_{kj}^i(x)\Gamma_{lm}^j(x))v^m dx^l.dx^k(\mathbb{L})$$

$$+ 2\Gamma_{mk}^i(x)dx^m.dv^k(\mathbb{L})\ ]\frac{\partial}{\partial v^i}$$

$$+ [dv^i.dv^j(\mathbb{L}) + \Gamma_{kl}^i(x)v^l\Gamma_{mr}^j(x)v^r dx^m.dx^k(\mathbb{L})$$

$$+ 2\Gamma_{ml}^i(x)v^l dx^m.dv^j(\mathbb{L})]\frac{\partial^2}{\partial v^i \partial v^j}.$$

Proof : For $V = (x, v)$ in TM, we compute the 2-jet $\tilde{p}_V(\mathbb{L}) = d^2\tau_V(V)(\mathbb{L})$ of τ_V at V, using (1). For all $\mathbb{L}$ in $\tau_V TM$, we get

$$\tilde{p}_V(\mathbb{L}) = \left[\frac{\partial \tau_V^i}{\partial V^m}(V)d^2V^m(\mathbb{L}) + \frac{\partial^2 \tau_V^i}{\partial V^m \partial V^k}(V)dV^m.dV^k(\mathbb{L})\right]\frac{\partial}{\partial v^i}$$

$$+ \left[\frac{\partial \tau_V^i}{\partial V^m}\frac{\partial \tau_V^j}{\partial V^k}dV^m.dV^k(\mathbb{L})\right]\frac{\partial^2}{\partial v^i \partial v^j}.$$

The first term is

$$\left[\frac{\partial \tau_V^i}{\partial v^m}(V)d^2v^m(\mathbb{L}) + \frac{\partial \tau_V^i}{\partial x^m}(V)d^2x^m(\mathbb{L}) + \frac{\partial^2 \tau_V^i}{\partial v^m \partial v^k}(V)dv^m.dv^k(\mathbb{L})\right.$$

$$\left. +\frac{\partial^2 \tau_V^i}{\partial x^m \partial x^k}(V)dx^m.dx^k(\mathbb{L}) + 2\frac{\partial^2 \tau_V^i}{\partial x^m \partial v^k}(V)dx^m.dv^k(\mathbb{L})\right] \frac{\partial}{\partial v^i}.$$

By proposition 6, it becomes

$$\left[d^2v^i(\mathbb{L}) + \Gamma_{ml}^i(x)v^l \; d^2x^m \; (\mathbb{L}) + \left[\frac{\partial \Gamma_{ml}^i}{\partial x^k}(x) + \Gamma_{kj}^i(x)\Gamma_{ml}^j(x)\right] v^l dx^m.dx^k(\mathbb{L})\right.$$

$$\left. + 2 \; \Gamma_{mk}^i(x) \; dx^m.dv^k(\mathbb{L})\right]\frac{\partial}{\partial v^i}.$$

The second term is

$$\left[\frac{\partial \tau_V^i}{\partial v^m}\frac{\partial \tau_V^j}{\partial v^k}dv^m.dv^k(\mathbb{L}) + \frac{\partial \tau_V^i}{\partial x^m}\frac{\partial \tau_V^j}{\partial x^k}dx^m.dx^k(\mathbb{L})\right.$$

$$\left. + 2 \; \frac{\partial \tau_V^i}{\partial x^m}\frac{\partial \tau_V^j}{\partial v^k}dx^m.dv^k(\mathbb{L})\right]\frac{\partial^2}{\partial v^i \partial v^j}.$$

By proposition 6, it becomes

$$\left[dv^i.dv^j(\mathbb{L}) + \Gamma_{kl}^i(x)v^l \Gamma_{mr}^j(x)v^r dx^k.dx^m(\mathbb{L}) + 2 \; \Gamma_{ml}^i(x)v^l dx^m.dv^j(\mathbb{L})\right]\frac{\partial^2}{\partial v^i \partial v^j}.$$

Adding both terms, we find the expression of $\tilde{p}_V$ in coordinates. That of p_V is obtained by taking the part of $\tilde{p}_V$ acting on first order vectors. It coincides with expression (9) for p. □

3.2 Examples

We now illustrate the notion of transport with some examples. We see that a transport is often only locally defined. Indeed, although it is not defined along a curve, its construction often requires a way of joining two points on M. If M is not simply convex, this can only be locally done. Otherwise, one may consider a smooth map which coincides with τ_V on a neighbourhood of V, and vanishes elsewhere (by using a partition of the unity). Actually, a transport will be used to describe the jumps of a semimartingale in TM. Then a locally defined transport would only describe small jumps. Nevertheless, since there is a finite number of large jumps, each one can be described separately, and there is no loss of generality due to the local existence of a transport.

Example 5. Parallel transport along geodesics $(\tau^{(//)})$
A connection is given on M, assumed to be simply convex. Let x and y be two points on M. Let $\tau_{yx}^{//} : T_yM \to T_xM$ be a parallel transport along the geodesic

joining y to x (or along $(I(s,y,x))_{s\in[0,1]}$, if I is an interpolation rule on M). The following map is a linear idempotent and strongly invertible transport on M, called parallel transport along geodesics

$$\forall V = (x,v) \in TM, \ \forall W = (y,w) \in TM, \quad \tau_x^{(//)}(W) = \tau_{yx}^{//}(w)$$

Indeed, for all x in M, $\tau_x : TM \to T_xM$. The smoothness (C^2 in particular) of $(V,W) \to \tau_x^{(//)}(W)$ follows from the smoothness of $\tau^{//}$. Property $\tau_{xx}^{//} = Id_{T_xM}$ gives idempotence (i) and linearity of the parallel transport gives linearity (ii) required in proposition 5.

We now prove strong invertibility. If (c_t) denotes the geodesic from y to x, the geodesic from x to y is $(\tilde{c}_t) = (c_{1-t})$. Then the parallel transport along $\tilde{c}_t$ is $\tilde{\tau}_{0t}^{//} = \tau_{1\ 1-t}^{//}$. Hence, $\tau_{xy}^{//} = \tilde{\tau}_{01}^{//} = \tau_{10}^{//} = (\tau_{01}^{//})^{-1} = (\tau_{yx}^{//})^{-1}$. So $\tau^{(//)}$ is invertible, and moreover strongly invertible since $\tau_y^{(//)}(x,.)^{-1} = (\tau_{xy}^{//})^{-1} = \tau_{yx}^{//} = \tau_x^{(//)}(y,.)^{-1}$.

Example 6. More general transports $(\tau^{(\beta)})$

More general examples can be constructed as follows with the parallel transport, or with any linear transport on M. Let β be a map on $TM \times TM$ such that β is smooth on a neighbourhood of the diagonal, and satisfies the properties:
$\forall (V,W) \in TM \times TM, \ \beta(V,W) \in T_xM, \ \beta(V,V) = 0$ and $d^2\beta(V,.)(V) = 0$. Then the following map is a transport on M:

$$\forall V = (x,v) \in TM, \ \forall W = (y,w) \in TM, \quad \tau_V^{(\beta)}(W) = \tau_{yx}^{//}(w) + \beta(V,W).$$

Indeed, we have, for all $V = (x,v)$ in TM, $\tau_V^{(\beta)} : TM \to T_xM$. The smoothness ($C^2$ in particular) of $(V,W) \to \tau_V^{(\beta)}(W)$ on the diagonal follows from the smoothness of $\tau^{//}$ and that of β. Let us compute the 2-jet of $\tau_V^{(\beta)}$ at V. We have $d^2\tau_V^{(\beta)}(V) = d^2\tau_x^{(//)}(V) + d^2\beta(V,.)(V)$. But $d^2\beta(V,.)(V) = 0$. Moreover, since $\tau^{(//)}$ is a transport, its 2-jet $d^2\tau_x^{(//)}(V)$ is a connection of order two on M. Consequently, $d^2\tau_V^{(\beta)}(V)$ is a connection of order two on M so $\tau^{(\beta)}$ is a transport on M.

For instance, if M is assumed to be a Riemannian connected manifold with metric g, denote by $\|v\|$ the length of a tangent vector v, that is, $\|v\| = g(v,v)^{\frac{1}{2}}$, and by $\delta(x,y)$ the distance between two points x and y on M, that is, the infimum of the lengths of all piecewise differentiable curves of class C^1 joining x to y. Then, it can be shown that (see [Ma03])

- *the map defined by $\tau_V^{(\beta_1)}(W) = \tau_{yx}^{//}(w) + \|v - \tau_{yx}^{//}(w)\|^3 v$ is a non idempotent, non basic and non linear transport on M.*
- *the map defined by $\tau_V^{(\beta_2)}(W) = \tau_{yx}^{//}(w) + \|v - \tau_{yx}^{//}(w)\|^3 \exp_x^{-1}(y)$ is an idempotent but non basic and non linear transport on M.*
- *the map defined by $\tau_V^{(\beta_3)}(W) = \tau_{yx}^{//}(w) + \delta^3(x,y) \exp_x^{-1}(y)$ is a basic but non linear transport on M.*

- the map defined by $\tau_V^{(\beta_4)}(W) = \tau_{yx}^{//}(w) + \delta^3(x,y)v$ is a linear but non basic transport on M.

Example 7. Transport from a connection rule $(\tau^{(\gamma)})$

Let γ be a connection rule on M. For all x in M, write γ_x for the map $\gamma(x,.) : M \to T_xM$ and suppose that, for all y in M, its differential at point y is invertible. In [Pic91], Picard defined a transport as follows. The following map is a linear invertible basic transport on M, called the transport from the connection rule γ

$$\forall V = (x,v) \in TM, \ \forall W = (y,w) \in TM, \qquad \tau_x^{(\gamma)}(W) = d\gamma_x(y)(w).$$

Indeed, for all x in M, $d\gamma_x(y) : T_yM \to T_{\gamma(x,y)}T_xM \simeq T_xM$, so $\tau_x^{(\gamma)} : TM \to T_xM$. The C^2 property for $(V,W) \to \tau_x^{(\gamma)}(W)$ follows from the connection rule being a C^2 mapping. Property $d\gamma_x(x) = Id_{T_xM}$ gives idempotence (i) and linearity of the differential gives linearity (ii) required in proposition 5. Moreover, $\tau^{(\gamma)}$ is clearly invertible but not necessarily strongly invertible.

Note that, if $\gamma = \gamma_0$ is the geodesic connection rule $(\gamma_0)_x(y) = \exp_x^{-1}(y)$, we get a transport which corresponds to the notion of Jacobi fields: $\tau_x^{(\gamma_0)}(W) = d(\gamma_0)_x(y)(w) = d\exp_x^{-1}(y)(w)$.

4 Stochastic calculus of order two with covariant jumps $(\overset{\triangle}{d\!\!I}^v)$

We are going to use Cohen's results in [Coh96-1], to define a stochastic integral with covariant jumps along a càdlàg semimartingale $(Y_t) = ((x_t, y_t))$. A transport τ on M allows to describe a covariant jump as follows. When (Y_t) jumps from Y_{t-} to Y_t in TM, its corresponding covariant jump is a jump in $T_{x_{t-}}M$, from y_{t-} to $\tau_{Y_{t-}}(Y_t)$. Therefore, the jump of the covariant integral of a $\overset{\triangle}{\tau}^* TM$-valued θ along (Y_t) is given by the function $\theta \circ \tau$, that is, the pull-back of θ by τ.

Definition 24. For every $\overset{\triangle}{\tau}^* TM$-valued form θ, the pull-back $\tau^*(\theta)$ of θ by τ is the $\overset{\triangle}{\tau}^* TM$-valued form defined by $\forall V \in TM, \forall W \in TM, \quad \tau^*(\theta)_V(W) = \theta_V(\tau_V(W))$

Note that, since the transport τ_V is T_xM-valued, only the part of $\tau^*(\theta)$ acting on this space is involved. The vector $\tau_V(W) \in T_xM$ is seen as an element of TM.

Definition 25. Let τ be a transport on M. For every càdlàg semimartingale $(Y_t) = ((x_t, y_t))$ in TM and every $\overset{\triangle}{\tau}^* TM$-valued predictable locally bounded

*process (θ_t) above (Y_{t-}), the stochastic integral with covariant jumps of (θ_t)
along (Y_t) is defined by*

$$\int \theta \stackrel{\triangle\nu}{dI} Y = \int \tau^*(\theta) \stackrel{\triangle}{dI} Y.$$

Using formula (3), one can write in an imbedding

$$\int \theta \stackrel{\triangle\nu}{dI} Y = \int \frac{\partial\theta_s}{\partial v^i}(Y_{s-})\frac{\partial\tau^i_{Y_{s-}}}{\partial Y^m}(Y_{s-})dY^m_s + \frac{1}{2}\int\left[\frac{\partial\theta_s}{\partial v^i}(Y_{s-})\frac{\partial^2\tau^i_{Y_{s-}}}{\partial Y^m\partial Y^k}(Y_{s-})\right.$$

$$\left.+\frac{\partial^2\theta_s}{\partial v^i\partial v^j}(Y_{s-})\frac{\partial\tau^i_{Y_{s-}}}{\partial Y^m}(Y_{s-})\frac{\partial\tau^j_{Y_{s-}}}{\partial Y^k}(Y_{s-})\right]d<(Y^m)^c,(Y^k)^c>_s$$

$$+\sum_{0<s\leqslant t}(\theta_s(\tau_{Y_{s-}}Y_s) - \frac{\partial\theta_s}{\partial v^i}(Y_{s-})\frac{\partial\tau^i_{Y_{s-}}}{\partial Y^m}(Y_{s-})\triangle Y^m_s).$$

For instance, if τ is the parallel transport along geodesics, or any linear and strongly invertible transport on M, the continuous part is given by the 2-jet $\tilde{p}$ in proposition 7, and one gets

$$\int \theta \stackrel{\triangle\nu}{dI} Y = \int \frac{\partial\theta_s}{\partial v^i}(Y_{s-})\left[\Gamma^i_{mj}(x_{s-})y^j_{s-}dx^m_s + dy^i_s\right]$$

$$+\frac{1}{2}\int\left[\frac{\partial\theta_s}{\partial v^i}(Y_{s-})(\frac{\partial\Gamma^i_{mr}}{\partial x^k} + \Gamma^i_{kl}.\Gamma^l_{mr})(x_{s-})y^r_{s-}\right.$$

$$\left.+\frac{\partial^2\theta_s}{\partial v^i\partial v^j}(Y_{s-})\Gamma^i_{mr}.\Gamma^j_{kl}(x_{s-})y^l_{s-}y^r_{s-}\right]d<(x^m)^c,(x^k)^c>_s$$

$$+\int\left[\frac{\partial\theta_s}{\partial v^i}(Y_{s-})+\frac{\partial^2\theta_s}{\partial v^i\partial v^k}(Y_{s-})y^l_{s-}\right]\Gamma^i_{mk}(x_{s-})d<(x^m)^c,(y^k)^c>_s$$

$$+\frac{1}{2}\int\frac{\partial^2\theta_s}{\partial v^i\partial v^j}(Y_{s-})d<(y^i)^c,(y^j)^c>_s + \sum_{0<s\leqslant t}[\theta_s(\tau^{//}_{x_s,x_{s-}}Y_s)$$

$$-\frac{\partial\theta_s}{\partial v^i}(Y_{s-})(\Gamma^i_{mk}(x_{s-})y^k_{s-}\triangle x^m_s + \triangle y^i_s)].$$

By the following proposition, definition 25 extends the covariant continuous calculus of order two described in definition 21.

Proposition 8 *If (Y_t) is a continuous semimartingale on TM, we get*

$$\int \theta \stackrel{\triangle\nu}{dI} Y = \int d^2\theta \, dI^\nu Y,$$

where dI^ν is defined with the connection $\tilde{p}$, that is $\int \theta \stackrel{\triangle\nu}{d} Y = \int \tilde{p}^*(d^2\theta)dIY$.

Proof : According to (4), if (Y_t) is continuous, we get $\int \theta \stackrel{\triangle\nu}{dI} Y = \int \tau^*(\theta) \stackrel{\triangle}{dI} Y = \int d^2(\tau^*(\theta))dIY$. Formula (1), applied for $W = V$, yields $d^2(\theta_V \circ \tau_V)(V) = \tilde{p}^*_V(d^2\theta_V(V))$. It follows $\int d^2(\tau^*(\theta))dIY = \int \tilde{p}^*(d^2\theta)dIY$ which is $\int d^2\theta dI^\nu Y$, by definition 21. $\square$

5 Application: Stratonovich and Itô calculi with covariant jumps

5.1 A Stratonovich calculus with covariant jumps ($\overset{\triangle\nu}{\partial}$)

We now apply the previous results to define a Stratonovich integral with co-variant jumps. On the one hand, the covariant calculus is described by the notion of transport τ on M. On the other hand, the Stratonovich calculus with jumps in TM requires an interpolation rule on TM. To get an integral with covariant jumps, we apply τ^* to the jump $\overset{\triangle}{\mathbb{I}}(\alpha)$.

A first definition: *Let τ be a transport on M and $\mathbb{I}$ an interpolation rule on TM. For every càdlàg semimartingale $(Y_t) = ((x_t, y_t))$ on TM and every first order form α on TM, the Stratonovich integral with covariant jumps of α along (Y_t) is*

$$\int \alpha_{Y_{t-}} \overset{\triangle\nu}{\partial} Y_t = \int \overset{\triangle}{\mathbb{I}}(\alpha)_{Y_{t-}} \overset{\triangle\nu}{d\mathbb{I}} Y_t,$$

where the jump of the integral, when (Y_t) jumps from Y_{t-} to Y_t, is

$$\tau^*(\overset{\triangle}{\mathbb{I}}(\alpha))_{Y_{t-}}(Y_t) = \overset{\triangle}{\mathbb{I}}(\alpha)_{Y_{t-}}(\tau_{Y_{t-}}(Y_t)) = \int_0^1 \alpha_{\mathbb{I}(s,(x_{t-},y_{t-}),(x_{t-},\tau_{Y_{t-}}(Y_t)))}$$
$$\times \dot{\mathbb{I}}(s,(x_{t-},y_{t-}),(x_{t-},\tau_{Y_{t-}}(Y_t)))ds$$

It is important to notice that this definition only involves the part of $\mathbb{I}$ in the fibre and not the one on M. The map $\mathbb{I}(s,(x,v),(x,u))$ can be seen as a map $(x, \mathbb{I}(s,v,u))$, with $\mathbb{I}(s,v,u) \in T_xM$ (written with an abuse of notation since it also depends on x and y.) Then, it would be natural to consider for $\mathbb{I}(s,v,w)$ the linear interpolation rule $\mathbb{I}(s,v,w) = s(w-v)+v$. In this case, we write $\mathbb{I}(s,(x,v),(x,w)) = (x, s(w-v)+v)$, and then we get $\dot{\mathbb{I}}(s,(x,v),(x,w)) = (0, w-v)$, where $w-v$ is seen as an element of $T_v(T_xM)$.

For this reason, the definition of the Stratonovich integral with covariant jumps does not really involve an interpolation rule. Therefore, we will keep the following definition.

Definition 26. *Let τ be a transport on M. For every càdlàg semimartingale $(Y_t) = ((x_t, y_t))$ on TM and every first order form α on TM, the Stratonovich integral with covariant jumps of α along (Y_t) is*

$$\int \alpha_{Y_{t-}} \overset{\triangle\nu}{\partial} Y_t = \int \overset{\triangle}{\mathbb{I}}(\alpha)_{Y_{t-}} \overset{\triangle\nu}{d\mathbb{I}} Y_t,$$

where the jump of the integral, when (Y_t) jumps from Y_{t-} to Y_t, is

$$\tau^*(\overset{\triangle}{\mathbb{I}}(\alpha))_{Y_{t-}}(Y_t) = \int_0^1 \alpha_{(x_{t-}, s[\tau_{Y_{t-}}(Y_t) - y_{t-}] + y_{t-})} \, ds \; (0, \tau_{Y_{t-}}(Y_t) - y_{t-}). \quad (20)$$

Proposition 9 *If (Y_t) is a continuous semimartingale on TM, the Stratonovich integral with covariant jumps gives*

$$\int \alpha \overset{\triangle\nu}{\partial} Y = \int d_s \alpha \; d\!I^\nu Y,$$

where $d\!I^\nu$ is defined with the connection $\tilde{p}$, that is $\int \alpha \overset{\triangle\nu}{\partial} Y = \int \tilde{p}^*(d_s \alpha) d\!I Y$.

Proof : According to proposition 8, if (Y_t) is continuous, we have $\int \alpha \overset{\triangle\nu}{\partial} Y$ $= \int \overset{\triangle}{\mathbb{I}}(\alpha) \, d\!I \overset{\triangle\nu}{} Y = \int d^2(\overset{\triangle}{\mathbb{I}}(\alpha)) \, d\!I^\nu Y$. Then, using (6), we get the result. Note that, if we consider the first definition with a more general interpolation rule, proposition 1 implies that the integral obtained in the continuous case does not depend on the choice of $\mathbb{I}$.$\square$

5.2 An Itô calculus with covariant jumps ($\overset{\triangle\nu}{d}$)

We now apply the previous results to define an Itô integral with covariant jumps. On the one hand, the covariant calculus is described by the notion of transport τ on M. On the other hand, the Itô calculus with jumps in TM requires a connection rule on TM. To get an integral with covariant jumps, we apply τ^* to the jump $\overset{\triangle}{\gamma}(\alpha)$.

A first definition: *Let τ be a transport on M and γ a connection rule on TM. For every càdlàg semimartingale $(Y_t) = ((x_t, y_t))$ on TM and every first order form α on TM, the Itô integral with covariant jumps of α along (Y_t) is*

$$\int \alpha_{Y_{t-}} \overset{\triangle\nu}{d} Y_t = \int \overset{\triangle}{\gamma}(\alpha)_{Y_{t-}} \, d\!I^\nu Y_t,$$

where the jump of the integral, when (Y_t) jumps from Y_{t-} to Y_t, is

$$\tau^*(\overset{\triangle}{\gamma}(\alpha))_{Y_{t-}}(Y_t) = \overset{\triangle}{\gamma}(\alpha)_{Y_{t-}}(\tau_{Y_{t-}}(Y_t)) = \alpha_{Y_{t-}} \gamma((x_{t-}, y_{t-}), (x_{t-}, \tau_{Y_{t-}}(Y_t))).$$

It is important to notice that this definition only involves the part of γ in the fibre and not the one on M. The map $\gamma((x, v)(x, u))$ can be seen as a map $(0, \gamma(v, u))$, with $\gamma(v, u) \in T_x M$ (written with an abuse of notation since it also depends on x and y.) Then, it would be natural to consider for $\gamma(v, w)$ the linear connection rule $\gamma(v, w) = w - v$. In this case, we write $\gamma((x, v), (x, w)) = (0, w - v)$, where $w - v$ is seen as an element of $T_v(T_x M)$.

For this reason, the definition of the Itô integral with covariant jumps does not really involve a connection rule. We recover the fact that, for the Itô covariant calculus, it suffices to have one connection, as in Norris' definition for the continuous case. One connection is p, the other one is a flat connection on the vertical space, derived from the linear connection rule. Therefore, we will keep the following definition.

Definition 27. *Let τ be a transport on M. For every càdlàg semimartingale $(Y_t) = ((x_t, y_t))$ on TM and every first order form α on TM, the Itô integral with covariant jumps of α along (Y_t) is*

$$\int \alpha_{Y_{t-}} \overset{\triangle v}{d} Y_t = \int \overset{\triangle}{\gamma}(\alpha)_{Y_{t-}} \overset{\triangle v}{dI} Y_t,$$

where the jump of the integral, when (Y_t) jumps from Y_{t-} to Y_t, is

$$\tau^*(\overset{\triangle}{\gamma}(\alpha))_{Y_{t-}}(Y_t) = \alpha_{Y_{t-}}(0, \tau_{Y_{t-}}(Y_t) - y_{t-}). \tag{21}$$

Proposition 10 *If (Y_t) is a continuous semimartingale on TM, the covariant Itô integral gives*

$$\int \alpha \overset{\triangle v}{d} Y = \int \alpha \, dI^V Y,$$

where dI^V is defined with the connection $\tilde{p}$, that is, $\int \alpha \overset{\triangle v}{d} Y = \int \tilde{p}^*(\alpha) dI Y$.

Remark 1. It can be shown that the connection of order two $\tilde{p}$, seen as a second order vector, has an intrinsic first order part. Then $\tilde{p}^*(\alpha) = \alpha \circ \tilde{p}$ corresponds, in the above formula, to the application of α to this first order part.

Proof : According to proposition 8, if (Y_t) is continuous, we have $\int \alpha \overset{\triangle v}{d}$
$Y = \int \overset{\triangle}{\gamma}(\alpha) \overset{\triangle v}{dI} Y = \int d^2 \overset{\triangle}{\gamma}(\alpha) \, dI^V Y$. Then, using (8) (with a flat G, in particular such that $G(\alpha) = \alpha$), we get the result. $\square$

Part III: Comparison of both stochastic calculi: covariant with jumps or with covariant jumps

We now compare both different approaches of parts I and II. To define Stratonovich and Itô covariant integrals, we integrated a first order form α along a càdlàg semimartingale (Y_t). We had to explain, on the one hand, how we transform α into a function on M describing the jump of the integral and, on the other hand, how we use the connection to pass to the covariant case. Actually, both definitions correspond to both possible orders for these steps. An important difference between these two approaches results in the different given tools for the covariant calculus: the first definition uses a connection whereas a transport is required for the second definition. On the one hand, every transport τ on M defines a connection p (of order one) on M by $p_V = d\tau_V(V)$. This connection yields a parallel transport $\tau^{//}$ along geodesics, but equality $\tau_V(W) = \tau_{yx}^{//}(w)$ does not necessarily hold. On the other hand, every connection q of order one on M defines a parallel transport $\tau^{//}$ along geodesics. It can be shown that we the connection q is recovered as by $q_V = d\tau_x^{//}(V)$. Thus, in general, there is no real equivalence between a connection and a transport, except when we consider the parallel transport along geodesics.

To do the comparison, the choice for the interpolation and connection rules of part I has to be coherent with that of part II. Recall that, in part II, interpolation and connection rules are supposed to be linear on the fibre when the point is fixed on the base. For instance, the geodesic interpolation and connection rules satisfy this property. Recall their expression

$$\mathbb{I}(s, V, W) = (\ \exp_x(s \exp_x^{-1} y)\ ,\ \tau_{0s}^{//}[s(\tau_{10}^{//}(w) - v) + v]\ ),$$
$$\gamma(V, W) = \dot{\mathbb{I}}(0, V, W),$$

where $(\tau_{0s}^{//})$ is the parallel transport along the geodesic joining x to y.

We will see that the covariant Itô integrals of part I and II can equal. On the contrary, there is a real difference of nature between the covariant Stratonovich integrals of part I and II.

6 Comparison of both Stratonovich covariant calculi

6.1 Comparison of the jumps

To compare both definitions for the Stratonovich covariant integrals along a càdlàg semimartingale (Y_t) on TM, it is enough to compare the jumps of the integrals. Let us have a look at the example of the parallel transport along geodesics, with $\mathbb{I}$ the geodesic interpolation rule on TM.

In part I, we have defined a covariant Stratonovich integral with jumps $\int \alpha\ \overset{\triangle}{\partial^{\mathcal{V}}} Y$. With the geodesic interpolation rule, its jump, when (Y_t) jumps from Y_{t-} to Y_t, is given by example 3, as

$$\overset{\triangle}{\mathbb{I}}(p^*\alpha)_{Y_{t-}}(Y_t) = \int_0^1 \alpha_{\mathbb{I}(s, Y_{t-}, Y_t)}(0, \tau_{x_{t-}, \pi\mathbb{I}(s)}^{//}(\tau_{x_t, x_{t-}}^{//} y_t - y_{t-}))\ ds \qquad (22)$$

In part II, we have defined a Stratonovich integral with covariant jumps $\int \alpha\ \overset{\triangle_v}{\partial} Y$. Its jump, when (Y_t) jumps from Y_{t-} to Y_t, is given by

$$(\tau^{//})^*(\overset{\triangle}{\mathbb{I}}(\alpha))_{Y_{t-}}(Y_t) = \int_0^1 \alpha_{(x_{t-},\mathbb{I}(s,Y_{t-},(x_{t-},\tau^{//}_{x_t,x_{t-}}y_t)))} \, ds \, (0, \tau^{//}_{x_t,x_{t-}} y_t - y_{t-}).$$

(23)

Note that, since $\mathbb{I}(s,v,\tau^{//}_{10}(w)) = s(\tau^{//}_{10}(w) - v) + v = \tau^{//}_{s0}\mathbb{I}(s,v,w)$, using the invertibility of $\tau^{//}$, one can write $\mathbb{I}(s,v,w) = \tau^{//}_{0s}\mathbb{I}(s,v,\tau^{//}_{10}(w))$, and so (23) can equivalently be written

$$(\tau^{//})^*(\overset{\triangle}{\mathbb{I}}(\alpha))_{Y_{t-}}(Y_t) = \int_0^1 \alpha_{(x_{t-},\tau^{//}_{\pi\mathbb{I}(s)},x_{t-}}\mathbb{I}(s,y_{t-},y_t))} \, ds \, (0, \tau^{//}_{x_t,x_{t-}} y_t - y_{t-}).$$

(24)

In (22), the integration of α along $\mathbb{I}$ simultaneously uses the part of $\mathbb{I}$ on M and the one in the fibres $T_{\pi\mathbb{I}(s)}M$ whereas (24) shows that the transport $\tau^{//}$ brings everything at $\pi\mathbb{I}(0,Y_{t-},Y_t) = x_{t-}$, so that α is integrated only along the part of $\mathbb{I}$ in the tangent space $T_{x_{t-}}M$. If we change the part of $\mathbb{I}$ on M, the second definition will not change but the first one will. As a consequence, both Stratonovich integrals differ in general.

They can agree only if, for all s, $\pi\mathbb{I}(s) = \pi\mathbb{I}(0) = x_{t-}$. It means that $x_t = x_{t-}$. We will see later that this corresponds to a vertical jump for (Y_t).

6.2 Comparison of the continuous Stratonovich integrals

As a consequence of the transfer principle, the definition of part I contains Norris' definition. With regard to the definition of part II, it cannot contain Norris'one, since it is not even a Stratonovich integral. Indeed, according to [Em89], a second order integral $\int \theta d\mathbb{I}Y$ is a Stratonovich integral if, by applying the application d_s to the first order part $R\theta$ of θ, we recover θ, that is, if we can write

$$\int \theta d\mathbb{I}Y = \int d_s(R\theta)d\mathbb{I}Y.$$

Here, we have $R(\tilde{p}^*(d_s\alpha)) = p^*(\alpha)$ but $d_s(p^*(\alpha))$ is not $\tilde{p}^*(d_s\alpha)$. On the one hand, $d_s(p^*(\alpha))$ is obtained by considering the differential $d\tau_V(W)$ at point $W = V$ and then applying, in a way, the differentiation d_s to the application $V \to p_V = d\tau_V(V)$, which depends on V in two ways. On the other hand, $\tilde{p}^*(d_s\alpha)$ is obtained by computing the 2-jet $d^2\tau_V(W)$ of the application τ_V, with V fixed, and then taking it at point $W = V$. Consequently, $\int \tilde{p}^*(d_s\alpha)d\mathbb{I}Y$ is not a Stratonovich integral and so the covariant "Stratonovich" integrals of part I and II keep on being different in the continuous case.

7 Comparison of both Itô covariant calculi

7.1 Comparison of the jumps

To compare both definitions given for the Itô covariant integral along a càdlàg semimartingale (Y_t) on TM, it is enough to compare the jumps of the integrals.

In part I, we have defined a covariant Itô integral with jumps $\int \alpha \, \overset{\triangle}{d^v} Y$. Its jump, when (Y_t) jumps from Y_{t-} to Y_t, is given by

$$\overset{\triangle}{\gamma} (p^*\alpha)_{Y_{t-}} (Y_t) = \alpha_{Y_{t-}} \, p_{Y_{t-}} (\gamma(Y_{t-}, Y_t)). \qquad (25)$$

In part II, we have defined an Itô integral with covariant jumps $\int \alpha \, d^{\overset{\triangle}{v}} Y$. Its jump, when (Y_t) jumps from Y_{t-} to Y_t, is given by

$$\tau^*(\overset{\triangle}{\gamma} (\alpha))_{Y_{t-}} (Y_t) = \alpha_{Y_{t-}} (0, \tau_{Y_{t-}} (y_t) - y_{t-}). \qquad (26)$$

Writing the equality of both jumps expresses a commutativity between τ and γ :

$$\forall V = (x, v), W = (y, w) \in TM, \; p_V(\gamma(V, W)) = \tau_V(W) - v \qquad (27)$$

According to the following proposition, this equality can be satisfied for any transport on M. Relation (27) explains how the transport can lift a connection rule on M into a connection rule on TM.

Theorem 1 *Let τ be a transport on M, and p the associated connection (defined by $p_V = d\tau_V(V)$). Let γ be a connection rule on M. Define the map γ as follows: $\forall V = (x, v), W = (y, w) \in TM$,*

$$\gamma(V, W) = (\gamma(x, y), p_V(\gamma(x, y), .)^{-1}(\tau_V(W) - v))$$

Then γ is a connection rule on TM, linked with τ by (27).

Remark 2. Seeing the tangent space $T_V TM$ at $V = (x, v)$ as a product space $T_x M \times T_v(T_x M)$, we write

$$p_V(U) = p_V(dx^i(U)\frac{\partial}{\partial x^i}, dv^i(U)\frac{\partial}{\partial v^i}).$$

This implies, using the linearity of p_V and (9), that

$$p_V(U) = dx^i(U)p_V(\frac{\partial}{\partial x^i}, 0) + dv^i(U)\frac{\partial}{\partial v^i}.$$

As a consequence, if the part $dx^i(U)\frac{\partial}{\partial x^i}$ of U on $T_x M$ and its vertical part $p_V(U)$ are known, we recover the part $dv^i(U)\frac{\partial}{\partial v^i}$ of U in $T_v(T_x M)$. In other words, for every U_x in $T_x M$, the map

$$p_V(U_x, .) : U_v \to p_V(U_x, 0) + U_v$$

is invertible, from $T_v(T_x M)$ to $T_v(T_x M)$, and one has

$$p_V(U_x, .)^{-1}(Z) = Z - p_V(U_x, 0). \qquad (28)$$

Proof : First note that γ and τ, so defined, satisfy (27). Let us verify if γ satisfies properties (i), (ii) and (iii) of the definition of a connection rule.

- (i). We have $\gamma(x,y) \in T_x M$, since γ is a connection rule on M. We also have
  $p_V(\gamma(x,y),.)^{-1}(\tau_V(W) - v) \in T_v(T_x M)$. Then we get $\boldsymbol{\gamma}(V,W) \in T_x M \times T_v(T_x M) \subset T_V TM$.

- (ii). For all V in TM, $\boldsymbol{\gamma}(V,V) = p_V(\gamma(x,x),.)^{-1}(\tau_V(V) - v)$, that is, using properties $\gamma(x,x) = 0$ and $\tau_V(V) = v$, $\boldsymbol{\gamma}(V,V) = p_V(0,.)^{-1}(0)$. By (28), we get $p_V(0,.)^{-1}(0) = 0 - p_V(0,0) = 0$ with the linearity of p_V. Then $\boldsymbol{\gamma}(V,V) = 0$.

- (iii). For all V and W in TM,
  $\frac{\partial}{\partial W}(\boldsymbol{\gamma}(V,W))|_{W=V}$

$$= (\frac{\partial}{\partial W}(\gamma(x,y))|_{W=V}, \frac{\partial}{\partial W}[p_V(\gamma(x,y),.)^{-1}(\tau_V(W) - v))|_{W=V}]). \quad (29)$$

Since γ is a connection rule on M, we have $\frac{\partial}{\partial W}(\gamma(x,y))|_{W=V} = \frac{\partial}{\partial y}(\gamma(x,y))|_{y=x} = Id_{T_x M}$. Therefore, we need to prove that $\frac{\partial}{\partial W}[p_V(\gamma(x,y),.)^{-1}(\tau_V(W) - v))|_{W=V}) = Id_{T_v(T_x M)}$.
For this, let us use (27). By differentiating with respect to W, we get

$$dp_V(\boldsymbol{\gamma}(V,V))(\frac{\partial \boldsymbol{\gamma}(V,W)}{\partial W}|_{W=V}) = d\tau_V(V).$$

But p_V is linear so $dp_V(\boldsymbol{\gamma}(V,V)) = dp_V(0) = p_V$. Recall that $d\tau_V(V) = p_V$. Then we get, for every $U = (U_x, U_v)$ in $T_V TM$, with $U_x \in T_x M$ and $U_v \in T_v(T_x M)$,

$$p_V(\frac{\partial \boldsymbol{\gamma}(V,W)}{\partial W}|_{W=V}(U_x, U_v)) = p_V(U_x, U_v),$$

that is, by (29) and the linearity of p_V, with the fact that it is a projection on $T_v(T_x M)$,

$$p_V(\frac{\partial \gamma(x,y)}{\partial y}|_{W=V}(U_x), 0) + \frac{\partial}{\partial W}(p_V(\gamma(x,y),.)^{-1}(\tau_V(W) - v))|_{W=V}(U_v)$$

$$= p_V(U_x, 0) + U_v.$$

But $\frac{\partial \gamma(x,y)}{\partial y}|_{W=V}(U_x) = U_x$, so $\frac{\partial}{\partial W}(p_V(\gamma(x,y),.)^{-1}(\tau_V(W)-v))|_{W=V})(U_v)$
$= U_v$, proving (iii), that is $\frac{\partial}{\partial W}(\boldsymbol{\gamma}(V,W))|_{W=V} = Id_{T_V TM}$. $\square$

Proposition 11 *Let τ be a transport on M and $\boldsymbol{\gamma}$ a connection rule on TM, both of them satisfying (27). Then we have an equivalence between the two following properties*

1. τ is a linear idempotent transport
2. γ satisfies

$$\forall V, W \in TM, \gamma(\bar{\lambda}V, \bar{\lambda}W) = \lambda\gamma(V, W), \tag{30}$$

$$\text{and } \forall v, w \in T_xM, \gamma((x, v), (x, w)) = (0, w - v). \tag{31}$$

Proof :

- $1 \Rightarrow 2$: Assume that τ is a linear idempotent transport. If τ and γ satisfy (27), γ can be written as in the previous proposition:

$$\gamma(V, W) = (\gamma(x, y) , p_V(\gamma(x, y), .)^{-1}(\tau_V(W) - v)).$$

Let us verify the first property for γ. We have

$$\gamma(\bar{\lambda}V, \bar{\lambda}W) = (\gamma(x, y) , p_{\bar{\lambda}V}(\gamma(x, y), .)^{-1}(\tau_{\bar{\lambda}V}(\bar{\lambda}W) - \lambda v)).$$

Using the linearity of τ, we get

$$\gamma(\bar{\lambda}V, \bar{\lambda}W) = (\gamma(x, y) , p_{\bar{\lambda}V}(\gamma(x, y), .)^{-1}[\lambda(\tau_V(W) - v)].$$

According to (28), it follows

$$p_{\bar{\lambda}V}(\gamma(x, y), .)^{-1}[\lambda(\tau_V(W) - v)] = \lambda(\tau_V(W) - v) - p_{\bar{\lambda}V}(\gamma(x, y), 0).$$

For every λ, one can write $(\gamma(x, y), 0) = (\gamma(x, y), \lambda 0) = \bar{\lambda}(\gamma(x, y), 0))$. Then $p_{\bar{\lambda}V}(\gamma(x, y), 0) = p_{\bar{\lambda}V}[\bar{\lambda}(\gamma(x, y), 0)] = \lambda p_V(\gamma(x, y), 0)$. It follows, using (28) again, that

$$p_{\bar{\lambda}V}(\gamma(x, y), .)^{-1}[\lambda(\tau_V(W) - v)] = \lambda(\tau_V(W) - v - p_V(\gamma(x, y), 0))$$
$$= \lambda p_V(\gamma(x, y), .)^{-1}(\tau_V(W) - v).$$

Then we get $\gamma(\bar{\lambda}V, \bar{\lambda}W) = \lambda\gamma(V, W)$.
Let us verify the second property for γ. We have

$$\gamma((x, v), (x, w)) = (\gamma(x, x) , p_{(x,v)}(\gamma(x, x), .)^{-1}(\tau_{(x,v)}(x, w) - v)).$$

Using the fact that τ is idempotent and by (28), we get

$$\gamma((x, v), (x, w)) = (0, p_{(x,v)}(0, .)^{-1}(w - v) = (0, w - v).$$

- $2 \Rightarrow 1$: Assume that γ satisfy both properties given by (30) and (31) . If τ and γ satisfy (27), the transport τ is written

$$\tau_V(W) = v + p_V(\gamma(V, W))$$

Let us verify that τ is linear. According to the first property for γ, we have

$$\tau_{\bar{\lambda}V}(\bar{\lambda}W) = \lambda v + p_{\bar{\lambda}V}(\gamma(\bar{\lambda}V, \bar{\lambda}W)) = \lambda v + p_{\bar{\lambda}V}(\lambda\gamma(V, W)).$$

Then, using the properties of the connection p, we get
$\tau_{\bar{\lambda}V}(\bar{\lambda}W) = \lambda v + \lambda p_V(\gamma(V,W)) = \lambda \tau_V(W)$, which yields the linearity for
τ. Let us verify that τ is idempotent. According to the second property
for γ, we have

$$\tau_{(x,v)}(x,w) = v + p_V(\gamma((x,v),(x,w))) = v + p_V(0, w - v).$$

But $p_V(0,U) = U$, so $\tau_{(x,v)}(x,w) = v + w - v = w$ and τ is idempotent. $\square$

A consequence of this proposition is the following theorem, yielding a
reciprocal result of theorem 1.

Theorem 2 *Let γ be a connection rule on TM, satisfying both properties
given by (30). Let p be a connection on M. We define τ as follows:*

$$\forall V = (x,v), W = (y,w) \in TM, \ \tau_V(W) = v + p_V(\gamma(V,W)). \tag{32}$$

*Then τ is a linear idempotent transport on M, such that, for every V in TM,
$p_V = d\tau_V(V)$. Moreover, γ and τ satisfy (27).*

Proof : Note that, for this definition of τ, $p_V(\gamma(V,W))$ is seen as an element
of T_xM.

The smoothness of τ comes from that of p_V and γ. By propositions 5
and 11, τ is an idempotent and linear transport on M. Moreover, we have
$d\tau_V(V) = dp_V(0) \circ d\gamma(V,.)(V)$. Since γ is a connection rule on TM, it satisfies $\gamma(V,.)(V) = Id_{T_VTM}$, and, by linearity of p_V, we have $dp_V(0) = p_V$, so
$d\tau_V(V) = p_V$. $\square$

Note that a more general result may be shown, yielding a general transport,
but it requires *a priori* other conditions on γ.

Remark 3. In the particular case of the parallel transport and the geodesic
connection rule, (27) is satisfied, so the Itô covariant integral with jumps of
part I coincides with the Itô integral with covariant jumps of part II. Note
that another way to find the jump of the Itô covariant integral $\int \alpha \ d^{\triangle} Y$,
following Norris' presentation with the horizontal lift, and Cohen's description
in [Coh96-1] for the jump of the horizontal lift, would be to write it as

$$\alpha_{Y_{t-}}(0, u_{t-}\triangle(u_t^{-1}y_t)) = \alpha_{Y_{t-}}(0, u_{t-}(u_t^{-1}y_t - u_{t-}^{-1}y_{t-}))$$
$$= \alpha_{Y_{t-}}(0, (u_{t-}u_t^{-1}y_t - y_{t-}))$$
$$= \alpha_{Y_{t-}}(0, \tau_{x_t,x_{t-}}'' y_t - y_{t-}),$$

which gives, for this example, the same jump as (27).

7.2 Comparison of the continuous Itô integrals

Recall that the definition of part II only requires one connection. Then the equality of definitions of part I and II in the continuous case, written, according to propositions 4 and 10,

$$\mathbb{G}(p^*\alpha) = \tilde{p}^*(\alpha)$$

implies in particular that also one connexion is used for the Itô covariant calculus of part I. It can be shown (see [Ma03]) that it then corresponds to Norris' definition.

8 Equality for vertical or horizontal semimartingales

We have introduced two covariant stochastic integrals along a semimartingale $(Y_t) = ((x_t, y_t))$. To call them "covariant", we expect from them that they correspond to the classical stochastic integration along (y_t) when (Y_t) is vertical, and that they vanish when (Y_t) is horizontal.

A càdlàg semimartingale $(Y_t) = ((x_t, y_t))$ on TM is said to be vertical if it stays in a vector space, that is if the semimartingale (x_t), projection of (Y_t) on M, satisfies: $\forall t, x_t = x_0$.

Proposition 12 *If $(Y_t) = ((x_t, y_t))$ is a vertical càdlàg semimartingale on TM, one has*

$$\int \alpha \stackrel{\triangle v}{\partial} Y = \int \alpha \stackrel{\triangle}{\partial^v} Y = \int \alpha \stackrel{\triangle}{\partial} y, \quad and \quad \int \alpha \stackrel{\triangle v}{d} Y$$

$$= \int \alpha \stackrel{\triangle}{d^v} Y = \int \alpha \stackrel{\triangle}{d} y.$$

Proof : If (Y_t) is a vertical semimartingale, then its part (x_t) on M is a constant semimartingale. In particular we have $x_t = x_{t-}$ for every jump of Y_t. With regard to the Stratonovich integral of part I, $x_t = x_{t-}$ implies $\pi\mathbb{I}(s, Y_{t-}, Y_t) = x_{t-}$ for all s. It follows that the curve $(\mathbb{I}(s, Y_{t-}, Y_t))$ is vertical, so its covariant derivative satisfies $p_{\mathbb{I}(s,Y_{t-},Y_t)}(\frac{\partial \mathbb{I}}{\partial s}(s, Y_{t-}, Y_t)) = \frac{\partial \mathbb{I}}{\partial s}(s, y_{t-}, y_t)$. Consequently $\int \alpha \stackrel{\triangle}{\partial^v} Y = \int \alpha \stackrel{\triangle}{\partial} y$. With regard to the Itô integral of part I, $x_t = x_{t-}$ implies $\pi\boldsymbol{\gamma}(Y_{t-}, Y_t) = 0$. Then $\boldsymbol{\gamma}(Y_{t-}, Y_t) = (0, \boldsymbol{\gamma}(y_{t-}, y_t))$ is vertical and so we get $p_{Y_{t-}}(\boldsymbol{\gamma}(Y_{t-}, Y_t)) = \boldsymbol{\gamma}(y_{t-}, y_t)$. It follows $\stackrel{\triangle}{\boldsymbol{\gamma}}(p^*\alpha)_{Y_{t-}}(Y_t) = \stackrel{\triangle}{\boldsymbol{\gamma}}(\alpha)_{y_{t-}}(y_t)$. Consequently $\int \alpha \stackrel{\triangle}{d^v} Y = \int \alpha \stackrel{\triangle}{\partial} y$. With regard to the integrals of part II, $x_t = x_{t-}$ implies $\tau_{Y_{t-}}(Y_t) = y_{t-}$. Then, for all θ in τ^*TM, $(\tau^{//})^*(\theta)_{Y_{t-}}(Y_t) = \theta_{y_{t-}}(y_t)$ and $\int \theta \stackrel{\triangle v}{d\mathbb{I}} Y = \int \theta \stackrel{\triangle}{d\mathbb{I}} y$. Apply this to $\theta = \stackrel{\triangle}{\mathbb{I}}(\alpha)$ and $\theta = \stackrel{\triangle}{\boldsymbol{\gamma}}(\alpha)$. $\square$

Let (Y_t) be a C^1 curve on TM and p a connection on M. Recall that (Y_t) is said to be horizontal if, for every t, $\frac{\partial Y_t}{\partial t}$ belongs to the horizontal space at Y_t, that is if $p_{Y_t}(\frac{\partial Y_t}{\partial t}) = 0$. Note that the notion of an horizontal curve depends on the connection p.

In the same way, the definition for a càdlàg semimartingale depends on the choice of a transport. Here, it is important to consider equivalent tools for the connection and the transport, so we write the definition for the parallel transport along geodesics.

We will say that a càdlàg semimartingale $(Y_t) = ((x_t, y_t))$ is horizontal if it satisfies

$$y_t = \tau^{//}_{x_{t-}, x_t} y_{t-}, \ \text{if } x_t \neq x_{t-}$$

$$\text{and } \ \partial^V Y_t = 0 \text{ if } x_t = x_{t-}.$$

Note that, for a rigorous definition, we would have to write (Y_t) as the solution of a covariant s.d.e. with jumps (studied in a forthcoming paper).

Proposition 13 *Let p be a connection on M, associated to the parallel transport along geodesics. If $(Y_t) = ((x_t, y_t))$ is an horizontal semimartingale on TM, all following covariant integrals vanish:*

$$\int \alpha \overset{\triangle_V}{\partial}\, Y = \int \alpha \overset{\triangle}{\partial}^V Y = \int \alpha \overset{\triangle_V}{d}\, Y = \int \alpha \overset{\triangle}{d}^V Y = 0.$$

Proof : If (Y_t) is horizontal, we have $\tau_{Y_{t-}}(Y_t) = y_{t-}$. The term $\tau_{Y_{t-}}(Y_t) - y_{t-}$ is in factor in all the expressions of the integrals when considering the example of parallel transport along geodesics. Since it vanishes, the jumps and so the integrals vanish too. $\square$

As a consequence, when we integrate the particular classes of vertical and horizontal semimartingales, the covariant integrals of parts I and II are equal.

9 Conclusion

The covariant stochastic calculi of part I and II both permit to recover the covariant calculus for a C^1 curve in TM. Moreover, they both coincide with vector integration when integrating a vertical semimartingale, and both vanish when integrating an horizontal semimartingale. Both calculi can recover the stochastic continuous covariant calculus of order two, involving a connection of order two, $d_s p$ for Stratonovich calculus, $\tilde{p}$ for Itô calculus. This extends in a way the Schwartz principle: like classical stochastic calculus, the covariant stochastic calculus on manifolds is a second order calculus.

The stochastic covariant calculus with jumps of part I is perfectly compatible with the Stratonovich and the Itô calculi, since it coincides with Norris' formalism when integrating a continuous semimartingale. Nevertheless, it does not include these covariant Stratonovich and Itô calculi in a second order covariant formalism.

This second order formalism is given by the stochastic calculus with covariant jumps of part II. It is geometrically more significant since it describes a covariant jump on the manifold. This second part uses the notion of transport, which can be seen as a connection "of order 0". Then we get a larger class of connections than connections of order one, with transports non necessarily linear, basic, idempotent, or invertible.

We saw that the calculus with covariant jumps is not well adapted to the Stratonovich calculus, but it is to the Itô calculus. In part II, we start with describing the covariant jump of a semimartingale, in order to describe the jump of a stochastic integral. This method is better adapted to the Itô integral. Indeed, in the non covariant case, the jump of an Itô integral is described from the one of the semimartingale, which is geometrically represented by a vector $\gamma(z_{t-}, z_t)$ in $T_{z_{t-}} U$. This is not the case for the Stratonovich calculus with jumps, for which only the jump of the integral is known, but does not come from a geometrical representation of the jump of the semimartingale itself.

This formalism can be applied to the study of covariant s.d.e. It yields theorems of discretization for covariant s.d.e with jumps and for s.d.e. with covariant jumps. This will be done in a next paper. The notions of transport and connection of order two are essential for the Itô covariant calculus. They will be further studied in a next paper too.

References

[Coh96-1] S. Cohen: Géométrie différentielle stochastique avec sauts I. Stoch. Stoch. Rep. **56** (1996), no. 3–4, 179–203.

[Coh96-2] S. Cohen: Géométrie différentielle stochastique avec sauts II. Discrétisation et applications des EDS avec sauts. Stoch. Stoch. Rep. **56** (1996), no. 3–4, 205–225.

[Elw82] K.D. Elworthy: Stochastic Differential Equations on Manifolds. LMS Lecture Note Series **70**. Cambridge University Press, Cambridge-New York, 1982.

[Em89] M. Emery: Stochastic Calculus in Manifolds. Universitext. Springer-Verlag, Berlin, 1989.

[Em90] M. Emery: On two transfer principles in stochastic differential geometry, Séminaire de Probabilités XXIV, 1988/89, 407–441, Lecture Notes in Math. **1426**, Springer, Berlin, 1990.

[KN63] S. Kobayashi, K. Nomizu: Foundations of Differential Geometry. Vol I. Interscience Publishers, John Wiley & Sons, New York-Lond, 1963.

[Ma03] L. Maillard-Teyssier: Calcul Stochastique Covariant à Sauts et Calcul Stochastique à Sauts Covariants . *Thèse de doctorat (16 décembre 2003) - Université de Versailles Saint Quentin en Yvelines.* "thèses-EN-ligne": http://tel.ccsd.cnrs.fr/

[Mey81] P.A. Meyer: Géométrie stochastique sans larmes. Séminaire de Probabilités XV, 44–102, Lecture Notes in Math. **850**, Springer, Berlin-New York, 1981.

[Mey82] P.A. Meyer: Géométrie différentielle stochastique II. Séminaire de Probabilités XVI, Supplément Géométrie Différentielle Stochastique, 165–207, Lecture
Notes in Math. **921**, Springer, Berlin-New York, 1982.

[Nor92] J.R. Norris: A complete differential formalism for stochastic calculus in
manifolds. Séminaire de Probabilités XXVI, 189–209, Lecture Notes in Math.
1526, Springer, Berlin, 1992.

[Pic91] J. Picard: Calcul stochastique avec sauts sur une variété. Séminaire de Probabilités XXV, 196–219, Lecture Notes in Math. **1485**, Springer, Berlin, 1991.

[Sch82] L. Schwartz: Géométrie différentielle du 2ème ordre, semi-martingales et
équations différentielles stochastiques sur une variété différentielle. Séminaire
de Probabilités XVI, Supplément Géométrie Différentielle Stochastique, 1–148,
Lecture Notes in Math. **921**, Springer, Berlin-New York, 1982.

[Spi79] M. Spivak: A Comprehensive Introduction to Differential Geometry. Vol. II.
Second edition. Publish or Perish, Inc., Wilmington, Del., 1979.

[YI73] K. Yano, S. Ishihara: Tangent and Cotangent Bundles: Differential Geometry.
Pure and Applied Mathematics, No. **16**. Marcel Dekker, Inc., New York, 1973.

SÉMINAIRE DE PROBABILITÉS ON THE WEB

The content of all volumes of the Séminaire de Probabilités is now available on the web:

- Volumes I to XXXVI can be consulted on the NUMDAM website

  http://www.numdam.org/numdam-bin/browse?j=SPS&sl=0

(the first 34 volumes are freely accessible; volumes XXXV and XXXVI are accessible by SpringerLink subscribers only);

- all volumes from XXXII to the present volume can also be consulted by subscribers to SpringerLink at

  http://www.springerlink.com/

A small database with keywords, classification and comments on the contributions to the Séminaire de Probabilités can be found at

  http://www-irma.u-strasbg.fr/irma/semproba/e_index.shtml

It is unfinished but already provides efficient search tools by themes or contents in the Séminaire volumes published in the sixties and seventies, as well as interesting retrospective comments (mostly due to P.A. Meyer).

NOTE TO CONTRIBUTORS

Contributors to the Séminaire are reminded that their articles should be formatted for the Springer Lecture Notes series.

Manuscripts should preferably be prepared with LaTeX version 2e, using the svmult.cls environment provided by Springer-Verlag for multi-authored LNM books. This file, together with instructions and examples, can be downloaded from the Springer website http://www.springer.com (first choose *Mathematics*, then *For Authors*, then *LaTeX for different book types*, and finally *Contributed books*).

Lecture Notes in Mathematics

For information about earlier volumes
please contact your bookseller or Springer
LNM Online archive: springerlink.com

Vol. 1788: A. Vasil'ev, Moduli of Families of Curves for Conformal and Quasiconformal Mappings (2002)

Vol. 1789: Y. Sommerhäuser, Yetter-Drinfel'd Hopf algebras over groups of prime order (2002)

Vol. 1790: X. Zhan, Matrix Inequalities (2002)

Vol. 1791: M. Knebusch, D. Zhang, Manis Valuations and Prüfer Extensions I: A new Chapter in Commutative Algebra (2002)

Vol. 1792: D. D. Ang, R. Gorenflo, V. K. Le, D. D. Trong, Moment Theory and Some Inverse Problems in Potential Theory and Heat Conduction (2002)

Vol. 1793: J. Cortés Monforte, Geometric, Control and Numerical Aspects of Nonholonomic Systems (2002)

Vol. 1794: N. Pytheas Fogg, Substitution in Dynamics, Arithmetics and Combinatorics. Editors: V. Berthé, S. Ferenczi, C. Mauduit, A. Siegel (2002)

Vol. 1795: H. Li, Filtered-Graded Transfer in Using Noncommutative Gröbner Bases (2002)

Vol. 1796: J.M. Melenk, hp-Finite Element Methods for Singular Perturbations (2002)

Vol. 1797: B. Schmidt, Characters and Cyclotomic Fields in Finite Geometry (2002)

Vol. 1798: W.M. Oliva, Geometric Mechanics (2002)

Vol. 1799: H. Pajot, Analytic Capacity, Rectifiability, Menger Curvature and the Cauchy Integral (2002)

Vol. 1800: O. Gabber, L. Ramero, Almost Ring Theory (2003)

Vol. 1801: J. Azéma, M. Émery, M. Ledoux, M. Yor (Eds.), Séminaire de Probabilités XXXVI (2003)

Vol. 1802: V. Capasso, E. Merzbach, B.G. Ivanoff, M. Dozzi, R. Dalang, T. Mountford, Topics in Spatial Stochastic Processes. Martina Franca, Italy 2001. Editor: E. Merzbach (2003)

Vol. 1803: G. Dolzmann, Variational Methods for Crystalline Microstructure – Analysis and Computation (2003)

Vol. 1804: I. Cherednik, Ya. Markov, R. Howe, G. Lusztig, Iwahori-Hecke Algebras and their Representation Theory. Martina Franca, Italy 1999. Editors: V. Baldoni, D. Barbasch (2003)

Vol. 1805: F. Cao, Geometric Curve Evolution and Image Processing (2003)

Vol. 1806: H. Broer, I. Hoveijn. G. Lunther, G. Vegter, Bifurcations in Hamiltonian Systems. Computing Singularities by Gröbner Bases (2003)

Vol. 1807: V. D. Milman, G. Schechtman (Eds.), Geometric Aspects of Functional Analysis. Israel Seminar 2000-2002 (2003)

Vol. 1808: W. Schindler, Measures with Symmetry Properties (2003)

Vol. 1809: O. Steinbach, Stability Estimates for Hybrid Coupled Domain Decomposition Methods (2003)

Vol. 1810: J. Wengenroth, Derived Functors in Functional Analysis (2003)

Vol. 1811: J. Stevens, Deformations of Singularities (2003)

Vol. 1812: L. Ambrosio, K. Deckelnick, G. Dziuk, M. Mimura, V. A. Solonnikov, H. M. Soner, Mathematical Aspects of Evolving Interfaces. Madeira, Funchal, Portugal 2000. Editors: P. Colli, J. F. Rodrigues (2003)

Vol. 1813: L. Ambrosio, L. A. Caffarelli, Y. Brenier, G. Buttazzo, C. Villani, Optimal Transportation and its Applications. Martina Franca, Italy 2001. Editors: L. A. Caffarelli, S. Salsa (2003)

Vol. 1814: P. Bank, F. Baudoin, H. Föllmer, L.C.G. Rogers, M. Soner, N. Touzi, Paris-Princeton Lectures on Mathematical Finance 2002 (2003)

Vol. 1815: A. M. Vershik (Ed.), Asymptotic Combinatorics with Applications to Mathematical Physics. St. Petersburg, Russia 2001 (2003)

Vol. 1816: S. Albeverio, W. Schachermayer, M. Talagrand, Lectures on Probability Theory and Statistics. Ecole d'Eté de Probabilités de Saint-Flour XXX-2000. Editor: P. Bernard (2003)

Vol. 1817: E. Koelink, W. Van Assche(Eds.), Orthogonal Polynomials and Special Functions. Leuven 2002 (2003)

Vol. 1818: M. Bildhauer, Convex Variational Problems with Linear, nearly Linear and/or Anisotropic Growth Conditions (2003)

Vol. 1819: D. Masser, Yu. V. Nesterenko, H. P. Schlickewei, W. M. Schmidt, M. Waldschmidt, Diophantine Approximation. Cetraro, Italy 2000. Editors: F. Amoroso, U. Zannier (2003)

Vol. 1820: F. Hiai, H. Kosaki, Means of Hilbert Space Operators (2003)

Vol. 1821: S. Teufel, Adiabatic Perturbation Theory in Quantum Dynamics (2003)

Vol. 1822: S.-N. Chow, R. Conti, R. Johnson, J. Mallet-Paret, R. Nussbaum, Dynamical Systems. Cetraro, Italy 2000. Editors: J. W. Macki, P. Zecca (2003)

Vol. 1823: A. M. Anile, W. Allegretto, C. Ringhofer, Mathematical Problems in Semiconductor Physics. Cetraro, Italy 1998. Editor: A. M. Anile (2003)

Vol. 1824: J. A. Navarro González, J. B. Sancho de Salas, $\mathscr{C}^\infty$ – Differentiable Spaces (2003)

Vol. 1825: J. H. Bramble, A. Cohen, W. Dahmen, Multiscale Problems and Methods in Numerical Simulations, Martina Franca, Italy 2001. Editor: C. Canuto (2003)

Vol. 1826: K. Dohmen, Improved Bonferroni Inequalities via Abstract Tubes. Inequalities and Identities of Inclusion-Exclusion Type. VIII, 113 p, 2003.

Vol. 1827: K. M. Pilgrim, Combinations of Complex Dynamical Systems. IX, 118 p, 2003.

Vol. 1828: D. J. Green, Gröbner Bases and the Computation of Group Cohomology. XII, 138 p, 2003.

Vol. 1829: E. Altman, B. Gaujal, A. Hordijk, Discrete-Event Control of Stochastic Networks: Multimodularity and Regularity. XIV, 313 p, 2003.

Vol. 1830: M. I. Gil', Operator Functions and Localization of Spectra. XIV, 256 p, 2003.

Vol. 1831: A. Connes, J. Cuntz, E. Guentner, N. Higson, J. E. Kaminker, Noncommutative Geometry, Martina Franca, Italy 2002. Editors: S. Doplicher, L. Longo (2004)

Vol. 1832: J. Azéma, M. Émery, M. Ledoux, M. Yor (Eds.), Séminaire de Probabilités XXXVII (2003)

Vol. 1833: D.-Q. Jiang, M. Qian, M.-P. Qian, Mathematical Theory of Nonequilibrium Steady States. On the Frontier of Probability and Dynamical Systems. IX, 280 p, 2004.

Vol. 1834: Yo. Yomdin, G. Comte, Tame Geometry with Application in Smooth Analysis. VIII, 186 p, 2004.

Vol. 1835: O.T. Izhboldin, B. Kahn, N.A. Karpenko, A. Vishik, Geometric Methods in the Algebraic Theory of Quadratic Forms. Summer School, Lens, 2000. Editor: J.-P. Tignol (2004)

Vol. 1836: C. Năstăsescu, F. Van Oystaeyen, Methods of Graded Rings. XIII, 304 p, 2004.

Vol. 1837: S. Tavaré, O. Zeitouni, Lectures on Probability Theory and Statistics. Ecole d'Eté de Probabilités de Saint-Flour XXXI-2001. Editor: J. Picard (2004)

Vol. 1838: A.J. Ganesh, N.W. O'Connell, D.J. Wischik, Big Queues. XII, 254 p, 2004.

Vol. 1839: R. Gohm, Noncommutative Stationary Processes. VIII, 170 p, 2004.

Vol. 1840: B. Tsirelson, W. Werner, Lectures on Probability Theory and Statistics. Ecole d'Eté de Probabilités de Saint-Flour XXXII-2002. Editor: J. Picard (2004)

Vol. 1841: W. Reichel, Uniqueness Theorems for Variational Problems by the Method of Transformation Groups (2004)

Vol. 1842: T. Johnsen, A.L. Knutsen, K3 Projective Models in Scrolls (2004)

Vol. 1843: B. Jefferies, Spectral Properties of Noncommuting Operators (2004)

Vol. 1844: K.F. Siburg, The Principle of Least Action in Geometry and Dynamics (2004)

Vol. 1845: Min Ho Lee, Mixed Automorphic Forms, Torus Bundles, and Jacobi Forms (2004)

Vol. 1846: H. Ammari, H. Kang, Reconstruction of Small Inhomogeneities from Boundary Measurements (2004)

Vol. 1847: T.R. Bielecki, T. Björk, M. Jeanblanc, M. Rutkowski, J.A. Scheinkman, W. Xiong, Paris-Princeton Lectures on Mathematical Finance 2003 (2004)

Vol. 1848: M. Abate, J. E. Fornaess, X. Huang, J. P. Rosay, A. Tumanov, Real Methods in Complex and CR Geometry, Martina Franca, Italy 2002. Editors: D. Zaitsev, G. Zampieri (2004)

Vol. 1849: Martin L. Brown, Heegner Modules and Elliptic Curves (2004)

Vol. 1850: V. D. Milman, G. Schechtman (Eds.), Geometric Aspects of Functional Analysis. Israel Seminar 2002-2003 (2004)

Vol. 1851: O. Catoni, Statistical Learning Theory and Stochastic Optimization (2004)

Vol. 1852: A.S. Kechris, B.D. Miller, Topics in Orbit Equivalence (2004)

Vol. 1853: Ch. Favre, M. Jonsson, The Valuative Tree (2004)

Vol. 1854: O. Saeki, Topology of Singular Fibers of Differential Maps (2004)

Vol. 1855: G. Da Prato, P.C. Kunstmann, I. Lasiecka, A. Lunardi, R. Schnaubelt, L. Weis, Functional Analytic Methods for Evolution Equations. Editors: M. Iannelli, R. Nagel, S. Piazzera (2004)

Vol. 1856: K. Back, T.R. Bielecki, C. Hipp, S. Peng, W. Schachermayer, Stochastic Methods in Finance, Bressanone/Brixen, Italy, 2003. Editors: M. Fritelli, W. Runggaldier (2004)

Vol. 1857: M. Émery, M. Ledoux, M. Yor (Eds.), Séminaire de Probabilités XXXVIII (2005)

Vol. 1858: A.S. Cherny, H.-J. Engelbert, Singular Stochastic Differential Equations (2005)

Vol. 1859: E. Letellier, Fourier Transforms of Invariant Functions on Finite Reductive Lie Algebras (2005)

Vol. 1860: A. Borisyuk, G.B. Ermentrout, A. Friedman, D. Terman, Tutorials in Mathematical Biosciences I. Mathematical Neurosciences (2005)

Vol. 1861: G. Benettin, J. Henrard, S. Kuksin, Hamiltonian Dynamics – Theory and Applications, Cetraro, Italy, 1999. Editor: A. Giorgilli (2005)

Vol. 1862: B. Helffer, F. Nier, Hypoelliptic Estimates and Spectral Theory for Fokker-Planck Operators and Witten Laplacians (2005)

Vol. 1863: H. Fürh, Abstract Harmonic Analysis of Continuous Wavelet Transforms (2005)

Vol. 1864: K. Efstathiou, Metamorphoses of Hamiltonian Systems with Symmetries (2005)

Vol. 1865: D. Applebaum, B.V. R. Bhat, J. Kustermans, J. M. Lindsay, Quantum Independent Increment Processes I. From Classical Probability to Quantum Stochastic Calculus. Editors: M. Schürmann, U. Franz (2005)

Vol. 1866: O.E. Barndorff-Nielsen, U. Franz, R. Gohm, B. Kümmerer, S. Thorbjønsen, Quantum Independent Increment Processes II. Structure of Quantum Lévy Processes, Classical Probability, and Physics. Editors: M. Schürmann, U. Franz, (2005)

Vol. 1867: J. Sneyd (Ed.), Tutorials in Mathematical Biosciences II. Mathematical Modeling of Calcium Dynamics and Signal Transduction. (2005)

Vol. 1868: J. Jorgenson, S. Lang, $Pos_n(R)$ and Eisenstein Sereies. (2005)

Vol. 1869: A. Dembo, T. Funaki, Lectures on Probability Theory and Statistics. Ecole d'Eté de Probabilités de Saint-Flour XXXIII-2003. Editor: J. Picard (2005)

Vol. 1870: V.I. Gurariy, W. Lusky, Geometry of Müntz Spaces and Related Questions. (2005)

Vol. 1871: P. Constantin, G. Gallavotti, A.V. Kazhikhov, Y. Meyer, S. Ukai, Mathematical Foundation of Turbulent Viscous Flows, Martina Franca, Italy, 2003. Editors: M. Cannone, T. Miyakawa (2006)

Vol. 1872: A. Friedman (Ed.), Tutorials in Mathematical Biosciences III. Cell Cycle, Proliferation, and Cancer (2006)

Vol. 1873: R. Mansuy, M. Yor, Random Times and Enlargements of Filtrations in a Brownian Setting (2006)

Vol. 1874: M. Émery, M. Yor (Eds.), In Memoriam Paul-André Meyer - Séminaire de Probabilités XXXIX (2006)

Vol. 1875: J. Pitman, Combinatorial Stochastic Processes. Ecole d'Eté de Probabilités de Saint-Flour XXXII-2002. Editor: J. Picard (2006)

Vol. 1876: H. Herrlich, Axiom of Choice (2006)

Vol. 1877: J. Steuding, Value Distributions of L-Functions (2006)

Vol. 1878: R. Cerf, The Wulff Crystal in Ising and Percolation Models, Ecole d'Eté de Probabilités de Saint-Flour XXXIV-2004. Editor: Jean Picard (2006)

Vol. 1879: G. Slade, The Lace Expansion and its Applications, Ecole d'Eté de Probabilités de Saint-Flour XXXIV-2004. Editor: Jean Picard (2006)

Vol. 1880: S. Attal, A. Joye, C.-A. Pillet, Open Quantum Systems I, The Hamiltonian Approach (2006)

Vol. 1881: S. Attal, A. Joye, C.-A. Pillet, Open Quantum Systems II, The Markovian Approach (2006)

Vol. 1882: S. Attal, A. Joye, C.-A. Pillet, Open Quantum Systems III, Recent Developments (2006)

Vol. 1883: W. Van Assche, F. Marcellàn (Eds.), Orthogonal Polynomials and Special Functions, Computation and Application (2006)

Vol. 1884: N. Hayashi, E.I. Kaikina, P.I. Naumkin, I.A. Shishmarev, Asymptotics for Dissipative Nonlinear Equations (2006)

Vol. 1885: A. Telcs, The Art of Random Walks (2006)

Recent Reprints and New Editions

Vol. 1471: M. Courtieu, A.A. Panchishkin, Non-Archimedean L-Functions and Arithmetical Siegel Modular Forms. – Second Edition (2003)

Vol. 1618: G. Pisier, Similarity Problems and Completely Bounded Maps. 1995 – Second, Expanded Edition (2001)

Vol. 1629: J.D. Moore, Lectures on Seiberg-Witten Invariants. 1997 – Second Edition (2001)

Vol. 1638: P. Vanhaecke, Integrable Systems in the realm of Algebraic Geometry. 1996 – Second Edition (2001)

Vol. 1702: J. Ma, J. Yong, Forward-Backward Stochastic Differential Equations and their Applications. 1999. – Corrected 3rd printing (2005)

Printing: Krips bv, Meppel
Binding: Stürtz, Würzburg